Bioelectrics

Hidenori Akiyama • Richard Heller
Editors

Bioelectrics

Springer

Editors
Hidenori Akiyama
Institute of Pulsed Power Science
Kumamoto University
Kumamoto, Japan

Richard Heller
Norfolk, Virginia
USA

ISBN 978-4-431-56093-7   ISBN 978-4-431-56095-1 (eBook)
DOI 10.1007/978-4-431-56095-1

Library of Congress Control Number: 2016950238

Printed on acid-free paper

This Springer imprint is published by Springer Nature
The registered company is Springer Japan KK

Preface

Bioelectrics is a new field encompassing both the science and technology of applying electrical stimuli to biological systems. Typical stimuli utilize what is called pulsed power, a pulse with high voltage and/or current with a pulse width from subnanoseconds to milliseconds. Pulsed power may be altered to such forms as pulsed electric fields, plasmas, shock waves, pulsed electromagnetic waves, and pulse light. Application targets include specific cells such as cancer cells and bacteria, more complex arrangements such as tissues and organs, and complex environments and ecosystems including plant and animal species. Elucidation of effects of pulsed power on biology has led to numerous applications in such fields as medical treatment and welfare, environment, food, agriculture and fisheries, and biotechnology.

Research integrating pulsed power science and technology with the biological and medical fields began in 2000. The Research Center for Bioelectrics, Old Dominion University, was established in 2002 by Dr. Karl Schoenbach; the Bioelectrics Research Center, Kumamoto University, was established in 2007 by Dr. Hidenori Akiyama. The International Bioelectrics Consortium was established in 2005 by three institutions, Old Dominion University (USA), Kumamoto University (Japan), and Karlsruhe Institute of Technology (Germany). By 2016, the consortium had grown to encompass participation by15 universities and institutions.

Military research has driven an increase in the size of pulsed power generators. Sandia National Laboratories' Z Machine, the largest pulsed power generator in the world, has stored energy capabilities of 20 MJ, output current of 26 MA, X-ray peak power of 350 TW, and diameter of about 33 m. Concurrent development of low- to no-maintenance pulsed power supply systems continue to expand industrial applications. Such systems require only moderate output power and a small footprint but do require high repetition rate and high-functioning pulsed power.

The rapid development of bioelectric science and technology leads to hope that, in the near future, bioelectrics will substantially contribute to such global concerns as cancer treatment, environmental maintenance, and food and energy supplies.

Chapter 1 contains an introduction to the application of bioelectrics. Electroporation, the electropermeabilization of cell membranes, is a fundamental and the most thoroughly studied bioelectric effect due to pulsed electric fields. This chapter provides background into the research history of this effect and introduces medical and biotechnological applications based on both reversible and irreversible electropermeabilization with millisecond to microsecond pulses. The Chap. 1 also demonstrates application of nanosecond- to picosecond-range pulses with pulsed electric fields ranging from a kV/cm to more than 100 kV/cm; these affect not only the cell plasma membrane but also subcellular structures.

Chapter 2 includes an introduction into pulsed power technology. Pulsed power is produced by transferring energy generally stored in capacitors and inductors to a load very quickly through switching devices. The chapter focuses on the basics of two key technologies, electric circuits and switches, tools used for pulsed power measurement, and delivery of electric pulses to biological tissues using antennas.

Chapter 3 includes discussions about unique electromagnetic agents ranging from cold plasma to electromagnetic radiation. Nonequilibrium plasmas in gases and liquids are known to produce reactive agents that contribute to biological effects including membrane disruption and apoptosis. These reactive agents include ultraviolet radiation, heat, reactive oxygen-based and nitrogen-based species, charged particles, and electric fields. The interactions between pulsed electromagnetic fields and biological systems are included.

Chapter 4 outlines biological responses to pulsed electric fields (PEFs), which exert profound effects on cells by interacting with the cell membrane and other cellular components. The initial discussion within this chapter is a description from multifaceted standpoints of interactions of PEFs with biological membranes, membrane pore formation, and their physiological significance; this is followed by a description of subcellular events induced by PEFs including their effects on the cytoskeleton and signal transduction. The chapter also provides a detailed description on irreversible electroporation and cell death by PEFs.

Chapter 5 covers the use in medical applications. Therapeutic applications of bioelectrics have been developed for a large number of medical conditions including cancer, wound healing, ischemia, cardiovascular disease, and diabetes. Utilization of bioelectrics-based medical techniques has recently surged as the potential of such techniques lies in a more holistic approach to fundamental biophysical mechanisms driving underlying success.

Chapter 6 includes a discussion of environmental applications such as food and biomass processing. PEF-based techniques have been demonstrated to be energy saving and persistent in efficacy for bacterial inactivation in wastewater and liquid food and for the eradication of cyanobacteria in surface waters, obviating harmful chemical usage. PEF treatment offers specific advantages for component extraction—low heat influx, low energy demand, and selectivity of compound release—which make PEF processing suitable for winemaking, for extraction of sugar from sugar beets and valuable components from fruits and vegetables, and in PEF downstream processing of microalgae.

Kumamoto, Japan Hidenori Akiyama
Norfolk, VA, USA Richard Heller

Contents

1 **Introduction** 1
Karl H. Schoenbach, Eberhard Neumann, Richard Heller, P. Thomas Vernier, Justin Teissie, and Stephen J. Beebe

2 **Pulsed Power Technology** 41
Hidenori Akiyama, Sunao Katsuki, Luis Redondo, Masahiro Akiyama, A.J.M. Pemen, T. Huiskamp, F.J.C.M. Beckers, E.J.M. van Heesch, G.J.J. Winands, S.J. Voeten, L. Zhen, J.W.M. van Bree, Shu Xiao, and Ross Petrella

3 **Special Electromagnetic Agents: From Cold Plasma to Pulsed Electromagnetic Radiation** 109
Petr Lukes, Hidenori Akiyama, Chunqi Jiang, Andrea Doria, Gian Piero Gallerano, Alfonsina Ramundo-Orlando, Stefania Romeo, Maria Rosaria Scarfì, and Olga Zeni

4 **Biological Responses** 155
Ken-ichi Yano, Lea Rems, Tadej Kotnik, Damijan Miklavčič, James C. Weaver, Kyle C. Smith, Reuben S. Son, Thiruvallur R. Gowrishankar, P. Thomas Vernier, Zachary A. Levine, Marie-Pierre Rols, Justin Teissie, Lluis M. Mir, Andrei G. Pakhomov, Peter Nick, Wolfgang Frey, David A. Dean, Keiko Morotomi-Yano, Robert E. Neal II, Suyashree Bhonsle, Rafael V. Davalos, and Stephen J. Beebe

5 **Medical Applications** 275
Richard Heller, Justin Teissie, Marie-Pierre Rols, Julie Gehl, Gregor Sersa, Lluis M. Mir, Robert E. Neal II, Suyashree Bhonsle, Rafael Davalos, Stephen Beebe, Barbara Hargrave, Richard Nuccitelli, Chunqi Jiang, Maja Cemazar, Youssef Tamzali, and Natasa Tozon

6 Environmental Applications, Food and Biomass Processing by Pulsed Electric Fields 389
Wolfgang Frey, Christian Gusbeth, Takashi Sakugawa, Martin Sack, Georg Mueller, Juergen Sigler, Eugene Vorobiev, Nikolai Lebovka, Ignacio Álvarez, Javier Raso, Loree C. Heller, Muhammad A. Malik, Christian Eing, and Justin Teissie

Index 477

Chapter 1
Introduction

Karl H. Schoenbach, Eberhard Neumann, Richard Heller, P. Thomas Vernier, Justin Teissie, and Stephen J. Beebe

Abstract Electroporation, the electropermeabilization of cell membranes, is a basic and most thoroughly studied bioelectric effect caused by pulsed electric fields. Section 1.1 gives a brief introduction into the research history of this effect and revisits the original chemical thermodynamic concept of membrane poration, which was introduced in 1982 in the context of the first electro-genetic reprogramming of cells with foreign DNA. Most of the applications of this effect have since been based on the use of millisecond and microsecond pulses with electric field amplitudes ranging from the upper tens of V/cm to several kV/cm. Section 1.2 provides a brief overview of the history in this important field of research and introduces medical and biotechnological applications, based on reversible and irreversible electropermeabilization with millisecond/microsecond pulses. More recently the pulse duration has been extended into the nanosecond and even picosecond range, with pulsed electric fields ranging from kV/cm to more than 100 kV/cm. The application of such ultrashort pulses has been shown to affect not only the plasma membrane of cells but also subcellular structures. The development of the research in this area and applications of nanosecond and picosecond bioelectric effects are covered in Sect. 1.3.

Keywords Electroporation • History of electroporation/electropermeabilization • Millisecond/microsecond pulse effects • Nanosecond/picosecond pulse effects • Medical and biotechnological applications

---

K.H. Schoenbach (✉) • R. Heller • P.T. Vernier • S.J. Beebe
Old Dominion University, Norfolk, VA, USA
e-mail: kschoenb@odu.edu

E. Neumann
University of Bielefeld, Bielefeld, Germany

J. Teissie
Université de Toulouse and Institute of Pharmacology and Structural Biology, Toulouse, France

H. Akiyama, R. Heller (eds.), *Bioelectrics*, DOI 10.1007/978-4-431-56095-1_1

1.1 History of the Membrane Electroporation Concept

Eberhard Neumann

The original chemical thermodynamic concept of membrane electroporation (MEP), introduced in 1982 in the context of the first genetic reprogramming of cells with foreign naked DNA by using electric pulses, is revisited. In addition, an advanced physical chemical analysis of MEP data correlations is outlined in terms of pore density statistics. A new strategy for a stepwise analysis of electroporation data correlations suggests starting with the low-field range, which directly yields the average single-pore polarizability and eventually the pore radii. Further, at a given applied electric field, the induced membrane fields in spheroidal membrane shells are dependent on the angular position. Consequently, experimental quantities, such as the extent of poration and the pore distribution constant, reflect averages of field-dependent cos squared terms. Remarkably, the "in-field" kinetics indicates that the pore density increases in two consecutive steps and includes the delayed Maxwell stress-induced shape changes of membrane shells. Both the rapid in-field kinetic modes and the longer-lasting post-field pore resealing phases indicate that transmembrane fluxes of small ions and larger molecules are coupled to the structural changes modulating the transport. Therefore, flux analysis uses the concept of time-dependent (interaction) flux coefficients to rationalize the characteristic sigmoid onsets and apparently stretched exponentials, qualifying "stretched" as "exponentials of exponentials."

1.1.1 Introduction

When it was found that electric field pulses together with naked foreign gene DNA can be used to genetically reprogram biological cells, the general interest in bioelectric phenomena and electric pulse techniques gained enormous impetus for field-controlled chemical cell manipulations. Historically, the field effect leading to uptake of biogenic agents like DNA was termed membrane "electroporation," short for "electric pore formation," in 1982 [1]. As such, the term electroporation intends to qualify the chemical structural basis for the observed permeability changes for small ions and (even poly-ionic) macromolecules that usually do not penetrate the lipid phase of membranes.

In more detail, the chemical thermodynamic concept for electroporation was already specified in the context of the first electrotransfer of naked foreign DNA into living mouse lyoma cells by the high-voltage pulse trains leading to stable reprogramming of the cell genome [1, 2]. Prior and complementary to functional electro-uptake, it had been found that electric pulses cause electro-release of cellular components, such as catecholamines, ATP, Ca-ions, and chromogranine proteins from isolated chromaffin granules of bovine adrenal medullae [3].

Similarly, in 1980 it was shown that aggregated dictyostelium cells exposed to a train of electric field pulses, in the post-field time phase, reorganize (facilitated by added Ca-ions) and form viable giant syncytia by (electroporative) cell–cell fusion (electrofusion) [4]. The post-field kinetics is generally slower and covers a longer time range of several seconds. This indicates that the (in-field) induced porous membrane states are (at zero field) structurally long lived as compared to the rapid process of in-field pore formation.

These early field effect data and the electroporation concept have been valued, among others, in Nature Methods [5] as seminal for the various biotechnological and medical applications; among them are the clinical applications of voltage pulses combined with bioactive agents in the clinical discipline of electrochemotherapy and gene electrotransfer. In this context, we appreciate the critical input of members of the laboratories of, initially, L. M. Mir, J. Teissie, M.-P. Rols, D. Miklavcic, G. Sersa, R. Heller, J. C. Weaver, Y. Chizmadzhev, B. Rubinsky, K. H. Schoenbach, I. Tsoneva, and later, U. Pliquett, M. Kotulska, W. Frei, J. Gehl, R. Boeckmann, H. Grubmueller, T. Vernier, M. Tarek, and many others.

It is recalled that the original "thermodynamic concept of membrane electroporation" was only sketched for the overall description of the dependence of the fractional extent of poration (and implicitly the kinetic constants) on the field pulse parameters in terms of pore reaction quantities like induced pore polarization volumes (and eventually pore radius). The present personal review on the early history of membrane electroporation (MEP) is therefore restricted to those selected examples where electroporation data were subjected to rigorous chemical kinetic and chemical thermodynamic analysis.

In the original thermodynamic concept, the mass action law for a two-state structural interaction scheme had been formulated, and the dependence of the chemical state distribution constant on physical state parameters like temperature, pressure, or electric field was given in the differential form of a general Van't Hoff relationship [1]. It is emphasized that, as such, chemical thermodynamics is a general physical chemical approach for the analysis of appropriate data correlations in terms of established thermodynamic laws. Thus, chemical thermodynamics is not an alternative theory competing with other physical theories.

Due to novel single-particle techniques and progress in analytical theory, however, the original thermodynamic approach to MEP has to be substantially refined. In particular, it is realized that the experimental electroporation data of cell ensembles (and thus of electropore ensembles) in homogeneous suspensions are "double-average values" of the polar angular position and of (position dependent) local electric field strength. It is outlined below how the experimental parameters like pore distribution constants and fractional pore densities can be explicitly expressed in terms of positional averages of the (centrosymmetric) electropore reactivity confined to spherical shells with position-dependent local fields. The inclusion of these, so far often disregarded, average aspects and of the field-dependent "membrane conductivity factor" should resolve obvious discrepancies in data interpretations. For example, the actual data assumed to reflect the fractional increase of the field effect appear to be better described when the exponent of the field parameter is linearly dependent on the electric field strength, E. However, the concept of induced polarization (of the pore water) clearly is associated with a square dependence (E^2).

1.1.2 Ensemble Kinetics of Electroporation

It is essential that there is a fundamental difference between the rate of opening and closing of a single-pore event, as compared to the chemical kinetics of ensembles of membrane pores. The rate parameters of ensemble kinetics are obtained from the analysis of the entire time-course Sig(t) of an appropriate signal that reflects—either direct or indirect—proportionality to the density of pores. Here, pore density refers to the number of pores in the shell volume of the closed membrane system of a lipid vesicle or a cell.

The access to electric pore formation is usually indirect. It has to be checked whether the observed transport kinetics are rate limited by the structural changes (in the lipid–water interfaces) underlying the statistics of pore opening and closing events. For instance, the time course of forward light scattering (of polarized light) can indicate the entrance of water into the lipid membrane, and also global shape changes concomitant with vesicle volume changes. The time course of conductance parameters can indicate field-induced transmembrane flux of small ions. Spectroscopic methods are used to indirectly monitor membrane transport of larger molecules like optical dyes or dye-labeled DNA.

Here, it is practical to differentiate between "in-field" responses and "post-field" responses like the slower pore ensemble resealing phases. In classical electroporation kinetics, the actual time-course Sig(t) of the respective signal is analyzed in terms of (exponential) normal modes. Each exponential mode is characterized by the relaxation time constant and the mode amplitude. It is shown below that flux analysis requires "time-dependent rate coefficients."

In particular, the relaxation kinetic data of unilamellar lipid vesicles [6] and of densely packed CHO cells [7] are instrumental for elucidating the mechanism of pore formation. Already the actual time courses of poration events show peculiarities such as delayed parallel modes. In the low-field strength range, pore formation is roughly mono-exponential with time. The (field-dependent) relaxation times are in the range of 5 μs (low field) down to about 0.5 μs (high field). The post-field relaxation covers a longer time range (stretched exponentials); the "zero-field" time constants are in the range of 1 ms to 10 ms. Even if the ion concentration outside is much smaller than the inside concentration, the post-field response indicates no net transport of ions to the outside. So, the pores are conductive but do not indicate measurable net outflux. These "rapid" pores appear to be permselective either for single file cation flux or for single file anion flux; they are thus associated with local

Nernst potentials counteracting the field-induced potentials. They were called type-1 pores (P1), estimated average pore radius of 0.5 ± 0.2 nm.

At field strengths higher than a "threshold field strength," the initial exponential phase (of P1 pores) appears to saturate, and at about 5 μs the signal slowly increases. At about 5 μs, a delayed second transport mode appears. The delay is particularly pronounced when adsorbed macromolecules are present or when shape changes due to Maxwell stress lead to net outflux of intra-particle salt solution. This special feature suggests that a second type of larger pores is formed at the expense of P1 pores. The in-field time constants for the P2 pores are estimated to be in the range of 100–30 μs, estimated pore radius of about 1.0 ± 0.2 nm. The post-field response is extremely stretched and can cover the time range of 10–100 ms and even several seconds [7]. In the case of voltage clamp (i.e., the membrane field is forced to be constant), ion fluxes do not reduce the transmembrane field and pores may expand up to pore radii of 10 nm and more [8–12]. In both cases, structural longevity of pore ensembles at zero field underlies (and rationalizes) the large post-field membrane transport. This is particularly apparent when the unavoidable adsorption of (macro) molecules leads to prolonged membrane insertion and thus to delayed (zero-field) dissociation from the membrane. In addition, at higher field strengths, electromechanical Maxwell stress leads to elongation of lipid vesicles and cells in the field direction and to release of intracellular liquid. Adsorbed molecules facilitate, by curvature effects, both electroporation and material transport as well as shape changes.

In brief, the thermodynamic ensemble approach provides a comprehensive formalism for quantitative analysis and characterization of electroporation data of cells and tissue. Together with results obtained from flux analysis, using the concept of time-dependent flux coefficients, the thermodynamic and kinetic information provides not only a general guideline for the analysis of data correlations but also for the goal-directed design of experiments and apparatus for the various biotechnological and medical electroporation treatments.

1.1.3 Molecular Chemical Schemes for Pore Formation

The chemical electro-thermodynamic concept of membrane electroporation (MEP) views the primary effect of the induced membrane electric field as directly acting on the interface between bulk water and the hydrated polar head groups of lipids, leading to the formation of hydrophobic and hydrophilic pores. The chemical process may be viewed as field-induced cooperative rearrangements of n lipids L_{aq} (hydrated polar head *groups*). *Thereby aqueous pores* L_n and $L_n\,(W)$ are formed by water (W) entrance., according to

$$nL_{aq} \rightleftarrows (\mathrm{L}_{aq})_n \overset{+W}{\rightleftharpoons} (L_{aq})_n(W) \tag{1.1}$$

The water in the cluster-like configuration $(L_{aq})_n$ (W) is polarized stronger in the induced transmembrane field (E_m) as compared to the smaller field E in the bulk water. Therefore, water entrance into the higher electric field of the membrane site contributes strongly to the thermodynamic stability of small aqueous hydrophobic (*HO*) and larger hydrophilic (*HI*) or inverted pores [13]. The observed longevity (larger pore life times) of the ensemble of local pore structures is due to the local cooperativity of the ordered lipids in the highly curved pore wall of a hydrophilic pore. The steps from the closed membrane sites $(L_{aq})_n$ (denoted *C*) to the various pore states had been cast in the scheme: C $\rightleftarrows$ HO $\rightleftharpoons$ HI.

Field-induced rotational motions of the polar lipids in the curved pore wall (of the *HI* pore) also rationalize the huge acceleration of lipid flip-flop and other intra-wall motions such as the translocation of phosphoryl inositol from the internal membrane monolayer to the outer monolayer. Electric pulses of low-field intensity but longer pulse duration facilitate, via electroporation, both endocytotic uptake of external particles and exocytotic release of intracellular components.

The structural feature of pore longevity is also instrumental for rationalizing some of the voltage pulse data for pulse train combination modes of high-voltage (HV) pulses and low-voltage (LV) pulses and the effects of a time interval between the pulses [14]. Viewed afterward, the originally applied "exponential field pulses" [1, 2], with the longer RC-circuit discharge times, combine the HV part of the initial time course with the LV part of the slower part.

As compared to the chemical thermodynamic approach, it appears that the majority of detailed (physical) electroporation theories primarily address the physical aspects like the electric polarization term of pore water, pore line tension, and membrane surface tension, membrane curvature and bending rigidity, and other factors in terms of "pore radius." They do not explicitly address the chemical free energy changes of rearrangements of the lipids and water in pore wall formation and pore resealing.

1.1.4 Global Electroporation Scheme

In order to understand the electroporation phenomena on the membrane level, it is essential to recall that an electric field (of a voltage pulse) acts (here via the larger induced membrane field) as a force on chemical structures, vectorial and simultaneously on all polar (ionic and dipolar) groups of the membrane components. Note, the induced membrane field originates from the mobile ions near the two surface sides of the dielectric membrane (Maxwell–Wagner ionic polarization). When macromolecules adsorb on the membrane surface, the adsorption complex necessarily is situated within the induced field across the "complex."

It is common experience that cellular electroporation data correlations can be satisfactory rationalized in terms of a global overall scheme for the structural transitions of pore formation and pore resealing viewed as cascades of field-sensitive closed membrane states (*C*) and a sequence of porous states (*P*):

$$(C) \overset{+W}{\rightleftharpoons} (P) \rightarrow X(Sig) \tag{1.2}$$

In scheme (1.2), the term $X(Sig)$ is to indicate that the preceding structural transition is necessarily coupled to transport phenomena and transport feeds back to the extent of transition. The transport of small ion fluxes leads to reduction of the induced transmembrane field; this reduces the extent of poration even if the external field remains constant. Under the special condition of low external salt concentration, oscillations in the membrane field and in the global shape are observed [12].

The thermodynamic overall distribution constant K_p (E) for the overall two-state scheme is defined as

$$K_p(E) = \frac{[P]}{[C]} = \frac{[P]}{[P]_{\max} - [P]} = \frac{f_p}{1 - f_p} \tag{1.3}$$

The thermodynamic field dependence is denoted by K_p (E). The state density of potential pore sites is given by $[\mathrm{C}] = [\mathrm{P}]_{\max} - [\mathrm{P}]$, where the maximum value $[P]$ max (at field infinity) is used as a reference for the pore state density $[P]$. The overall fraction of pore states is given by

$$f_p(E) = \frac{[\mathrm{P}]}{[\mathrm{P}]_{\max}} = \frac{K_p(E)}{1 + K_p(E)} \tag{1.4}$$

The numerical value of f_p (E) is limited to the range $0 \le f_p \le 1$. (The previous restricted assumption of $K \ll 1$, reducing Eq. (1.4) to $\mathrm{f_p(E)} = \mathrm{K_p(E)}$, is justified only for the low-field range [6, 7].) The analysis of the field dependence of f_p (E) requires to first express the fractional extent of poration as f_p (E_m), i.e., in terms of the induced local membrane field E_m.

1.1.5 *Chemical Energetics of Field Effects*

The various poration phenomena are thermodynamically rationalized in differential form as dependence of K_p on the generalized thermodynamic forces in terms of a generalized Van't Hoff relationship (1.1). The total differential d ln $K(p,T,E)$ is expressed as partial change dT in the Kelvin temperature T, partial change dp in the pressure p, and partial change dE_m in the electric field strength E_m of the "locally active" electric field induced by the external applied field E. For a two-state structural equilibrium transition, Eq. (1.2), the general thermodynamic expression reads

$$RT\, d\ell nK(p,\, T,\, E) = \Delta_r H^o_{p,T}\, dt - \Delta_r V^o_{\mathrm{E},T}\, dp + \Delta_{\mathrm{r}} M^o_{p,T}\, dE_m \tag{1.5}$$

In Eq. (1.5), $R = k_B N_A$ is the gas constant, k_B the Boltzmann constant, and N_A the Loschmidt–Avogadro constant. Equation 1.5 covers the overall electroporation process in terms of the standard value $\Delta_r M^0 = M^0(P) - M^0(C)$ of the (molar) reaction dipole moment, "sonoporation" with the standard value of the reaction volume $\Delta_r V^0$, and "thermo-poration" and thermal aspects of laser "opto-poration" with the standard reaction enthalpy $\Delta_r H^0$. The standard value of the total molar reaction energy is $\Delta_r H^0(E_m) = \Delta_r G^0(E_m) + T\Delta_r S^0(E_m)$ and includes the standard value $\Delta_r G^0(E_m)$ of the reversible work potential and the standard value $\Delta_r S^0(E_m)$ of the molar reaction entropy at constant p,T.

The field effect distribution constant $K_p\,(E_m)$ is determined by the standard value of the Legendre-transformed Gibbs reaction energy $\Delta_r \hat{G}^0(E_m)$ in the field E_m. Using the transformation definition $\hat{G}\,(E_m) = G(0) - E_m M$, where G(0) is the ordinary Gibbs reaction energy (at $E = 0$) and M the projection of the total electric moment vector $\boldsymbol{M}$ onto the direction of the external field vector $\boldsymbol{E}$, the standard value is expressed as

$$\Delta_r \hat{G}^0(E_m) = \Delta_r G^0(0) - E_m \Delta_r M^0$$

Hence, in terms of the induced local field E_m at constant p,T, Eq. (1.5) is specified by

$$d\ell nK_p(E_m)/dE_m = \Delta_r M^0_{pp}/RT = \langle \Delta_r m_{pp} \rangle / k_B T \tag{1.6}$$

In Eq. (1.6), $\langle \Delta_r m_{pp} \rangle = \Delta_r M^0_{pp}/N_A$ is the average "molecular reaction moment." For field-induced polarization, it is given by $\langle \Delta_r m_{pp} \rangle = \langle \Delta_r \alpha_{pp} \rangle E_m$, where the reaction pore polarizability can be expressed as $\langle \Delta_r \alpha_{pp} \rangle = \nu_{pp} \varepsilon_0 (\varepsilon_{pp} - \varepsilon_{lip}), \varepsilon_0$ being the dielectric permittivity of the vacuum. The "pore polarization volume" ν_{pp} includes the pore wall and the aqueous inner part. The dielectric term is an effective dielectric constant (order of 50) for the pore volume ν_{pp}, and ε_{lip} (=2.3) is the dielectric constant of the lipid phase.

The cylindrical pore polarization model specifies the pore polarization volume as $\nu_{pp} = d_m \pi\, r^2_{\mathrm{out}}$ with the membrane thickness $d_m = 5$ nm and the "outer pore radius" r_{out}. The inner ring radius r_{in} specifies the aqueous volume as $\nu_{aq} = d_m \pi\, r^2_{in}$; thus the volume of the pore wall ring is $\nu_{pw} = d_m \pi (r^2_{\mathrm{out}} - r^2_{\mathrm{in}})$. Integration of Eq. (1.6) in the limits $E_m = 0$ and E_m, yields

$$K_p(E_m) = K_0 \exp[\langle \Delta\alpha_{pp} \rangle E_m^2 / 2k_B T] \tag{1.7}$$

In Eq. (1.7), $K_0 = K_p$ $(E = 0)$ is the apparent distribution constant for occasional fluctuative pore opening–closing events at zero-applied field. Usually K_0 is in the order of 10^{-4} or even smaller. At zero-applied field, $K_0 = f_0/(1 - f_0)$ refers to the zero-field fraction $f_0 = f_p(E = 0)$ of fluctuative pore events. In living cells f_0 is finite due to the membrane potential. Because of the very small values of K_0, we may use the approximation $f_0 = K_0$. It is outlined below that the primary parameter obtained from thermodynamic data analysis is the "θ - average reaction polarizability" $\langle \Delta_r \alpha_{pp} \rangle$.

1.1.6 The Induced Potential Difference

The data of classical lipid membrane electroporation suggest that the externally applied electric field E induces the larger membrane field E_m that, in turn, "operates" on the membrane structures. Obviously, E_m is parallel to E, and in spherical membrane shells, the curvature conditions the asymmetry of the field effect on the lipid head groups, different on the anode side and the cathode side.

For porated membranes, the local induced field E_m determines the current density vector j_m for the cross-membrane ion fluxes (of both, the cations and the anions): $j_m = \lambda_m(-\nabla\varphi_m) = \lambda_m E_m$, where λ_m is the conductivity of the membrane (referring to all conductive pores). Because of continuous misconceptions in the literature, some further details have to be refined. In brief, for spherical membrane shells in conductive media, after rapid buildup of both, the ion concentration polarization and the interfacial charge separation polarization (Maxwell-Wagner), the *stationary value* $\Delta\varphi_{\text{ind}}(\theta, E)_{ss}$ of the polarization buildup kinetics $\Delta\varphi_{\text{ind}}(t) = \Delta\varphi_{ind}(\theta, E)_{ss}(1 - \exp[-t/\tau_{pol}])$ at the polar angle θ to the direction of the external field E (Fig. 1.1) is given by

$$\Delta\varphi_{\text{ind}}(\theta, E_m)_{ss} = -g \cdot a \cdot E \cdot f_\lambda |\cos\theta| \tag{1.8}$$

In Eq. (1.8), $a = a_{out}$ is the outer radius of the spherical membrane shell. The inner radius is given by $a_{in} = a - d_m$, membrane thickness d_m. Note, the polar angle for spherical geometry is confined to the range $0° \leq \theta \leq 180°$ (or $0 \leq \theta \leq \pi$), thus $1 \geq |\cos\theta| \geq 0$. For ultra-short pulses (ps, ns) or for the high frequency range (>10 Hz) of a.c. pulses, the Maxwell-Wagner limit of $g = 3/2$ applies. For short pulses (μs, ms and longer) or for the low frequency range of a.c. pulses, the Dukhin g-term *g(Du)* applies. This term contains the Dukhin factor, which for spherical particles is given by $Du_{sp} = \lambda_{surf}/a_{sp}\,\lambda_{sx}$, where a_{sp} is the outer particle radius, $\lambda_{surf} \approx 1\text{-}3 \times 10^{-9}$ S the surface conductance (linear surface conductivity of O'Konski) and λ_{ex} the medium conductivity. The *g(Du)* term thus accounts for the coupling of the polarization ion fluxes to the ionic content of the suspending medium.

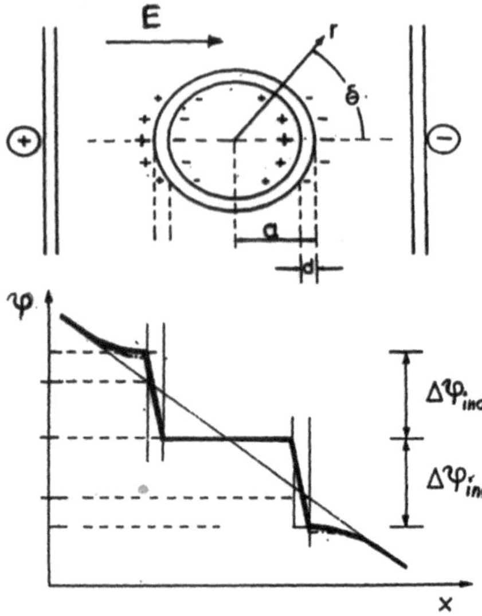

$$\vec{E} = -\vec{\nabla}\varphi\,,\ \Delta\varphi_{ind} = -Ff_\lambda(E\cdot a)$$

$$\Delta\varphi_{ind}^{\ominus} = -\frac{3}{2}E\cdot a\cdot f_\lambda\cdot|\cos\Theta|$$

$$E_m^{\ominus} = -\frac{\Delta\varphi_{ind}^{\ominus}}{d} = \frac{3}{2}\cdot\frac{a}{d}\cdot E\cdot f_\lambda\cdot|\cos\Theta|$$

$$f_\lambda = 1-\lambda_m\cdot\frac{a}{2d\lambda_{ex}} \qquad (\lambda_{in} >> \lambda_{ex})$$

Fig. 1.1 Scheme for the membrane field amplification by interfacial charge separation polarization (Maxwell–Wagner). Cross section through the center point of the spherical shell (centrosymmetric case) of outer radius $r = a$. The electric potential $\varphi(x)$ through the center point as a function of the x-coordinate for the presence of the dielectric sphere as compared to the absence of the object. The field-induced potential differences $\Delta\varphi_{ind}$ are indicated for the anode (*left*) membrane side and for the cathode (*right*) side. The field amplification by the ratio (a/d_m) is specified for the purely dielectric case of $F = g = 3/2$. For the anionic cell wall structures of bacteria and yeast cells and other walled cells (glycocalix shells of up to 200 nm thickness), the Dukhin g-term, $g(Du)$, applies. Note, at finite ionic strength, $g(Du) < 3/2$; see the text

Further, the validity of Eq. (1.8) is restricted to larger objects like cells ($d_m \ll a$) and to very low membrane conductivity λ_m, i.e., $\lambda_m \ll \lambda_{ex}$, λ_{in}, where λ_{ex} is the external medium conductivity and λ_{in} the conductivity of the cell interior. For this case, the conductivity factor for spherical shells in ionic media reduces to [9, 15, 16]

$$f_\lambda = 1 - \lambda_m \frac{2\lambda_{ex} + \lambda_{in}}{2\lambda_{ex}\lambda_{in}d_m/a} \tag{1.9}$$

Note that Eq. (1.8) is consistent with the Maxwell definition of the vector **E** as the negative gradient ($E = -\nabla\varphi$) of the electric potential φ. Since the induced membrane potentials are dependent on position, experimental ensemble properties of spherical shells reflect averages of the contributions of all ring positions (i.e., all polar angles θ of the spherical shell).

The induced membrane field $E_m(\theta)$ at the polar angle θ in the direction of **E** (Cartesian coordinate system) is given in terms of membrane voltage $U_m(\theta) = -\Delta\varphi(\theta)$ by

$$E_m(\theta) = \frac{U_m(\theta)}{d_m} = \frac{-\Delta\varphi(\theta)}{d_m} = E_m(0)|\cos\theta| \tag{1.10}$$

In Eq. (1.10), $E_m(0)$ is the (maximum) membrane field at the pole caps at $\theta = 0°$ and at 180°. In the (required) notation of the absolute $|\cos\theta|$, there is no change in sign when going from the "cathode" pole cap, in the range $1 \leq \cos\theta \leq 0$, to the anode one in the angular range $0 \geq \cos\theta \geq (-1)$. Hence Eq. (1.10) is unrestrictive, i.e., generally applicable for the description of current flows *in the direction* of the respective E vector, through the two electroporated hemispheres of a spherical membrane shell in an external field.

1.1.7 Contribution of Natural Membrane Potential

Living cell membranes have a finite natural membrane potential (difference) $\Delta\varphi_{\text{nat}}$. In electrophysiology, the membrane potential is defined as $\Delta\varphi = \varphi(i) - \varphi(\text{o})$, where the outside potential is taken as the reference $\varphi(\text{o}) = 0$. In many cells, the membrane potential is dominated by the Nernst potential $\Delta\varphi(K^+)$ for K^+-ions. Due to the dominant contribution of permselectivity for K^+, we may use the approximation $\Delta\varphi_{\text{nat}} = \Delta\varphi\ (K^+)$. Typically, $\Delta\varphi_{\text{nat}} = -70$ mV. The contribution to $\Delta\varphi_m$ of the natural membrane potential (difference) $\Delta\varphi_{\text{nat}}$ is readily incorporated. Due to the natural law of additivity of electric potentials, the total term $\Delta\varphi_{\text{m}}$ (θ) at the angular ring position θ is given by $\Delta\varphi_m(\theta) = \Delta\varphi_{\text{ind}}(\theta, E)_{ss} - \Delta\varphi_{nat}\Delta\ sgn(\cos\theta)$, where $sgn(\cos\theta) = \cos\theta/|\cos\theta| = \pm 1$ is the sign of $\cos\theta$. Note, $\Delta\varphi_{\text{nat}}$ is independent of the angle θ *(and of E)*.

Insertion of Eq. (1.8) yields [16]

$$\Delta\varphi_m(\theta) = -g \cdot a \cdot E \cdot f_\lambda |\cos\theta| - \Delta\varphi_{nat} sgn(\cos\theta) \tag{1.11}$$

As required, the total membrane potential in the absence of E_{ind}, but viewed in the direction of the external field vector E, is given by $\Delta\varphi_m = \Delta\varphi_{\text{nat}} sgn(\cos\theta)$.

The geometrical amplification factor a/d_m of a larger spherical shell, for instance, radius $a = 5$ μm and $d_m = 5$ nm is as large as $a/d_m = 10^3$. It is this geometrical amplification which rationalizes that comparatively small external fields in the range of $E = 1$ kV/cm are amplified to yield the large field strength $E_m = 10^3$ kV/cm, which then has such high (electroporative) power on the membrane structure.

1.1.8 Positional Averages

It is recalled that the membrane potentials, which are induced by the external field, are position dependent. For proper data analysis, it is instrumental to realize that the respective experimental electroporation parameters of homogeneous cell suspensions reflect the angular average of the different local poration reactivities (pore densities) at the polar angle θ.

The overall membrane field E_m induced by the applied field E can be expressed in terms of the positional average $\langle|\cos\theta|\rangle$. Since the field is poratively active as E^2, the effective field strength is

$$E_m = \langle E_m^2(\theta)\rangle^{1/2} = \left(E_m^2(0)\langle|\cos^2\theta|\rangle\right)^{1/2} = g(a/d_m)E\,f_\lambda(\langle|\cos^2\theta|\rangle)^{1/2} \quad (1.12)$$

Note, for charged surfaces and pulses of duration larger than μs or low frequency a.c. pulses, the g factor has to be calculated from the Dukhin factor with the respective medium conductivity. For ultra-short pulses or high frequency a.c. pulses, the geometric factor is $g = 3/2$.

On the same line, the distribution constant K_p (E_m) is given by $\langle K(\theta)\rangle$. Therefore, analogous to the expression in Eq. (1.7), the average distribution constant is expressed $\langle K(\theta)\rangle$ as

$$\langle K(\theta)\rangle/K_0 = \exp\left[b\langle\cos^2\theta(E_m)\rangle E^2 f_\lambda^2\right] \quad (1.13)$$

The quantity b is given by

$$b = \frac{\langle\Delta\alpha_{pp}\rangle}{2k_BT}g^2(a/d_m)^2 \quad (1.14)$$

The function $\langle\cos^2\theta(E_m)\rangle$ represents the field-dependent reactivity average and is limited to the range $1/3 \le \langle\cos^2\theta(E_m)\rangle \le 1$ for $E = 0$ and the virtual "E-infinity," respectively.

Comparing Eq. (1.7) with Eq. (1.13), the experimental quantity K_p (E) is explicitly given as

$$K_p(E) = K_0\exp\left[\mathrm{b}\langle\cos^2\theta\ (\mathrm{E}^2\mathrm{f}_\lambda^2)\rangle\rangle\right] \quad (1.15)$$

It is recalled that K_0 contains the "zero-field" standard value of Δ_rG^0 containing the sum of the chemical structural contribution of pore fluctuations, the interfacial parameters of pore line tension, and membrane surface tension, membrane curvature, and bending rigidity [9].

Using $f_p(E) = \langle f_p(\theta)\rangle$, the experimental pore density fraction f_p (E) is, in line with Eq. (1.4), specified as the polar angle average quantity:

$$f_p(E) = \frac{K_p(E)}{1 + K_p(E)} = \frac{K_0 \exp[b\langle \cos^2\theta(E_m)\rangle E^2 f_\lambda^2]}{1 + K_0 \exp[b\langle \cos^2\theta(E_m)\rangle E^2 f_\lambda^2]} \tag{1.16}$$

At the half-point of the fractional electroporation, we have $f_p = 0.5$; the respective half-point field strength is $E_{0.5}$. Because at $f_p = 0.5$, the half-point value of K_p is $K_p(E_{0.5}) = 1$. For this case, we have in logarithmic form $\ell n K_0 = -b\langle \cos^2\theta(E_{0.5})\rangle E_{0.5}^2 f_\lambda^2$. As a practical estimate, we may use $\langle \cos^2\theta(E_{0.5})\rangle f_\lambda^2 = 1/2$, including $f_\lambda = 1$. Hence $K_0 = \exp[-(b/2) \cdot E_{0.5}^2]$ may be estimated and, later in the analysis, subjected to iterative refinement.

1.1.9 Experimental Data Correlations

If the measured quantity *Sig* (*E*), e.g., signal due to uptake of a dye or of DNA or RNA, can be judged to reflect proportionality to the (pore) fraction f_p (*E*), see Eq. (1.3), the practical approximation is defined as

$$f_p(E) = Sig(E)/Sig(\max) = [P(E)]/[P]_{\max} \tag{1.17}$$

Indeed, the majority of measured field dependencies of *Sig* (*E*) look as if they reflect Eq. (1.16), starting sigmoid exponentially and continuing toward saturation. Due to the usually very low value of K_0, the experimental data correlation *Sig* (*E*) appears as an "onset E^2 scale-shifted" exponential. This characteristic feature deceives a "threshold field strength" E_{th}, operationally denoted as the field strength range where the Sig (E_{th}) becomes "visible" under the given experimental conditions. In single-pulse experiments, the value of E_{th} is dependent on the pulse length (and the observation time).

It is noted that, at the start of the data analysis, the term *Sig* (max) is at first a virtual quantity similar to P(max) and later accessible by iterative fitting.

In any case, the evaluation of the characteristic system parameters from the experimental data is rather involved and has to proceed in several steps. The first step is to explore the so-called "low-field strength range" of the data correlation *Sig* (*E*) as a function of the applied field strength *E* by using the small-field approximations $f_\lambda{}^2 = 1$ and $\langle \cos^2\theta(E)\rangle = 1/3$. Recalling Eq. (1.15) and using Eq. (1.17), the initial part (up to about $f_p = 0.2$) of the experimental data correlations *Sig* (*E*) can be analyzed in the explicit form:

$$K_p(E) = K_0 \exp\left[b' E^2\right] = Sig(E)/Sig(\max) \tag{1.18}$$

Practically, the low-field data are evaluated according to

$$\ell n|Sig(E)| = \ell n|Sig(\max)K_0| + b' E^2 \tag{1.19}$$

The slope yields the (experimental) term $b' = (1/3)\, b$ and the intercept (at $E = 0$) is $\ell n|Sig(\max)K_0|$. The term Sig(max) can be determined from the product Sig(max) K_0 if an estimate of K_0 is used. The value of K_0 may be estimated in two ways. The sigmoidal data function Sig (E) allows the estimation of a turning point range at the half-point field strength $E(\text{at } f_p = 0.5) = E_{0.5}$ by using the maximum of the derivative $d|Sig(E)|/dE$. If we now apply the slope value b', K_0 is estimated from the relationship $\ell n\, K_0 = -(b/2)\, E_{0.5}^2$, using the estimate $\langle \cos^2\theta(E_m) \rangle = 1/2$ as a mean value for this initial range of the field strength. Since $Sig(\max) = 2\, Sig(E_{0.5})$, K_0 can also be estimated from the intercept $\ell n|Sig(\max)K_0|$ by inserting a rough visual estimate of Sig(max).

In the next step, the entire "zero-membrane conductivity transition curve," i.e., for the case $f_\lambda{}^2 = 1$, is calculated according to $f_p(E) = K_p(E)/(1 + K_p(E))$ using the estimates for K_0, the slope value b, and the exact function $\langle \cos^2\theta(E_m) \rangle$ in the limits of 1/3 and 1. The differences between this curve and the actual data points yield the conductivity factor as a function of the applied field strength. From the value of $\langle \Delta\alpha_{pp} \rangle$, eventual pore polarization volume and pore radius may be calculated.

1.1.10 Flux Coefficient Integrals

The newly introduced concept of time-dependent flux coefficient functions has turned out to be instrumental for proper flux analysis [16] of the post-field conductance relaxations (resealing curves) reflecting transport through a decreasing number of pores. It is practical to start the analysis of the measured signals, for instance, conductance relaxations g(t) with the introduction of reduced signals like signal ratios Y(t). For example, the time course of the signal ratio $Y(t) = (g(t) - g(0))/g(0)$ $= (I(t) - I(0))/I(0)$ of the after-field currents (I) or conductances (g), relative to the zero time values $I(0)$ and $g(0)$ before pulse application, looks like an increasing "stretched exponential" of the Kohlrausch type. However, the nonequilibrium concept of a flux coefficient function k(t) provides an expression in terms of an "exponential of an exponential." Explicit, instead of the simple exponent term kt, where k is a constant flux coefficient, the exponent term in the respective $Y(t)$-function is an integral $\int k(t)dt$, such that $Y(t) = Y(\max) \exp[-\int k(t)dt]$. For instance, if in the simplest case of monoexponential pore resealing during the efflux phase, the efflux coefficient function is given by, $k_R(t) = k_0(t_p, E)\exp[-t/\tau_R]$, where $k_0(t_p, E)$ is the initial value of the flux coefficient at the end of the pulse (at t_p and E) and τ_R is the ensemble time constant for pore resealing. Note, at this time point the new time scale starts at $t = 0$ for the post-field efflux function $Y(t)$. The integral is given by

$$\int k_R(t)dt = k_0(t_p, E)\tau_R[1 - \exp(-t/\tau_R)]\} \quad (1.20)$$

Insertion of Eq. (1.20) yields the integral function [7, 9]:

$$Y(t) = Y(\max)\{1 - \exp\{-k_0(t_p, E)\tau_R[1 - \exp(-t/\tau_R)]\}\} \quad (1.21)$$

In Eq. (1.21), $Y(\max)$ refers to the virtual maximum value for the case of complete equilibration between the intracellular ion concentration and that of the external medium. The stationary value $Y(ss) = Y(t \to \infty) \leq Y(\max)$ may be estimated by extrapolation (parallel to the abscissa). It refers to the stationary value of resealing (before concentration equilibration) and is given by

$$Y(ss) = Y(\max)(1 - \exp[-k_0(t_p(E)\tau_R)]) \quad (1.22)$$

In terms of pore fraction, the ensemble resealing of the porous area in the absence of the field is given by $f_p(t) = f_p(E)\exp[-t/\tau_R]$.

This procedure of analysis was, for instance, successfully applied for the resealing phase of densely packed CHO cells. As rationalized with Eq. (1.21), the measured transport curves are therefore exponentials of exponentials. The actually measured curves can deceive stretched exponentials (of the Kohlrausch type). As seen in the case of exponential decrease in the pore fraction, the post-field kinetics provides mechanistic details of the long-lived electroporated membrane states. This analytical framework has been used to obtain the values of k_0 (E, t_E) as a function of the field strength E and of the pulse duration t_E, respectively, and of the (field-independent) time constant τ_R of the resealing process [7].

Complementary to efflux of ions into the outer cell compartment, influx of a dye into a cell, for instance, is analyzed in a similar way [11]. If the measured signals $Sig(t)$ can be expressed as the fraction $y(t) = Sig(t)/Sig(ss)$, where the stationary signal is given by the amplitude, $Sig(ss) = S(t \to \infty)$. The proper differential equation describing the signal increase is formally analogous to the linear form of $dy(t)/d(t)$:

$$\frac{d\,y(t)}{dt} = -k_p(t)[y(t) - y(ss)] \quad (1.23)$$

where k_p(t) is the influx coefficient function and the stationary term, $y(ss) = Sig(t \to \infty)/Sig(ss) = 1$. Integration of Eq. (1.23) yields the integral flux equation for the stretched increasing (or the delayed increasing) time course:

$$y(t) = y(ss)\left(1 - \int k_p(t)dt\right) \quad (1.24)$$

In the linear case, the flux coefficients for the (apparently stretched) increase are $k_p(t) = k_0\exp[-t/\tau_p]$ and, for the delayed increase, $k_p(t) = k(ss)\exp[1 - t/\tau_p]$, respectively.

1.1.11 Elementary Kinetics of Electroporation

Since pore formation is a chemical structural process involving lipid reorganizations and interfacial water molecules, see Eq. (1.1), the kinetic analysis uses "chemical" rate equations. In terms of the scheme in Eq. (1.2), the individual rate equation for a particular relaxation mode of the increase in the pore density is given by

$$\frac{d[P(t)]}{dt} = -k_p(t)[P(t) - P(ss)] \tag{1.25}$$

In Eq. (1.24), *P(ss)* is the conventional relaxation amplitude and the term, $k_p(t)$ represents the reaction flux coefficient function. It replaces the normal relaxation rate term $1/\tau_p = k_p + k_{-p}$, where *kp* is the ensemble poration rate coefficient and *k-p* is the pore closing ensemble rate coefficient. In the case of a sigmoid onset of the integral relaxation curve, we may use the membrane conditioning coefficient (dimple formation preceding the actual pore formation process) in the form of $k_p(t) = k(ss)\exp[1 - t/\tau_{cond}]$.

Here, *k(ss)* is the constant stationary coefficient and τ_{cond} the ensemble time constant of the respective conditioning process. We may start with the assumption of proportionality between the measured quantity Sig(t) and [P(t)]. Consequently the integrated rate equation is given by

$$Sig(t) = Sig(\max)\{1 - \exp[-k(ss)(t - \tau_{cond}(1 - \exp[-t/\tau_{cond}]))]\} \tag{1.26}$$

In Eq. (1.26), the—at first—virtual term $Sig(\max)$ is given by the measured stationary value

$$Sig(ss) = Sig(\max)\left(1 - \exp[-k(ss)\tau_{cond}]\right) \tag{1.27}$$

Analogous to K_p, the field dependence of the rate coefficient is given by the Arrhenius-like equation: $k(E) = k(0)\exp[-G^*(E)/RT]$, respectively, for the on-rate coefficient and the off-rate coefficient, where $G^*(E)$ is the respective field-dependent Gibbs activation free energy [12].

In each case, however, it must be checked which equation must be applied and whether existing equations have to be modified or expanded [12], as dictated by proper physical chemical reasoning along the fundamental laws of flux nonequilibrium thermodynamics.

1.2 Millisecond and Microsecond Pulse Effects: An Introduction

Richard Heller, P. Thomas Vernier, and Justin Teissie

Electric fields can be used to elicit specific responses from biological cells. Research exploring these effects has been ongoing since the 1960s [17–22]. The electric fields are typically applied as a series of short-duration pulses and are easily obtained by applying a voltage pulse between a set of two conducting electrodes in contact with the sample. The time course and the magnitude of the field are directly controlled by the voltage generator and the electrode geometry. The voltage is obtained by charging a capacitor that is discharged under the control of a trigger in an electronic circuit that controls the decay. This can be an exponential decay or a square pulse of a controlled duration. Pulse duration (or decay) is from micro- to milliseconds (see details in Chap. 2). The specific effect on the cell(s) exposed to the field is dependent on the pulse characteristics (amplitude, duration, and number). With micro-millisecond pulse durations, the effect is predominately on the cell membrane and can result in reversible or irreversible permeabilization. Reversible permeabilization also known as electroporation and electrotransfer can be used to move molecules from one side of the membrane to the other, including large molecules such as proteins and nucleic acids. When pulse characteristics are chosen, appropriately cells will typically survive the reversible permeabilization process. Irreversible permeabilization (electroporation) is done to induce cell death. This approach can be used to effectively ablate tissue such as solid tumors without causing damage to vital structures. Another result of the application of pulse electric field is fusion between two cells that are in contact when the field is applied.

1.2.1 History

There is evidence through scientific history of the use of electric fields in biomedical applications. This includes evidence from the first century that Largus utilized electric fields as a treatment for taste and headaches [23]. In the eighteenth century, several scientists utilized electric fields to evaluate the effects on animals, human skin, muscle contractions, and basic responses of humans following electrical stimulation [24–28]. In the twentieth century, work is more focused on the effects on cells and the potential uses of these fields. In 1959, Pauly and Schwan reported on an electrophysical model of a cell [29]. Coster reported on the ability to use electricity to punch through biological membranes [17], and Hamilton and Sale demonstrated how electric pulses could cause damage to cell membranes [30].

Following these breakthroughs in the 1960s, work began to evaluate utilizing electric fields to manipulate cells and cell membranes. In the 1970s several breakthroughs occurred. The work during this decade focused on the membrane effects following brief exposure to intense electric fields. Neumann and Rosenheck in 1972 reported on permeabilization of vesicular membranes [3]. In 1974, Zimmermann et al. reported on increasing membrane conductance [22]. In the late 1970s, focus shifted more toward membrane breakdown and how this could be utilized. Kinosita and Tsong demonstrated the electroporation of erythrocytes in 1977 [31]. Electrical stimulation and fusion of plant protoplasts were accomplished in 1979 by Senda et al. [32]. This was followed in 1981 with Scheurich and Zimmermann successfully fusing different plant protoplasts [33]. Transient electroporation in phospholipid vesicles was demonstrated by Teissie and Tsong in 1981 [34]. Gordon and Seglen demonstrated reversible permeabilization to accomplish small-molecule uptake in 1982 [35]. In that same year, Neumann et al. demonstrated that genes could be delivered to mammalian cells using electric pulses [1]. Also in 1982, Teissie et al. demonstrated electrofusion of fibroblasts [36]. Potter reported on a standard method for in vitro gene electrotransfer in 1984 [37]. Sugar, Neumann, and Chizmadzhev reported on a model of aqueous pore in the late 1970s early 1980s [13, 38]. The work performed in the 1970s and 1980s formed the foundation for the current description of both electroporation and electrofusion. In addition, the studies performed during this era made critical contributions to the basic description of the interaction between electric pulses and biological membranes. This confirmed that intense electric fields could be used to destabilize membranes to facilitate transport and that this effect could be reversible, preserving the cell viability.

In the late 1980s, the focus shifted to utilizing electric fields for specific applications. Electrofusion was utilized to facilitate the production of monoclonal antibodies [39]. In 1986, Okino and Mohri reported on using pulse electric fields to deliver chemotherapeutic agent (bleomycin) to a tumor in a mouse model [40]. Mir et al. optimized this approach demonstrating the potential as an anticancer therapy [41]. This approach was later expanded to include other chemotherapeutic agents, calcium delivery, electrode devices, and administration routes [42–44]. This work was translated into clinical use with the first clinical results being reported by Mir et al. in 1991 [45]. The in vivo delivery of genes was first reported in 1991 by Titomirov purpose; while the delivery was done in vivo, the expansion and detection were done in vitro [46]. The first complete in vivo delivery of plasmid DNA was accomplished in 1996 by Heller et al. [47]. The first delivery to a solid tumor was accomplished in 1998 by Rols and Teissie [48] and to muscle by Aihara et al. [49] The delivery of DNA has also been translated to clinical use with the first clinical trial being reported in 2008 by Daud et al. [50] The use of pulse electric fields has also been used for transdermal delivery as demonstrated by Prausnitz and Weaver in 1993 [51, 52]. Electric fields can also be used to directly ablate tumors by performing irreversible electroporation. This was initially demonstrated by Davalos, Mir, and Rubinsky in 2005 [53]. In the late 1980s and early 1990s, Grasso et al. demonstrated that electric pulses could be used to fuse cells directly to tissue both in vitro and in vivo [54, 55].

1.2.2 *Physical Effects on Membranes (Experiments and Modeling)*

The first effect of applying a voltage pulse on a conducting sample is Joule heating. The temperature of the sample is increased. This affects the thermodynamic properties of the membrane. The amount of heating is dependent on the number of pulses and the time between pulses. The shorter the time between pulses, the more of an effect on heating [56].

The second effect is to induce electromechanical stress on the cells. The molecular organization of the membrane is altered. Charged species are submitted to an electrophoretic drift. The domain organization in the membrane is changed [57].

Cells can be considered as "spherical capacitors." The membrane is indeed a dielectric thin layer. The external field induces a charge loading of the membrane that is position dependent due to the vectorial character of an electric field. This charging time ranges from submicrosecond to a few microseconds. As a result, the transmembrane potential is heterogeneously modified during the application of the field pulse [58].

Details are in Sects. 4.1, 4.2, and 4.4.

1.2.3 *Primary Biological Effects (Cytoplasmic Content Changes and Swelling)*

The most dramatic consequence is the observation of a membrane alteration to a permeabilized state. Polar compounds that cannot freely cross the membrane under normal resting conditions are observed to freely diffuse across it after application of pulses with appropriate characteristics. This new organization can be reversible or irreversible [3].

Under reversible conditions, the cytoplasmic content is going to be changed. Due to the leakage, secondary messengers leave the cytoplasm. At the same time, new species will flow inside such as Ca2+. This results in a dramatic alteration of the cell organization. Organelles (nucleus) and cytoskeleton are the main targets of these changes. An additional change is that the cell undergoes osmotic swelling. Those primary biological effects are secondary consequences of the physical trigger (the electric field pulse). These changes are transient as the membrane is going to reseal and recover its characteristic-specific permeability. Through an active process (dependent on the energetic reserves of the cell), the cytoplasm will be brought back to normal [59–61].

Under irreversible conditions, the membrane cannot recover and cell death follows. When cells are aligned properly during the administration of pulses, fusion may occur. This will result in a multinucleated cell. Cells can survive this process [62].

Details are in Sects.. 4.4 and 4.9.

1.2.4 Secondary Biological Effects (Fusogenicity, Apoptosis, etc.)

Even when membrane permeabilization is reversible, the cytoplasmic alteration may result in a cascade of reactions involving the organelles. In these circumstances, it is possible that the cell will undergo apoptosis.

Membrane electropermeabilization is a complex multimolecular event affecting the interfacial properties of the cell surface. When cells are brought into contact during (or after) the electric pulse application, membrane merging between the two partners is observed as well as their content mixing. This is the direct evidence that membrane fusion is mediated by the electric pulse application.

Details are in Sects.. 4.4, 4.6, 4.7, and 4.9.

1.2.5 Medical Applications

A major application of administration of pulse electric fields is to deliver molecules with therapeutic or prophylactic potential to the interior of cells. Cell membrane is an important protective barrier for cells. However, the membrane can also be a barrier preventing therapeutic molecules from reaching their intracellular targets. In the case of chemotherapeutic agents, the influx or efflux of the drug can be influenced by the presence or absence of specific receptors or transport proteins that move the drug out of the cell. The use of pulse electric fields can bypass this process by giving the agent a more direct path into the cytosol. By allowing uploading of hydrophilic compounds to the cytoplasm, electric pulses give access to their cytoplasmic target to drugs that can only very poorly cross the plasma membrane spontaneously. In addition, delivery in this manner will typically result in higher concentrations of the drug within the cell thereby reducing the impact of transport proteins moving drug out of the cell. Lower drug concentrations can be used to obtain the clinical effect minimizing the secondary effects of chemotherapy. This approach has been named electrochemotherapy (ECT) and has been under development since the middle of the 1980s and is now a routine practice in Europe, where the equipment is approved by the regulatory agencies [63].

Delivery of nucleic acids is another major application of reversible permeabilization and its use has been steadily growing. Nucleic acids are significantly larger than drugs and cannot easily pass through the cell membrane. The administration of pulse electric field with the appropriate characteristics provides a means to move nucleic acid into the cell even within a tissue. Immune responses against the protein coded by the transferred plasmid can be induced resulting in efficient delivery of DNA vaccines. Another promising approach is to boost the immune expression by the expression of cytokines. Common target tissues are the skin, tumor, or muscle that are easily accessible for the electrodes and therefore for a calibrated field pulse delivery. The approach has been successfully translated from laboratory and preclinical models to both human and veterinary applications [50, 64–68].

Membrane fusion can be used for the production of hybridomas for the production of monoclonal antibodies [39]. It can also be used to make phenotypical changes within cells.

Pulse electric fields can also be used to have direct effect on cells and tissues. Using pulse characteristics that result in irreversible permeabilization can result in death of exposed cells [52]. This can be used to ablate solid tumors. In this case, the pulse electric field directly causes the destruction of the tissue instead of using the fields to deliver the effector molecule. The irreversible approach can be done in a manner that does not significantly elevate temperature which reduces or eliminates effects on vital structures. This clinical technology is now on the market.

This is described in detail in Chap. 5.

1.2.6 Biotechnological Applications

Membrane permeabilization through the use of pulse electric fields can be utilized to extract substances from cells. This can be utilized to enhance yields of products being collected from cells acting as "factories" for specific substances (yeast, microalgae, plants). This can be used to enhance olive oil production, lipid extraction, anthocyanin extraction from red cabbage, extraction of juice, or wine production. The technology has advanced to the point that electroextraction is now used at the industrial level through the development of flow process treating hundreds of liters per hour. A suspension of cells is pumped through a chamber, where electric pulses are delivered at high frequencies [69].

Irreversible permeabilization of bacteria and yeasts is now used in the food industry for cold sterilization preserving the taste of the electrotreated product. Here again flow processes are routinely used. A similar approach is under development to treat wastewater without the use of chemicals [70].

Gene engineering by electrotransfer is used routinely in applications for the obtention of GMO. A major advantage is that gene transfer is obtained on walled species (yeast, bacteria) not on protoplasts. This makes regeneration easier [71–74].

Membrane fusion between protoplasts can be used for the production of hybrid cells [75].

This is detailed in Chap. 6.

1.3 Nanosecond and Picosecond Pulse Effects: An Introduction

Karl H. Schoenbach and Stephen J. Beebe

1.3.1 From Classical Plasma Membrane Poration to Subcellular Membrane Effects

The effect of intense pulsed electric fields on biological cells and tissues has been the topic of research since the late 1950s. Intense means that the electric field is of sufficient magnitude to cause instant, nonlinear changes in cell membranes. The first paper reporting the reversible breakdown of cell membranes when electric fields are applied was published in 1958 [76]. The first report on the increase in permeability of the plasma membrane of a biological cell, an effect subsequently named "electroporation," appeared in 1972 [3]. The electric fields that are required to achieve electroporation depend on the duration of the applied pulse. Typical pulses range from tens of milliseconds with amplitudes of several 100 V/cm to pulses of a few microseconds and several kV/cm.

More recently, the pulse duration range has been shortened into the nanosecond range. The effects of such short pulses have been shown to reach into the cell interior [77]. Pulse durations are as brief as several nanoseconds, with pulse amplitudes as high as 300 kV/cm for short pulses [78]. A new field of research opens when the pulse duration is decreased even further, into the subnanosecond range. By applying such ultrashort pulses, it will also become possible to use wideband antennas, rather than direct-contact electrodes, to deliver the pulses to tissue [79].

Fig. 1.2 The effect of 7 μs long pulses with 1.1 kV/cm field amplitude (*left*) and that with 60 ns at 60 kV/cm amplitude (*right*) on cells. The electrical parameters were chosen such that the electrical energy for both cases is identical [85], with permission

The prediction that ultrashort pulses cause intracellular effects was based on a simple analytical, passive, and linear electrical model of the cell. In this model, the cell membranes, plasma membrane and subcellular membranes, were considered as capacitors and the cytosol and the medium inside the organelles as resistors [77, 80]. Passive and linear mean that changes in the properties of the cell structures, such as electroporation, are not considered. The assumption holds strictly, therefore, only for electric field amplitudes below those required for electroporation or nanoporation. The results of this model indicated that for pulses on the order of or less than the charging time constant of the plasma cell membrane, τ_m, the applied electric field reaches into the interior of the cell and therefore affects cell substructures and possibly cell functions. Since typical charging time constants of cells with dimensions of 10 μm are on the order of hundred nanoseconds, submicrosecond pulses were expected to cause different effects on cells than pulses with durations of microseconds and longer.

After the onset of poration, this simple linear, passive element approach is no longer applicable to describe electric field–cell interactions. The membrane becomes then an "active" cell element, with variable resistivity and variable permeability. Modeling of cells with "active" membranes had been a topic of multiple publications [81–84]. A detailed discussion on modeling by J. Weaver can be found in Sect. 4.2. Figure 1.2 shows the results of a continuum model [85, 86]. Here, the poration of the plasma membrane and subcellular membranes was compared for pulses of 7-μs and 60-ns duration. The electric field of the pulses was adjusted such that the energy density in both cases was identical. For 60-ns long pulses, (left) the cell membranes, plasma and internal, are fully exposed to the applied 60-kV/cm pulse, clearly demonstrating that nanosecond pulses allow us to affect subcellular membrane potentials. When the cell was exposed to a long pulse (7 μs) at 1.1 kV/cm, effects were only seen on the plasma membrane. The cell interior is shielded.

Under the conditions mentioned above (assuming that all membranes are equal, and the conductivity of organelle interiors and cytosol is the same), the voltage across the organelle membranes never exceeds the voltage across the plasma membrane. However, if we deviate from these assumptions and take into account that cell and organelle membranes differ electrically as do their interiors, it is possible to construct scenarios where the voltage across the organelle membranes exceeds that of the plasma membrane in a certain spectral range. This has been shown by using a simple equivalent circuit but assuming a lower capacitance of subcellular membranes (thicker membranes) than that of the plasma membrane [77, 87]. Similar results have been reported using a more elaborate continuum model, by Kotnik and Miklavcic [88]. Also, a cell membrane of low curvature requires a larger voltage to be electroporated than a highly curved membrane of small vesicles [89]. Consequently, under certain conditions, poration of subcellular membranes could be more likely than plasma membrane poration.

The first experimental study on the effect of nanosecond pulses on the intracellular structures showed that such pulses affect membranes of subcellular structures [77]. Human eosinophils were loaded with calcein-AM (calcein-

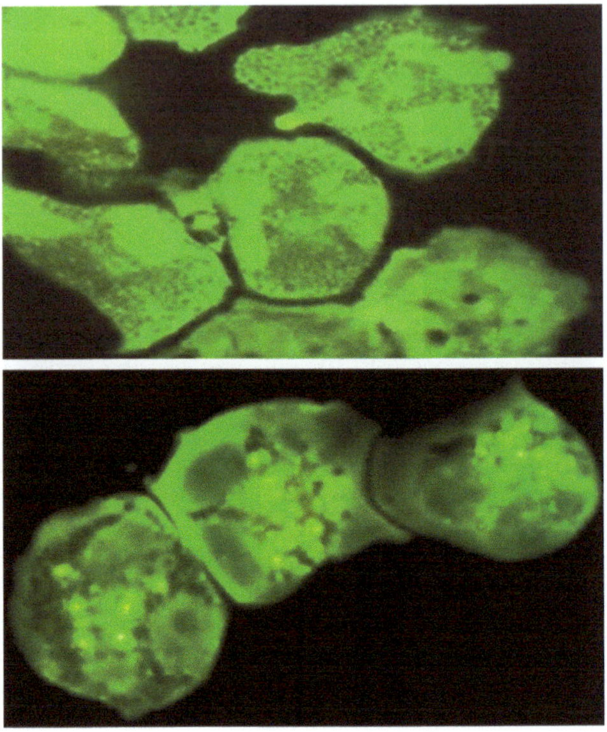

Fig. 1.3 Eosinophils (white blood cells) before (*top*) and after (*bottom*) the application of 60 ns pulses with electric fields of 50 kV/cm—photograph on bottom shows that inner structures have opened and taken up dyes—shown as "sparklers" [77] © [2001] John Wiley & Sons, Inc

acetoxymethylester), an anionic fluorochrome that enters cells freely and becomes trapped in the cytoplasm by an intact cell surface membrane following removal of the AM group. The granules in the eosinophils stay unlabeled because the cytosolic calcein is impermeant to the granular membrane. When 60-ns long pulses with an electric field amplitude of 50 kV/cm and higher were applied to the eosinophils suspended in Hanks Balanced Salt Solution, the granules, which were dark (nonfluorescent) before pulsing, began to fluoresce brightly (Fig. 1.3). This is a strong evidence for the breaching of the granule membranes and ionic binding of free calcein from the cytosol to the cationic granule components. On the other hand, the retention of the cytoplasmic calcein staining indicated that the surface (outer membrane) was not electroporated in such a way that it became permeable for these ions.

1.3.1.1 Nanoporation of Membranes

There is definitely an effect of nanosecond pulsed electric fields (nsPEFs) on the plasma membrane. Laser stroboscopy with a temporal resolution of 5 ns showed a rapid increase in membrane conductance in only a few nanoseconds when a 60-ns, 100-kV/cm pulse was applied [90]. The increase is assumed to be due to the formation of nanopores. This concept of nanoporation, the creation of a high density of "nanopores," had been introduced by Joshi [81] and by Weaver [86, 91]. These are pores of such small diameters (1.5 nm) that they become permeable only for small ions. Using whole-cell patch clamp to measure plasma membrane conductance changes resulting from nanosecond pulsed electric field (nsPEF) exposure, such nanopores have been observed following nsPEF exposure [92]. In Sect. 4.5 by A. Pakhomov, nanoporation is discussed in more detail. Molecular dynamics simulation is a method which is particularly suitable to study these nanopores. The importance of this method is in the visualization of membrane effects on timescale of nanoseconds, the inclusion of complex underlying physics, and the determination of critical electric fields for pore formation [93–96]. T. Vernier provides an overview on "molecular dynamics in electropermeabilization" in Sect. 4.3.

1.3.1.2 Scaling of Nanosecond Pulse Effects

Studies in the nanosecond and even in the near subnanosecond (700 ps) range indicate that nanosecond bioelectric effects, S, scale with the product of electric field, E, times pulse duration, τ, times the square root of the number of pulses, $N : S = S\left(E\tau N^{1/2}\right)$ [97]. Such a scaling law can be understood by assuming that the intensity of the bioelectric effect is proportional to the total number of ions passing through the cell membrane or, in other words, depends primarily on the current density (times the pulse duration). The square root dependence on the pulse number in in vitro studies can be considered the result of the thermally initiated statistical rotation of cells with respect to the electric field direction. It needs to be noted that this scaling law for nanosecond pulses has been derived for monopolar pulses. For pulses of different shape, it will need to be modified. Recent measurements of the diminished calcium uptake of cells when bipolar pulses are applied [98] could possibly be explained by drift as the major transport process for the ions through the nanoporated plasma membrane. The "reverse drift" model for bipolar pulses indicates a reduced importance of diffusion for nanosecond pulses and allows estimates on the recovery time of membranes [99].

1.3.2 Biological Effects

Whereas the previous section introduced biophysical effects of nanosecond pulses, especially effects on plasma membranes, the following will give a brief overview on biological studies of nanosecond pulsed electric fields: intracellular signaling, physiological functions, and apoptosis or regulated cell death (RCD) induction. In early experiments (1997–2005), new biological applications for nanosecond pulse power technology, which had been used for decades mainly for military purposes, were realized. This was of much interest because organisms had not evolved to respond to such brief and intense electric fields, which did not exist in nature. Thus, cells must use mechanisms that evolved for responses to other stimuli, most likely other stresses, especially when electric field intensities are high. (This aspect will be presented in a chapter devoted to biological responses.) In addition, given that nsPEFs had been shown to kill bacteria [100, 101], it seemed to be possible that this technology has application for cancer therapy. Therefore, mechanisms for cell death induction were of specific interest.

A variety of effects in living systems both in vitro and in vivo have been observed in response to nsPEFs depending on the pulse duration, electric field amplitude, and the number of pulses. With pulse repetition rates of one to two pulses per second, elevated temperatures could be circumvented, and responses were strictly due to nonthermal, electric fields. Since submicrosecond pulses were hypothesized to induce intracellular effects [77], early analyses were focused on effects on intracellular membrane permeability, calcium mobilization, and typical responses indicative of apoptosis, including phosphatidylserine externalization, caspase activation, cytochrome *c* release, and/or DNA damage as well as cell viability. These early studies revealed a number of fundamental observations about how cells responded to nsPEFs.

As might be expected, longer pulses with higher electric fields were found to be more effective for all cell responses than shorter pulses with lower electric fields, and essentially all cell responses were electric field and pulse number dependent. As pulse durations were shortened, higher electric fields were required for given effects seen at longer pulse durations. Furthermore, for many nsPEF effects, including ethidium homodimer uptake (HL60, Jurkat), phosphatidylserine externalization (Jurkat), caspase activation (HL60, Jurkat), and cytochrome *c* release (Jurkat), numbers of "sparkler" granules per cell and viability for human eosinophils did not scale with energy density, which is consistent with nonthermal responses. In general, for cell death and apoptosis induction, longer pulses and/or higher electric fields were required, while reversible effects on plasma membranes and calcium mobilization could be observed at lower electric fields. These fundamentals will be reiterated in the sections below.

1.3.2.1 Calcium Mobilization and Intracellular Calcium Release

Most of the research in the lower range of pulse amplitudes was focused on its effect on the release of intracellular free calcium. Calcium is known as a ubiquitous second messenger molecule that regulates a number of responses in cell signaling, including modulation of metabolism, gene transcription, secretion of neurotransmitter and hormone, muscle contraction, and proliferation and apoptosis regulation, among others. Intracellular calcium is primarily stored in the endoplasmic reticulum (ER) (α-granules in platelets) with lower levels in mitochondria. Nevertheless, several mitochondrial enzymes are calcium dependent, and there is continuous calcium cycling between the two organelles. Calcium is not only important for life, but calcium-overloaded mitochondria are a harbinger for cell death [102, 103].

One of the earliest demonstrations of submicrosecond pulses on calcium mobilization was carried out with human polymorphonuclear leukocytes (PMNs). When the PMNs, crawling over a coverslip between electrodes, were exposed to a 300-ns pulse with an electric field strength of 15 kV/cm, apparent random fluctuations in intracellular calcium levels abruptly changed to cellular coordinated spikes in intracellular calcium with cessation of crawling followed by slowly decreasing levels in intracellular calcium when cells become mobile again 7–10 min after the pulse [104]. The immobilization phase was electric field dependent. Lowering the electric field allowed cells to recover mobility and fluctuations in intracellular calcium more quickly. This mobility effect had been previously observed when aquatic organisms, such as brine shrimp [61] or hydrozoans [105], were subjected to submicrosecond pulsed electric fields; however, this was the first demonstration of reversible coordinate effects on calcium dynamics and cell function.

The origin of this calcium release was not determined, but was later shown in other cell types to be from extracellular sources through permeabilized plasma membranes as well as from intracellular sources such as the endoplasmic reticulum (ER). In experiments where cells were treated with nsPEF in the absence of extracellular calcium, elevation of intracellular calcium was observed coming from intracellular stores [104–109]. One reason for the observed calcium emission from intracellular stores was the nanoporation of the subcellular membranes, in line with the initial assumption that the high electric fields generated by the nanosecond pulses cause permeabilization of these membranes. Another explanation for how nsPEFs induce calcium release from the ER is that the nsPEFs mimic a ligand signal that triggers receptors on internal membranes, thus causing calcium to be released from the internal stores into the cytoplasm [109].

NsPEFs also cause platelet calcium release and aggregation in human platelet-rich plasma and washed platelets [110], an effect which is now successfully used in wound healing. Activation and aggregation of platelets with nsPEFs will avoid adverse effects of adding exogenous thrombin for wound healing. This topic will be presented in greater detail in Sect. 5.5 by S. Beebe and B. Hargrave.

1.3.2.2 Regulated/Programmed Cell Death

During development and homeostasis, cells that are no longer necessary, worn out, or damaged during cell division are eliminated by a programmed cell death (PCD) mechanism called apoptosis that is carried out by a genetically regulated program (s). The term "programmed" in PCD is specifically used to define a regulated cell death (RCD) that is part of a developmental program or used to preserve physiological homeostasis. In contrast RCD, while genetically programmed, can be "influenced," at least partially, by specific pharmacologic or genetic interventions [111] and as it turns out "influenced" by electrical interventions [112, 113]. These developmental and homeostatic programs are initiated by extrinsic mechanisms activated by cell death receptors on plasma membranes (extrinsic apoptosis) or by intrinsic mechanisms activated by internal signals in response to intracellular errors, stress, or defects (intrinsic apoptosis). Under these circumstances apoptotic cell death is anti-inflammatory and immunologically silent, protecting the organism from inflammatory and autoimmune complications. In vivo apoptosis was not readily observed under normal physiological circumstances even though millions of cells are dying by the minute. This is because apoptotic cells are phagocytized before their plasma membranes break and intracellular products that induce inflammatory reactions are contained and inhibited. Thus, the final stages of apoptosis take place inside phagocytes in these homeostatic occurrences.

This apoptotic process is different than necrosis, which has, until recently, been classically characterized as accidental cell death due to catastrophic, abrupt, and irreversible membrane permeabilization and characterized by inflammation and immunogenic recognition. Necrosis is now known to be programmed under certain circumstances and classified as programmed necrosis, one of several regulated cell death (RCD) mechanisms. While defects in apoptosis were first known to cause diseases, regulated necrosis is now known to play roles in physiological (embryonic development) and pathological (ischemic conditions) scenarios [111]. A later chapter is dedicated to mechanisms of RCD including apoptosis. However, at the beginning of experimentation with nsPEFs, around 1998, apoptosis was the only known mechanism of RCD (necrosis was not known to be programmed), and studies were focused to determine if nsPEFs induced apoptosis as a means for causing losses in cell viability.

Early experiments to investigate cell death were carried out with HL-60 cells and Jurkat cells for in vitro studies and fibrosarcoma tumors for ex vivo and in vivo studies using pulse-forming networks with coaxial cable or a strip line giving pulse durations of 10, 60, or 300 ns with electric fields as high as 280 kV/cm (28.0 MV/m) [108, 112, 113]. Another study was carried out in Jurkat and rat glioma cells using a MOSFET-based, inductive-adder pulse generator with pulse durations of 10 ns and electric fields of 28 or 40 kV/cm (2.8 or 4.0 MV/m) [114]. Generally, these studies showed that apoptosis was induced by ultrashort pulses when the electric field amplitude exceeded a threshold value. All of these studies showed differences in plasma membrane behavior between nsPEFs and classical electroporation pulses and used phosphatidylserine (PS) externalization and caspase activation as apoptosis markers.

These early studies on cell death induction demonstrated several fundamental aspect of how cells responded to nsPEFs, stimuli that were not present during evolution of the species. As might be expected, some cells were less sensitive to nsPEFs. For example, unlike the Jurkat cells, rat glioma C6 cells [75] and 3 T3–L1 pre-adipocytes [113], which normally grow attached to surfaces, were found to be highly resistant to the same pulses and pulse sequences. In other experiments where cell survival was explored after ultrashort pulse application, adherent cells required higher electric fields than nonadherent cells for cell death induction [115]. Generally, rapid progression of apoptosis induced by nsPEFs was quite different from that obtained with other apoptotic stimuli, such as UV light and toxic chemicals, which require hours for apoptosis markers to appear. However, the kinetics of ultrashort pulse-induced apoptosis depends on the pulse duration. Shorter pulses result in slower apoptosis progression than longer pulses for the same electrical energy density [113]. Likewise, it is possible for shorter pulse durations to induce apoptosis by increasing electric fields or pulse numbers. For pulse regimens with higher pulse numbers, no systematic studies have been conducted to determine the effect of pulse repetition rate. However, these early studies, like many of later ones, used pulse conditions with relatively low repetition rates so that the electric field was the primary determinant or the only variable, because high repletion rates resulted in increases in temperature.

Since the initial discovery that nsPEFs trigger an apoptosis signaling pathway, there have been significant efforts made to determine the exact mechanism of this process. One of the possible reasons for nsPEF-induced apoptosis was suggested to be extensive intracellular calcium elevation [116]. Calcium at high concentration is known to induce apoptosis [116]. Another possible mechanism for apoptosis induction with ultrashort electric pulses was suggested by Weaver [2003]. The electric field of nanosecond pulses correlated to the high current density in the cytoplasm will affect subcellular membranes, e.g., mitochondrial membranes. It is known that biochemically induced apoptosis involves the mitochondrial permeability transition pore (mPTP) complex [113] and the mitochondrial membrane voltage-dependent anion channels [117, 118]. A hypothesis is that ultrashort pulses change the transmembrane voltage at mitochondrial membrane sites, which leads to an opening of the mPTP, inducing apoptosis [119].

More recent nsPEF studies experimentally demonstrate the importance of influxes of extracellular calcium and dissipation of the mitochondria membrane potential ($\Delta\Psi$m) for cell death [120–122]. However, calcium was necessary, but not sufficient to induce cell death; cell death was coincident with a decrease in the mitochondrial membrane potential, $\Delta\Psi$m, in a calcium-dependent manner. The calcium dependence for dissipation of $\Delta\Psi$m is not consistent with a permeabilization event of the inner mitochondrial membrane, but is consistent with nsPEF-induced changes in the transmembrane voltage at mitochondrial membrane sites in the presence of elevated calcium, which leads to the opening of the mPTP and apoptotic cell death. This is a significant deviation from what is expected for nsPEF effects on intracellular membranes such as that observed for calcium release from the ER. Additional studies of nsPEF effects on mitochondria and $\Delta\Psi$m will be necessary to further analyze this hypothesis.

More recently, a determining role for calcium in cell death was observed using HeLa cells [123]. Calcium was shown to not only be important for cell death but to determine what cell death mechanisms were activated in some cells. While non-apoptotic mechanisms of nsPEF-induced cell death were known [124, 125], no specific mechanisms were defined. However, in HeLa cells, calcium was shown to be required for a type of regulated necrosis identified by formation of poly (ADP-ribose) or PAR. While these cells could express apoptosis markers (caspase activation), they underwent PAR-associated regulated cell necrosis, which was suggested to be a preferential cell death mechanism. These authors also showed that nsPEF-induced cell death was cell type specific, since HEK 293 cells and K562 cells underwent PAR-associated regulated necrosis in the presence of calcium, while Jurkat cells underwent apoptosis regardless of the presence or absence of calcium.

The production of reactive oxygen species (ROS) and subsequent DNA damage has also been proposed as a possible effect causing apoptosis [126]. NsPEFs do increase intracellular ROS, but it is not clear that this has significant effects on cell viability [127]. There is evidence that nsPEF stimulation with multiple, intense pulses causes damage to DNA or other critical proteins. Studies by comet assay have shown that significant DNA damage does occur very quickly after treatment (90-s posttreatment) with nsPEFs [128]. Some DNA repair was observed when permeabilization of cells for the comet assay was delayed for an hour. At this time it is not known if DNA damage is a primary (direct) effect of intense nanosecond electric fields or a secondary effect caused by membrane permeabilization and/or mediated by apoptosis-inducing factor (AIF) released from mitochondria or following caspase cleavage of DNA. A more detailed discussion on cell death mechanisms is given in Sect. 4.10 by S. Beebe.

Since earth life forms did not evolve in the presence of nsPEFs, it was of interest to determine what cellular systems would function in response to nsPEFs. It has been shown that cellular response to nsPEFs is distinct from those induced by previously known forms of cellular stress such as the protein-unfolding response, UV radiation, and heat [129]. A more specific discussion of these finding will be presented in the chapter by K Yano.

The research on cell death caused by nanosecond pulses has triggered studies on its application for cancer treatment. First studies [112] indicated the potential of such pulses for the controlled destruction of tumor tissue. Tumors were grown in flanks of mice and removed and treated ex vivo. In the first rudimentary experiments treating tumors in vivo using electrodes with two needles, tumor growth was decreased by 60 %. When treatment regimens and electrode designs were optimized, B16f10 melanoma [130] and hepatocellular tumors in mice [131] and rats [132] were eliminated by 75–100 %. These early studies have now led to the development of therapeutic methods for cancer treatment, particularly treatment of skin cancer [133–135]. This topic is described in detail in Sect. 5.12 by R. Nuccitelli.

1.3.3 New Research Directions

1.3.3.1 From Nanosecond to Picosecond Pulses

By reducing the duration of electric pulses from microseconds into the nanosecond range, the electric field–cell interactions shift increasingly from the plasma (cell) membrane to subcellular structures. Yet another domain of pulsed electric field interactions with cell structures and functions opens when the pulse duration is reduced to values such that membrane charging becomes negligible, and direct electric field-molecular effects determine biological mechanisms. For dominance of such effects, the pulse duration needs to be less than the dielectric relaxation time of the cytoplasm. For mammalian cells, this holds for pulse duration of less than 1 ns.

The practical reason for entering this new field of bioelectrics by moving into the subnanosecond range is the possible use of antennas as pulse delivery systems instead of needle electrodes as well as the opportunity to observe direct electric field effects on biological mechanisms. These antenna systems are mainly based on the use of a prolate spheroidal reflector, where the pulse is launched from one focal point and reflected into a second focal point [77]. The use of needle or plate electrodes in therapeutic applications which rely on electroporation [136] or nanosecond pulsed electric fields (nsPEF) [133] requires that the electrodes are brought into close contact with the treated tissue. This limits the application to treatments of tissue close to the skin or surface of the body or by using invasive procedures or minimally invasive procedures by laparoscopy to treat internal organs. The use of antennas, on the other hand, would allow one to apply such electric fields to tissues (tumors) that are not easily accessible with needles. Also, focusing electrical energy on the target would reduce the damage to the tissue layers surrounding the target and the skin.

1.3.3.2 From Membranes to Proteins

Most, if not all, effects of electric fields are analyzed in terms of electrical charging of plasma membranes and/or subcellular membranes. Since nsPEFs are extensions of classical electroporation, which was developed for permeabilization of plasma membranes for drug or nucleotide delivery to cells, the focus with submicrosecond pulses has remained on lipid bilayers for electric field effects. Although cellular plasma membranes include integral and peripheral proteins, these structures have generally been ignored because of their complexity compared to lipid bilayers. Only a few experimental analyses have approached the subject of electric field effects on proteins.

Initial studies with submicrosecond pulses demonstrated effects that breached membranous intracellular vesicles in human eosinophils [77]. Yet caution was exercised not to assume all cell responses would be exclusively on membranes. The intracellular effects were referred to as "intracellular electromanipulation" instead of "intracellular electroporation/permeabilization." Beyond consideration for membrane effects, nsPEFs have been considered for possible effect on DNA [112, 115, 133]. However it is presently unclear if observed effects on DNA damage are due to direct, primary electric field effects or if they are secondary to intracellular assaults on upstream biological mechanisms that lead to DNA damage. More specifically, it is debated whether nsPEFs deliver sufficient energy, power, or high-frequency components to disrupt molecular bonding such as hydrogen bonding that contributes to the structure and stability of DNA or such bonding in proteins. However, there are several simulation studies using molecular dynamics (MD) suggesting that nsPEFs can affect protein structure and at least one experimental study showing that nsPEFs can directly affect the catalytic activity of a protein kinase.

Using MD, high-intensity nsPEFs were found to cause a marked structural rearrangement along the major geometrical axis of the protein myoglobin [137]. In other MD studies, nsPEFs caused the soybean hydrophobic protein, a well-studied food protein, to unfold and lose all secondary structure [138]. Similar conclusions were reached for static and oscillating effects of nanosecond pulses that destabilized the structure of the insulin β-chain (30 amino acids) [139]. In a study with the human aquaporin four channel, nsPEFs were shown to switch the dipolar orientation of the histidine-201 residue [140]. However, it needs to be noted that except for the last study where 50- and 100-ns pulses at 650 kV/cm were applied, the assumed pulsed electric field amplitudes were in several MV/cm range, although for shorter pulse durations. Such pulse amplitudes are not realistic when considering the state of present pulse power technology.

However, both direct and indirect experimental evidence indicate that effects on proteins are likely occurring at much lower electric fields. NsPEFs were shown to have direct effects on the phosphotransferase activity of the catalytic subunit of the cAMP-dependent protein kinase (PKA) [122]. NsPEFs caused a 41 % and 45 % loss in catalytic activity with one and ten pulses at 60 ns, 60 kV/cm, respectively, and a 55 % and 77 % activity loss for one and ten pulses at 300 ns, 26 kV/cm, respectively. Given that the one pulse and ten pulse conditions between 60 ns, 60 kV/cm and 300 ns, 26 kV/cm exhibited similar energy densities (1.7 J/cc), the inhibitory effects on kinase structure/function appear to be, at least in part, not to scale with energy density. This is consistent with effects of nsPEFs on nsPEF-induced cytochrome *c* release and caspase activity in intact cells [112, 113]. These studies further indicate that nsPEF-induced inactivation of kinase activity is due, at least in part, to charging effects transferred to the enzyme rather than energy deposition.

Also, observations that nsPEFs induce Ca^{2+}-dependent dissipation of the mitochondria membrane potential (ΔΨm), which is enhanced when high-frequency components are present in fast rise-fall waveforms, suggest events that are not due to permeabilization of the inner mitochondrial membrane; plasma membrane poration (and thereby intracellular membranes) is not Ca^{2+} dependent. Since all Ca^{2+}-mediated events require proteins, it is possible that nsPEFs affect a protein(s) or protein complex that changes the transmembrane voltage, thereby leading to opening the mitochondrial permeability transition pore (mPTP) [81, 85]. Effects of nsPEFs on proteins open a new paradigm for electric field effects on cell structures and functions.

1.3.3.3 Thermally Assisted Electroeffects

Thermal effects, although generally avoided in bioelectric studies, have their place in bioelectrics, even for pulses short compared to the dielectric relaxation time of the membrane. Simultaneously pulsing and heating cells, either by utilizing the Joule heat generated through high repetition rate pulsing [141] or using pulse-independent methods such as microwave heating, allow us to reduce the electrical energy required for inducing cell death by a considerable amount. This has been shown recently in a study where 200-ps long pulses were applied to a slightly heated cell suspension [142] Cell death, as recorded by means of trypan blue uptake, increased considerably for heating above physiological temperature. The results indicate that thermal assistance of electro-technologies, such as tumor ablation and gene therapy, will allow us to reduce the electrical energy (pulse amplitude) and consequently reduce tissue damage.

1.3.3.4 Nonthermal Plasmas for Environmental and Medical Applications

A fascinating new field of research using pulsed electric fields to generate nonthermal plasma is plasma medicine [143, 144]. Plasma reactors for environmental and medical applications have found their place in "bioelectrics." Again, as in the field of bioelectrics as discussed in this chapter, nanosecond pulses play a major role. Experimental studies with nonequilibrium gliding gas discharges have shown their potential for environmental and medical applications, such as chemical and bacterial decontamination [145].

References

1. Neumann, E., Schaefer-Ridder, M., Wang, Y., Hofschneider, P.H.: Gene transfer into mouse lyoma cells by electroporation in high electric fields. EMBO J. **1**, 841–845 (1982)
2. Wong, T.K., Neumann, E.: Electric field mediated gene transfer. Biophys. Biochem. Res. Commun. **107**, 584–587 (1972)
3. Neumann, E., Rosenheck, K.: Permeability changes induced by electric impulses in vesicular membranes. J. Membr. Biol. **10**, 279–290 (1972)
4. Neumann, E., Gerisch, G., Opatz, K.: Cell fusion induced by high electric impulses applied to Dictostelium. Naturwissenschaften **67**, 414–415 (1980)
5. Eisenstein, M.: A look back: a shock to the system. Nat. Methods **3**, 66 (2006)
6. Griese, T., Kakorin, S., Neumann, E.: Conductometric and electrooptic relaxation spectrometry of lipid vesicle electroporation at high fields. Phys. Chem. Chem. Phys. **4**, 1217–1227 (2002)
7. Schmeer, M., Seipp, T., Pliquett, U., Kakorin, S., Neumann, E.: Mechanism the conductivity changes caused by membrane electroporation of CHO cell-pellets. Phys. Chem. Chem. Phys. **6**, 5564–5574 (2004)
8. Pliquett, U., Gallo, S., Hui, S.W., Gusbeth, C., Neumann, E.: Local a transient structural changes in stratum corneum at high electric fields: Contribution of Joule heating. Bioelectrochemistry **67**, 37–46 (2005)
9. Neumann, E., Kakorin, S.: Electroporation of curved lipid membranes in ionic strength gradients. Biophys. Chem. **85**, 249–271 (2000)

10. Kakorin, S., Redeker, E., Neumann, E.: Electroporative deformation of salt filled vesicles. Eur. Biophys. J. **27**, 43–53 (1998)
11. Neumann, E., Tönsing, K., Kakorin, S., Budde, P., Frey, J.: Mechanism of electroporative dye uptake by mouse B cells. Biophys. J. **74**, 98–108 (1998)
12. Dimitrov, V., Kakorin, S., Neumann, E.: Transient Oscillation of shape and membrane conductivity changes by field pulse-induced electroporation in nano-sized phospholipid vesicles. Phys. Chem. Chem. Phys. **15**, 6303–6632 (2013)
13. Abidor, I. G., Arakelyan, V. B., Chernomordik, L. V., Chizmadzhev, Y. A., Pastuchenko, V. P., Tarasevich, M. R.: Electric Breakdown of Bilayer Lipid Membrane. I. The main experimental facts and their theoretical discussion, Bioelectrochem. Bioenerg. **6**, 37–52. (1979)
14. Andre, F.M., Gehl, J., Sersa, G., Preat, V., Hojman, P., Eriksen, J., Golzio, M., Cemazar, M., Pavselj, N., Rols, M.-P., Miklavcic, D., Neumann, E., Teissie, J., Mir, L.M.: Efficiency of high- and low voltage pulse combinations for gene electrotransfer in muscle liver, tumor, and skin. Hum. Gene Ther. **19**, 1261–1271 (2008)
15. Neumann, E.: The relaxation hysteresis of membrane electroporation. In: Neumann, E., Sowers, A.E., Jordan, C.A. (eds.) Electroporation and Electrofusion in Cell Biology, pp. 61–82. Plenum Press, New York (1989)
16. Neumann, E., Kakorin, S., Toensing, K.: Fundamentals of electroporative delivery of drugs and genes. Bioelectrochem. Bioenerg. **48**, 3–1 (1999)
17. Coster, H.G.: A quantitative analysis of the voltage-current relationships of fixed charge membranes and the associated property of "punch-through". Biophys. J. **5**(5), 669–686 (1965)
18. Sale, A.J.H., Hamilton, W.A.: Effects of high electric fields on microorganisms: I. Killing of bacteria and yeasts. Biochim. Biophys. Acta Gen. Subj. **148**(3), 781–788 (1967)
19. Sale, A.J.H., Hamilton, W.A.: Effects of high electric fields on micro-organisms: III. Lysis of erythrocytes and protoplasts. Biochim. Biophys. Acta Biomembr. **163**(1), 37–43 (1968)
20. Pohl, H.A., Crane, J.S.: Dielectrophoresis of cells. Biophys. J. **11**(9), 711–727 (1971)
21. Crowley, J.M.: Electrical breakdown of bimolecular lipid membranes as an electromechanical instability. Biophys. J. **13**(7), 711–724 (1973)
22. Zimmermann, U., Pilwat, G., Riemann, F.: Dielectric breakdown of cell membranes. Biophys. J. **14**(11), 881–899 (1974)
23. Largus, S. 47. Compositiones Medicae. Quoted in Kellaway, P.: The part played by electric fish in the early history of bioelectricity and electrotherapy. Bull. Hist. Med. **20**, 112–137 (1946)
24. Jallabert, J.: Experiences sur l'electricite avec quelques conjectures sur la cause de ses effets. http://catalogue.bnf.fr/ark:/12148/cb393088488 (1746)
25. Franklin, B: Letters to Peter Collison (1749). In: Labaree, L. W. (ed.) The Papers of Benjamin Franklin, vol. 3, January 1, 1745, through June 30, 1750, pp 391–393. Yale University Press, New Haven (1961)
26. Nollet, J. A.: Recherches sur les causes particulieres des phénoménes électriques, et sur les effets nuisibles ou avantageux qu'on peut en attendre. Nouv. éd. Paris, Guerin u. Delatour (1754)
27. Galvani, L.: Aloysii Galvani De viribus electricitatis in motu musculari commentarius, Ex Typographia Instituti Scientiarium (1791)
28. Mauduyt de la Varenne, P.J.C. Mémoire sur les différentes manières d'administrer l'électricité, et observations sur les effets qu'elles ont produits. Paris. l'Imprimerie Royale. (1784)
29. Pauly, H., Schwan, H.P.: Impedance of a suspension of ball-shaped particles with a shell; a model for the dielectric behavior of cell suspensions and protein solutions. Z Naturforsch B **14B**, 125–131 (1959)
30. Hamilton, W.A., Sale, A.J.H.: Effects of high electric fields on microorganisms II. Mechanism of action of the lethal effect. Biochim. Biophys. Acta **148**, 789–800 (1967)

31. Kinosita Jr., K., Tsong, T.Y.: Hemolysis of human erythrocytes by transient electric field. Proc. Natl. Acad. Sci. U. S. A. **74**, 1923–1927 (1977)
32. Senda, M., Takeda, J., Abe, S., Nakamura, T.: Induction of cell fusion of plant protoplasts by electrical stimulation. Plant Cell Physiol. **20**, 1441–1443 (1979)
33. Scheurich, P., Zimmermann, U.: Electrically stimulated fusion of different plant cell protoplasts: mesophyll cell and guard cell protoplasts of *Vicia faba*. Plant Physiol. **67**, 849–853 (1981)
34. Teissie, J., Tsong, T.Y.: Electric field induced transient pores in phospholipid bilayer vesicles. Biochemistry **20**, 1548–1554 (1981)
35. Gordon, P.B., Seglen, P.O.: Autophagic sequestration of [^{14}C]sucrose, introduced into rat hepatocytes by reversible electro-permeabilization. Exp. Cell Res. **142**, 1–14 (1982)
36. Teissie, J., Knutson, V.P., Tsong, T.Y., Lane, M.D.: Electric pulse-induced fusion of 3T3 cells in monolayer culture. Science **216**, 537–538 (1982)
37. Potter, H., Weir, L., Leder, P.: Enhancer-dependent expression of human kappa immunoglobulin genes introduced into mouse pre-B lymphocytes by electroporation. Proc. Natl. Acad. Sci. U. S. A. **81**, 7161–7165 (1984)
38. Sugar, I.P., Neumann, E.: Stochastic model for electric field-induced membrane pores. Electroporation. Biophys. Chem. **19**, 211–225 (1984)
39. Lo, M.M., Tsong, T.Y., Conrad, M.K., Strittmatter, S.M., Hester, L.D., Snyder, S.H.: Monoclonal antibody production by receptor-mediated electrically induced cell fusion. Nature **310**(5980), 792–794 (1984)
40. Okino, M., Mohri, H.: Effects of a high-voltage electrical impulse and an anticancer drug on in vivo growing tumors. Jap. J. Cancer Res. **78**, 1319–1321 (1987)
41. Mir, L.M., Banoun, H., Paoletti, C.: Introduction of definite amounts of nonpermeant molecules into living cells after electropermeabilization: direct access to the cytosol. Exp. Cell Res. **175**, 15–25 (1988)
42. Sersa, G., Cemazar, M., Miklavcic, D.: Antitumor effectiveness of electrochemotherapy with cis-diamminedichloroplatinum(II) in mice. Cancer Res. **55**, 3450–3455 (1995)
43. Gilbert, R., Jaroszeski, M.J., Heller, R.: Novel electrode designs for electrochemotherapy. Biochemica Biophysica Acta **1334**, 9–14 (1997)
44. Frandsen, S.K., Gissel, H., Hojman, P., Tramm, T., Eriksen, J., Gehl, J.: Direct therapeutic applications of calcium electroporation to effectively induce tumor necrosis. Cancer Res. **72**, 1336–1341 (2012)
45. Mir, L.M., Belehradek, M., Domenge, C., Orlowski, S., Poddevin, B., Belehradek Jr., J., Schwaab, G., Luboinski, B., Paoletti, C.: Electrochemotherapy, a new antitumor treatment: first clinical trial. C. R. Acad. Sci. III **313**, 613–618 (1991)
46. Titomirov, A.V., Sukharev, S., Kistanova, E.: In vivo electroporation and stable transformation of skin cells of newborn mice by plasmid DNA. Biochim. Biophys. Acta **1088**, 131–134 (1991)
47. Heller, R., Jaroszeski, M., Atkin, A., Moradpour, D., Gilbert, R., Wands, J., Nicolau, C.: In vivo gene electroinjection and expression in rat liver. FEBS Lett. **389**, 225–228 (1996)
48. Rols, M.P., Delteil, C., Golzio, M., Dumond, P., Cros, S., Teissie, J.: In vivo electrically mediated protein and gene transfer in murine melanoma. Nat. Biotechnol. **16**, 168–171 (1998)
49. Aihara, H., Miyazaki, J.: Gene transfer into muscle by electroporation in vivo. Nat. Biotechnol. **16**, 867–870 (1998)
50. Daud, A.I., DeConti, R.C., Andrews, S., Urbas, P., Riker, A.I., Sondak, V.K., Munster, P.N., Sullivan, D.M., Ugen, K.E., Messina, J.L., Heller, R.: Phase I trial of interleukin-12 plasmid electroporation in patients with metastatic melanoma. J. Clin. Oncol. **26**, 5896–5903 (2008)
51. Prausnitz, M.R., Bose, V.G., Langer, R., Weaver, J.C.: Electroporation of mammalian skin: a mechanism to enhance transdermal drug delivery. Proc. Natl. Acad. Sci. U. S. A. **90**(22), 10504–10508 (1993)
52. Prausnitz, M.R., Lau, B.S., Milano, C.D., Conner, S., Langer, R., Weaver, J.C.: A quantitative study of electroporation showing a plateau in net molecular transport. Biophys. J. **65**, 414–422 (1993)

53. Davalos, R.V., Mir, L., Rubinsky, B.: Tissue ablation with irreversible electroporation. Ann. Biomed. Eng. **33**(2), 223–231 (2005)
54. Grasso, R.J., Heller, R., Cooley, J.C., Haller, E.M.: Electrofusion of individual animal cells directly to intact corneal epithelial tissue. Biochim. Biophys. Acta **980**(1), 9–14 (1989)
55. Heller, R., Grasso, R.J.: Transfer of human membrane surface components by incorporating individual human cells into intact animal tissue by cell-tissue electrofusion *in vivo*. Biochim. Biophys. Acta **1024**, 185–188 (1990)
56. Garcia, P.A., Davalos, R.V., Miklavcic, D.: A numerical investigation of the electric and thermal cell kill distributions in electroporation-based therapies in tissue. PLoS One **9**(8), e103083 (2014)
57. Poo, M., Robinson, K.R.: Electrophoresis of concanavalin A receptors along embryonic muscle cell membrane. Nature **265**(5595), 602–605 (1977)
58. Pucihar, G., Miklavcic, D., Kotnik, T.: A time-dependent numerical model of transmembrane voltage inducement and electroporation of irregularly shaped cells. IEEE Trans. Biomed. Eng. **56**(5), 1491–1501 (2009)
59. Rols, M.P., Delteil, C., Golzio, M., Teissié, J.: Control by ATP and ADP of voltage-induced mammalian-cell-membrane permeabilization, gene transfer and resulting expression. Eur. J. Biochem. **254**(2), 382–388 (1998)
60. Pucihar, G., Kotnik, T., Miklavcic, D., Teissié, J.: Kinetics of transmembrane transport of small molecules into electropermeabilized cells. Biophys. J. **95**(6), 2837–2848 (2008)
61. Huynh, C., Roth, D., Ward, D.M., Kaplan, J., Andrews, N.W.: Defective lysosomal exocytosis and plasma membrane repair in Chediak-Higashi/beige cells. Proc. Natl. Acad. Sci. U. S. A. **101**(48), 16795–16800 (2004)
62. Neil, G.A., Zimmermann, U.: Electrofusion. Methods Enzymol. **220**, 174–196 (1993)
63. NICE interventional procedures guidance [IPG478] (2014)
64. Broderick, K.E., Humeau, L.M.: Electroporation-enhanced delivery of nucleic acid vaccines. Expert Rev. Vaccines **14**(2), 195–204 (2015)
65. Pavlin, D., Cemazar, M., Sersa, G., Tozon, N.: IL-12 based gene therapy in veterinary medicine. J. Transl. Med. **10**, 234 (2012)
66. Heller, L.C., Heller, R.: In vivo electroporation for gene therapy. Hum. Gene Ther. **17**(9), 890–897 (2006)
67. Heller, R., Heller, L.C.: Gene electrotransfer clinical trials. Adv. Genet. **89**, 235–262 (2015)
68. Young, J.L., Dean, D.A.: Electroporation-mediated gene delivery. Adv. Genet. **89**, 49–88 (2015)
69. Puértolas, E1., Luengo, E., Álvarez, I., Raso, J.: Improving mass transfer to soften tissues by pulsed electric fields: fundamentals and applications. Annu. Rev. Food Sci. Technol. **3**, 263–282 (2012)
70. Rieder, A., Schwartz, T., Schön-Hölz, K., Marten, S.M., Süss, J., Gusbeth, C., Kohnen, W., Swoboda, W., Obst, U., Frey, W.: Molecular monitoring of inactivation efficiencies of bacteria during pulsed electric field treatment of clinical wastewater. J. Appl. Microbiol. **105**(6), 2035–2045 (2008)
71. Dower, W.J., Miller, J.F., Ragsdale, C.W.: High efficiency transformation of E. coli by high voltage electroporation. Nucleic Acids Res. **16**(13), 6127–6145 (1988)
72. Miller, J.F., Dower, W.J., Tompkins, L.S.: High-voltage electroporation of bacteria: genetic transformation of Campylobacter jejuni with plasmid DNA. Proc. Natl. Acad. Sci. U. S. A. **85**(3), 856–860 (1988)
73. Meilhoc, E1., Masson, J.M., Teissié, J.: High efficiency transformation of intact yeast cells by electric field pulses. Biotechnology (N Y) **8**(3), 223–227 (1990)
74. Brown, L.E., Sprecher, S.L., Keller, L.R.: Introduction of exogenous DNA into Chlamydomonas reinhardtii by electroporation. Mol. Cell. Biol. **11**(4), 2328–2332 (1991)
75. Trick, H.N., Bates, G.W.: Electrofusion of plant protoplasts. Selection and screening for somatic hybrids of Nicotiana. Methods Mol. Biol. **55**, 165–179 (1995)

76. Staempfli, R.: Reversible breakdown of the excitable membrane of a ranvier node. An. Acad. Bras. Ciências **30**, 57 (1958)
77. Schoenbach, K.H., Beebe, S.J., Buescher, E.S.: Intracellular effect of ultrashort electrical pulses. J. Bioelectromagnetics **22**, 440–448 (2001)
78. Kolb, J.F., Kono, S., Schoenbach, K.H.: Nanosecond pulsed electric field generators for the study of subcellular effects. Bioelectromagnetics J. **27**, 172–187 (2006)
79. Xiao, S., Altunc, S., Kumar, P., Baum, C.E., Schoenbach, K.H.: A reflector antenna for focusing in the near field. IEEE Antennas Wirel. Propag. Lett. **9**, 12–15 (2010)
80. Schoenbach, K.H., Joshi, R.P., Kolb, J.F., Chen, N., Stacey, M., Blackmore, P.F., Buescher, E.S., Beebe, S.J.: Ultrashort electrical pulses open a new gateway into biological cells. Proc. IEEE **92**, 1122–1137 (2004)
81. Joshi, R.P., Hu, Q., Schoenbach, K.H.: Modeling studies of cell response to ultrashort high-intensity electric fields – implications for intracellular manipulation. IEEE Trans. Plasma Sci. **32**, 1677–1686 (2004)
82. Schoenbach, K.H., Hargrave, B., Joshi, R.P., Kolb, J.F., Osgood, C., Nuccitelli, R., Pakhomov, A., Swanson, R.J., Stacey, M., White, J.A., Xiao, S., Zhang, J., Beebe, S.J., Blackmore, P.E., Buescher, E.S.: Bioelectric effects of nanosecond pulses. IEEE Trans. Diel. Electr. Insul. **14**, 1088–1119 (2007)
83. Smith, K.C., Gowrishankar, T.R., Esser, A.T., Stewart, D.A., Weaver, J.C.: The spatially distributed dynamic transmembrane voltage of cells and organelles due to 10 ns pulses: meshed transport networks. IEEE Trans. Plasma Sci. **34**, 1394–1404 (2006)
84. Stewart, D.A., Gowrishankar, T.R., Weaver, J.C.: Three dimensional transport lattice model for describing action potentials in axons stimulated by external electrodes. Bioelectrochemistry **69**, 88–93 (2006)
85. Weaver, J.C.: Harvard-MIT Division of Health Sciences and Technology, Cambridge, MA 02139, USA, private communication (2006)
86. Gowrishankar, T.R., Esser, A.T., Vasilkoski, Z., Smith, K.C., Weaver, J.C.: Microdosimetry for conventional and supra-electroporation in cells with organelles. BBRC **341**, 1266–1276 (2006)
87. Schoenbach, K.H., Katsuki, S., Stark, R.H., Buescher, E.S., Beebe, S.J.: Bioelectrics – new applications for pulsed power technology. IEEE Trans. Plasma Sci. **30**, 293–300 (2002)
88. Kotnik, T., Miklavcic, D.: Theoretical evaluation of voltage inducement on internal membranes of biological cells exposed to electric fields. Biophys. J. **90**, 480–491 (2006)
89. Kakorin, E., Redeker, E., Neumann, E.: Electroporative deformation of salt filled lipid vesicles. Eur Biophys **J27**, 43–53 (1998)
90. Frey, W., White, J.A., Price, R.O., Blackmore, P.F., Joshi, R.P., Nuccitelli, R., Beebe, S.J., Schoenbach, K.H., Kolb, J.F.: Plasma membrane voltage changes during nanosecond pulsed electric field exposure. Biophys. J. **90**, 3608–3615 (2006)
91. Vasilkoski, Z., Esser, A.T., Gowrishankar, T.R., Weaver, J.C.: Membrane electroporation: the absolute rate equation and nanosecond time scale pore creation. Phys. Rev. **E74**, 021904-1–021904-12 (2006)
92. Pakhomov, A.G., Shevin, R., White, J., Kolb, J.F., Pakhomova, O.N., Joshi, R.P., Schoenbach, K.H.: Membrane permeabilization and cell damage by ultrashort electric field shocks. Arch. Biochem. Biophys. **465**(1), 109–118 (2007)
93. Halestrap, A.P., McStay, G.P., Clarce, S.J.: The permeability transition pore complex: another view. Biochimie **84**, 153–166 (2002)
94. Tieleman, D., Leontiadou, H., Mark, A.E., Marrink, S.J.: Simulation of pore formation in lipid bilayers by mechanical stress and electric fields. J. Am. Chem. Soc. **125**, 6382–6383 (2003)
95. Vernier, P.T., Ziegler, M.J., Sun, Y., Gundersen, M.A., Tieleman, T.P.: Nanopore-facilitated, voltage-driven translocation in lipid bilayers – in cells and in silico. Phys. Biol. **3**, 233–247 (2005)

96. Vernier, P.T., Ziegler, M.J., Sun, Y., Chang, W.V., Gundersen, M.A., Tieleman, D.P.: Nanopore formation and phosphatidylserine externalization in a phospholipid bilayer at high transmembrane potential. J. Am. Chem. Soc. **128**, 6288–6289 (2006)
97. Schoenbach, K.H., Baum, C.E., Joshi, R.P., Beebe, S.J.: A scaling law for membrane permeabilization with nanopulses. Special issue on Bioelect. IEEE Trans. Dielect. Elect. Insulation. **16**, 1224–1235 (2009a)
98. Pakhomov, A.G., Semenov, I., Xiao, S., Pakhomova, O.N., Gregory, B., Schoenbach, K.H., Ullery, J.C., Beier, H.T., Rajulapati, S.R., Ibey, B.L.: Cancellation of cellular responses to nanoelectroporation by reversing the stimulus polarity. Cell. Mol. Life Sci. **71**(22), 4431–4441 (2014)
99. Schoenbach, K.H., Pakhomov, A.G., Semenov, I., Xiao, S., Pakhomova, O.N., Ibey, B.L.: Ion transport into cells exposed to monopolar and bipolar nanosecond pulses. Bioelectrochemistry **103**, 44–51 (2015)
100. Schoenbach, K.H., Peterkin, F.E., Alden, R.W., Beebe, S.J.: The effect of pulsed electric fields on biological cells: experiments and applications. Trans. Plasma Sci. **25**, 284–292 (1997)
101. Schoenbach, K.H., Joshi, R.P., Stark, R.H., Dobbs, F., Beebe, S.J.: Bacterial decontamination of liquids with pulsed electric fields. IEEE Trans Dielect. Electr. Insul. **7**, 637–645 (2000)
102. Berridge, M.J., Bootman, M.D., Lipp, P.: Calcium – a life and death signal. Nature **395**, 645–648 (1998)
103. Susin, S.A., Zamzami, N., Kroemer, G.: Mitochondria as regulators of apoptosis: doubt no more. Biochim. Biophys. Acta **1366**, 151–165 (1998)
104. Buescher, E.S., Schoenbach, K.H.: Effects of submicrosecond, high intensity pulsed electric fields on living cells – intracellular electromanipulation. IEEE Trans Dielect. Electr. Insul. **10**, 788–794 (2003)
105. Ghazala, A., Schoenbach, K.H.: Biofouling prevention with pulsed electric fields. IEEE Trans. Plasma Sci. **28**, 115–121 (2000)
106. Vernier, P.T., Sun, Y., Marcu, L., Salemi, S., Craft, C.M., Gundersen, M.A.: Calcium bursts induced by nanosecond electrical pulses. BBRC **310**, 286–295 (2003)
107. Vernier, P.T., Sun, Y., Marcu, L., Craft, C.M., Gundersen, M.A.: Nanoelectropulse-induced phosphatidylserine translocation. Biophys. J. **86**, 4040–4048 (2004)
108. Beebe, S.J., White, J.A., Blackmore, P.F., Deng, Y., Somers, K., Schoenbach, K.H.: Diverse effects of nanosecond pulsed electric fields on cells and tissues. DNA Cell Biol. **22**, 785–796 (2003)
109. White, J.A., Blackmore, P.F., Schoenbach, K.H., Beebe, S.J.: Stimulation of capacitive calcium entry in HL-60 cells by nanosecond pulsed electric fields (nsPEF). J. Biol. Chem. **279**, 22964–22972 (2004)
110. Zhang, J., Blackmore, P.F., Hargrave, B.Y., Xiao, S., Beebe, S.J., Schoenbach, K.H.: The characteristics of nanosecond pulsed electrical field stimulation on platelet aggregation in vitro. Arch. Biochem. Biophys. **471**, 240–248 (2008)
111. Galluzzi, L., Bravo-San Pedro, J.M., Vitale, I., Aaronson, S.A., Abrams, J.M., Adam, D., et al.: Essential versus accessory aspects of cell death: recommendations of the NCCD 2015. Cell Death Differ. (2014). doi:10.1038/cdd.2014.137
112. Beebe, S.J., Fox, P.M., Rec, L.C., Somers, K., Stark, R.H., Schoenbach, K.H.: Nanosecond Pulsed Electric Field (nsPEF) effects on cells and tissues: apoptosis induction and tumor growth inhibition. IEEE Trans. Plasma Sci. **30**, 286–292 (2002)
113. Beebe, S.J., Fox, P.M., Rec, L.J., Willis, L.K., Schoenbach, K.H.: Nanosecond, high intensity pulsed electric fields induce apoptosis in human cells. FASEB J. **17**, 1493 (2003)
114. Vernier, P.T., Li, A., Marcu, L., Craft, C.M., Gundersen, M.A.: Ultrashort pulsed electric fields induce membrane phospholipid translocation and caspase activation: differential sensitivities of jurkat T lymphoblasts and rat glioma C6 cells. IEEE Trans. Diel. Electr. Insul. **10**, 795–809 (2003)

115. Stacey, M., Stickley, J., Fox, P., Statler, V., Schoenbach, K.H., Beebe, S.J., Buescher, S.: Differential effects in cells exposed to ultra-short, high intensity electric fields: cell survival, DNA damage, and cell cycle analysis. Mutat. Res. **542**, 65–75 (2003)
116. Müller, A., Günther, D., Düx, F., Naumann, M., Meyer, T.F., Rudel, T.: Neisseria porin (PorB) causes rapid calcium influx in target cells and induces apoptosis by the activation of cysteine proteases. EMBO J. **18**, 339–352 (1999)
117. Savill, J., Haslett, C.: Granulocyte clearance by apoptosis in the resolution of inflammation. Semin. Cell Biol. **6**, 385–360 (1995)
118. Tsujimot, Y., Shimizu, S.: The voltage-dependent anion channel: an essential player in apoptosis. Biochimie **84**, 187–193 (2002)
119. Weaver, J.C.: Electroporation of biological membranes from multicellular to nanoscales. IEEE Trans. Diel. Electr. Insul. **10**, 754–768 (2003)
120. Beebe, S.J., Chen, Y.J., Sain, N.M., Schoenbach, K.H., Xiao, S.: Transient features in nanosecond pulsed electric fields differentially modulate mitochondria and viability. PLoS One **7**, e51349 (2012)
121. Beebe, S.J., Sain, N.M., Ren, W.: Induction of cell death mechanisms and apoptosis by nanosecond pulsed electric fields (nsPEFs). Cells **2**, 136–162 (2013)
122. Beebe, S.J.: Considering effects of nanosecond pulsed electric fields on proteins. Bioelectrochemistry **S1567–5394**(14), 00130–00133 (2014). doi:10.1016/j.bioelechem.2014.08.014
123. Morotomi-Yano, K., Akiyama, H., Yano, K.: Different involvement of extracellular calcium in two modes of cell death induced by nanosecond pulsed electric fields. Arch. Biochem. Biophys. **555–556**, 47–54 (2014)
124. Ford, W.E., Ren, W., Blackmore, P.F., Schoenbach, K.H., Beebe, S.J.: Nanosecond pulsed electric fields stimulate apoptosis without release of pro-apoptotic factors from mitochondria in B16f10 melanoma. Arch. Biochem. Biophys. **497**, 82–89 (2010)
125. Ren, W., Beebe, S.J.: An apoptosis targeted stimulus with nanosecond pulsed electric fields (nsPEFs) in E4 squamous cell carcinoma. Apoptosis **16**, 382–393 (2011)
126. Walker III, K., Pakhomova, O.N., Kolb, J.F., Schoenbach, K.H., Stuck, B.E., Murphy, M.R., Pakhomov, A.G.: Oxygen enhances lethal effect of high-intensity, ultrashort electrical pulses. Bioelectromagnetics J. **27**, 221–225 (2006)
127. Pakhomova, O.N1., Khorokhorina, V.A., Bowman, A.M., Rodaitė-Riševičienė, R., Saulis, G., Xiao, S., Pakhomov, A.G.: Oxidative effects of nanosecond pulsed electric field exposure in cells and cell-free media. Arch. Biochem. Biophys. **527**(1), 55–64 (2012)
128. Nuccitelli, R., Chen, X., Pakhomov, A.G., Baldwin, W.H., Sheikh, S., Pomicter, J.L., Ren, W., Osgood, C., Swanson, R.J., Kolb, J.F., Beebe, S.J., Schoenbach, K.H.: A new pulsed electric field therapy for melanoma disrupts the tumor's blood supply and causes complete remission without recurrence. Int. J. Cancer **125**, 438–445 (2009)
129. Morotomi-Yano, K., Oyadomari, S., Akiyama, H., Yano, K.: Nanosecond pulsed electric fields act as a novel cellular stress that induces translational suppression accompanied by eIF2α phosphorylation and 4E-BP1 dephosphorylation. Exp. Cell Res. **318**, 1733–1744 (2012)
130. Chen, X., Kolb, J.F., Swanson, R.J., Schoenbach, K.H., Beebe, S.J.: Apoptosis initiation and angiogenesis inhibition: melanoma targets for nanosecond pulsed electric fields. Pigment Cell Melanoma Res. **23**, 554–563 (2010)
131. Chen, X., Zhuang, J., Kolb, J.F., Schoenbach, K.H., Beebe, S.J.: Long term survival of mice with hepatocellular carcinoma after pulse power ablation with nanosecond pulsed electric fields. Technol. Cancer Res. Treat. **11**, 83–93 (2012)
132. Chen, R., Sain, N.M., Harlow, K.T., Chen, Y.J., Shires, P.K., Heller, R., Beebe, S.J.: A protective effect after clearance of orthotopic rat hepatocellular carcinoma by nanosecond pulsed electric fields. Eur. J. Cancer **50**, 2705–2713 (2014)

133. Nuccitelli, R.L., Pliquett, U., Chen, X., Ford, W., Swanson, R.J., Beebe, S.J., Kolb, J.F., Schoenbach, K.H.: Nanosecond Pulsed electric fields cause melanomas to self-destruct. Biochem. Biophys. Res. Commun. (BBRC) **343**, 351–360 (2006)
134. Nuccitelli, R., Tran, K., Athos, B., Kreis, M., Nuccitelli, P., Chang, K.S., Epstein Jr., E.H., Tang, J.Y.: Nanoelectroablation therapy for murine basal cell carcinoma. Biochem. Biophys. Res. Commun. **424**, 446–450 (2012)
135. Nuccitelli, R., Wood, R., Kreis, M., Athos, B., Huynh, J., Lui, K., Nuccitelli, P., Epstein Jr., E.H.: First-in-human trial of nanoelectroablation therapy for basal cell carcinoma: proof of method. Exp. Dermatol. **23**, 135–137 (2014)
136. Hofmann, G.A.: Instrumentation and electrodes for in vivo electroporation. In: Jaroszewski, M. J., Heller, R., Gilbert, R. (eds.) Electrochemotherapy, Electrogenetherapy, and Transdermal Drug Delivery, pp. 37–61. Humana Press, Totowa (2000)
137. Marracino, P., Apollonio, F., Liberti, M., d'Inzeo, G., Amadei, A.: Effect of high exogenous electric pulses on protein conformation: myoglobin as a case study. J. Phys. Chem. B **117**(8), 2273–2279 (2013)
138. Singh, A., Orsat, V., Raghavan, V.: Soybean hydrophobic protein response to external electric field: a molecular modeling approach. Biomolecules **3**, 168–179 (2013)
139. Budi, A., Legge, F.S., Treutlein, H., Yarovsky, I.: Electric field effects on insulin chain-B conformation. J. Phys. Chem. B **109**, 22641–22648 (2005)
140. Reale, R., English, N.J., Garate, J.A., Marracino, P., Liberti, M., Apollonio, F.: Human aquaporin 4 gating dynamics under and after nanosecond-scale static and alternating electric-field impulses: a molecular dynamics study of field effects and relaxation. J. Chem. Phys. **139**(20), 205101 (2013)
141. Xiao, S., Guo, S., Vasyl, N., Heller, R., Schoenbach, K.H.: Subnanosecond electrical pulses cause membrane permeabilization and cell death. IEEE Trans. Biomed. Eng. **58**, 1239–1245 (2011)
142. Camp, J.T., Jing, Y., Zhuang, J., Juergen, K., Beebe, S.J., Song, J., Joshi, R.P., Xiao, S., Schoenbach, K.H.: Cell death induced by subnanosecond pulsed electric fields at elevated temperatures. IEEE Trans. Plasma Sci **40**(10), 2334–2347 (2012)
143. Jiang, C., Chen, M.T., Gorur, A., Schaudinn, C., Jaramillo, D.E., Costerton, J.W., Sedghizadeh, P.P., Vernier, P.T., Gundersen, M.A.: Atmospheric-pressure cold plasma for endodontic disinfection. IEEE Trans. Plasma Sci. **37**, 1190–1195 (2009)
144. Kong, M.G., Groesen, G., Morfill, G., Nosenko, T., Shimizu, T., van Dijk, J., Zimmermann, J. L.: Plasma medicine: an introductory review. New J. Phys. **11**, 115012 (2009) (35 pp)
145. Malik, M.A., Jiang, C., Dhali, S.K., Heller, R., Schoenbach, K.H.: Coupled sliding discharges: a scalable nonthermal plasma system utilizing positive and negative streamers on opposite sides of a dielectric layer. Plasma Chem. Plasma Process. (2014). doi:10.1007/s11090-014-9528-2

Chapter 2
Pulsed Power Technology

Hidenori Akiyama, Sunao Katsuki, Luis Redondo, Masahiro Akiyama, A.J.M. Pemen, T. Huiskamp, F.J.C.M. Beckers, E.J.M. van Heesch, G.J.J. Winands, S.J. Voeten, L. Zhen, J.W.M. van Bree, Shu Xiao, and Ross Petrella

Abstract Pulsed power refers to the science and technology of accumulating energy over a relatively long period of time and releasing it as a high-power pulse composed of high voltage and current over a short period of time; as such, it has extremely high power but moderately low energy. Pulsed power is produced by transferring energy generally stored in capacitors and inductors to a load very quickly through switching devices. Applications of pulsed power continue expansion into fields including the environment, recycling, energy, defense, material processing, medical treatment, plasma medicine, and food and agriculture.

Building upon the development of pulsed power generators which offer both high repetition and performance, scientists are now able to investigate effects of pulsed power on living organisms, and their research has expanded to encompass a new field known as bioelectrics. Section 2.1 summarizes pulsed power technology

H. Akiyama (✉) • S. Katsuki
Kumamoto University, Kumamoto, Japan
e-mail: akiyama@cs.kumamoto-u.ac.jp

L. Redondo
Lisbon Engineering Superior Institute, Lisbon, Portugal

M. Akiyama
Iwate University, Iwate, Japan

A.J.M. Pemen • T. Huiskamp • F.J.C.M. Beckers • E.J.M. van Heesch
Eindhoven University of Technology, Eindhoven, The Netherlands

G.J.J. Winands
Demcon, Eindhoven, The Netherlands

S.J. Voeten
Moog-Bradford, Heerle, The Netherlands

L. Zhen
Zhejiang University, Hangzhou, China

J.W.M. van Bree
Vabrema, Eindhoven, The Netherlands

S. Xiao • R. Petrella
Old Dominion University, Norfolk, USA

H. Akiyama, R. Heller (eds.), *Bioelectrics*, DOI 10.1007/978-4-431-56095-1_2

with a focus on this new field. Section 2.2 summarizes the basics of electric circuits, while Sect. 2.3 discusses pulsed power generators utilized for bioelectrics. Section 2.4 describes switches as a key technology. Measurement tools of pulsed power are shown in Sect. 2.5, and delivery of electric pulses to biological tissues using antennas is described in Sect. 2.6.

Keywords Pulsed power • Switch • Measurement • Antenna • Bioelectrics

2.1 Introduction

Hidenori Akiyama

Pulsed power refers to the science and technology of accumulating energy over a relatively long period of time and releasing it as a high-power pulse over a short period of time. Though pulsed power reaches 1 GW, which corresponds to the output of a power plant, the energy in the case of 100 ns pulse width is only 100 J, which is small enough to heat 1 cm^3 water to only about 24 °C. Therefore, while the power of the pulses is extremely high, their energy is moderately low.

Pulsed power is produced by transferring a primary stored energy to a load quickly. Though chemical energy sources such as batteries or explosives or kinetic energy sources such as generators may be used as the primary stored electrical energy, in fact, capacitors and inductors are the most common methods. When the primary stored electrical energy is transferred to a secondary stored energy through a switching device, the pulse width decreases and the power increases while maintaining the same energy. The third and fourth stored energies are used to obtain a shorter pulse width and a higher output power.

Applications of pulsed power are expanding into disparate fields such as the environment (concrete and other recycling methods, algae treatment), energy (oil production from algae, improvement of engines), defense (electromagnetic wave and particle beams), material processing (extreme ultraviolet light source and high-pressure processing), medical treatment (drug delivery and plasma medicine), and food and agriculture (pulse range and plant factories). Recently, the development of high-repetition and high-performance pulsed power generators has enabled scientists to investigate effects of pulsed power on living organisms. This has led to a new field of pulsed power application on living organisms, which is known as bioelectrics.

This chapter summarizes pulsed power technology focusing on bioelectrics. Section 2.1 summarizes pulsed power technology with a focus on this new field. Section 2.2 summarizes the basics of electric circuits, while Sect. 2.3 describes switches as a key technology; both are in preparation to understand Sect. 2.4, which discusses pulsed power generators utilized for bioelectrics. Measurement tools of pulsed power are shown in Sect. 2.5, and delivery of electric pulses to biological tissues using antennas is described in Sect. 2.6.

2.2 Basis of Electric Circuit

Sunao Katsuki

Generation of pulsed power utilizes the transient energy flow in a relatively simple circuit consisting of passive elements and a transmission line. In some bioelectric applications, the main concern is satisfactory voltage flattop. Here, transient phenomena of elemental circuits, voltage multiplication, transmission lines, and pulse-forming networks are described as a basis of the generation and handling of pulsed power which will be described later.

2.2.1 *Resistors, Inductors, and Capacitors*

2.2.1.1 Voltage–Current Relationships

Pulsed power circuits can be analyzed in terms of four elements: resistors, capacitors, inductors, and ideal switches, as shown in Fig. 2.1. These elements are two-terminal passive components, which are characterized by two quantities: the current through the device and the voltage between the terminals. The action of the circuit element is described by the relationship between voltage and current.

A switch is either an open circuit (zero current at any voltage) or a closed circuit (zero voltage at any current). In pulsed voltage circuits, a closing switch is an open circuit for times $t < 0$ and a short circuit for $t > 0$. An opening switch has the inverse properties.

A resistor contains material that impedes the flow of electrons via collisions. The flow of current is proportional to the driving voltage

$$I = V/R, \tag{2.1}$$

where I is in amperes, V in volts, and R is the resistance in ohms. Energy is transferred from flowing electrons to the resistive material. With the polarity shown in Eq. 2.1, electrons flow to the bottom of the resistor. Each electron absorbs an energy $eV0$ from the driving circuit during its transit through the resistor. This

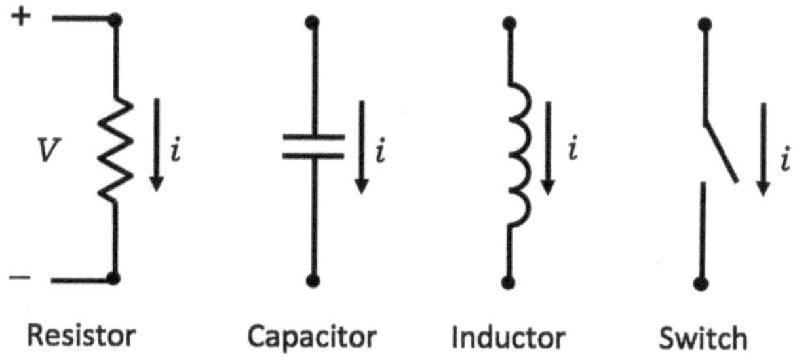

Fig. 2.1 Symbols and polarity convention for common circuit elements

energy acts to accelerate the electrons between collisions. They emerge from the top of the resistor with low velocity because most of the energy gained was transferred to the material as heat. The power deposited is

$$P = VI = \frac{V^2}{R} = I^2R \tag{2.2}$$

Basic capacitor geometry is of two conducting plates separated by a dielectric. The voltage between the plates is proportional to the stored charge on the plates and the geometry of the capacitor:

$$V = \frac{Q}{C} \tag{2.3}$$

The quantity V is in volts, Q in coulombs, and C in farads. Neglecting fringing fields, the capacitance of the parallel plate geometry can be expressed as

$$C = \frac{\varepsilon_0 \varepsilon_r A}{d} \tag{2.4}$$

where $\varepsilon_0 = 8.85 \times 10^{-12}$ and ε_r is the relative dielectric constant of the material between the plates, A is the plate area in square meters, and d is the plate separation in meters. The values of most plastic materials range between 2 and 5, whereas those of strong dielectric materials such as barium titanate are on the order of 10,000. The current through a capacitor is the time rate of change of the stored charge:

$$I = \frac{dQ}{dt} = C\frac{dv}{dt} \tag{2.5}$$

The capacitor contains a region of electric field. The inductor is configured to produce a magnetic field. The most common geometry is solenoidal winding. The magnetic flux linking the windings is proportional to the current in the winding, such as

$$\Phi = LI \tag{2.6}$$

where L is a constant dependent on inductor geometry. Inductance is measured in Henries. The voltage across the terminals is proportional to the time rate of change of magnetic flux. Therefore,

$$V = \frac{d\Phi}{dt} = L\frac{dI}{dt} \tag{2.7}$$

2.2.1.2 Electrical Energy Storage

A resistor converts electrical energy to thermal. No stored electrical energy remains in a resistor in the absence of a voltage supply. Conversely, capacitors and inductors, known as reactive elements, store electrical energy in the form of electric and magnetic fields, respectively. No average energy dissipation exists in a reactive element. Energy stored in a capacitor in the form of electric fields is

$$U_c = \frac{1}{2}CV^2 \tag{2.8}$$

Magnetic energy stored in an inductor is

$$U_c = \frac{1}{2}LI^2 \tag{2.9}$$

2.2.2 *Basic Circuits for Pulsed Power [1, 2]*

2.2.2.1 Capacitor–Resistor Circuit

Figure 2.2 shows a familiar circuit combining a resistor, capacitor, and switch. This is the simplest model for a pulsed voltage circuit; electrical energy is stored in a capacitor and then dumped into a load resistor via a switch. Continuity of current around the circuit combined with Eqs. 2.1 and 2.5 implies the following differential equation for the load voltage after switching:

$$C\frac{dV}{dt} + \frac{V}{R} = 0 \tag{2.10}$$

The solution plotted in Fig. 2.2 for switching at $t = 0$ is

$$V(t) = V_0 \ \exp(-t/RC) \tag{2.11}$$

where V_0 the initial charge voltage. The product RC is the characteristic time for the transfer of energy from a capacitor to a resistor in the absence of inductance. A circuit with a resistor, capacitor, voltage source, and switch is shown in Fig. 2.3. This circuit models the capacitor charged by a DC voltage source through a resistor. Continuity of current and the *V-I* relations of the components, the time variation for load voltage, plotted in Fig. 2.3, is

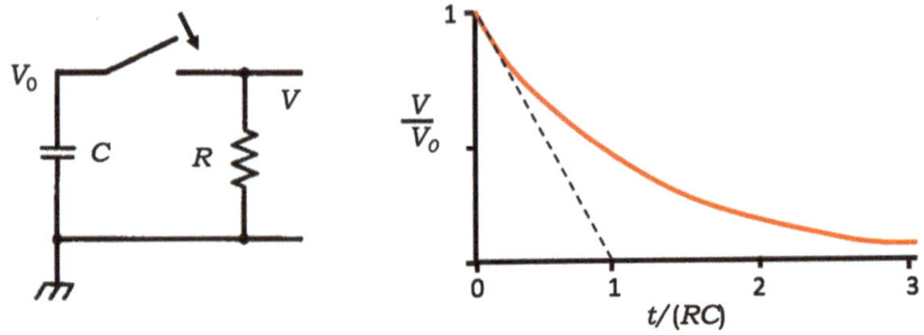

Fig. 2.2 Time-dependent load voltage in a switched RC circuit

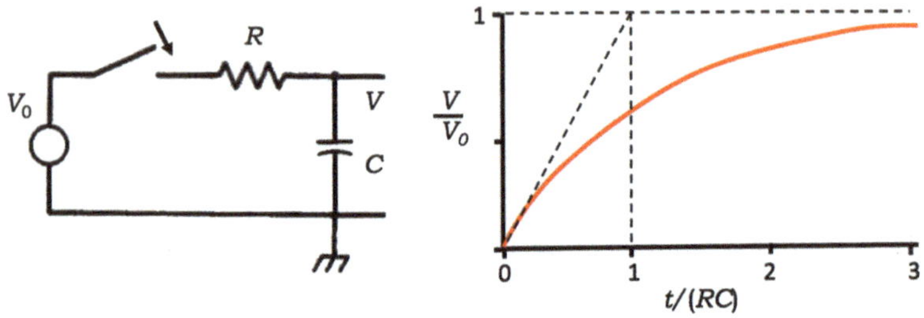

Fig. 2.3 Time-dependent load voltage at capacitor from a pulse generator with series resistance

$$V(t) = V_0\left[1 - \exp\left(-\frac{t}{RC}\right)\right] \tag{2.12}$$

2.2.2.2 Inductor–Resistor Circuit

This is the simplest model for a pulsed current circuit on the basis of an inductive energy storage system. Electrical energy is stored in an inductor as a magnetic field and then released into a load resistor via a switch. Applying Kirchhoff's voltage law combined with Eqs. 2.1 and 2.7 leads to the following differential equation for the load current after switching:

$$L\frac{dI}{dt} + RI = 0 \tag{2.13}$$

The time variation for circuit current, plotted in Fig. 2.4, is

$$I(t) = I_0\exp\left(-\frac{t}{L/R}\right) \tag{2.14}$$

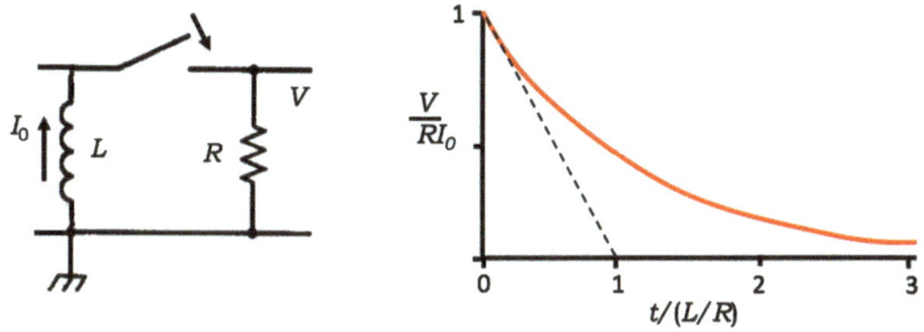

Fig. 2.4 Time-dependent load voltage in a switched LR circuit

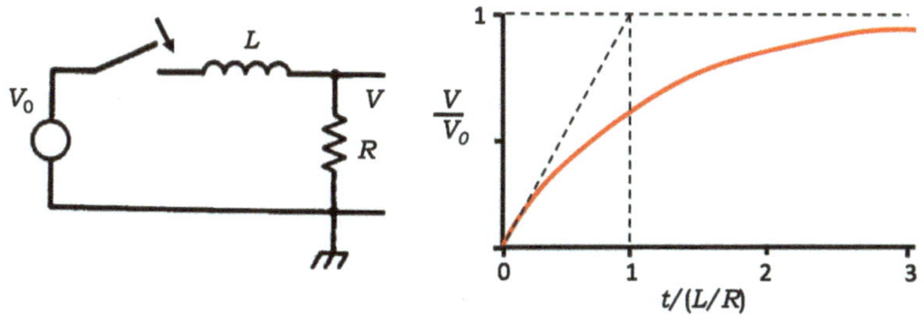

Fig. 2.5 Time-dependent load voltage from a pulse generator with a series inductance

where I_0 is an initial current in the inductor. This initial current must be provided by a separate circuit in advance. The voltage on the load resistor is $i(t)R$.

Figure 2.5 shows a circuit with an inductor, resistor, voltage source, and switch, modeling the output region of a pulsed voltage generator under significant inductance of leads and load. A rapid risetime for power to the load is typically desirable. Current flow initiation time to the load is limited by parasitic inductance. Kirchhoff's voltage law after switching is given as

$$V_0 = L\frac{dI}{dt} + RI \tag{2.15}$$

The time variation for load voltage, plotted in Fig. 2.5, is

$$V(t) = V_0\left[1 - \exp\left(-\frac{t}{L/R}\right)\right] \tag{2.16}$$

The L/R time determines the speed at which current and voltage can be induced to the load. The 10–90 % risetime for the voltage pulse is $2.2(L/R)$. For instance, where a fast pulse generator used to drive a 10 Ω load, the total inductance of the load circuit would have to be less than 23 nH if the risetime was to be less than 5 ns.

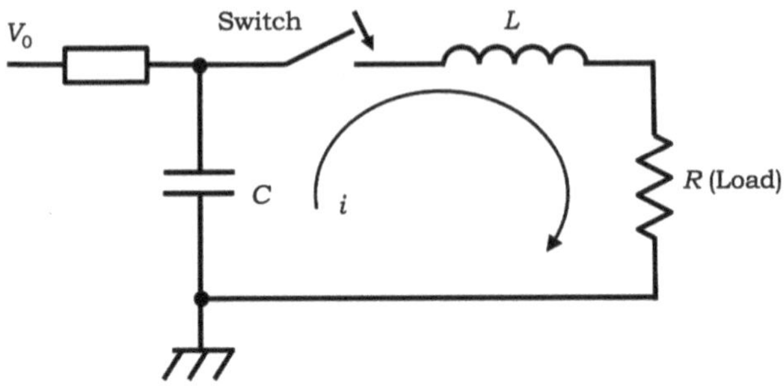

Fig. 2.6 CLR circuit (capacitive energy storage circuit)

2.2.2.3 CLR Circuit: Capacitive Energy Storage Circuit

All pulsed voltage circuits have an energy storage element where electrical energy is contained in the form of electric or magnetic fields. The energy is transferred by a fast switch to a load. The speed of transfer is limited by parasitic inductance or capacitance in the circuit. The voltage pulse waveform is determined by the configuration of the energy storage element and the nature of the load. The circuit produces a variation in time of the voltage.

The simplest electrical energy storage device is a single capacitor. The voltage modulator shown in Fig. 2.6 consists of a capacitor charged to voltage V_0 and a closing switch to transfer the energy. The energy is deposited in a load resistor, R. The flow of current involved in the transfer generates magnetic fields, so the effects of series inductance L must be included in the circuit. Setting the loop voltage equal to zero gives

$$L\frac{dI}{dt}+RI+\int\frac{I}{C}dt=0 \tag{2.17}$$

The second condition follows from the fact that, immediately after switching, the total capacitor voltage appears across the inductor rather than the resistor. The solution of Eq. 2.17 is usually written in three different forms depending on the values of the following parameters:

$$\omega_0=1/\sqrt{LC}, \tag{2.18}$$

$$\beta=\frac{R}{2L}, \tag{2.19}$$

$$\omega_1=\sqrt{\omega_0^2-\beta^2},\ \ \omega_2=\sqrt{\beta^2-\omega_0^2},\ \text{and}\ \delta=\tan^{-1}(\beta/\omega_1) \tag{2.20}$$

The solution with $\beta<\omega_0$, which is termed an under-damped circuit, is plotted as curve (a) in Fig. 2.7. The circuit behavior is oscillatory. Energy is transferred back and forth between the inductor and capacitor at approximately the characteristic frequency, ω_0. Slight energy loss occurs at each oscillation, determined by the damping parameter. The time-dependent current is

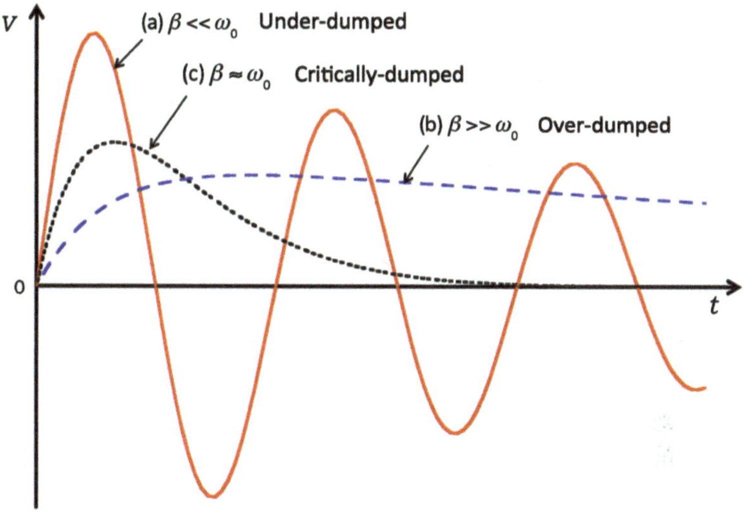

Fig. 2.7 Voltage on a load driven by a switched capacitor with series inductance as a function of β/ω_0: L and C are constant, load resistance R varied; (**a**) $\beta << \omega_0$ under-dumped, (**b**) $\beta >> \omega_0$ over-dumped, and (**c**) $\beta = \omega_0$ critically dumped

$$\begin{aligned} I(t) &= \left(\frac{CV_0}{\cos\delta}\right)\exp(-\beta t)[\omega_1 \sin(\omega_1 t - \delta) + \beta\cos(\omega_1 t - \delta)] \\ &= CV_0\omega_0 \exp(-\beta t)\sin(\omega_0 t) \quad (\beta << \omega_0) \end{aligned} \tag{2.21}$$

Conversely, when $\beta > \omega_0$, the circuit is termed over-damped. As indicated in curve (b) in Fig. 2.7, the circuit is dominated by resistance and does not oscillate. The monopolar voltage pulse on the load rises in a time of approximately L/R and decays exponentially over time RC. The current following switching in an over-damped circuit is

$$i(t) = \left[CV_0\left(\beta^2 - \omega_2^2\right)/2\omega_2\right][\exp(\omega_2 t) - \exp(-\omega_2 t)]\exp(-\beta t) \quad (\beta >> \omega_0) \tag{2.22}$$

Finally, an LRC circuit when $\beta = \omega_0$ or $R_C = 2\sqrt{L/C}$ is termed critically damped.
The current for a critically damped circuit is

$$i(t) = \beta CV_0(\beta t)\exp(-\beta t) \tag{2.23}$$

Time-dependent load voltage is plotted as curve (c) in Fig. 2.7. The curves (a), (b), and (c) in Fig. 2.7 share identical values for L and C, while those of R differ. Note that the transfer of energy from the capacitor to the load resistor is accomplished most rapidly under the critically damped circuit. Thus, the power extracted from a pulsed circuit is maximum when $R = 2\sqrt{L/C}$, which is called the characteristic impedance of the circuit. Energy transfer is optimized when the load resistance is

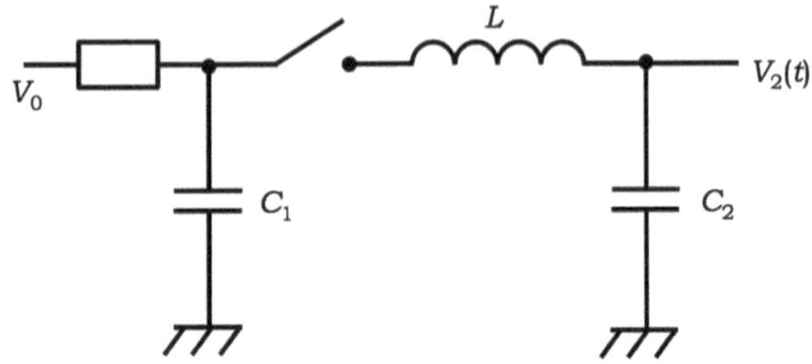

Fig. 2.8 CLC energy transfer circuit

matched to the modulator impedance. The peak power flow in a critically damped circuit occurs at time $t = 1/\beta$. The maximum voltage at this time is $V_{max} = 0.74\,V_0$, the maximum current is $0.37\,V_0/\sqrt{L/C}$, and the maximum power in the load is $0.27\,V_0/\sqrt{L/C}$. The maximum power from this simple pulsed power generator is limited by the parasitic inductance of the circuit.

2.2.2.4 CLC Energy Transfer Circuit

Transfer of energy between capacitors forms the basis of most pulsed power generators. A circuit model for the transfer is illustrated in Fig. 2.8. A capacitor is charged to voltage V_0, and then energy is switched through an inductance to a second capacitor by closing switch. The inductance may be introduced purposely or may represent the inevitable parasitic inductance associated with current flow. Current in the circuit is described by the equation

$$\int \frac{I}{C_1} dt + L\frac{dI}{dt} + \int \frac{I}{C_2} dt = 0 \tag{2.24}$$

If the initial conditions are $I(0) = 0$ and $V_c = V_0$, the solution of Eq. 2.24 is

$$I(t) = \frac{V_0}{\omega L} \sin(\omega t) \tag{2.25}$$

where

$$\omega = \sqrt{LC_1C_2/(C_1 + C_2)} \tag{2.26}$$

The quantities of interest are the time-dependent charge voltages on the two capacitors:

$$V_1(t) = V_0 - \int Idt/C_1, \quad V_2(t) = \int Idt/C_2 \tag{2.27}$$

Substituting Eqs. 2.25 into 2.27, we find

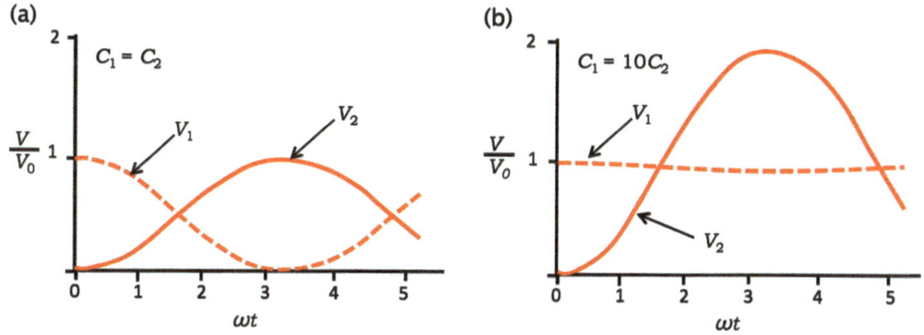

Fig. 2.9 Voltage waveforms on storage capacitor and charged capacitor in a CLC circuit. (**a**) Efficient energy transfer mode with $C_1 = C_2$ and (**b**) voltage gain mode with $C_2 = 0.1C_1$

$$V_1 = V_0\left[1 - \left(\frac{C_2}{C_1 + C_2}\right)(1 - \cos(\omega t))\right], \tag{2.28}$$

and

$$V_2 = V_0\frac{C_1}{C_1 + C_2}(1 - \cos(\omega t)). \tag{2.29}$$

Waveforms are plotted in Fig. 2.9 for $C_2 = C_1$ (a) and $C_2 << C_1$ (b). The first case is the optimum choice for a high-efficiency power compression circuit. A complete transfer of energy from the first to the second capacitor occurs at time $t = \omega/\pi$. In the second case, though the energy transfer is inefficient, the second capacitor is driven to a charge voltage twice that of the first. For this reason, the circuit of Fig. 2.8 is often called the peaking capacitor circuit.

2.2.3 Impulse Generators [2, 3, 4, 5]

The level of the voltage generated by a single-capacitor modulator is limited by both the power supply and the tolerance of the capacitor; in addition, insulation must be taken care of carefully at levels above 30 kV when you work in atmospheric air. Impulse generators enable production of pulsed voltages exceeding 100 kV relatively easily even with the use of inexpensive low-voltage power supply and low-voltage capacitors. Recently, pulsed power modulators have utilized semiconductor power device operation voltages below 5 kV. An impulse generator configuration is often used to generate 50 kV even with the use of semiconductor switches. Impulse generators consist of a number of capacitors charged in parallel to moderate voltage levels. The capacitors are switched to a series configuration by simultaneously triggered closing switches. The voltages of the capacitors add in the

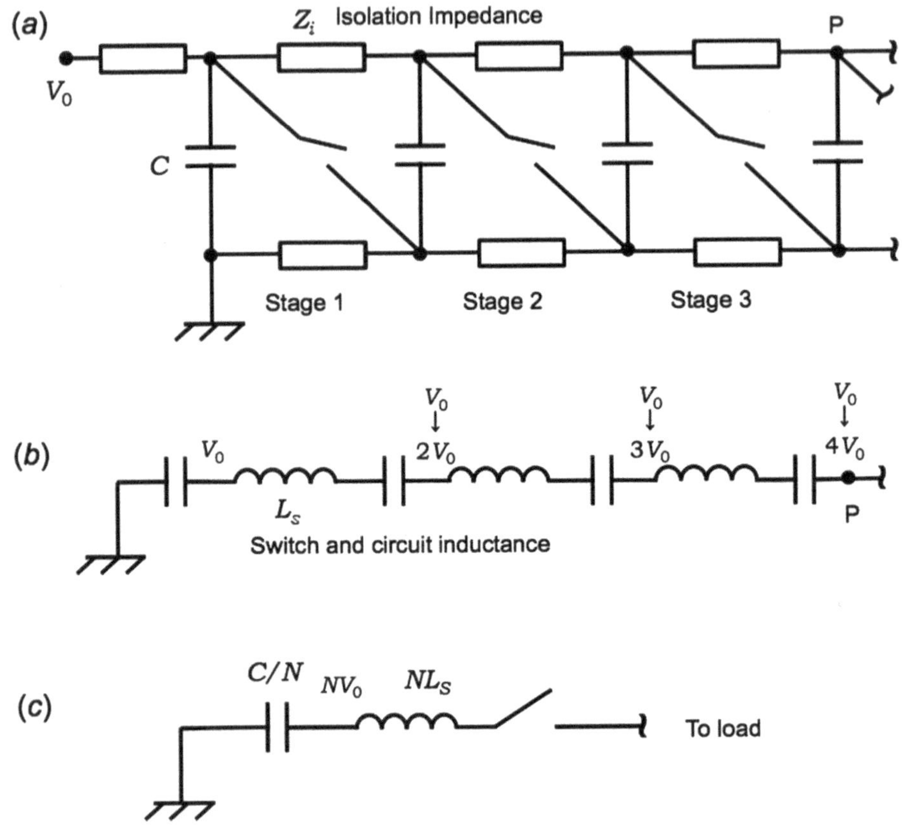

Fig. 2.10 Circuit diagram of a typical Marx generator and its equivalent circuit with all switches on

series configuration. Impulse generators that produce submicrosecond pulses require less insulation than a single capacitor with a steady-state charge at the full output voltage. Below, two widely used circuits are discussed: the Marx generator and the LC generator.

2.2.3.1 Marx Generator

A typical configuration for a Marx generator is illustrated in Fig. 2.10a. The charging current is carried to the capacitors through isolation resistors (or inductors) that act as open circuits during the fast output pulse. The capacitor stack is interrupted by high-voltage closing switches.

The circuit configuration immediately after switch closing is shown in Fig. 2.10b. The voltage across the load rapidly increases from 0 to NV_0, where N is the number of switches and V_0 is the magnitude of charge voltage on each capacitor. The series inductance shown arises mainly from the narrow discharge channels in the spark gaps. If voltage variations on timescales less than the dimension of the generator divided by the speed of light in the insulating medium are ignored, the inductances and capacitors can be lumped together as shown in Fig. 2.10c. The Marx generator in the high-voltage phase is equivalent to a single-capacitor modulator. Output to a resistive load is described by the equations of Sect. 2.2.2.3.

The total series inductance is proportional to the number of switches, while the series capacitance is C/N. Therefore, the characteristic generator impedance is proportional to N. This implies difficulty in designing high-voltage Marx generators with low characteristic impedance. High-energy density capacitors and short connections help lower inductance, but the main limitation arises from the fact that the discharge current must flow through inductive spark gap switches.

One favorable feature of the Marx generator is that triggering all switches actively is unnecessary: shorts in the spark gaps at the low-voltage end of the stack can be compensated with by overvoltage on the remaining gaps. Furthermore, the trigger electrodes of the spark gaps can be connected by circuit elements so that a trigger wave propagates rapidly through the generator.

2.2.3.2 LC Generator

The LC generator, illustrated in Fig. 2.11, is more difficult to trigger than the Marx generator but has lower characteristic impedance for equivalent output voltage. As in the Marx generator, a stack of capacitors is charged slowly in a parallel configuration by a voltage supply. The main difference is that the switches are external to the main power flow circuit. Transition from a parallel to a series configuration is accomplished in the following way: half of the capacitors are connected to external switched circuits with a series inductance, and when the switches are triggered, each LC circuit begins a harmonic oscillation. After one-half cycle, the polarity on the switched capacitors is reversed. At this time, the voltages of all capacitors add in series, as shown in Fig. 2.11b. The voltage at the output varies as

$$V_c = 2NV_0\left[1 - \cos\left(\sqrt{\frac{2}{lc}}t\right)\right] \tag{2.30}$$

where C is the capacitance of a single capacitor, N is the number of switches, and V_0 is the magnitude of the charge voltage.

The load must be isolated from the generator by a high-voltage switch during the mode change from parallel to series. It can be actively triggered near the peak

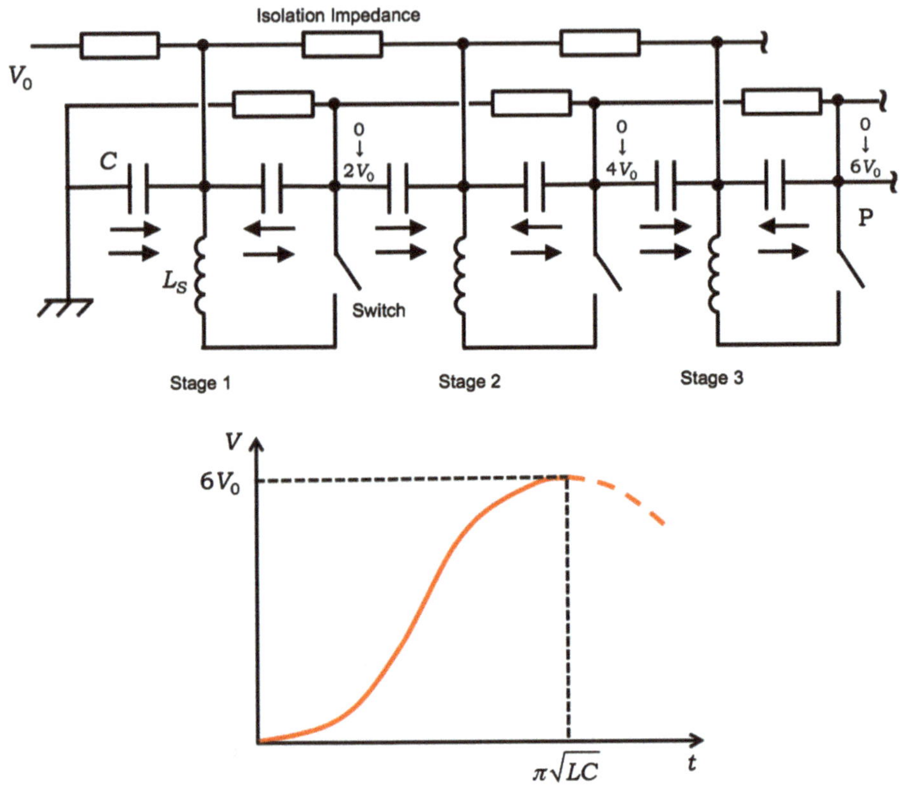

Fig. 2.11 Circuit diagram of a typical LC generator showing voltage at point P following switching through the external inductor

output voltage. Energy transfer to the load should take place at t = $\sqrt{LC}/2$ so that no energy remains in the external inductors. The transfer must be rapid compared to the mode change time or energy will return to the inductors, reducing generator efficiency.

The equivalent circuit for the LC generator in the series state is similar to that of the Marx generator (Fig. 2.10c), with the main difference being that the series inductances are reduced by elimination of the switches. Three main disadvantages of the LC generator compared to the Marx generator are that the switching sequence is more complex, a low-inductance output switch is required, and the circuit remains at high voltage for a longer time. Because triggering one reversing circuit does not result in an overvoltage on the spark gaps of other sections, all switches in an LC generator must be actively fired with strong, synchronized trigger pulses.

2.2.4 *Transmission Line [1, 2, 3, 4, 5]*

2.2.4.1 Transmission Line Theory

Most applications for pulse modulators require a constant-voltage pulse. The critically damped waveform is the closest a modulator with a single capacitor and inductor can approach constant voltage. Better waveforms can be generated by modulators with multiple elements; such circuits are called pulse-forming networks (PFNs), whose transmission line is the continuous limit of a PFN. Here, we analyze transmission lines by a lumped element description rather than the direct solution of the Maxwell equations. Discrete element PFNs are treated in Sect. 2.2.5.

We will concentrate on the coaxial transmission line, with properties of other common geometries listed in Table 2.1. The coaxial line consists of an inner conducting cylinder (radius R_i) and a grounded outer cylinder (R_o) separated by a medium with dielectric constant ε_r. Assumed is linear magnetic permeability, μ. Figure 2.12a shows a sectional view of a line divided into differential elements of length Δz. Each element has a capacitance between the inner and outer conductors proportional to the length of the element Δz. If C is the capacitance per length, the capacitance of an element is $C\Delta z$. Magnetic fields are produced by current flow along the inner conductor. Each differential element also has a series inductance, $L\Delta z$, where L is the inductance per unit length. The circuit model of Fig. 2.12b can be applied as a model of the transmission line. The quantities C and L for cylindrical geometry are given by

Table 2.1 Typical transmission lines and their capacitance, inductance, and impedance

Coaxial transmission line

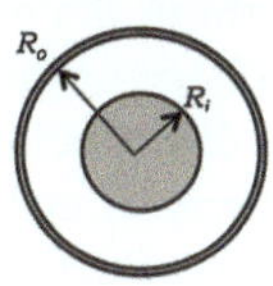

$$C = 2\pi\varepsilon / \ln(R_0 / R_i)$$
$$L = \frac{\mu}{2\pi}\ln(R_0 / R_i)$$
$$Z_0 = \frac{\sqrt{\mu/\varepsilon}}{2\pi}\ln(R_0 / R_i)$$

Isolated parallel plates (*d* << *D*)

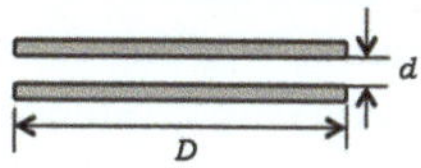

$$C = \varepsilon D / d$$
$$L = \mu d / D$$
$$Z_0 = \sqrt{\mu/\varepsilon}\,\frac{d}{D}$$

Two-wire transmission line

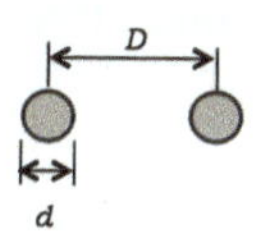

$$C = \pi\varepsilon / \cosh^{-1}(D / d)$$
$$L = \mu / \pi \cosh^{-1}(D / d)$$
$$Z_0 = \frac{\sqrt{\mu/\varepsilon}}{\pi}\cosh^{-1}(D / d)$$

Strip line (*d* << *D*)

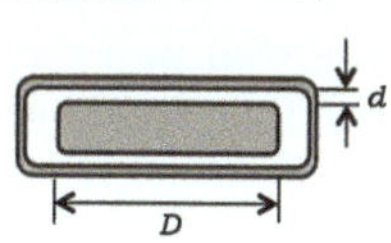

$$C = 2\varepsilon D / d$$
$$L = \mu d / 2D$$
$$Z_0 = \sqrt{\mu/\varepsilon}\,\frac{d}{2D}$$

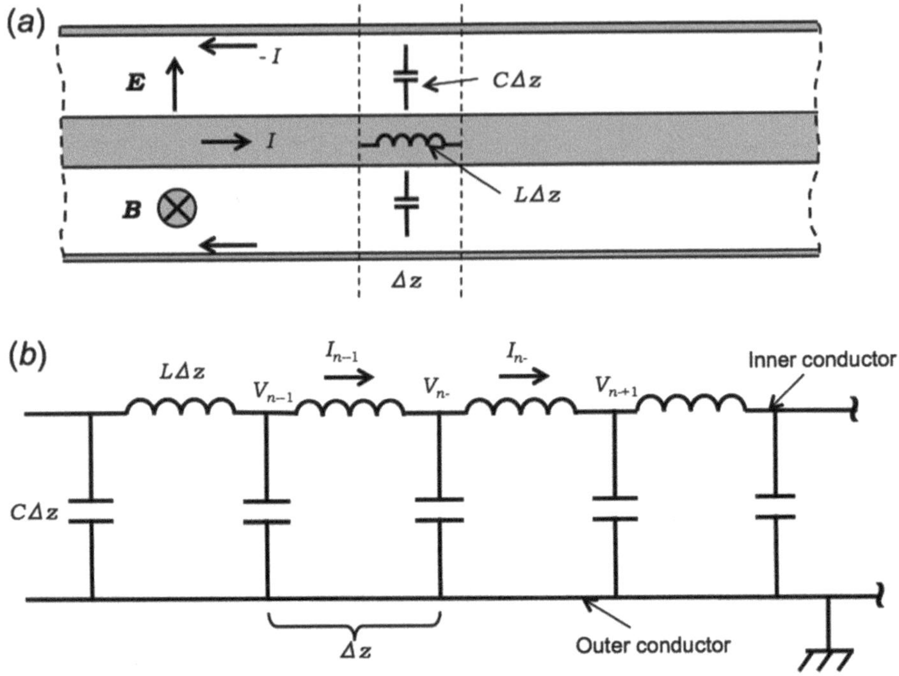

Fig. 2.12 Coaxial transmission line. (**a**) Physical basis for lumped circuit model. (**b**) Lumped circuit element analogue of a coaxial transmission line

$$C = 2\pi\epsilon/\ln(R_o/R_i) \tag{2.31}$$

$$L = \frac{\mu}{2\pi}\ln(R_o/R_i) \tag{2.32}$$

The outer conductor may be considered an ideal ground; voltage variations occur along the inner conductor. The two quantities of interest that determine the fields in the line are the current flow along the inner conductor and the voltage of the inner conductor with respect to the grounded outer conductor. Voltage differences along the inner conductor are inductive. Differences in current along the inner conductor result from displacement current between the inner and outer conductors. The current returns along the outer conductor to complete the circuit. The balance between these effects determines the relationship between voltage and current and the propagation velocity of the pulse. The voltage at point n is equal to the total current that has flowed into the point divided by the capacitance. Using the sign convention shown, this statement is expressed by

$$-C\Delta z\left(\frac{\partial V_n}{\partial t}\right) = I_n - I_{n-1} \tag{2.33}$$

The relationship is a partial differential equation because the change in voltage with time is viewed at a constant position. The difference in voltage between two points is the inductive voltage between them:

$$-L\Delta z\left(\frac{\partial I_n}{\partial t}\right) = V_n - V_{n-1} \tag{2.34}$$

When Δz is small, the discrete voltages approach a continuous function, $V(z, t)$. The voltage difference between two points can be approximated in terms of this function as continuous partial differential equations

$$\frac{\partial V}{\partial z} = -L\frac{\partial I}{\partial t} \tag{2.35}$$

$$\frac{\partial I}{\partial z} = -C\frac{\partial V}{\partial t} \tag{2.36}$$

The two above equations are called telegraphist's equations and can be combined to give wave equations for V and I of the form

$$\frac{\partial^2 V}{\partial z^2} = (LC)\frac{\partial^2 V}{\partial t^2} \tag{2.37}$$

Similarly,

$$\frac{\partial^2 I}{\partial z^2} = (LC)\frac{\partial^2 I}{\partial t^2} \tag{2.38}$$

Equations 2.37 and 2.38 are mathematical expressions of the properties of a transmission line. It can easily be verified that any 3 functions of the forms

$$V(z,\ t) = F\left(t \pm \frac{z}{v}\right) \tag{2.39}$$

$$I(z,\ t) = G\left(t \pm \frac{z}{v}\right) \tag{2.40}$$

are solutions of Eqs. 2.37 and 2.38, respectively, if

$$v = 1/\sqrt{LC} \tag{2.41}$$

Equations 2.39 and 2.40 indicate that voltage or current pulses propagate in the transmission line at velocity v without a change in shape. Pulses can travel in either the $+z$ or $-z$ direction depending on input conditions. The velocity of propagation in the coaxial transmission line can be found by substituting Eqs. 2.31 and 2.32 into 2.41:

$$v = 1/\sqrt{\varepsilon\mu} \tag{2.42}$$

This velocity is the speed of light in the medium; that is, geometric factors in C and L cancel for all transmission lines, meaning that the propagation velocity is determined only by the properties of the medium filling the line.

Inspection of Eqs. 2.35 and 2.36 shows that voltage is linearly proportional to the current at all points in the line, independent of the functional form of the pulse shape. In other words,

$$V = IZ_0 \tag{2.43}$$

where Z_0 is called the characteristic impedance of the line. Its value depends on the geometry of the line. The characteristic impedance for a coaxial transmission line is

$$Z_0 = \sqrt{L/C} = \frac{\sqrt{\mu/\varepsilon}}{2\pi}\ln\left(\frac{R_o}{R_i}\right) \tag{2.44}$$

2.2.4.2 Reflection and Matching

Let us consider a stepped voltage pulse with amplitude of V_0 traveling in a transmission line whose end is terminated by a load with an impedance of Z_L, as shown Fig. 2.13a. The pulse reaches the load and is reflected by the load unless the load impedance is equal to the characteristic impedance of the line, Z_0. Figure 2.13b shows a snapshot of the voltage distribution consisting of one traveling to the load and the other traveling back to the transmission line. The voltage amplitude of the reflected pulse is

$$V_r = \rho V_0 \tag{2.45}$$

where ρ is a reflection coefficient given by

$$\rho = \frac{Z_L - Z_0}{Z_L + Z_0} \tag{2.46}$$

The load voltage is the sum of the incident and the reflected pulses, as

$$V_L = (1+\rho)V_0 \tag{2.47}$$

Equation 2.46 indicates that the condition of $Z_L = Z_0$ results in $\rho = 0$, i.e., no reflection results, and all electromagnetic energy is absorbed in the load. This situation of no reflection is called a matching between load and transmission line.

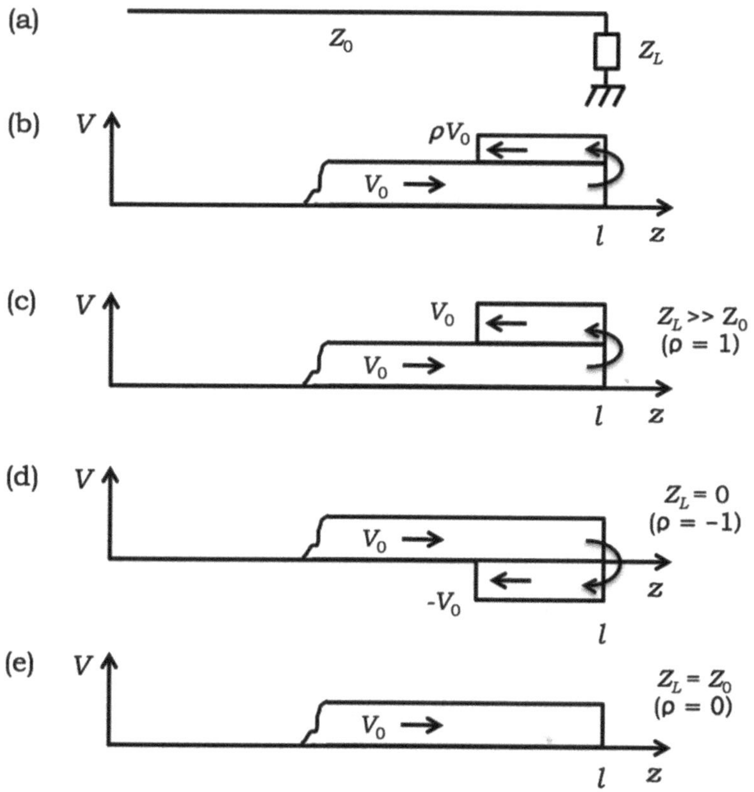

Fig. 2.13 Reflection of a step-functioned voltage pulse at the load. (**a**) Transmission line (Z_0) terminated by load (Z_L). (**b–e**) Reflection of a voltage pulse at load under various loads: (**c**) $Z_L >> Z_0$, (**d**) $Z_L = 0$, and (**e**) $Z_L = Z_0$

Figure 2.13c, e shows the reflection at the special cases in (c) Z_r is much larger than Z_0, (d) $Z_r = 0$, and (e) $Z_r = Z_0$, respectively.

2.2.5 *Pulse-Forming Network*

Transmission lines are well suited for output pulse lengths in the range 5 ns $< \Delta t_p <$ 200 ns but are impractical for pulse lengths above 1 μs. Discrete element circuits are usually used for long pulse lengths as they achieve better output waveforms than the critically damped circuit due to their use of more capacitors and inductors. Discrete element circuits providing a shaped waveform are called pulse-forming networks. The derivations of Sect. 2.2.4 suggest that the circuit of Fig. 2.14 can provide a pulse with an approximately constant voltage. A transmission line is simulated by a

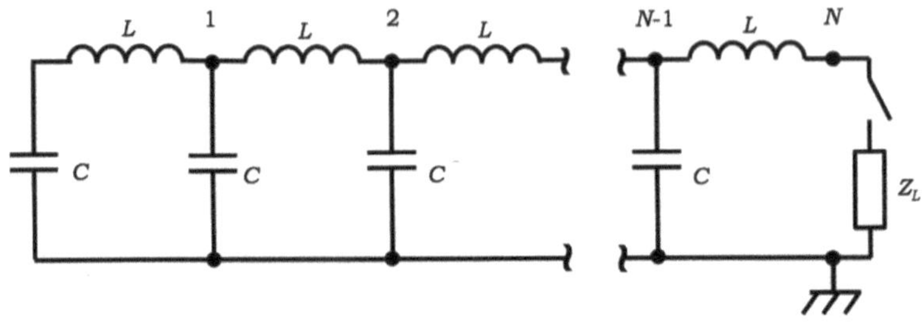

Fig. 2.14 Pulse-forming network; lumped element approximation of a transmission line

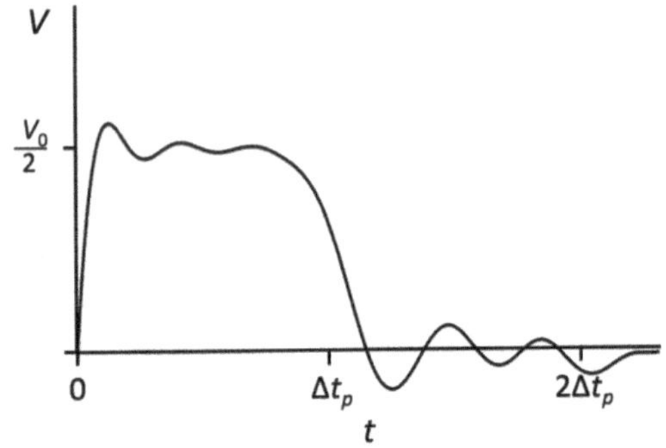

Fig. 2.15 Load voltage, five-element PFN discharged a matched load

finite number N of inductor–capacitor units. Following the derivation of Sect. 2.2.4, the resistance of a matched load is

$$Z_0 = \sqrt{L/C} \tag{2.48}$$

The quantities L and C are the inductance and capacitance of discrete elements. We shall call Z_0 the impedance of the PFN. The single transit time of an electromagnetic pulse through the network is approximately $N\sqrt{LC}$. The output voltage pulse has average magnitude $V_0/2$ and duration of

$$\Delta t_p = 2N\sqrt{LC} \tag{2.49}$$

The output pulse of a five-element network into a matched resistive load is shown in Fig. 2.15. Although the general features are as expected, there exist a substantial overshoot at the beginning of the pulse and an undershoot at the end in addition to voltage oscillations during the pulse.

For most applications, the main concern is a good voltage flattop; gradual voltage variation on the rise and fall of the pulse can be tolerated. A large number of elements are needed in the circuit of Fig. 2.14 for a relatively constant output voltage.

2.3 Pulsed Power Generators

Hidenori Akiyama and Masahiro Akiyama

2.3.1 Pulse-Forming Line

A simple pulsed power generator is composed of a pulse-forming line using distributed constant circuits. A coaxial line, which consists of inner and outer conductors along with an insulator between these conductors, is shown in Fig. 2.16. The coaxial line is charged by a high-voltage DC source through the resistor with a resistance much higher than the characteristic impedance, $Z_0\left(=\sqrt{L/C}\right)$, where L and C are the inductance and the capacitance per unit length. The charging voltage, V_0, appears on the inner conductor along the coaxial line.

Figure 2.17 shows the voltage change on the inner conductor at the times of $t = 0,\ 1/2v_0,\ 1/v_0,\ 31/2v_0$ after initiation of the switch S. Following initiation of S at $t = 0$, the transmission line is expressed by wave equations, and right- and left-going voltage waves with the voltage of $V_0/2$ appear. Wave velocity is expressed as $v_0 = 1/\sqrt{LC}$. One wave, represented here by dots, moves left and is reflected at the left edge of the transmission line since the reflection coefficient expressed by $\rho_r = (R_0 - Z_0)/(R_0 + Z_0)$ becomes 1 assuming $R_0 >> Z_0$. The other wave, represented here by oblique lines, moves right; all energy moving right is absorbed by the resistor since $\rho_r \cong 0$ assuming $R = Z_0$.

Voltage waveform at load R is shown in Fig. 2.18. A pulse voltage with amplitude of $V_0/2$ and pulse width of $21/v_0$ appears, since $R = Z_0$. This pulse-forming line is the simplest pulsed power generator.

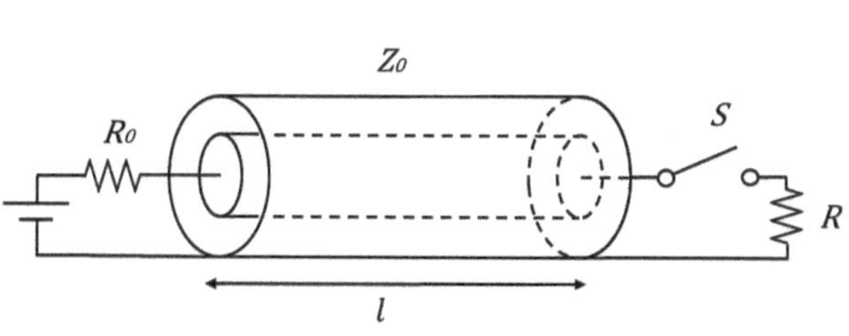

Fig. 2.16 A simple pulsed power generator composed of a pulse-forming line

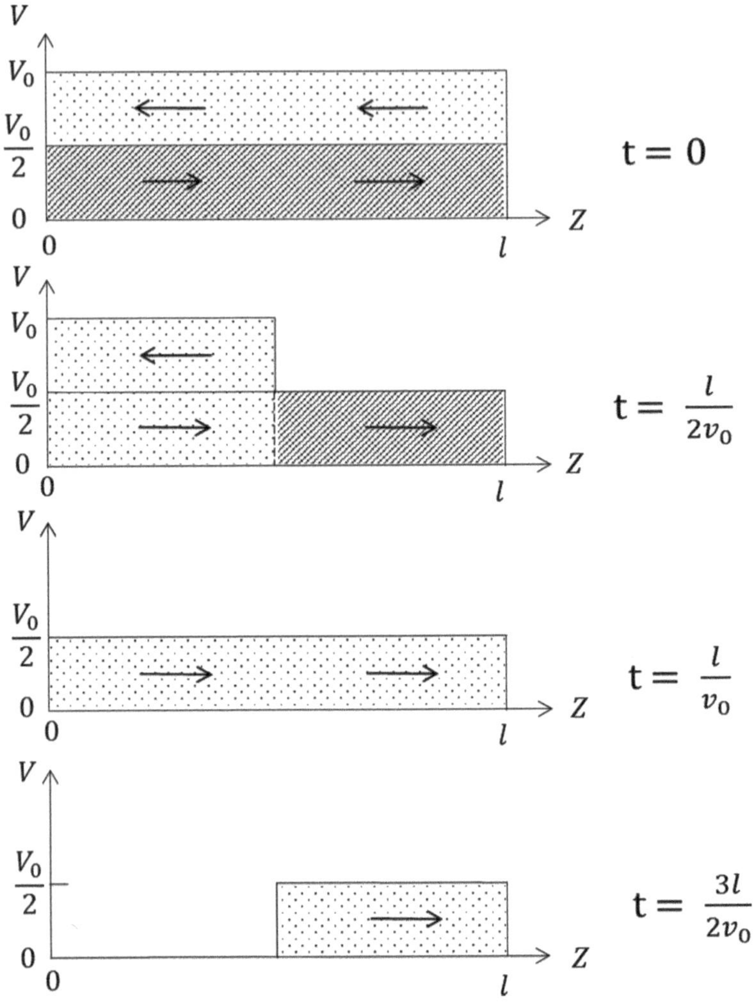

Fig. 2.17 Voltage change on the inner conductor at the different time after initiation of the switch S

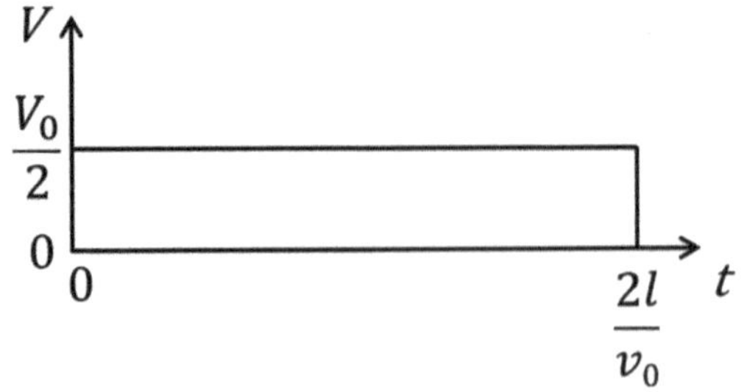

Fig. 2.18 Voltage waveform at load R

2.3.2 Blumlein Line

The pulse-forming line proposed by Blumlein is shown in Fig. 2.19. The Blumlein line is composed of three long and narrow metal strips, A, B, and C; a closing switch, S_c, between the center and lower strips; and the load, R. The characteristic impedances between B and A (or C) are Z_0 along l, and between A and C are $2Z_0$ along l_1. The load resistance, R, is selected as $2Z_0$. The center strip is charged to V_0 through a large resistance, R_0, and the inductance L. After the switch, S_c, initiates at $t = 0$, the waves propagate between A and B and between B and C as shown in Fig. 2.19. The left-going wave propagating between A and B meets high impedance and is reflected with the same polarity. The left-going wave propagating between B and C meets S_c initiation and is reflected with the reverse polarity. The right-going waves between A and B and between B and C are reflected with the same polarity since $\rho_r \cong 1$. These phenomena exist until $t = l/v_0$ after initiation of the switch S_c. A voltage difference between A and C begins to occur at $t = l/v_0$, and after this time, waves both between A and B and between B and C propagate to the space between A and C along l_1. The characteristic impedance between A and C along l becomes $2Z_0$, which corresponds to the characteristic impedance between A and C along l_1. Therefore, no reflection occurs at the right edge of B after $t = l/v_0$; instead, the right-going wave is absorbed by the load resistance, R, since $2Z_0 = R$. The voltage V_{AB} between A and B and V_{BC} between B and C is summarized in Fig. 2.20, which also shows the voltage V_{AC} between A and C calculated by adding V_{AB} and V_{BC}.

The voltage waveform at load R is shown in Fig. 2.21. The pulse voltage with amplitude of V_0 and pulse width of $2l/v_0$ appear. Though the pulse width is the same as that in Fig. 2.18, the voltage is double.

2.3.3 Magnetic Pulse Compression

Figure 2.22 shows a schematic diagram of a pulsed power generator using a magnetic pulse compression circuit. After the capacitor, C_0, is charged to V_0, insulated-gate bipolar transistors (IGBTs) are initiated. Since the saturable inductor, L_0, has a large value, IGBT current remains low during its initiation. Therefore, switching loss calculated from the IGBT voltage and current is minimized. Figure 2.23 shows waveforms of IGBT voltage and current. The inductor, L_0, assumes

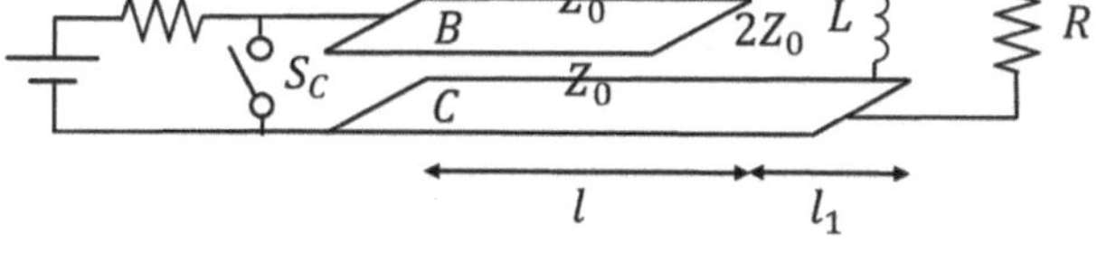

Fig. 2.19 The pulse-forming line proposed by Blumlein

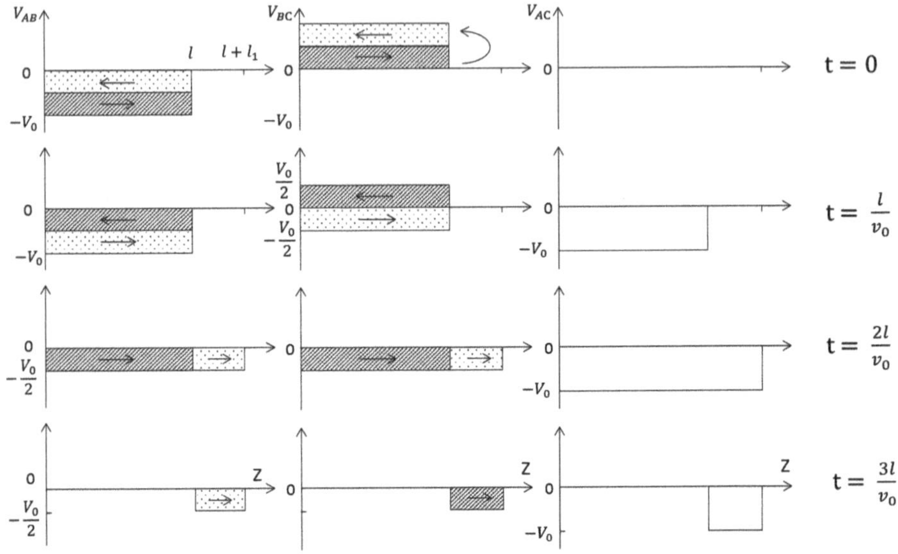

Fig. 2.20 The voltage V_{AB} between A and B and V_{BC} between B and C are summarized, and the voltage V_{AC} between A and C calculated by adding V_{AB} and V_{BC}

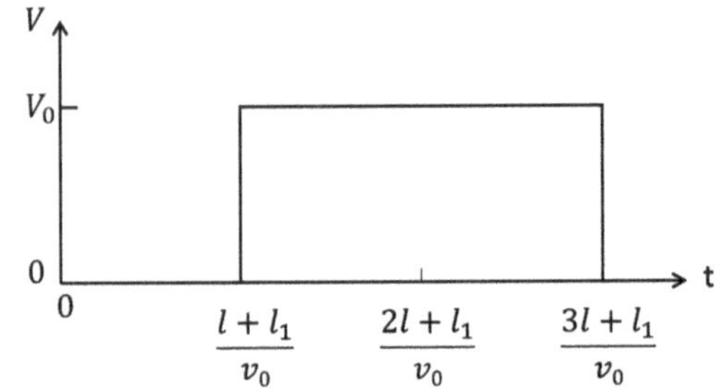

Fig. 2.21 The voltage waveform at load *R*

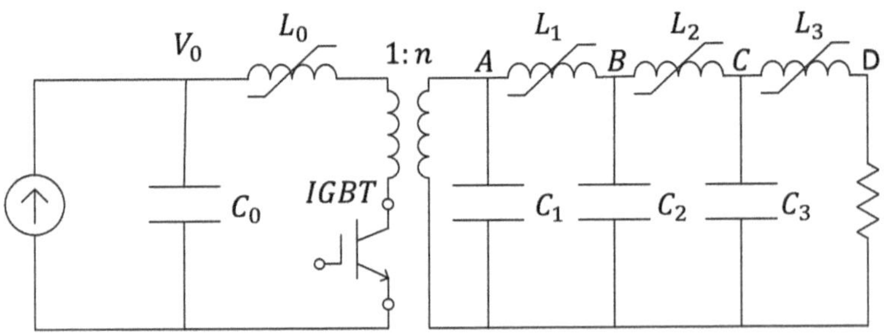

Fig. 2.22 Schematic diagram of a pulsed power generator using a magnetic pulse compression circuit

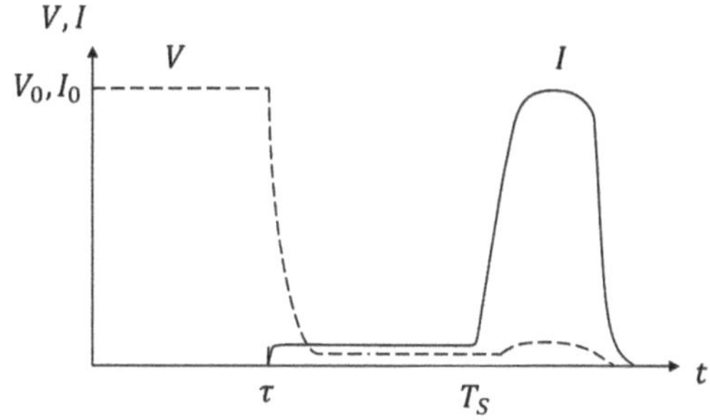

Fig. 2.23 Waveforms of IGBT voltage and current

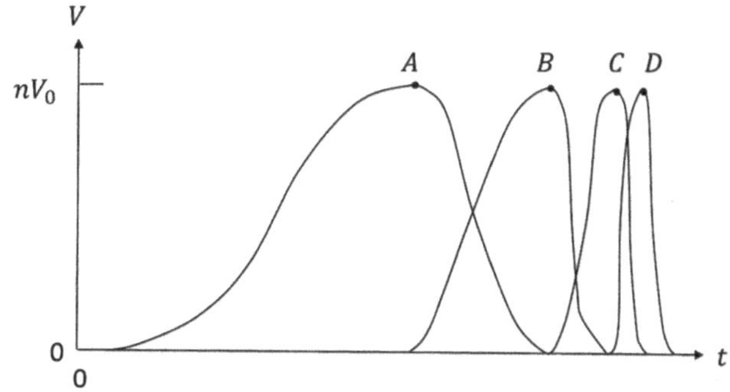

Fig. 2.24 When the capacitor, C_1, is charged to nV_0, where n is the amplification factor of the transformer, L_{1u} shifts to L_{1s} and, concurrently, the stored energy of C_1 transfers to C_2. Subsequently, the energy transfers from C_2 to C_3 and then from C_3 to the load occur

a low value at a finite time, T_s, after IGBT initiation. Detailed behavior of the saturable inductor will be explained below in Figs. 2.25 and 2.26; presently, it is sufficient to know that the inductance of saturable inductor changes from a large value, L_{0u}, of unsaturation to a small value, L_{0s}, of saturation after T_s. Since the main current, I, at IGBTs begins to flow at T_s after IGBTs are initiated at $t = 0$ as shown in Fig. 2.23, the switching loss of IGBTs is minimized by the existence of the saturable inductor.

When the capacitor, C_1, is charged to nV_0, where n is the amplification factor of the transformer, L_{1u} shifts to L_{1s} and, concurrently, the stored energy of C_1 transfers to C_2. Subsequently, the energy transfer from C_2 to C_3 occurs concurrently with the shift of L_{2u} to L_{2s}. The energy transfer from C_3 to the load occurs in the same way. These energy transfers are summarized in Fig. 2.24. When the voltage at point A in Fig. 2.22 becomes nV_0, the value of L_1 becomes small and the voltage at B becomes nV_0. In the same way, the voltages at C and D become nV_0. No backward current from C_2 to C_1 occurs since L_{1u}, which appears by the shift from saturation to unsaturation just after the energy transfer from C_1 to C_2, is much larger than L_{2s}. When capacitances of C_1, C_2, and C_3 reach the same value, voltages at A, B, C, and D become nV_0 without considering loss. The risetime of voltage decreases gradually as shown in Fig. 2.24, when $L_{1s} \gg L_{2s} \gg L_{3s}$. The fall time at D is dependent on load resistance.

Saturable inductors are used as the closing switches shown in Fig. 2.22. Figure 2.25 shows a B–H curve of the magnetic material used to make a saturable

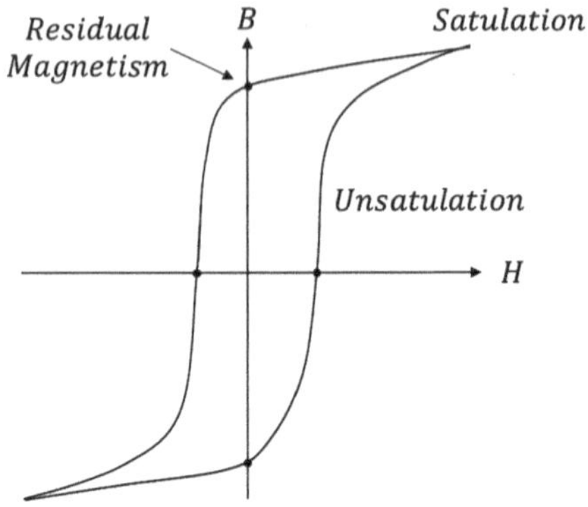

Fig. 2.25 *B–H* curve of the magnetic material used to make a saturable inductor

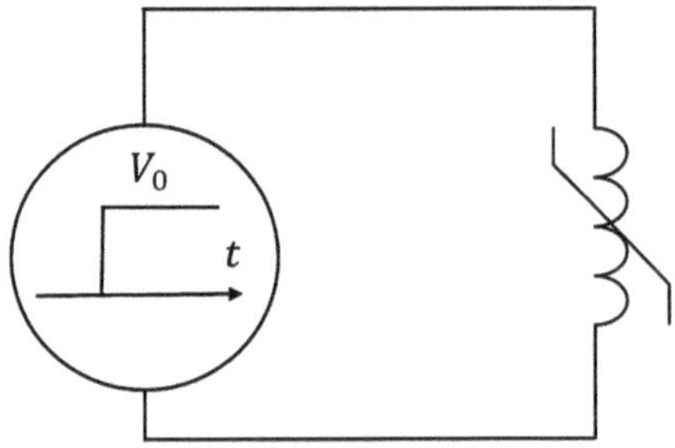

Fig. 2.26 The case in which a step function with voltage of V_0 is applied to saturable inductor

inductor. The saturable inductor sits near the residual magnetism (negative) on the B–H curve. The magnetic permeability during unsaturation is far larger than that during saturation. Since inductance is proportional to magnetic permeability, the inductance during saturation is far smaller than that during unsaturation.

How T_s is determined is of importance. Figure 2.26 shows a case in which a step function with voltage of V_0 is applied to saturable inductance. The following equation is obtained from $\partial(N\varnothing)/\partial t = V_0$:

$$\frac{1}{N}\int_0^{T_s} V_0 dt = \frac{T_s V_0}{N} = \int_S B \cdot dS = 2B_s A, \tag{2.50}$$

where N, φ, B_s, and A are winding number, magnetic flux, saturation magnetic flux density, and cross section of the magnetic material. The time T_s, at which the magnetic material changes from unsaturation to saturation states, is expressed as

$$T_s = \frac{2B_s AN}{V_0}. \tag{2.51}$$

The time T_s is adjusted by N, B_s, and A at the constant applied voltage V_0. In the circuit of Fig. 2.22, T_s is a half period of sine wave as shown in Fig. 2.24.

2.3.4 Inductive Energy Storage Using Opening Switch

Figure 2.27 shows a typical pulsed power generator using inductive energy storage and an opening switch. The capacitor, C_0, which is charged to V_0, is discharged through a pulse transformer, PT, and SiC-MOSFET as a power device, and the capacitor, C_1, is charged to $10 V_0$ through the saturable transformer, ST, and the diode, D. The total winding ratio of PT and ST is 10. After C_1 is charged, ST is saturated, and the backward current flows through D since the carrier stored into D by the forward current moves by reverse voltage. After a short time of backward current, a depletion layer is formed in D, and the current is interrupted quickly. A high voltage at D, V_D, is produced by $L\partial I_D/\partial t$, which corresponds to RI_D, where L is the inductance of the saturated ST and R is the equivalent resistance of D. Figure 2.28 shows waveforms of I_D and V_D. The positive and negative currents show the forward and backward current at D, and the high voltage at D is produced between the peak backward current and zero current. This diode is called a fast recovery diode.

2.3.5 Generator Using Power Semiconductor Device

Both the rise and fall times of switches and switching loss have decreased with the development of power semiconductor devices used as switches. Pulsed power generation is possible by direct switching of a DC power source with high voltage and current. Figure 2.29 shows a schematic circuit composed of a DC power source, SiC-MOSFET module, and load. The rise and fall times are several tens of nanoseconds.

Figure 2.30 shows a schematic circuit of a fully solid-state Marx-type pulsed power generator. The voltage level, polarity, pulse width, and repetition rate can be changed by a series connection of full-bridge switch-capacitor cells (SCCs). The operation of this generator is explained in Fig. 2.31a–c. The capacitors C_1, C_2, and C_3 are charged to V_0 as shown in (a), where C_3 is charged through S_1, S_2, and S_3; the diode at B_3, D_2; the diode at B_2, D_1; and the diode at B_1.

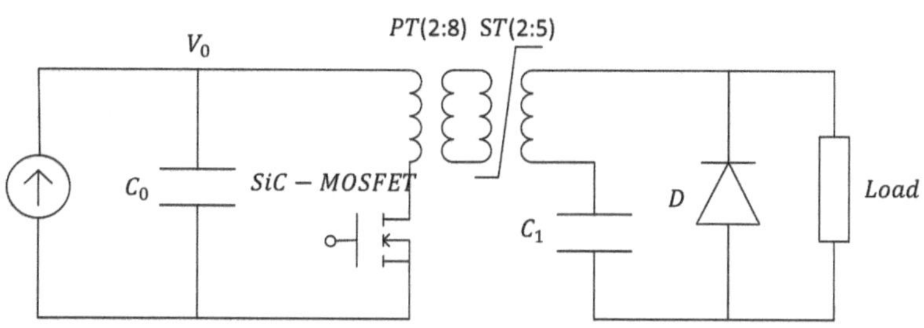

Fig. 2.27 The typical pulsed power generator using inductive energy storage and an opening switch

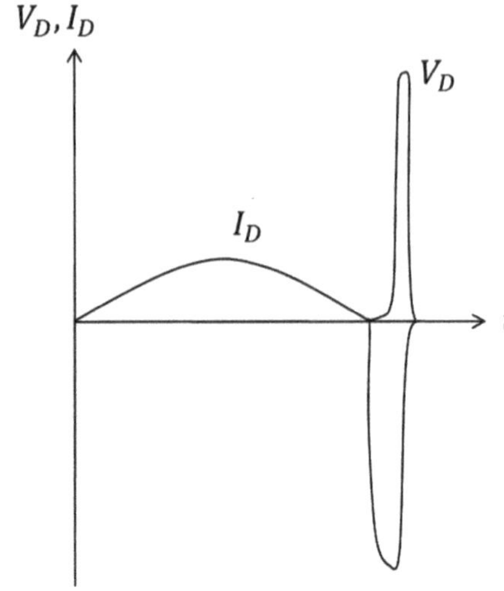

Fig. 2.28 Waveforms of I_D and V_D

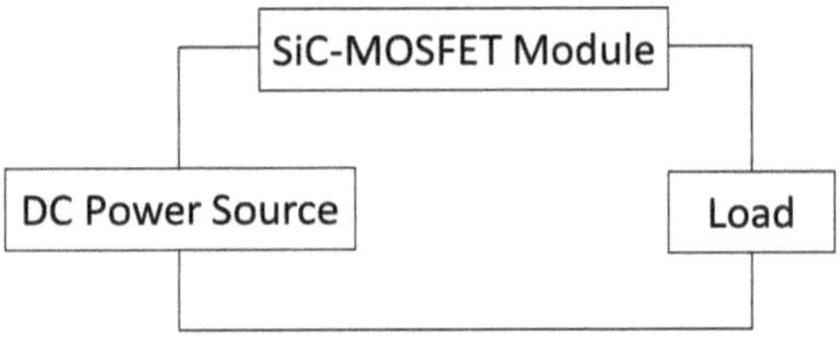

Fig. 2.29 The schematic circuit composed of a DC power source, SiC-MOSFET module, and load

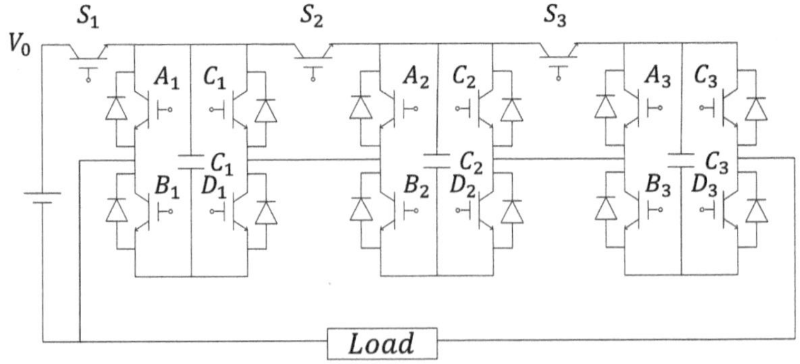

Fig. 2.30 The schematic circuit of a fully solid-state Marx-type pulsed power generator

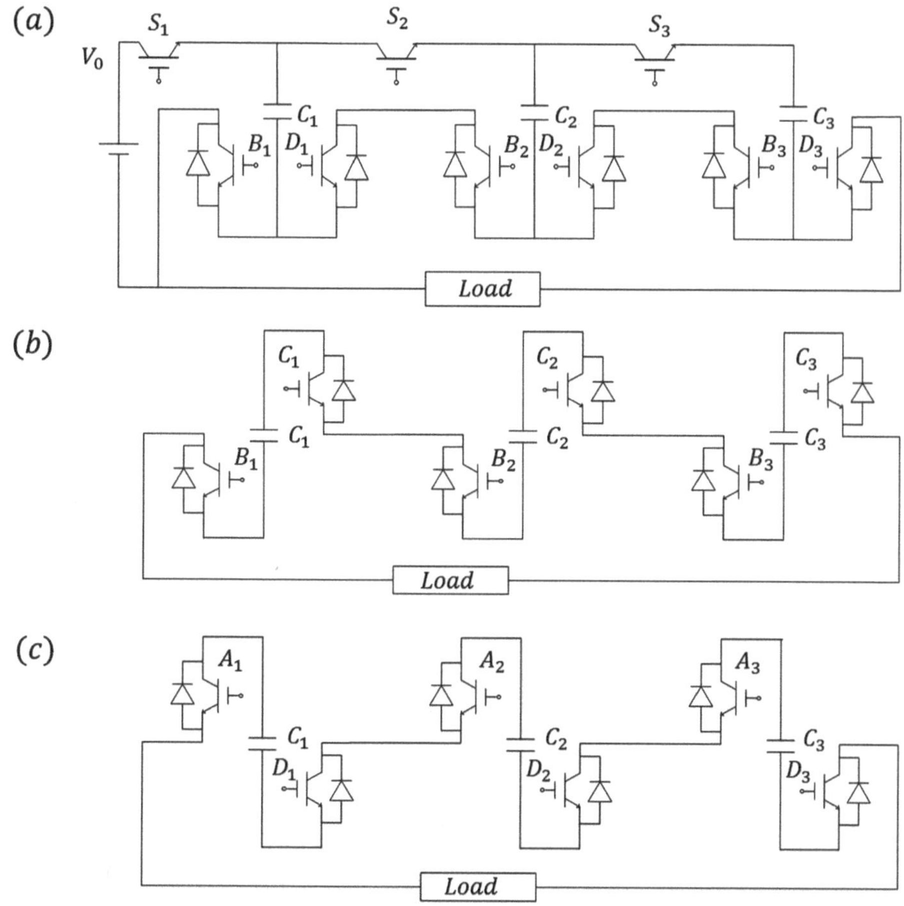

Fig. 2.31 The operation of a fully solid-state Marx-type pulsed power generator is explained: (**a**) the capacitors C_1, C_2, and C_3 are charged to V_0, (**b**) the voltage of 3 V_0 is applied to the load, (**c**) the voltage of $-3\,V_0$ is applied to the load. The load voltage of-2 V_0 appears by turning off A_3 and turning on B_3

When only B_s and C_s are initiated just after charging of C_s, as shown in Fig. 2.31b, the voltage of $3\,V_0$ is applied to the load. When only A_s and D_s are initiated, as shown in (c), the voltage of -3 V_0 is applied to the load. The load voltage of $-2\,V_0$ appears by turning off A_3 and turning on B_3 in (c).

2.3.6 Conclusions

Pulsed power generators with extremely high power have been developed for military and big science applications. Since moderate pulsed power is suitable for bioelectrics, the use of present power switching devices rather than gap switches is sufficient to generate pulsed power. Therefore, reliable, maintenance-free, and highly repetitive pulsed power generators which are completely solid-state have been developed.

Pulse-forming and Blumlein lines have been used for comparatively large generator using gap switches. These lines are utilized in the bioelectric field for small generators using power switching semiconductors. Pulsed power generators using magnetic pulse compression circuits have become popular both for research in the bioelectric field and for industrial applications of pulsed power. Pulsed power generators using inductive energy storage and opening semiconductor switches are able to generate pulsed power with a nanosecond or subnanosecond pulse width. Pulsed power generators using power semiconductor devices are advancing rapidly due to the recent remarkable development of power semiconductor switching devices.

2.4 Switches

Luis Redondo

This chapter reviews the main characteristics of the most used switches in pulse generators for bioelectric applications. The switch is a fundamental component in the pulse generators performance and a key factor for achieving the best results. The generator output characteristics, pulse width, waveform, voltage and current peaks, power, and frequency, depend greatly on the switch behavior. As a single switch cannot be used in all the pulse generator topologies for different applications, it is essential to understand which can be used in each situation, depending on the output characteristics of the pulse applied to the load.

2.4.1 Magnetic Switches

Magnetic switches, also known as saturable inductors, are passive switches, having no control gate to trigger. Instead they consist of a ferromagnetic core inductor (or coupled inductors), comprising of significant inductance variation, between saturated and unsaturated ferromagnetic core conditions, due to core nonlinear permeability behavior. These switches can operate at high-repetition (kHz) and power (kW) rates, due to their small losses, short recovery time, and no moving parts, giving long lifetime [6].

A typical B–H curve of a magnetic switch is shown in Fig. 2.32, where the optimum switching characteristics are obtain if the external circuit imposes a flux

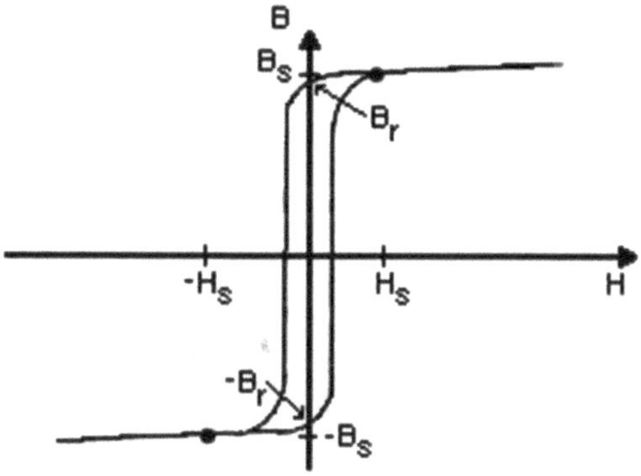

Fig. 2.32 *B–H* loop from a typical magnetic switch material, where B_s is the saturation flux density, H_s the corresponding magnetizing field intensity, and B_R the remanent flux density

swing, $\Delta B \approx B_s - (-B_R)$ between the negative remanent flux density, B_R, into the positive saturation flux, B_S [6].

The two magnetic switch key operating conditions can be define as an *off*-state condition with unsaturated magnetic material, with switch inductance given by

$$L_u = \frac{\mu_0 \mu_{ru} A_m N^2}{l}, \tag{2.52}$$

where μ_0 and μ_{ru} are the free space and unsaturated relative permeability of the core, respectively, A_m is the magnetic material cross-section area, N is the number of conductor turns, and l is the magnetic field path length and an *on*-state condition with saturated magnetic material, with switch inductance given by

$$L_s = \frac{\mu_0 \mu_{rs} A_m N^2}{l}, \tag{2.53}$$

where μ_{rs} is the saturated relative permeability of the core. The difference between the *on* and *off* states of the magnetic switch depends on the fact that a special type of magnetic material is used, which has the ratio between (2.52) and (2.53) of several thousands to tens of thousands, resulting in a large impedance difference.

Considering the B–H loop in Fig. 2.32 and switch operation, with the magnetic core, initially, biased at a negative saturation, $-B_s$. As voltage is applied across the magnetic switch, there is a flux change in the core in order to hold the voltage, the bias magnetizing point of the core move to where the slope is very steep. The relationship between the voltage-time product and the flux density change can be expressed as

$$\int_0^{T_s} v(t)dt = NA_m \Delta B, \tag{2.54}$$

where T_s is time to saturation, $v(t)$ is the applied voltage, and ΔB is the magnetic flux density swing. Initially, little current flows through the switch and the device is seen as having relatively large impedance. As the voltage continues to be applied, the bias point moves further up the B–H curve until the positive saturation flux is reached, B_s. When the core saturates, it can no longer provide the flux change necessary to hold the voltage, and the impedance decreases dramatically, as in a close switch [6].

Before next operating cycle, the core is reset to the negative saturation flux, $-B_s$, by applying a DC current through the main, or auxiliary, winding in a direction that provides a magnetizing field, $-H_s$, opposite to that induced by normal operating conditions [7, 8].

The main characteristics required for magnetic cores, used as magnetic switch, include [6] (a) large B_s to reduce size, (b) large μ_{ru} and unity μ_{rs}, (c) saturation that occurs drastically, hard core, (d) low loss, (e) low-frequency dispersion, (f) the ration between B_s and B_r that is close to one, and (g) very low magnetostriction, to reduce mechanical stress on the core.

2.4.2 Semiconductor Opening Switches

Semiconductor opening switches (SOS) are modified PIN high-voltage diodes, using a $P^+PN^-N^+$ structure with gradual doped P layer, not presented in PIN diodes, which can generate nanosecond or subnanoseconds kV and kA range pulses with kHz operation in inductive energy storage circuits [9–11].

For proper operation, first, carriers need to be pumped by a current in the forward (anode to cathode) direction, I_F, during hundreds of ns time range and then pulsed rapidly (dozens of ns) with a current in the reverse direction, I_R, until the carriers are completely discharged (Fig. 2.33). The interruption of the reverse current can be very fast and this effect is used as an opening switch [12].

The opening, or reverse recovery of SOS diodes, differs from traditional PIN diodes, since current interruption occurs mostly in the narrower P doped layer, not

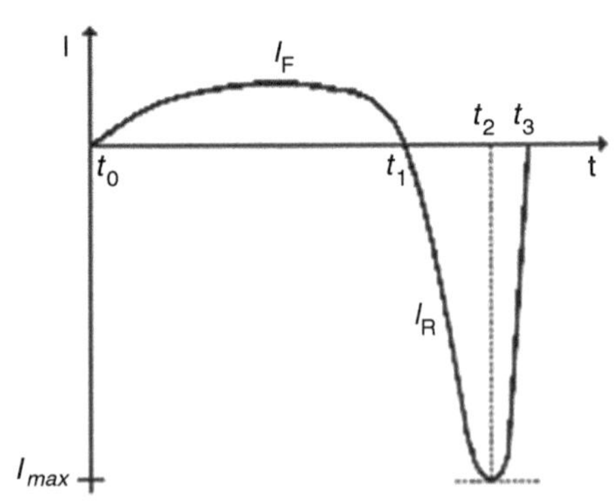

Fig. 2.33 SOS diode reverse blocking behavior

in the long drift region. Combined with dense excess plasma in the PN junction, the SOS can switch high-density currents in nanosecond opening times [10, 11].

SOS are used correctly in inductive storage circuits where they are pumped and pulsed via an inductor, L, and connected in parallel with a load, R_L. For optimum conditions of pumping time and current density in the junction, charge conservation can be achieved [12] as

$$\int_{t_0}^{t_1} I_F dt = -\int_{t_1}^{t_2} I_R dt \tag{2.55}$$

However, for reaching high values of the energy compression coefficient, it is necessary to reduce the excess plasma concentration that leads to recombination losses, which can be achieved by having a short pulse of forward pumping or by decreasing the pumping current density.

The output peak voltage, $V_{L\max}$, generated is correlated to the diode turn-off time between t_2 and t_3 and to the reverse peak current, $I_{\max}$, by

$$V_{L_{\max}} \approx L\frac{|I_{\max}|}{t_3 - t_2} \approx |I_{\max}|R_L \tag{2.56}$$

where the abrupt current interruption induces a very short pulse of voltage across the switch [12].

For a given charge at the junction and for its given turn-off time, it follows from (2.55) and (2.56) that, in order to increase the load peak voltage, it is desirable to increase $|I_{\max}|$ by increasing the rate of the diode reverse pulsing, dI_R/dt. The minimum possible length of the pulse is determined by the current cutoff time, which depends on the load. The current cutoff time decreased with decreasing pumping time. SOS operation can be optimized by maximizing the value of charge extracted by reverse current and in achieving a minimum time of the current cutoff [10].

Other SOS devices use unique diffusion techniques to achieve ultrafast switching times, such as the Drift Step Recovery Device (DSRD), the Inverse Recovery Diode (IRD), and the Delayed Ionization Devices (DID). These devices are capable of producing pulses ranging from 1 to 10 kV, with risetimes between 100 ps and 1 ns. Also, the PIN diode-based Silicon Avalanche Sharpener (SAS) is a sharpening device that shortens the pulse risetime [11].

2.4.3 Metal–Oxide–Semiconductor Technology Transistors

A metal–oxide–semiconductor field-effect transistor (MOSFET) is a unipolar, majority carrier device controlled by the v_{GS} voltage applied at the gate-source

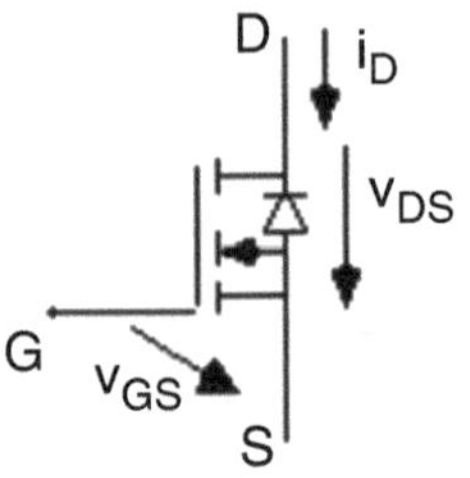

Fig. 2.34 MOSFET electronic symbol

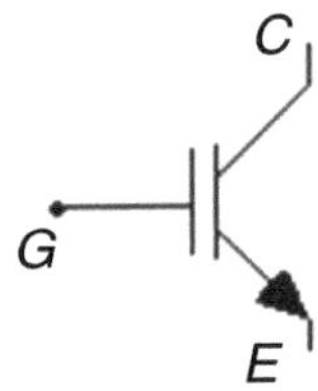

Fig. 2.35 IGBT common symbol

terminals. The N channel enhancement mode device is more common because of higher mobility of electrons (Fig. 2.34 and 2.35) [11, 13, 14]. The MOSFET switch speed is mostly due to the absence of carrier recombination since in N-type MOSFETs, the carriers are almost only electrons. Hence, this device has the capabilities for high-speed switching in the nanosecond or subnanosecond range, with low switching loss, within the rating of a few W to a few KW [15, 16].

The device has a slow intrinsic reverse source-drain diode, $v_{DS}<0$, and for high-speed applications often requires bypassing with external fast recovery diodes. Being a voltage-controlled device, the gate circuit impedance is extremely high. However, during fast turn-on and turn-off, the gate needs a current pulse to charge and discharge, respectively, the effective gate-source capacitance.

The switching times can be in the order of some nanoseconds or lower and can be affected by an external input resistance R_G value, used to damp the input RLC circuit. Considering a R_G value, by lowering it the input capacitance can be charged faster in order to fasten the turn-on and turn-off of the device, which decreases the commutation losses, but the EMI generated is greater due to the higher di/dt generated. Sometimes, switch-on t_{on} and off t_{off} times can be tailored using different resistive paths, limiting the charging current and the output voltage rise or fall rate, at the expense of increased switching losses.

During *on*-state the MOSFET behaves almost like a resistor R_{DSon}, with conduction drop $V_{DSon} \approx R_{DSon} I_D$ and positive temperature coefficient. Since the *on*-resistance increases with voltage rating, the device is very lossy at high current, limiting its voltage to 1200 V and hundreds of amperes. The resistance has positive temperature coefficient and therefore enables easy paralleling of devices. The MOSFET goes to *off*-state with $v_{GS} \leq 0\ V$, with $v_{DS}>0$ [11, 13, 14].

Insulated-gate bipolar transistors (IGBTs) are MOS technology-based semiconductor devices, which reduce *on*-state voltage drop, in comparison to MOSFETs, by adding an extra P^+ semiconductor layer in the drain side (punch-through PT-IGBT) or by substituting the N^+ by a P^+ layer (non-punch-through-NPT or reverse blocking IGBT, RB-IGBT) [11, 13, 14]. The P^+ layer injects holes in the drift layer reducing voltage drop by conductivity modulation.

Due to the injected minority recombination into the base, IGBTs are relatively slow turned-off devices and exhibit tail currents, in relation to MOSFETs. However, they can withstand much higher hold-off voltages (up to 6.5 kV) and currents (up to 3 kA) than MOSFETs [11, 13, 14].

IGBTs are turn-on and turn-off using a drive signal v_{GE} with similar properties as the MOSFETs. Most considerations about the MOSFET drive and protections are still valid for IGBTs.

2.4.4 Thyristors and Turn-Off Thyristor Devices

High-voltage high-power thyristors, or Silicon-Controlled Rectifiers (SCR), are PNIPN structures that can be gated on by injecting a relatively small gate current into the device. In pulsed power applications, thyristors can be used in capacitor discharging circuits and resonant circuits, since they can be easily turn-on gate controlled [11, 13, 14]. Thyristor turn-off behavior is not gate controlled, being similar to the PIN diode and of little use in pulsed power.

Thyristors can block voltages near 12 kV (the silicon limit is near 15 kV) and conduct currents up to 8 kA.

Modern thyristor structures like the Light-Activated Silicon-Controlled Rectifier (LASCR) or the Light-Triggered Thyristors (LTT) are optically triggered (Fig. 2.36) using an optical fiber and can be valuable in high-voltage applications to enable reducing stray capacitances [11].

Gate turn-off thyristors (GTO) are distributed gate thyristor structures, roughly equivalent to an NPN-PNP bipolar transistor association [11, 13, 14], having turn-*on* and turn-*off* gate capability, which enable blocking voltages up to 6 kV and forward currents up to 6 kA.

The GTO is much slower than MOSFET or IGBTs, since it is a bipolar carrier device, and it needs high turn-*off* gate currents, only slightly lower than their anode currents, meaning their turn-*off* current gain is small (between 2 and 10). Therefore, the GTO use in pulse power is usually limited to the generation of relatively long prepulses to be applied, through transformers, in sharpening devices such as the SOS diode. *Symmetrical* GTO structures are even slower than asymmetrical,

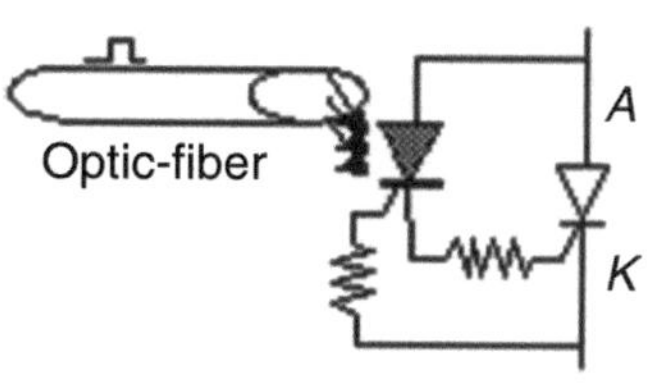

Fig. 2.36 LTT operating principle

although they can block both forward and reverse voltages (Symmetrical Gate Controlled Thyristor—SGCT).

Some GTO-derived devices, such as the integrated gate-commutated thyristor (IGCT), incorporate distributed gate drives in close association with the GTO, being a single device thus improving reliability [17]. This approach tries to circumvent the gate wiring inductances and enables accurate control of the turn-*on* speed.

Some GTO devices integrating MOSFET-based distributed gate turn-*off* drivers, with very low gate stray inductances, are sometimes called MOS turn-off thyristors (MTO) [11, 14]. They have the potential to block up to 9 kV and can be faster than the GTO itself, although they need hybrid gate drives to turn-*on*.

The MOS-controlled thyristors (MCT) [11, 14] are silicon fully integrated devices that try to improve GTOs, using integrated both turn-*on* and turn-*off* MOS devices. They can have lower *on*-state losses than an IGBT, current densities higher than IGBTs, working temperatures up to 200 °C, and block forward voltages of few kV. Switching speed is higher than in the GTO or even in the higher voltage IGBTs.

2.4.5 Power JFETs and Derived Devices

Power junction field-effect transistors (Fig. 2.37), or static induction transistors (SIT), are short-channel depletion-type junction field-effect devices operating in the prepunch through zone [11, 14]. Being unipolar devices, based on the flow of majority carriers (electrons) only, with very low stray capacitances, they have low switching losses and the potential to switch at very high frequencies, even higher than MOSFETs, up to the THz band.

The SIT is a normally *on* device (with $v_{GS} = 0$ it is in the *on*-state) that turns *off* by applying negative gate-source voltages v_{GS}, up to several tens of volts.

Unlike SCRs, GTOs, MOSFETs, or IGBTs, SIT normally *on* property enables its use in high-speed pulse generation circuits similar to those of SOS diodes, the main difference being the controlled turn-*off* with a negative gate voltage.

A SIT-derived structure, which is being used in pulsed power, is the static induction thyristor (SITh) also called field-controlled thyristor (FCTh) or field-controlled diode (FCD). It includes a P+ layer (instead of the SIT drain N+ layer) to obtain a PN junction (hence the name pinched diode). Present SITh devices block 4 kV and can switch peak currents in the kA range, with long lifetimes. The SITh is

Fig. 2.37 Power JFET symbol

being used for pulse generation as an opening switch in inductive circuits much like the SOS diode (the SITh behaves as a controlled diode) [11].

Much attention is being devoted to new SITs based on silicon carbide (SiC) structures and their derived structures, which can switch voltages in the kV range and current near the kA, specially the 4H–SiC SIT cascodes, which are expected to attain 8–10 kV blocking voltages. Other semiconductor materials such as gallium arsenide, GaAs; gallium nitride, GaN; and superjunction structures (SJ) are being investigated to build SJ power MOSFET structures, super trench MOSFETs (STM), vertical deep trench (Resurf DMOS—VTR-DMOS), SJ DMOS and UMOS, SITs, and SIT-derived structures. The GaAs SIThs are expected to be faster devices than existing power semiconductors [18, 19].

2.4.6 Conclusions

The most typical switching devices such as the magnetic switch, SOS diodes, IGBTs MOSFETs, and the emerging devices like SITh based on Si or SiC technology and the SiC devices such as JFETs were described. As the performance of these devices continues to improve, it is expected they will replace the more conventional switching devices for high-voltage applications.

The future of solid-state pulse power is linked to the capability of making semiconductor-based high-voltage generators with superior properties in compactness, low weight, low cost, high efficiency, modularity, and very important portability to deal with future applications near the consumer.

2.5 Measurements

A.J.M. Pemen, T. Huiskamp, F.J.C.M. Beckers, E.J.M. van Heesch,
G.J.J. Winands, S.J. Voeten, L. Zhen, and J.W.M. van Bree

2.5.1 Electrical Measurements

During the execution of a PEF treatment, it is important that the parameters of the applied electric field can be established reliably. Important pulse parameters are the amplitude, the pulse shape, the pulse duration, the pulse repetition frequency, and the energy of the pulse. These parameters must be determined with accurate voltage

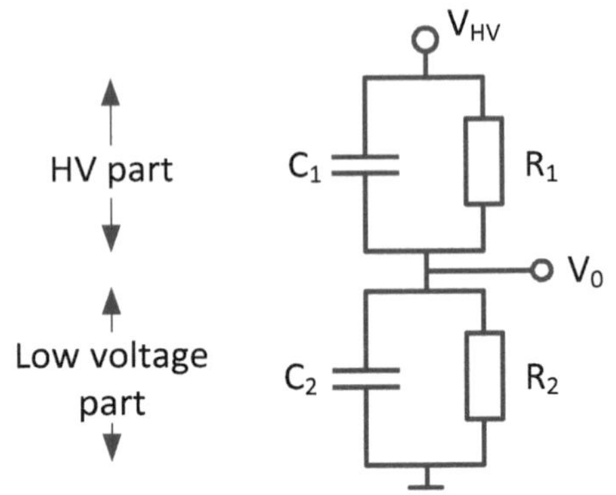

Fig. 2.38 General concept of a divider for high-voltage pulse measurements

and current measurements. With the measured voltage $v(t)$ and current $i(t)$, the energy $e(t)$ in a pulse can be determined by $e(t) = \int_0^T v(t)i(t)dt$. This section gives an overview of the concepts for pulsed high-voltage and pulsed current measurements. To be able to measure fast, high-amplitude voltage and current waveforms in an environment with large EM fields, much attention has to be paid to the measuring systems from EMC point of view.

2.5.1.1 High-Voltage Pulse Measurements

For measuring high-voltage pulses, several systems can be used, like capacitive dividers, resistive dividers, differentiating-integrating (D–I) circuits, and D-dot sensors. The general concept of a high-voltage pulse measurement system is shown in Fig. 2.38, and the corresponding transfer function is given by Eq. 2.57. The measured voltage V_0 relates to the high-voltage V_{HV} by the ratio of the impedance of the top part (or HV part) of the divider Z_1 and the total impedance $Z_1 + Z_2$, where Z_2 is the impedance of the bottom (or low-voltage) part of the divider. This concept can be configured in several ways; all options will be discussed below and are summarized in Table 2.2:

$$\frac{V_0}{V_{HV}} = \frac{Z_2}{Z_1 + Z_2} = \frac{R_2(1 + j\omega\tau_1)}{R_1(1 + j\omega\tau_2) + R_2(1 + j\omega\tau_1)}, \text{ with } \tau_1 = R_1C_1, \text{ and } \tau_2 = R_2C_2 \tag{2.57}$$

Resistive dividers, the case that only resistors R_1 and R_2 are used, are mainly used for DC or low-frequency HV measurements. For pulse measurements they are not so suitable, since due to the skin effect of the resistors, the transfer function

Table 2.2 Configurations of the voltage divider as in Fig. 2.38

Choice	Name	Behavior	Low ω	High ω
Only R_1 and R_2	Resistive divider	Not suitable for high frequencies	Also DC	Low
$\tau_1 = \tau_2$	Mixed divider	Wide bandwidth	Also DC	Up to resonance frequency
Only C_1 and C_2	Capacitive divider	Good at high ω	No DC	Up to resonance frequency
C_1, C_2, and R_2	Loaded divider	Good at high ω	No DC, cutoff at $\omega\tau_2 < 1$	Up to resonance frequency
C_1, C_2, and R_2, C_1 is very small	Differentiating, or D-dot sensor	Very good at high ω, measures derivative of V_{HV}	No DC, cutoff at $\omega\tau_2 < 1$	Up to resonance frequency

nonlinearly depends on the frequency. Also many resistances must be applied in series to hold the high-voltage level. This makes such dividers large and very sensitive for stray capacitance and inductance, which hampers the bandwidth.

A mixed divider has a much wider bandwidth as compared to a resistive divider, and it is also capable of measuring DC voltage. However, as for the resistive divider, the size of a mixed divider is generally large, making it sensitive for stray capacitance and inductance. The bandwidth is limited by the resonance frequency caused by the inductance and the total capacitance of the system. To some extent, effects of stray capacitance can be compensated by proper tuning of the values of C_1 and C_2. An example of a wide bandwidth divider is given in [20]. Various commercial HV probes are available, based on the mixed divider concept. An example is given in [21].

A capacitive divider is very suitable to measure pulses with bandwidths of over 100 MHz. Capacitive dividers can be compactly constructed, resulting in low inductance. A capacitive divider is generally loaded by the input impedance of a measuring instrument (represented by R_2 in the case of Fig. 2.38). This impedance is typically 50 Ω or 1 MΩ and results in a cutoff frequency ω_c as given by Eq. 2.58. For frequencies below ω_c, the system measures the derivative of V_{HV}. For frequencies above ω_c, the system acts as a true divider, with a divider ratio that depends of the ratio of the two capacitances. For a capacitive divider, the cutoff frequency is generally chosen to be low (<100 kHz) by applying relative large values for capacitances C_1 and C_2:

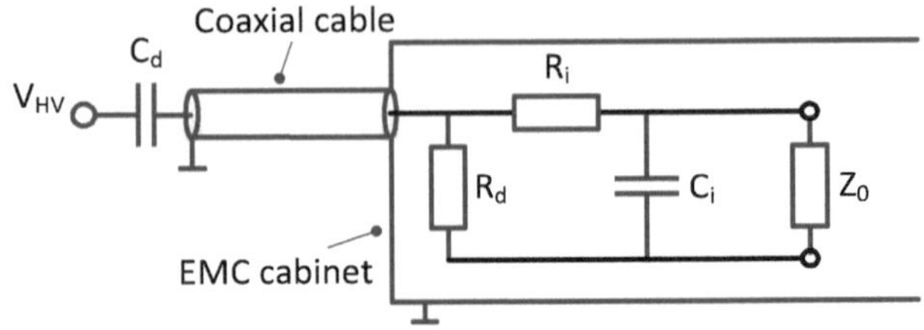

Fig. 2.39 Basic principle of a D–I measuring system

$$\frac{V_0}{V_{HV}} = \frac{j\omega R_2 C_1}{1 + j\omega R_2 (C_1 + C_2)}$$

$$\frac{V_0}{V_{HV}} = \begin{cases} j\omega R_2 C_1 \;\; \text{for } \omega \le \omega_c = \dfrac{1}{R_2(C_1+C_2)} \\ \dfrac{C_1}{C_1+C_2} \;\; \text{for } \omega \ge \omega_c = \dfrac{1}{R_2(C_1+C_2)} \end{cases} \tag{2.58}$$

For the situation that the measuring capacitor C_1 is small, such that the cutoff frequency ω_c is above the frequency range to be measured, the sensor has an output signal V_0 that corresponds to the derivative of the high-voltage VHV. In this case, the measured signal needs to be integrated, e.g., by a passive RC integrator. We generally refer to this situation as a differentiating-integrating (or D–I) measuring system (Fig. 2.39). Interesting examples of D–I systems are given in [22–29].

In a D–I system, a differentiated signal is transported via a coaxial cable to the passive RC input section of an integrator at the wall of an EMC cabinet. As a result of the differentiation, the high-frequency components of the signal are amplified with respect to the lower frequencies. The integrator restores the original waveform, but with a much smaller amplitude. Now the amplitude of the high frequencies is decreased with respect to the lower ones. A disadvantage of a D–I system is the inability to measure DC signals. The advantages of a D–I system are:

- A major advantage is the high signal-to-noise ratio, since the high-frequency noise that couples in via the coaxial cable will also be suppressed by the integrator. In other words, the integrator acts as a filter for the noise.
- The possibility to apply large amplitude signals to the sensors and a very wide dynamic range.
- The bandwidth of the sensors (can be as large as several hundred MHz).
- Differentiating sensors do not load the system.
- The sensors can be constructed with excellent EMC properties.
- Only one high-voltage component is required (capacitance C_1).

For a voltage measurement, a capacitive sensor (with capacitance C_i) is used to determine the voltage waveform V_{HV}. The output of this sensor induces a voltage across the resistor R_d equal to $R_d C_d \cdot dV_{HV}/dt$. The value of R_d must be the same as the characteristic impedance of the coaxial cable, which is typically 50 Ω. The signal over R_d is integrated by the passive R_i–C_i network. The total transfer

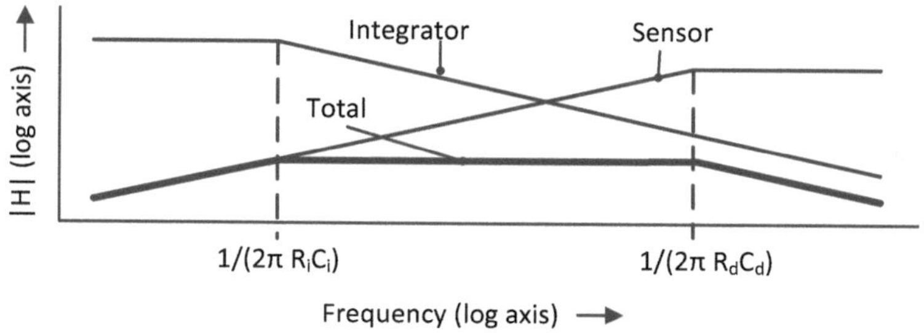

Fig. 2.40 Bode plot of the magnitude of the transfer function of a D–I measuring system

function of the D–I system as shown in Fig. 2.39 can be approximated (ignoring parasitic effects) by Eq. 2.59:

$$V_{HV}(t) = \frac{1}{R_d C_d}\left[R_i C_i V_m(t) + \left(1 + \frac{R_i}{Z_0}\right)\int V_m(t)dt\right] \tag{2.59}$$

where $V_m(t)$ is the waveform that can be recorded on an oscilloscope with input impedance Z_0 (typically 50 Ω). The effects of the parasitic inductances and capacitances of the sensors have been ignored during the derivation of Eq. 2.59. In general, these parasitic effects deteriorate the high-frequency response of the sensors. A bode plot of the magnitude of the transfer function of a D–I system is given in Fig. 2.40. The D–I system acts as a divider in the frequency range between the two cutoff frequencies $1/(2\pi R_i C_i)$ and $1/(2\pi R_d C_d)$. For frequencies below the low-frequency cutoff point, the system measures the derivative of V_{HV}. Therefore, the low-frequency part of the system response needs to be numerically corrected by the second term in Eq. 2.59. The high-frequency cutoff point must be higher than the highest frequency component in the signal to be measured. The resonance frequency of the system must be higher than the high-frequency cutoff point.

As the capacitive voltage sensor is usually a small electrode viewing a high-voltage conductor, calculating the value of C_d is not sufficiently accurate. Instead, the sensor has to be calibrated. The value of $(R_d C_d)^{-1}$ can be considered as the calibration factor of the D–I system. It can be determined by comparing the output signal of the D–I system with that of a calibrated voltage probe. Preferably, the calibration is carried out at a number of frequencies. This can be done by adding self-inductance between the output and the load of the pulse source. Preferably, the calibration is also performed at a number of voltage levels. As a control, the current through a known resistive load can be measured. Multiplication of this current with the value of the resistors indicates the voltage to be measured.

When a very high bandwidth measuring system is required (several 100 of MHz up to several GHz), the measuring capacitance C_1 must be very small in size and consequently has a very small capacitance value. A D-dot probe is a good choice for such measurements. The D-dot probe is a capacitive-coupled electrode that measures the voltage created by the displacement current through the capacitance from

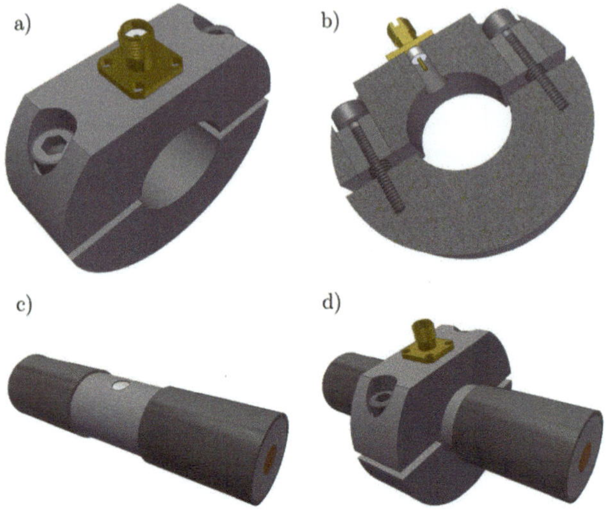

Fig. 2.41 (**a**) Drawing of the D-dot sensor. (**b**) Cut-away drawing of the D-dot sensor. (**c**) Drawing of the SA24272 cable with a hole in the outer conductor. (**d**) D-dot sensor mounted on the SA24272 cable

the high-voltage electrode and the D-dot sensor electrode. This current is the derivative of the voltage on the high-voltage electrode; thus the sensor measures purely the derivative of V_{HV}. Consequently, the high-voltage signal must be reconstructed by integration. Given the high bandwidth electronic integration is difficult. Therefore, the signal is usually numerically integrated. An example of a >4 GHz D-dot sensor for HV pulse measurements is given in the next section.

2.5.1.2 Example of a Very Wide Bandwidth D-Dot Probe [30]

At the Eindhoven University of Technology, nanosecond pulse technology has been developed which can deliver very short (0.5–10 ns) high-voltage (±0–50 kV) pulses with a very short risetime (<200 ps). The design, first implementation, and final realization of the pulse technology are given in [31, 32, 35]. The nanosecond pulse generator is connected to its load by a SA24272 coaxial cable, and D-dot sensors have been integrated with this cable for very wide bandwidth high-voltage pulse measurements. This solution can be applied at any coaxial connection between a pulse source and a load (e.g., PEF electrodes).

To obtain a very high bandwidth, a very small electrode (with very small capacitance) is used as D-dot sensor [32, 33]. The D-dot sensor is shown in Fig. 2.41a. The electrode that picks up the capacitive current from the high-voltage electrode is the tip of an SMA bulkhead connector (shown clearly in Fig. 2.41b). Figure 2.41c shows the hole made in the solid outer conductor of the SA24272 cable for the capacitive coupling between the inner conductor of the cable and the D-dot sensor electrode. Finally, Fig. 2.41d shows how the D-dot sensor is clamped onto the SA24272 cable.

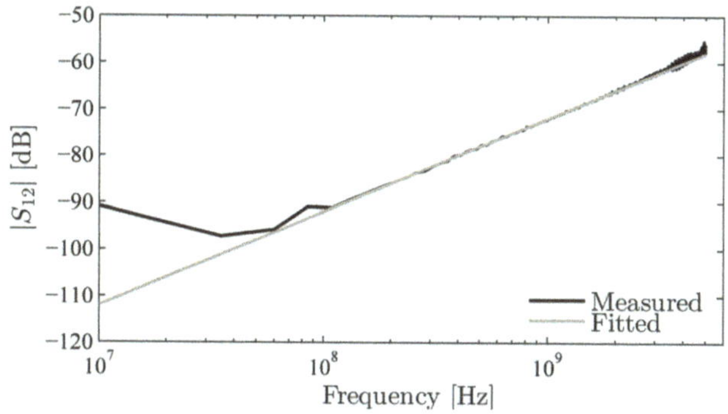

Fig. 2.42 Magnitude of S_{12} of the D-dot sensor measurement and the fitted values. Note that the measured results below 100 MHz are limited by the signal-to-noise ratio of the network analyzer

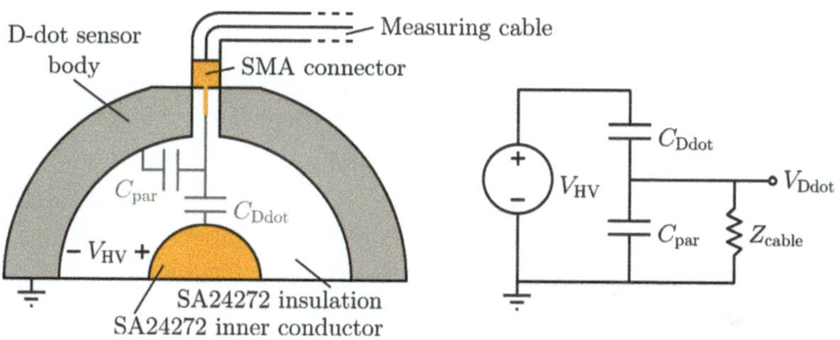

Fig. 2.43 Cut-away drawing of the D-dot sensor (compare with Fig. 2.41b) and its equivalent circuit

The D-dot sensor is calibrated by two ways: a network analyzer and a reed-relay pulse generator. The D-dot sensor was calibrated before the SA24272 cable was installed into the nanosecond pulse generator. For calibration, N-connectors were connected to both sides of the SA24272 cable. The N-connectors and the SMA connector of the D-dot sensor were connected to the network analyzer. The transfer function of the D-dot sensor is now determined by measuring the S_{12}-parameters of the setup.

Figure 2.42 shows the results of the measurement. We measured the S-parameters of the D-dot sensor up to 4 GHz. The result shows that the D-dot sensor is indeed a good differentiator up to this frequency. The lower-frequency limit of the network analyzer is 10 MHz and it has a usable signal-to-noise ratio of about 90 dB. This explains the results below 100 MHz in Fig. 2.42. Figure 2.43 shows the electrical equivalent circuit of the D-dot sensor system. Here, V_{Ddot} is the voltage from the D-dot sensor, V_{HV} is the voltage on the inner conductor of the SA24272 cable, Z_{cable} is the impedance of the measuring cable, C_{par} is the parasitic capacitance from the D-dot sensor electrode to ground, and C_{Ddot} is the capacitance

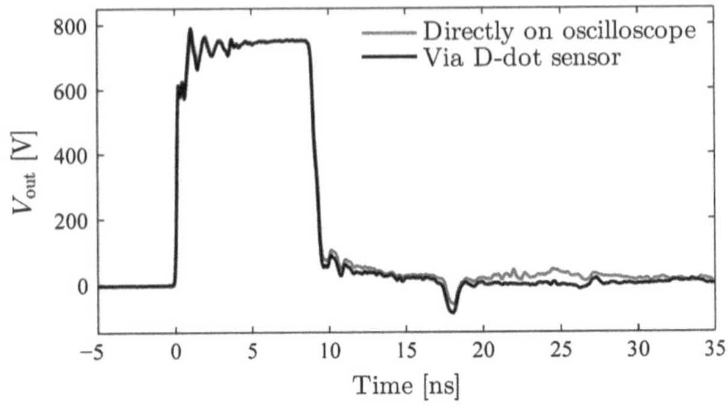

Fig. 2.44 Output voltage when the reed-relay pulse generator is connected directly to the oscilloscope (at a lower voltage) and when the same pulse is measured by the D-dot sensor

from the D-dot sensor electrode to the inner conductor of the SA24272 cable. This system has a transfer function as in Eq. 2.60:

$$\begin{aligned}\frac{V_{\text{Ddot}}}{V_{HV}} &= \frac{j\omega Z_{\text{cable}} C_{\text{Ddot}}}{1 + j\omega Z_{\text{cable}}\left(C_{\text{Ddot}} + C_{\text{par}}\right)},\\ \frac{V_{\text{Ddot}}}{V_{HV}} &= j\omega Z_{\text{cable}} C_{\text{Ddot}} \text{ when } j\omega Z_{\text{cable}}\left(C_{\text{Ddot}} + C_{\text{par}}\right) \ll 1\end{aligned} \tag{2.60}$$

The parasitic capacitance C_{par} is typically much larger than C_{Ddot} and will cause a deviation from the fitted transfer function at higher frequencies. However, for the frequency range, we are interested in (up to 4 GHz) the term $j\omega Z_{\text{cable}}(C_{\text{Ddot}} + C_{\text{par}})$ which is small and can therefore be neglected in the denominator in Eq. 2.60. We are then left with an ideal differentiator. From this simplified transfer function, we fitted a transfer function, which is shown in Fig. 2.42. With Z_{cable} as 50 Ω, we found a value of around 0.8 fF for C_{Ddot}.

To verify the fitted transfer function, we tested the D-dot sensor with a reed-relay pulse generator. This is a Blumlein pulse generator, which is switched by a mercury wetted reed relay [34]. This pulse generator is able to generate very fast pulses with a risetime of several hundred picoseconds and a voltage amplitude of up to around 750 V. We did one measurement at a charging voltage of 5 V with the output of the reed-relay pulse generator connected directly to an oscilloscope. Then, we generated the same pulse at a higher voltage, but now connected to one end of the SA24272 cable. A 0–4 GHz 50 Ω load was connected at the other end of the SA24272 cable. The output of the D-dot sensor was captured by the oscilloscope and integrated using the inverse of the fitted transfer function from Fig. 2.42. Figure 2.44 shows the result together with the scaled result of the 5 V measurement directly on the oscilloscope. The figure shows that we can measure correctly with the D-dot sensor.

After connecting the SA24272 cable with the implemented D-dot sensor to our (sub)nanosecond high-voltage pulse generator, the output pulses can be measured for all possible settings of the pulse generator. The results are shown in Fig. 2.45 and show that 200 ps risetime pulses can successfully be measured at pulse durations from 0.5 to 10 ns and pulse magnitudes from ±5 to 40 kV [35].

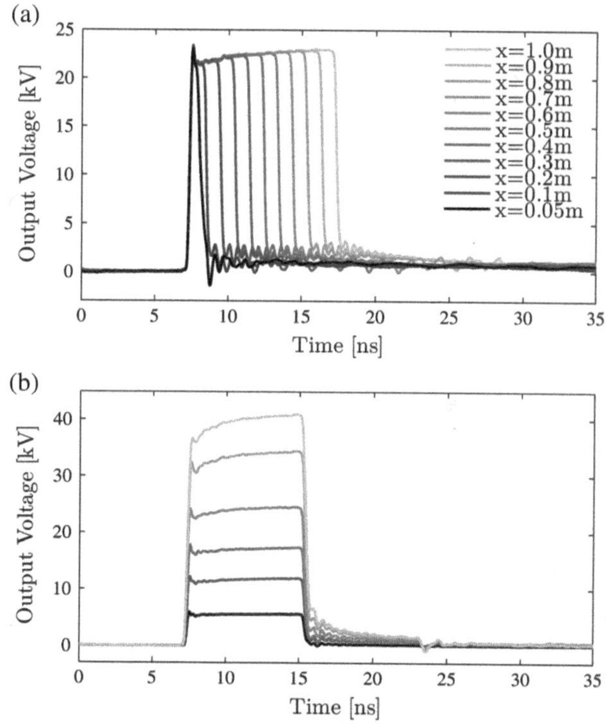

Fig. 2.45 (**a**) The output voltage of the pulse generator for different lengths of the PFL. The pulse duration can successfully be varied between 0.5 and10 ns. (**b**) The output voltage of the pulse generator for different amplitudes at pulse duration of 8 ns

2.5.1.3 Pulsed Current Measurements

A current sensor (Fig. 2.46) is based on Faraday's induction law. The pulsed current to be measured produces a changing magnetic field, which is enclosed by one or more windings of the sensor. By induction, a voltage is induced in the windings. This voltage is proportional to the derivative of the enclosed magnetic flux, the permeability of the magnetic core of the sensor, and the dimensions and the number of windings of the sensor. When the current sensor is loaded by a measuring resistor R_1, the output voltage V_0 of the sensor is given by Eq. 2.61, where I is the current to be measured, and M and L are respectively the mutual and the self-inductance of the sensor:

$$\frac{V_0}{I} = \frac{j\omega R_1 M}{R_1 + j\omega L}$$
$$\frac{V_0}{I} = \begin{cases} j\omega M & \text{for } \omega \leq \omega_c = R_1/(\omega L) \\ \dfrac{R_1 M}{L} & \text{for } \omega \geq \omega_c = R_1/(\omega L) \end{cases} \tag{2.61}$$

For the situation that the self-inductance L of the sensor is large, such that the cutoff frequency ω_c is low (few to a few 100 Hz), the sensor has an output signal V_0 that linearly corresponds to the current to be measured. In this case, the sensor has a

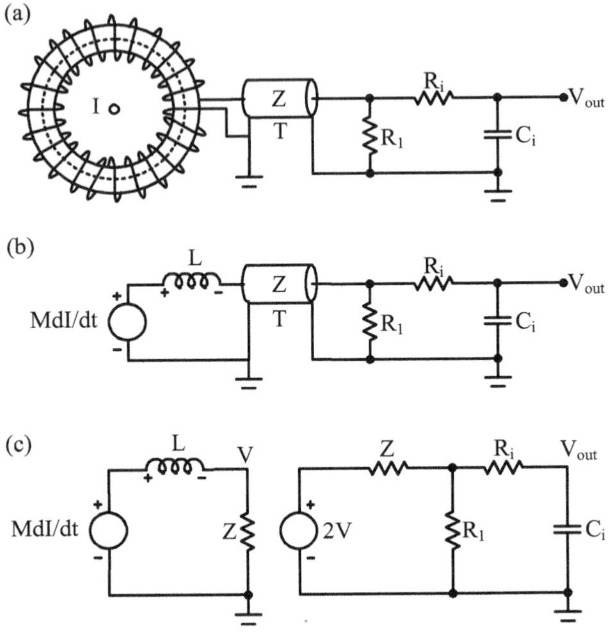

Fig. 2.46 (**a**) Schematic diagram of a current probe (Rogowski coil) as part of a D–I system; (**b**) equivalent circuit; (**c**) simplified circuit model, where the transmission line *T* is represented by a resistor *Z* and by a voltage source 2 V in series with a resistor *Z* at both sides, respectively [39]

magnetic core (thus the relative permeability of the core of the sensor $\mu_r > 1$), and we speak of a self-integrating sensor. Various commercial current probes are available, based on the self-integrating concept. An example is given in [36]. For the situation that the self-inductance *L* of the sensor is small, such that the cutoff frequency ω_c is above the frequency range to be measured, the sensor has an output signal V_0 that corresponds to the derivative of the current to be measured. In this case, the sensor generally has an air core and we speak of a Rogowski coil [37]. Advantages of an air core-Rogowski coil are the large bandwidth, its linearity, and that no saturation occurs at too high magnitude of the current. The measured signal needs to be integrated, e.g., by a passive RC integrator as in Fig. 2.46.

Fast pulsed currents are conveniently measured by Rogowski coils. The Rogowski coil gives an output voltage equal to $M{\cdot}dI/dt$, where *M* is the mutual inductance between the coil and the conductor passing through it. As was the case for the voltage sensor, the current sensor must also be calibrated since the value of *M* cannot be calculated with sufficient accuracy. In this section, we give the example of the calibration of the current sensor embedded into Vabrema's Mitoplicator [38]. This current sensor was calibrated by comparing the obtained signal with that of a calibrated, commercial current monitor.

The Mitoplicator (Fig. 2.47a) is a device for noncontact exposure of cells to nsPEF. The cells can be treated in their native cell culture medium and culture vessel. A single-turn loop antenna is designed that can contain a cell culture vessel. The electric field within the loop is created by magnetic induction and depends on the time derivative of the current through the loop. To generate the required field strength of 10 kV/cm with duration of ~6 ns, a fast high-voltage, low-inductance

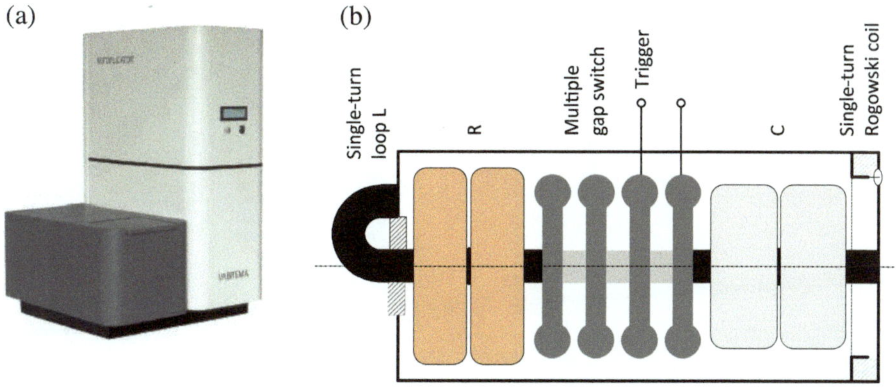

Fig. 2.47 (**a**) Mitoplicator, a device to generate intense nanosecond-duration electric fields to expose cells in their native culture medium and vessel (the Mitoplicator is commercial available from Vabrema, Netherlands). (**b**) RLC circuit to generate a high di/dt in the single-turn loop

RLC circuit is used (Fig. 2.47b). A fast-rising current of 100 A/ns is achieved in this circuit.

The device is equipped with a single-turn Rogowski coil to measure the current through the circuit (Fig. 2.47b). The air-cored Rogowski coil is included in the bottom flange of the device to measure the current without affecting the operation of the circuit. A single-turn sensor is used to obtain the desired high bandwidth (>100 MHz).

Figure 2.46 shows the schematic diagram of the current probe used [39]. The single-turn Rogowski coil is connected to an integrator via a transmission line T, which is terminated with a resistor R_1 (value is equal to the characteristic impedance of T, which is 50 Ω). The integrator with resistor R_i and capacitor C_i is connected to an oscilloscope with 1 MΩ input impedance. Figure 2.46b shows the equivalent circuit of the current probe where M is the mutual inductance between the current I and the coil and L represents the self-inductance of the coil. Representing the transmission line T by its characteristic impedance Z at the left side of T and by a voltage source 2 V in series with Z at the right side of T, one can derive the simplified circuit model as shown in Fig. 2.46c, where Z and V are the characteristic impedance of T and the voltage across T at the left side, respectively. The transfer function for this equivalent circuit is given by Eq. 2.62:

$$I = kV_0 + k\tau_1 \frac{dV_0}{dt} + \frac{k}{\tau_2}\int V_0 dt, \text{ when } R_iC_i >> \frac{L}{Z}, \text{ and } R_1 \approx\sim Z \qquad (2.62)$$

In the above equation, k $\approx R_iC_i/M$, $\tau_1 \approx L/Z$, and $\tau_2 \approx R_iC_i$. There are three terms at the right side of Eq. 2.62. The first term represents the ideal system; the second term is a correction factor for the high frequency; the third term is a correction factor for low frequency. To obtain the actual current, first the obtained signal V_0 must be corrected by Eq. 2.63. Then, by multiplying the corrected $V_{\text{corrected}}$ and the coefficient k, one can derive the actual current I:

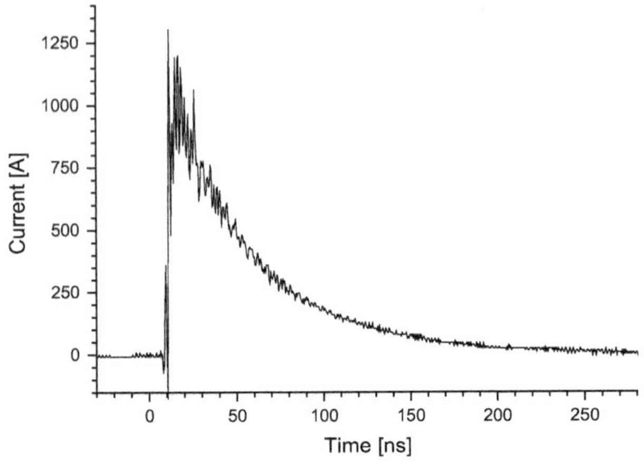

Fig. 2.48 Measured current in the Mitoplicator

$$I = kV_{\text{corrected}} = k\left[V_0 + \tau_1 \frac{dV_0}{dt} + \frac{1}{\tau_2}\int V_0 dt\right] \tag{2.63}$$

The self-inductance L and mutual inductance M of the single-turn Rogowski coil are estimated to be about 0.12 nH. Thus τ_1 and the second term in Eq. 2.63 were ignored when the signal V_0 was corrected. The calibration of the designed current probe was done by comparing the obtained signal with that of a calibrated current monitor, a Pearson 6600 probe (bandwidth 120 MHz). From this, a reliable value for k can be obtained (τ_1 is ignored, and τ_2 is known from the design of the integrator). A typical result of a calibrated current measurement is given in Fig. 2.48. With a peak current of 1250 A and a risetime of 5 ns, the *di*/*dt* is ~250 A/ns. This corresponds to an induced peak electric field of ~ 10 kV/cm with a duration of ~5 ns.

2.5.1.4 Electromagnetic Compatibility and Measuring Systems

EMC, electromagnetic compatibility, is an important issue when dealing with pulsed voltage and current measurements. In [24, 27, 40] details of a simple and proven EMC approach are given. Sources of interference are generally time-varying magnetic fields produced by the interfering equipment. As a result of these magnetic fields, interference currents (or common mode currents) are induced on the cables in a system (in our case the coaxial cables connecting the sensors with the measuring equipment). So an efficient EMC strategy must focus on minimizing the coupling of these interference currents with the signals to be measured.

First we need to make a distinction: common mode current and differential mode current (Fig. 2.49). An interference current is called common mode (CM) current I_{CM} because it is the net current in a specified multiconductor structure (such as a coaxial cable or a power cord). The CM circuit is always there and generally large.

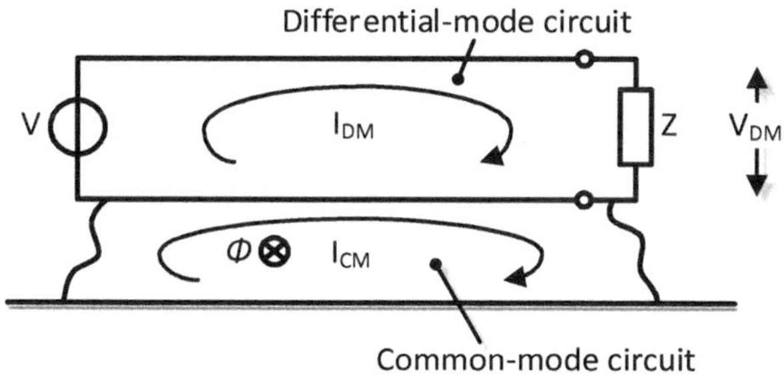

Fig. 2.49 Definition of common mode (*CM*) and differential mode (*DM*) signals. The DM circuit is the intended signal circuit. The CM circuit carries a CM current, or interference current, and is the net current through the two conductors, driven by an external interference flux Φ

Most often, interference is caused by CM currents that are generated by the enclosed time-varying magnetic flux of an interference source. A CM current is distinguished from differential mode (DM) currents I_{DM}. These are the balanced currents that follow a two-conductor structure such as a coaxial cable or power cord. The DM current causes a DM voltage V_{DM} over the impedance Z, which, e.g., can be the input of a measuring instrument. The transfer impedance Z_t is defined as the coupling between the CM current and the DM voltage as $V_{DM} = I_{CM}.Z_t$. Ideally, the voltage over Z is equal to the source voltage V (the signal to be measured). In practice, however, via the transfer impedance, the CM current can interfere with the DM or measurement voltage as $V_{DM} = V + I_{CM}.Z_t$.

The basis of an effective EMC strategy is an EMC wall, a low transfer impedance, and correct routing of interference currents. To avoid interference it is crucial that interference currents do not pass the EMC wall. An EMC wall can be realized by means of an EMC cabinet, a metallic enclosure/cabinet that contains all measuring equipment such as integrators and an oscilloscope (Fig. 2.39). An EMC cabinet does not have to be completely closed, such as a Faraday cage: its main function is to force interference currents on cables to take a path that does not cross this wall, see Fig. 2.50. Of course, cables (coaxial) must cross the wall to transport the measuring signals to the oscilloscope. But they are firmly connected by their braid to the EMC wall. This prevents interference currents from flowing over or in the vicinity of the sensitive measuring equipment, thus protecting this equipment from unwanted interference.

The EMC wall is generally connected to a grounding system (by a connection either via a grounding wire or via the (large) parasitic capacitance of the EMC wall to ground). In this way the interference currents follow paths as shown in Fig. 2.50. When the transfer impedance of the cable and its connector is sufficiently low, the interference currents cause minimal interference voltage across the differential mode cable terminals on the interior side of the EMC wall. These terminals can be the input to, e.g., measuring equipment. Generally the transfer impedance of coaxial cables with a double outer braid, like RG223 and RG214, is sufficiently low. The transfer impedance of connectors strongly depends on their connection to

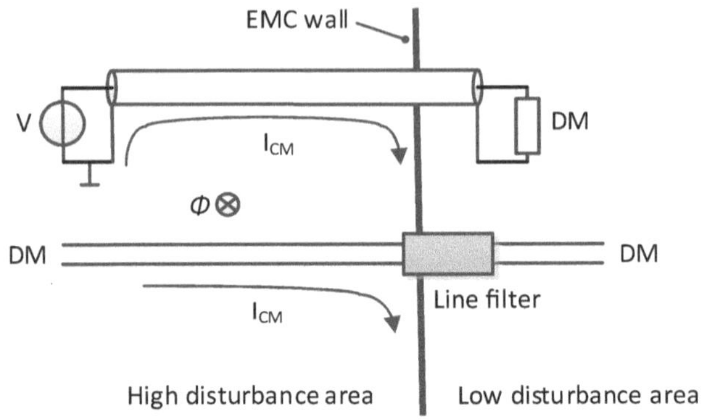

Fig. 2.50 Concept of EMC wall for effective reduction of interference

their contra connector. N-type or SMA-type connectors have proper screw connections and thus a low transfer impedance. BNC-type connectors have much higher transfer impedance due to their two-point connections. Figure 2.50 shows various CM and DM systems in combination with an EMC wall. Most CM sources occur due to time-varying magnetic flux in the space surrounding the DM circuits.

Proper EMC is also aided by using a D–I measuring system. In a D–I system, it is the large differentiated signal that is transported via the coaxial cable to the passive RC input section of an integrator at the wall of an EMC cabinet (Fig. 2.39). The integrator restores the original waveform but it also acts as an effective EMC filter. Large common mode currents are allowed to flow on the signal cables from sensor to integrator and back via grounding systems. A sufficiently low coupling impedance between the common mode currents and the measuring equipment leaves the measuring electronics undisturbed.

2.6 Delivery of Electric Pulses to Biological Tissues Using Antennas

Shu Xiao and Ross Petrella

2.6.1 Introduction

Intense electric fields with durations from milliseconds to nanoseconds interact with cells through charging membranes, for which the conductivities of the extracellular medium and intracellular medium primarily determine the charging time constants of the cell components. For shorter pulses in the subnanosecond range, the permittivities of various cell components, rather than their conductivities, determine the electric field distribution in the cell [41]. The electric field acts directly on membranes in the dielectric phase, rather than through charging. Thereby, the mechanism of interaction is different. The electric field intensity required for inducing biological effects depends on the end-goal applications. The critical electric field for stimulating pain nociceptors with subnanosecond pulses (e.g., 200 ps) can be extrapolated as 1 MV/m [42] and may be decreased to 20–50 kV/m if the pulses are applied at a high-repetition rate, suggested by Jiang and Cooper [43]. For cell killing, the electric field intensity in general is higher than the conditions for stimulation. For example, electric pulses of 800 and 200 ps have been studied in vitro on melanoma and hepatoma cells [44, 45]. The magnitude of the electric field needs to be approximately 2MV/m or higher in order to increase cell membrane permeability [44]. The temperature increase due to the high-repetition rate (10 kHz) and prolonged exposure (up to 10 min) can no longer be ignored.

Delivering subnanosecond pulses to tissues in the form of electromagnetic fields can be done by antennas, which is a step forward from using needles for delivering pulses of milliseconds to nanoseconds [46]. Previous studies have shown the promise of heating or imaging deeply situated lesions by narrowband microwave or RF wave antennas [47]. Clinical use of RF heating for hyperthermia is available [48]. By contrast, ultrawideband (UWB) pulses in biomedical applications are still in the research phase. In the meantime, UWB pulses (10 ps–1 ns) were shown to cause fewer hot spots than narrowband RF waves [49]. The study of pulsed radar also showed a better penetration into amorphous materials because the pulse has a wide-frequency spectrum [50]. In addition, the spatial distribution of the pulse can be small, offering high resolution for medical treatments. For example, in brain white matter that has a dielectric permittivity of 37, the spatial width of a 200-ps pulse is 1 cm. Despite all these advantages, there are still no UWB antennas available for clinical treatments.

Radiating UWB pulses relies on antennas evolved from narrowband types, such as the resistively loaded dipole antenna [51], antennas that contain multiple-resonance structures [52], and frequency-independent structures that can support transverse electromagnetic (TEM) waves, such as bicone or conical waveguides [53]. There is no single antenna that can work for all targets. When targeting tissues at various depths, the antenna is different. Here we give two options for subcutaneous targets and deep targets. A dielectric rod waveguide fed by conical TEM structure is used for shallow-depth subcutaneous targets, whereas a reflector antenna fed by a wave launcher is used for reaching deep targets. We start with the discussion of the general principle of radiation. It is important to point out that radiation consists of different terms with different dependence of distance. As the fields enter a biological medium, the transmittance largely depends on the field polarization and the difference of dielectric permittivity between tissue and air. It is important to utilize the right polarization to maximize the transmittance by choosing a certain antenna configuration and in the meantime place the antenna in a matching medium. The dielectric rod antenna in this sense can incorporate a desirable dielectric rod for sending the field into a tissue. For deep targets, we show the penetration of pulses in a human brain and the formation of a focal spot, but the challenge of tissue attenuation is discussed. Finally, we conclude this chapter with some biological implications with UWB antennas.

2.6.2 Fundamental Discussion of Radiation

Electromagnetic radiation is the result of changing the speed of moving charges. Along the trajectory of the charge, it creates the Lienard–Wiechert potentials, which contain the magnetic vector potential function $\vec{A}$ related to the current and the electric scalar potential function φ related to the charge alone. The electric field resulted from the potentials is

$$\vec{E} = -\frac{\partial \vec{A}}{\partial t} - \nabla \varphi \tag{2.64}$$

The Maxwell equations allow the calculation of electric field and magnetic field. The electric field for a moving charge is given as

$$\vec{\mathrm{E}}\left(\vec{r}, t\right) = \frac{q}{4\pi\epsilon_0}\left[\frac{\left(1-\frac{v^2}{c^2}\right)\left(\widehat{R}_q - \frac{\vec{v}}{c}\right)}{{R_q}^2\left(1-\frac{\widehat{R}_q\cdot\vec{v}}{c}\right)^3} + \frac{\widehat{R}_q \times \left[\left(\widehat{R}_q - \frac{\vec{v}}{c}\right)\times\frac{d\vec{v}}{dt}\right]}{c^2 R_q\left(1-\frac{\widehat{R}_q\cdot\vec{v}}{c}\right)^3}\right] \tag{2.65}$$

where q is the charge, $\vec{R}_q$ is the distance from the charge to the observing point P (Fig. 2.51), c is the speed of light, ε_0 is the free space permittivity, and $\vec{v}$ is the velocity of the charge. The first term in the bracket depends on the velocity vector and the second term is dominated by the acceleration. Accordingly, their electric fields decrease from the charge in the manner of $1/{R_q}^2$ or $1/R_q$. In the near distance from the charge, the $1/{R_q}^2$ term dominates, but the $1/R_q$ term dominates in the far distance. The direction of speed with respect to the direction of an observer at $\vec{r}$ has

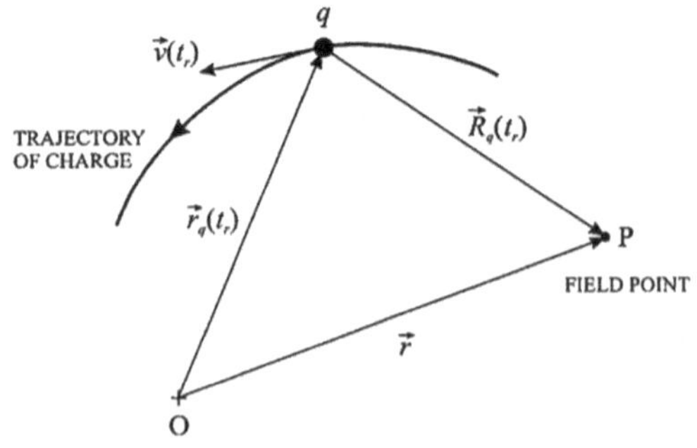

Fig. 2.51 Electromagnetic fields from a moving charge q [54]

an impact on the field. For example, in a linear dipole, the speed of moving charges is aligned with the direction of the dipole axis, which results in zero radiation ($1/R_q$ term) on the axis, whereas in a small coil, the speed is perpendicular to the axis, which results in the largest radiation for the observer on the axis.

As the radiated fields from single charges are dependent on the speed and acceleration, a series of moving charges, i.e., current, are responsible for the radiation. Generally, the current distribution in the antenna is known, so the calculation of radiated fields could start from the vector potential in the frequency domain by integrating the current distribution over the antenna structure, followed by the curl operation to obtain the magnetic field intensity. Further, the electric field is the curl of the magnetic field. Laplace transformation is then conducted to obtain the time-domain expression of electric fields [55]. This procedure is straightforward for simple antenna structures such as a linear dipole, but requires more analysis when a more complicated antenna structure is involved. Such a case can be seen in aperture antennas (e.g., reflector antennas or horn antennas), which are often used for focused radiation. On the aperture plane (Fig. 2.52), the current is not the real current anymore, but according to the Huygens–Fresnel principle, it is replaced by the equivalent current such as magnetic current or electric current. They are determined by the tangential components of electric fields ($-\text{n} \times \vec{E}$) or magnetic fields ($\text{n} \times \vec{H}$), which are established by the field structure of the antennas. Knowing the equivalent current distributions, the radiated fields can be calculated from the vector potential. As the equivalent current is set up by the feed structure at the aperture plane, the vector potential integration has a dependence on the aperture geometry as well as the design of the feed structure. One example is in a reflector antenna, and the field on the aperture is set up by a spherical TEM waveguide

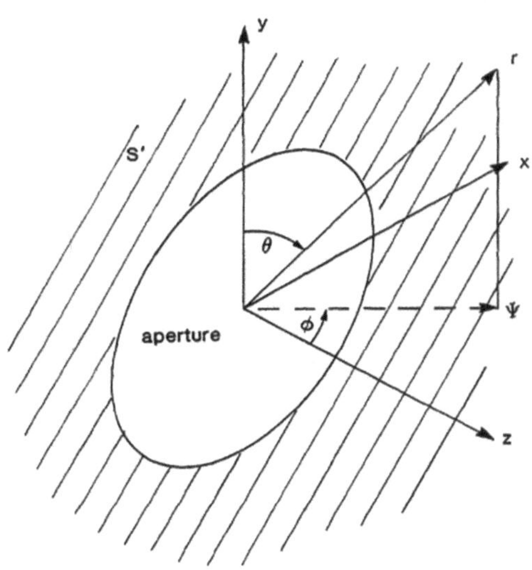

Fig. 2.52 Electromagnetic fields from a source plane S' [55]

originating from the first focal point. The tangential component of the electric field on the aperture plane is known, so the field radiated at a point on the axis is

$$E_y(\vec{R}_0, t) = E_0\left\{\alpha^{(1)}_{y,y}\frac{a^2}{R_0 c}\frac{\partial}{\partial t}f\left(t-\frac{R_0}{c}\right)+\alpha^{(2)}_{y,y}\frac{a^2}{R_0{}^2}f\left(t-\frac{R_0}{c}\right)\right\} \tag{2.66}$$

where R_0 is the distance of an observer on the axis from the aperture center. The input pulse has a time dependence of $f(t)$, but has a propagation delay $t - R_0/c$. Similar to the single-charge case, the field has a dependence of the input pulse voltage and the time derivative of the pulse. Specifically, for the field term near the antenna, $f(t)$ is dominant and the time derivative term $\partial f(t)/\partial t$ is dominant for the field far from it. The coefficients $\alpha^{(1\text{ or }2)}{}_{y,y}$ depict the integral of the tangent electric fields over the aperture plane. In general, homogenous, unidirectionally aligned "currents" from a uniform illumination result in larger coefficients and hence the electric field. The aperture size preferably is comparable to the observing distance, so the ratios of $\alpha^{(1\text{ or }2)}{}_{y,y}$ over R_0 or $R_0{}^2$ are not small. One implication from these expressions is that for biomedical targets that are in the near distance, both $f(t)$ and $\partial f/\partial t$ terms are important, suggesting both the amplitude of the pulses and the risetime contribute to the high fields at the target.

Once the electromagnetic waves are detached from antennas, they propagate in the free space through the induction process, i.e., the electric field inducing magnetic field, magnetic field inducing electric field, and so on. Even when the source is shut off, the fields continue to propagate toward the larger distance at the speed of light.

2.6.3 Entry of Electromagnetic Waves to Tissue

In the tissue, the physics of the propagation of electromagnetic waves is that they induce currents by mobilizing the dipole moments of tissue molecules, which reradiates electromagnetic waves in various directions, including back to the medium that the antenna is in (reflected waves) and those forward toward the tissue (refracted waves). The response of tissue dipole moments to the excitation of incident fields varies as frequency changes, which results in a different reradiation efficiency. A difference in phase exists between the incident field and the reradiated field. As a UWB pulse has a broadband spectrum, the overall phase differences are manifested in the broadening of the pulse width in the time domain. The response of dielectrics to wave excitation can be described by a complex dielectric permittivity $\varepsilon_0\varepsilon(\omega)$, which has a real part which accounts for the frequency-dependent phase shift with respect to the original field and an imaginary part that accounts for the damping effects resisting the displacement of the dipole moment. One can use a Debye single dispersion equation to describe $\varepsilon(\omega)$ [56]

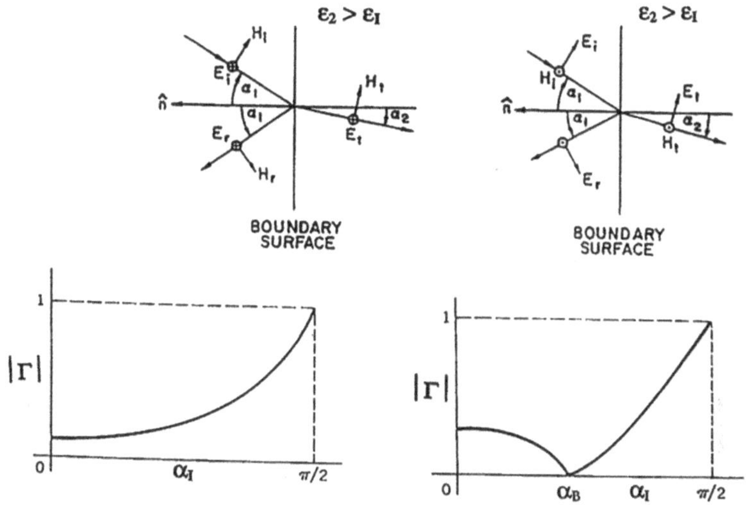

Fig. 2.53 The reflection coefficient for a wave that propagates from one medium to a tissue [57]

$$\varepsilon(\omega) = \varepsilon_\infty + (\varepsilon_r - \varepsilon_\infty)/(1 + j\omega\tau) \tag{2.67}$$

ε_∞ is the dielectric constant at infinite frequency and ε_r is the dielectric constant at low frequency and τ is the relaxation time constant. For simple dielectrics, such as water, τ is on the order of 10 ps (the frequency corresponding to 16 GHz). For pulses much longer than 10 ps or frequencies much lower than 16 GHz, $\omega\tau << 1$, the complex dielectric permittivity becomes real,

$$\varepsilon(\omega) = \varepsilon_r \tag{2.68}$$

In addition to the dielectric response, tissues have finite conductivity σ due to free charges or ions, which results in Joule heating and attenuation of the fields. So the overall complex dielectric permittivity can be written as

$$\varepsilon(\omega) = \varepsilon_r - j\sigma/\varepsilon_0\omega \tag{2.69}$$

At the air–tissue interface, the tangential electric field is continuous, i.e., the addition of incident field and reflected field should be equal to the refracted field, which governs the sign and amplitude of the transmittance and reflectance. Largely, three factors contribute to the transmittance: dielectric permittivities of the two media (ε_1 and ε_2), the angle of incidence, and the polarization of the electric fields. Their impact on the reflection coefficient can be seen in the plot of reflection coefficient versus the incident angle (Fig. 2.53).

If the wave is linearly polarized perpendicular to the tissue surface ($\widehat{n}$), a minimal reflectance is shown when the incident angle is 0. In this case, the reflectance is entirely limited by the dielectric constant:

$$\Gamma = \frac{\sqrt{\varepsilon_1} - \sqrt{\varepsilon_2}}{\sqrt{\varepsilon_1} + \sqrt{\varepsilon_2}} \tag{2.70}$$

Note that Eq. 2.7 can be negative in sign, suggesting that a reflected wave reverses its polarity at the air–tissue surface as tissues have a higher dielectric constant than air. Since the tissue electrical property (ε_2) is static and can't be changed, it is important to maximize the transmittance of waves, so a matching medium with a dielectric constant close to the tissue is desirable to minimize the difference in the dielectric constants.

When the wave is linearly polarized in parallel to the tissue surface ($\widehat{n}$), the reflectance can be reduced to zero if the wave's incident angle is the Brewster angle, allowing for a maximal entry of waves. When the incident angle α_1 becomes 0, the magnitude of reflectance equates the expression given by Eq. 2.7, which is the same as for perpendicular polarization. For waves to enter the tissue, preferably they should be polarized such that the incident angle is zero. In this case the reflectance is again determined by the two dielectric permittivities.

2.6.4 Delivery of Electric Pulses to Shallow Region

We have considered the factors that may allow us to maximize the transmittance of waves into tissues. Because the wave decreases over distance, tissue targets are ideally placed close to the antennas. The angle for the wave to enter the tissue can be chosen 0°, which means the electric field exiting from the antenna is linearly polarized with respect to the tissue. In that regard, antennas such as a TEM horn antenna or a tapered notch antenna can be used. It appears that a coaxial cable might even work. When a TEM mode is propagated in the cable, the electric field at the coax opening is mainly tangential. By making contact with the tissue, an open-ended coaxial cable seemingly could deliver the pulses to a certain depth.

Such a case was modeled by placing a coaxial cable on top of a tissue slab with $\varepsilon_r = 38$ and $\sigma = 1.18$ S/m. The cable was fed with a subnanosecond pulse for a peak voltage of 1 V. The electric field distribution on the axis inside a tissue is shown in Fig. 2.54. The electric field near the surface within the first 1–2 mm is very large, but the field decreases rapidly, faster than the inverse of distance. So, if one just wishes to expose the epidermis, a coaxial cable will suffice. But to a target in deeper region, since the field is small already, it may not be sufficient to induce any biological effects. However, if the field on the surface is too high, the surface tissue may be at risk. To give some perspective on how the electric field must penetrate biological tissue, we can consider the human skin. Skin thickness varies by location but it is reasonable to say that most of it is between 2- and 3-mm thick. Additionally, a sizeable skin melanoma can reasonably be 4 mm beneath that. Comparing this assessment to the relationship between field and depth (Fig. 2.54), it becomes apparent that the coaxial cable is only effective at the surface of the tissue. Why

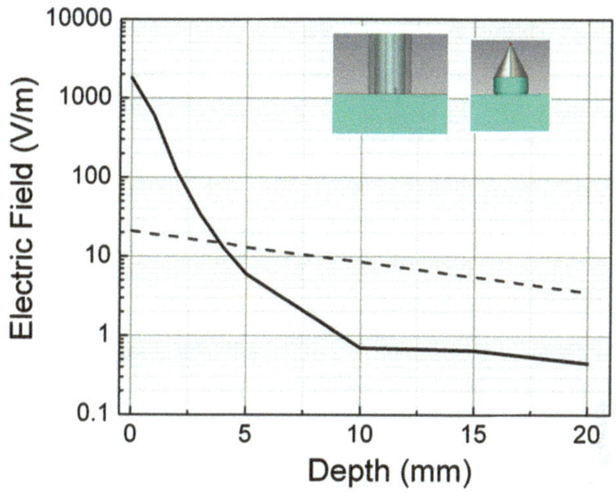

Fig. 2.54 The electric field decreases as the depth increases. The depth starts from the tissue surface. The *solid line* is the field on the axis of the coaxial cable but inside the tissue ($\varepsilon_r = 38$ and $\sigma = 1.18$ S/m). The coaxial cable has a radius of 0.45 mm for the inner conductor and the inner radius of 1.3 mm for the outer conductor. The *dashed line* is the field on the axis of the dielectric rod antenna but inside the tissue. The dielectric rod antenna has a radius of 1 cm and the dielectric constant $\varepsilon_r = 20$. The input voltage for both cases is 1 V

does the electric field decrease so fast? One can examine Eq. 2.1. The field has a dependence on the coefficients $\alpha_{y,y}^{(1)}$ and $\alpha_{y,y}^{(2)}$, which contain the aperture size and tangential field distribution of that aperture. For a coaxial cable with a small aperture (smaller than the observing distance R_0), the integration of the "surface current" over the aperture results in small coefficients. Furthermore, the field is azimuthally symmetrical about the axis, which leads to the cancelation of the transverse field on the axis, leaving only the axial field.

To allow the field to penetrate deeper, we use a cylindrical dielectric waveguide fed by a conical waveguide. First of all, using a rod with a high dielectric constant, the diameter of the waveguide can be small as long as it is still comparable to the spatial width of the pulse, allowing the establishment of a hybrid wave (HE), for which there is practically no cutoff frequency. This favors the transmission of UWB pulses. For HE waves, the energy is largely confined to the center of the rod. Secondly, the electric fields for HE waves are polarized in the cross section of the dielectric waveguide (as is the tissue surface), thereby minimizing the reflection as discussed previously. Finally, the dielectric constant of the waveguide can be chosen as a value close to that of tissue, avoiding the abrupt transition of dielectric constant from air to the tissue. To setup an HE wave in the dielectric rod, one can use a hollow conical waveguide wrapped around a conical dielectric rod. The apex of the conical waveguide is where the electric pulses are fed. The electric field distribution for a dielectric rod antenna is also plotted in Fig. 2.54. The field on the surface in this case is smaller than that of coaxial cable, but it decreases slower than the coaxial cable. So in the deeper region, the field can still be on the same order as that on the surface. Such distribution will then allow us to create a sufficiently high field in the subcutaneous region, and in the meantime, the field on the skin surface may not be damaged.

2.6.5 *Delivery of Electric Pulses to Deep Region*

To deliver pulses to deep targets, such as the ones in the range of 6–8 cm, the use of our previous antenna is insufficient. To this end, a large-aperture reflector antenna that has two focal points can be used. From the first focal point, a wave-launching antenna is placed and the spherical waves are emitted. The waves that will converge after scattering from the reflector surface are called impulse, while the other part of the wave being emitted in a diverging fashion is called prepulse. Two typical wave-launching antennas are shown (Fig. 2.55). One relies on a pair of conical wave-guides, feeding the reflector waves backward, which results in a converging impulse at the second focal point. This pulse has the opposite polarity to the prepulse (which is the diverging pulse). An example of such antenna was designed by Baum et al. [46, 58] and further studied at Old Dominion University [59, 60]. The second can be a cone that feeds the reflector waves forward, so the electric fields at the focus for both prepulse and impulse have the same polarity. In a study conducted at Kumamoto University [61], this type of antenna was used for radiating 800-MHz burst signals. Three radiating structures were compared: a sleeve antenna, a brown antenna, and a discone antenna (Fig. 2.56). The simulation results showed that burst electromagnetic waves are focused around the second focal point. The discone antenna outperforms the others, creating the narrowest confined region of electric fields.

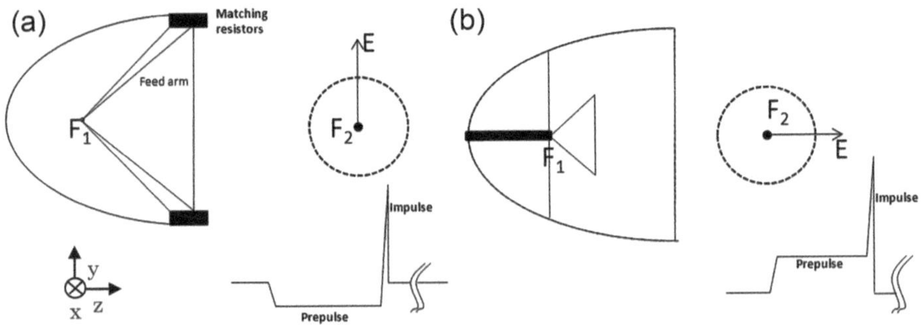

Fig. 2.55 An ellipsoidal reflector focuses radiation for reaching deep target. The polarization of the electric field at the second focus depends on the feed structure at the first focal point. (**a**) Conical plates are used to feed spherical TEM waves to the reflector, resulting in a vertically polarized electric field at the second focus; (**b**) a cone on a ground plane is used to feed a TEM wave to the reflector, resulting in a horizontally polarized electric field. In the drawings, a step input pulse is used

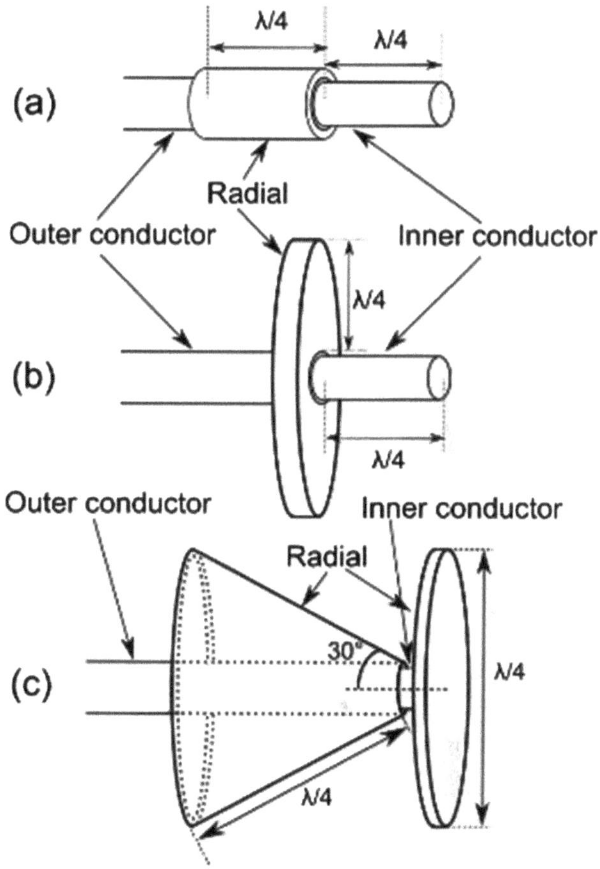

Fig. 2.56 The configurations of radiation antenna at the first focal point. (**a**) Sleeve antenna; (**b**)

In the case where the pulse is directly delivered to the brain by an impulse radiating antenna (Fig. 2.57), a simulation using the CST HUGO model was conducted to determine the temporal and spatial distribution of the pulsed electric fields in a human brain. In the human head model used in this study, the dielectric constants and conductivities of the tissues are bone ($\varepsilon_r = 11.78$, $\sigma = 0.28$), gray matter ($\varepsilon_r = 50$, $\sigma = 1.39$), fat tissue ($\varepsilon_r = 5.35$, $\sigma = 0.08$), nervus opticus ($\sigma = 30.87$, $\sigma = 0.84$), white matter ($\varepsilon_r = 37$, $\sigma = 0.91$), and skeletal muscle ($\varepsilon_r = 55.33$, $\sigma = 1.44$). These tissue dielectric properties are the values chosen at 1.8 GHz. In the frequency range of 0.9–5 GHz, the tissue properties stay relatively flat as the bulk tissue dispersion is mainly dominated by the water contained in the tissue. For example, white matter has $\varepsilon_r = 38.89$ and $\sigma = 0.59$ at 0.9 GHz and $\varepsilon_r = 33.44$ and $\sigma = 2.86$ at 5 GHz. The dispersion of the dielectrics in general broadens the pulse waveforms and lowers the pulse intensity; the results of the simulation without dispersion loss therefore serve as the best scenario estimate of the pulse delivery in terms of pulse intensity and focal spot size.

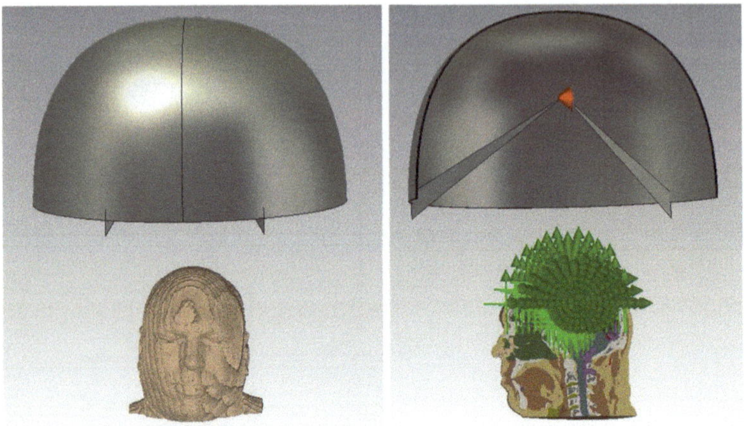

Fig. 2.57 An impulse radiating antenna is used to focus the radiation to the brain

The pulse waveform at the geometric focus in the brain (6-cm deep from the top of the skull) only has the y component, and the x and z components are negligible [62]. Figure 2.58 shows the propagation of subnanosecond pulses in the head. The intensity of the electric fields is shown as the linear isoline in the clamped range 0.1–0.3 V/m, meaning any values that are greater than 0.3 are clamped to 0.3. At 1.8 ns, the prepulse has already reached the brain. At the same time, the prepulse in the face and back side of the head has traveled faster and farther than the prepulse in the brain. At 3.1 ns, the impulse appears in the brain along the axis of the antenna. It gradually reduces in size and, at 3.7 ns, appears to be at its smallest. As the wave penetrates deeper, the intensity decreases. The isoline plot clearly indicates that there is no resonance in the brain except the impulse. In the side view of the wave propagation (Fig. 2.58), at 3.1 ns, the impulse actually consists of two parts indicated by two bright spots. In the time domain, the corresponding impulse waveform is bipolar, which is the time derivative of a Gaussian pulse. At 3.7 ns, the wave converges, which is consistent with the side view. A top view of the wave propagation is also shown in Fig. 2.58. The view is taken at a depth of 8 cm from the top of the head. Figure 2.59 shows the field distribution along various radii from the skin. The largest field is on the axis and the fields on the two sides seem symmetrical about the axis. They all decrease due to the strong absorption in the brain. Despite the fact that the field decreases as the penetration depth increases, the field values converge at 7–8 cm, slightly deeper than the geometric focus (6 cm in depth). The inhomogeneity of the brain does not seem to create any difference in the arrival time of the spherical waves along different radii, but the strong attenuation in brain tissue prevents the wave from reaching the highest amplitude near the focal point. Still, the focal volume can be estimated as $1 \times 2 \times 1$ cm in the xyz directions and is situated in the white-matter region. As white matter has a dielectric constant of 37, the spatial width of such pulse is 1 cm, which is consistent with the simulation result for an input pulse of 200 ps (Fig. 2.58). It is therefore reasonable to

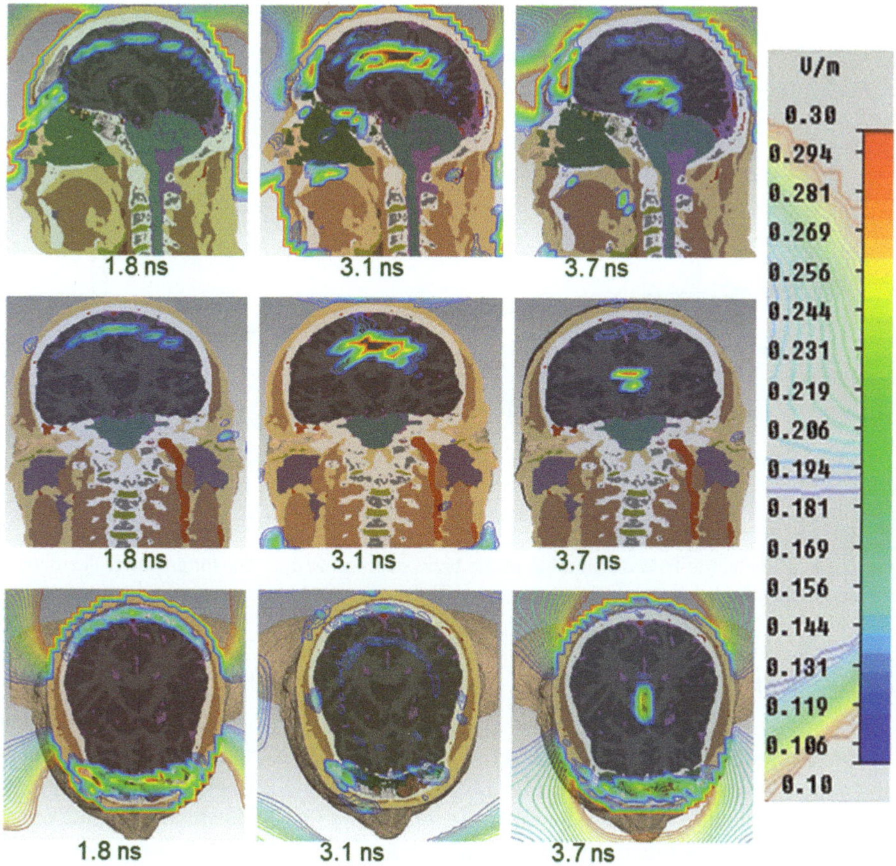

Fig. 2.58 The isoline plots of electric field distribution in the human brain exposed to a reflector antenna at various times with a clamp range of 0.1–0.3 V/m along x-axis (*side view*), y-axis (*back view*), and z-axis (*top view*) [62]. The input voltage is 1 V

predict the focal spot size simply by estimating the pulse spatial width in the dielectric where the focal point is to be created.

Although the field is confined at the focal point, it attenuates quite severely on its way to the focal point, resulting in a monotonically decreasing trend from the surface. So the spatial field distribution is not satisfactory. To improve the field distribution, one can introduce a second antenna, which can be a conical plate, in addition to the reflector antenna [63]. While the radiation from the reflector antenna is converging, the radiation from the conical plate is diverging. The input pulses for these two antennas can be set to be opposite in polarity, so that the fields in the shallow region are canceled, and in deeper regions, the field is largely dominated by the radiation from the reflector antenna. A deep, localized focal region may be formed, but the caveat is that the overall field intensity is reduced due to the cancelation.

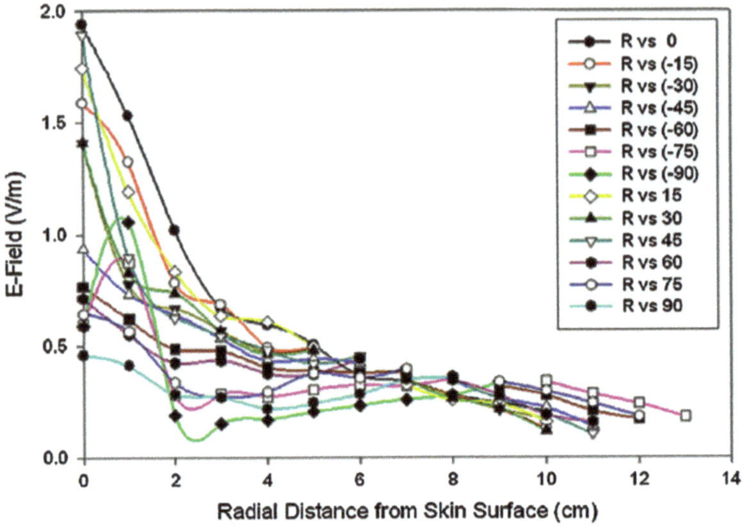

Fig. 2.59 The electric field distribution in the brain exposed to the reflector antenna. The data is shown along different paths at various angles [62]

2.6.6 Experimental Results of Reflector Antennas

Before the antennas are applied to actual biological targets, they were studied in air and water. Two experimental results are shown here: one is the impulse radiation in air and the other is the pulse burst in water. In Fig. 2.60, a subnanosecond pulse (Fig. 2.60a) was fed to the impulse antenna and the pulse at the focal point was shown in Fig. 2.60b. The prepulse is wider than the impulse since the prepulse is just the replica of the input pulse and the impulse is proportional to the time derivative of the feed pulse shown in Fig. 2.60a. The feed pulse, which was directly measured from the pulse generator, was differentiated to obtain dV/dt (dashed line in Fig. 2.60b). Apparently, the impulse part of the electric field is overlapped with dV/dt, which confirms the prediction of Eq. 2.3. The separation in time between these two pulses, defined as cleartime, is approximately 1 ns. The postpulse after 4.5 ns may contain both diffractions from various parts of the antenna (reflector rim, edge of the conical plate, etc.) and the resonance signal of the sensor itself. The amplitudes of both prepulse and impulse are plotted in y and z axes in Fig. 2.61. The prepulse decreases in the z direction, with $1/z$ dependence. The power density of the impulse, which is proportional to the square of the electric field, is focused in both lateral and axial directions. The distribution of power density has a full width at half-maximum (FWHM) width of 32 cm in the axial direction and a FWHM width of 10 cm in the lateral direction. The experimental results agree very well with the modeled results. This is true especially near the focal point in both the axial and the lateral electric field distributions. At an observing distance closer to the reflector

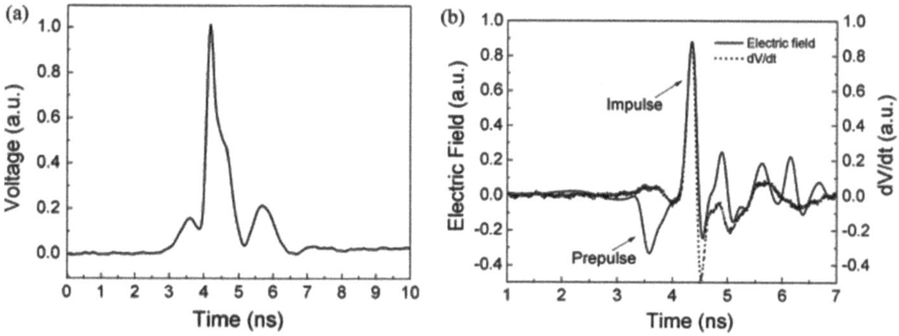

Fig. 2.60 The measured electric pulse from the pulse generator that was fed into the antenna. (**b**) The measured electric field at the second focal point (*solid line*). The *dashed line* is dV/dt, and V is the feed pulse shown in (**a**). The peak of dV/dt was normalized to the impulse amplitude [59]

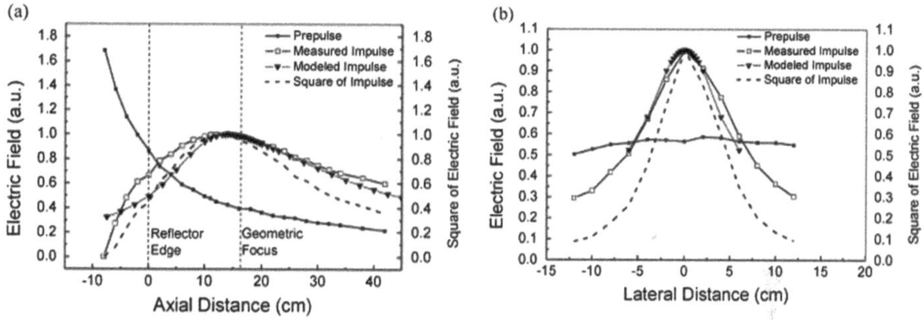

Fig. 2.61 The measured electric field (x-directed) distribution in the axial direction (z-axis). The prepulse, the impulse, and the square of impulse are shown. The curve with down-triangle symbols is the modeled impulse distribution in the z-axis. (**b**) The measured electric field (x-directed) distribution in the lateral direction (y-axis). The prepulse, the impulse, and the square of impulse are shown. The curve with *down-triangle* symbols is the modeled impulse distribution in the y-axis [59]

edge, we observe a slight discrepancy between the modeled field and measured field, which could be due to the fact that the sensor may pick up other electric field components besides the x-directed field and the model only gives the x-directed field.

Figure 2.62 shows a distribution map of electric field intensity in the case of three feeding antennas. The burst electromagnetic waves reflect on the reflector and are focused around the second focal point. The results were obtained when the antenna and reflector were submerged in water. The focal area, which is higher than half maximum around the second focal point, is approximately 14 cm^2. The FWHM (full width at half maximum) of the radial direction is narrower than that of the longitudinal direction. Figure 2.62 shows the electric field intensity distribution in the transverse and longitudinal directions. In the longitudinal direction (Fig. 2.62a),

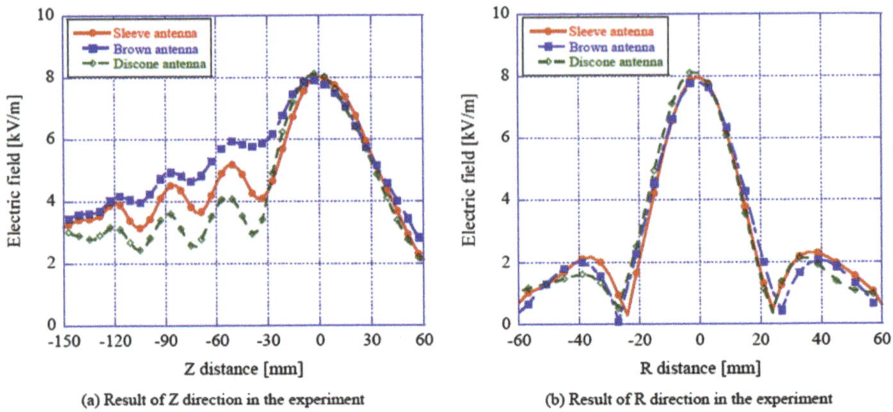

Fig. 2.62 Distribution of the electric field in the experiment using the sleeve antenna, the brown antenna, and the discone antenna: (**a**) longitudinal direction; (**b**) radial direction. The input peak voltage was 320 V [61]

the FWHM of the discone antenna is shortest. In the transverse direction (Fig. 2.62b), the FWHMs have almost identical values.

2.6.7 Concluding Remarks

The delivery of electric pulses to biological tissue can be implemented through the use of antennas for either shallow or deeper target regions. The heterogeneity of tissue composition does not seem to affect the operating principle of a reflector mirror, so the field is confined near the focal point. However, there is a major challenge, which is the tissue attenuation, making the field monotonically decreasing from the surface to a small field at the focal point. Using the brain as an example, while the pulse can be locally confined in a region of 1 cm^3 at a depth of 8 cm, the field intensity is only 0.3 V/m for a 1 V pulse input, which suggests that pulses of 170 kV need to be fed into the antenna in order to meet the estimated stimulation threshold of 50 kV/m. Such a high-amplitude pulse generator becomes technically challenging and harder to be incorporated with the antenna system. It is easier to focus the field on a shallow target with the dielectric rod antenna. The electric field is in the range of 5 V/m at a depth of 1 cm, which is the motor cortex region in the brain. This means an input voltage of 10 kV may be sufficient for an effective stimulation to meet the stimulation threshold of 50 kV/m. But this conclusion is based on the assumption of a biological threshold of 50 kV/m, which has not strictly been proven for subnanosecond pulses. Therefore, more biological experiments are called for in order to provide the guidance for antenna and pulsed power design.

References

1. Paul, W.: Smith, Transient Electronics, Pulsed Circuit Technology. Wiley, Chichester (2002)
2. Humphries Jr., S.: Principles of Charged Particle Acceleration. Wiley, New York (1990)
3. Bluhm, H.: Pulsed Power Systems, Principle and Applications. Springer, Berlin (2006)
4. Mesyats, G.A.: Pulsed Power. Springer, New York (2005)
5. Akiyama, H. (ed.): IEEJ EEText, Kodenatsu Parusu Pawa Kougaku, Ohmsha, (in Japanese) (2003)
6. Choi, J.: Introduction of the magnetic pulse compressor (MPC) – fundamental review and practical application. J. Electr. Eng. Technol. **5**(3), 484–492 (2010)
7. Barrett, D.M.: Core reset considerations in magnetic pulse compression networks, Pulsed Power Conference, 1995. Digest of technical papers. Tenth IEEE International, vol. 2, no., pp.1160,1165 vol. 2, (3–6 July 1995) doi: 10.1109/PPC.1995.599771
8. Deyu Wang, Weiyang Wu, Da Li, Liqiao Wang: Compact Magnetic Compression Repetitive Pulsed Power Generator Based on IGBT. Electrical Machines and Systems, 2008. ICEMS 2008. International Conference on, pp. 1255–1258, 17–20 Oct 2008
9. Mankowski, J., Kristiansen, M.: A review of short pulse generator technology. IEEE Trans. Plasma Sci. **2**(1), 102–108 (2000)
10. Lyubutin, S.K., Mesyats, G.A., Rukin, S.N., Slovikovskii, B.G., Turov, A.M.: New solid state opening switches for repetitive pulsed power technology, High-Power Particle Beams, 1996 11th International Conference on, vol. 1, no., pp.135,138 (10–14 June 1996)
11. Redondo, L., Silva, F.A.: Solid state pulsed power electronics. In: Rashid, M. et al. (eds.) Power Electronics Handbook 3rd edn. Butterworth-Hinemann Publishing, Elsevier, USA, ISBN # 9780123820365, chapter 26, pp 669–710 (2010)
12. Kesar, A.S., Merensky, L.M., Ogranovich, M., Kardo-Sysoev, A.F., Shmilovitz, D.: 6-kV, 130-ps rise-time pulsed-power circuit featuring cascaded compression by fast recovery and avalanche diodes. Electron. Lett. **49**(24), 1539–1540 (2013)
13. Mohan, N., Undeland, T., Robbins, W.: Power Electronics: Converters, Applications and Design, 2nd edn. Wiley, New York (1995)
14. Rashid, M.H. (ed.): Power Electronics Handbook, 2nd edn. Academic, Elsevier, San Diego (2007). ISBN 10:0-12-088479-8. ISBN 13:978-0-12-088479-7
15. Baker, R.J., Johnson, B.P.: Applying the Marx Bank circuit configuration to power mosfets. Electron. Lett. **29**(1), 56–57 (1993)
16. Baker, R.J., Ward, S.T.: Designing nanosecond high voltage pulse generators using power MOSFETs. Electron. Lett. **30**(20), 1634–1635 (1994)
17. Welleman, A., Waldmeyer, J., Ramezani, E.: Solid state switches for pulse power modulators. In: Proc. Linear Particle Accelerator Conf., pp. 707–709 (2002)
18. Jiang, W., et al.: Compact solid-state switched pulsed power and its applications. Proc. IEEE **92**(7), 1180–1196 (2004)
19. Mazumder, S.K., Sarkar, T.: SiC based optically-gated high-power solid-state switch for pulsed-power application. Mater. Sci. Forum **600–603**, 1195–1198 (2008)
20. Racz, B., Patocs, A.: Fast high-voltage resistive pulse divider. Meas. Sci. Technol. **3**, 926 (1992)
21. www.highvoltageprobes.com (as of 30 Dec 2014)
22. Winands, G.J.J.: Efficient Streamer Plasma Generation. PhD Thesis Eindhoven University of Technology (2007)
23. van Deursen, A.P.J., Gulickx, P.F.M., van der Laan, P.C.T.: A Current and Voltage Sensor in One Unit. 8th International Symposium on High Voltage Engineering, Yokohama (1993)
24. van Deursen, A.P.J., Smulders, H.W.M., de Graaff, R.A.A.: Differentiating/integrating measurement setup applied to railway environment. IEEE Trans. Instrum. Meas. **55**, 316–326 (2006)

25. van Heesch, E.J.M., van Deursen, A.P.J., van Houten, M.A., Jacobs, G.A.P., Kersten, W.F.J., van der Laan, P.C.T.: Field Tests and Response of the D/I H.V. Measuring System. Sixth International Symposium on High Voltage Engineering, New Orleans (1989)
26. van Heesch, E.J.M., van Rooij, J.N.A.M., Noij, R.G., van der Laan, P.C.T.: A new current and voltage measuring system; tests in a 150 kV and 400 kV GIS. Proc. 5th Int. Symp. High Voltage Eng. **3**, 73.06 (1987)
27. van Houten, M.A.: Electromagnetic Compatibility in High-Voltage Engineering. PhD thesis, Eindhoven University of Technology (1990)
28. Keller, R.: Wideband high voltage probe. Rev. Sci. Instrum. **35**, 1057–1059 (1964)
29. Smulders, H.W.M., de Graaff, R.A.A., Janssen, M.F.P., van Alphen, G.: Measurement systems for AC traction power supply systems. Int. Conf. Railw. Traction Syst. Capri Proc. **2**, 139–159 (2001)
30. Huiskamp, T., Beckers, F.J.C.M., van Heesch, E.J.M., Pemen, A.J.M.: First implementation of a subnanosecond rise time, variable pulse duration, variable amplitude, repetitive, high-voltage pulse source. IEEE Trans. Plasma Sci. **42**(3), 859–867 (2014)
31. Huiskamp, T., Voeten, S.J., van Heesch, E.J.M., Pemen, A.J.M.: Design of a subnanosecond rise time, variable pulse duration, variable amplitude, repetitive, high-voltage pulse source. IEEE Trans. Plasma Sci. **42**(1), 127–137 (2014)
32. Lorusso, A., Nassisi, V., Siciliano, M.: Fast capacitive probe for electromagnetic pulse diagnostic. Rev. Sci. Instrum. **79**(6), 064702 (2008)
33. Voeten, S.J.: Matching High Voltage Pulsed Power Technologies. Ph.D. dissertation, Dept. Electr. Eng., Eindhoven Univ. Technol., Eindhoven (2013)
34. Smith, P.: Transient Electronics: Pulsed Circuit Technology. Wiley, New York (2002)
35. Huiskamp, T., van Heesch, E.J.M., Pemen, A.J.M.: Final Implementation of a Subnanosecond Rise Time, Variable Pulse Duration, Variable Amplitude, Repetitive, High-Voltage Pulse Source. Accepted for IEEE Trans. Plasma Sci., on line 8 Dec 2014
36. http://www.pearsonelectronics.com (as of 30 Dec 2014)
37. Rogowski, W., Steinhaus, W.: Die Messung der magnetischen Spannung. Arch. Elektrotechnik **1**(Pt.4), 141–150 (1912)
38. van Bree, J.W.M., Geysen, J.J.G., van Heesch, E.J.M., Pemen, A.J.M.: Novel nanosecond pulsed electric field device for noncontact treatment of cells in native culture conditions. IEEE Trans. Plasma Sci. **41**(10), 2654–2658 (2013)
39. Zhen, L.: Multiple-Switch Pulsed Power Generation Based on a Transmission Line Transformer. PhD Thesis Eindhoven University of Technology (2008)
40. Smulders, H.W.M., van Heesch, E.J.M., van Paassen, S.V.B.: Pulsed power corona discharges for air pollution control. IEEE Trans. Plasma Sci. **26**, 1476–1484 (1998)
41. Schoenbach, K.H., Xiao, S., Joshi, R.P., Camp, J.T., Heeren, T., Kolb, J.F., Beebe, S.J.: The effect of intense subnanosecond electrical pulses on biological cells. IEEE Trans. Plasma Sci. **36**, 414–422 (2008)
42. Rogers, W.R., Merritt, J.H., Comeaux Jr., J.A., Kuhnel, C.T., Moreland, D.F., Teltschik, D.G., Lucas, J.H., Murphy, M.R.: Strength duration curve for an electrically excitable tissue extended down to near 1 nanosecond. IEEE Trans. Plasma Sci. **32**, 1587–1599 (2004)
43. Jiang, N., Cooper, B.Y.: Frequency-dependent interaction of ultrashort E-fields with nociceptor membranes and proteins. Bioelectromagnetics **32**, 148–163 (2011)
44. Xiao, S., Guo, S., Nesin, V., Heller, R., Schoenbach, K.H.: Subnanosecond electric pulses cause membrane permeabilization and cell death. IEEE Trans. Plasma Sci. **58**, 1239–1245 (2011)
45. Camp, J.T., Jing, Y., Zhuang, J., Kolb, J.F., Beebe, S.J., Song, J., Joshi, R.P., Xiao, S., Schoenbach, K.H.: Cell death induced by subnanosecond pulsed electric fields at elevated temperatures. IEEE Trans. Plasma Sci. **40**(10), 2334–2347 (2012)
46. Baum, C.E.: Focal waveform of a prolate-spheroidal impulseradiating antenna (IRA). Radio Sci. **42**, RS6S27 (2007)

47. Trefna, H.D., Vrba, J., Persson, M.: Time-reversal focusing in microwave hyperthermia for deep-seated tumors. Phys. Med. Biol. **55**, 2167–2185 (2010)
48. Wust, P., Hildebrandt, B., Sreenivasa, G., Rau, B., Gellermann, J., Riess, H., Felix, R., Schlag, P.M.: Hyperthermia in combined treatment of cancer. Lancet Oncol. **3**(8), 487–497 (2002)
49. Converse, M., Bond, J.E., Veen, B.D., Hagness, S.C.: A computational study of ultra-wide-band versus narrowband microwave hyperthermia for breast cancer treatment. IEEE Trans. Microwave Theory Tech. **54**(5), 2169–2180 (2006)
50. Yarovoy, A.G., Ligthart, L.P., Matuzas, J., Levitas, B.: UWB radar for human being detection. IEEE Aerosp. Electron. Syst. Mag. **21**, 22–26 (2006)
51. Miller, E.K.: Chapter 5: Time-Domain Measurements in Electromagnetics, pp. 122. Van Nostrand Reinhold Company Inc., New York (1986)
52. Allen, B., Dohler, M., Okon, E.E., Malik. W.Q., Brown, A.K., Edwards, D.J.: Chapter 7: Ultra-Wideband Antennas and Propagation for Communications, Radar and Imaging. Wiley, Chichester (2007)
53. Wiesbeck, W., Adamiuk, G., Sturm, C.: Basic properties and design principles of UWB antennas. Proc. IEEE **97**(2), 372–385 (2009)
54. Smith, G.S.: Teaching antenna radiation from a time-domain perspective. Am. J. Phys. **69**(3), 288 (2001)
55. Baum, C.E.: Focused Aperture Antennas. Sensor and Simulation Notes 306, (1987)
56. Grimnes, S., Martinsen, O.G.: Bioimpedance and Bioelectricity Basics. Academic Press, London (2000)
57. Barnes, F.S., Greenebaum, B.: Handbook of Biological Effects of Electromagnetic Fields. Introduction by C. Polk. CRC press, Boca Raton (2006)
58. Kumar, P., Baum, C.E., Altunc, S., Buchenauer, J., Xiao, S., Christodoulou, C.G., Schamiloglu, E., Schoenbach, K.H.: A hyperband antenna to launch and focus fast high-voltage pulses onto biological targets. IEEE Trans. Microwave Theory Tech. **59**, 1090–1101 (2011)
59. Xiao, S., Altunc, S., Kumar, P., Baum, C.E., Schoenbach, K.H.: A reflector antenna for focusing in the near field. IEEE Antennas Wirel. Propag. Lett. **9**, 12–15 (2010)
60. Bajracharya, C., Xiao, S., Baum, C.E., Schoenbach, K.H.: Target detection with impulse radiating antenna. IEEE Antennas Wirel. Propag. Lett. **10**, 496–499 (2011)
61. Ishizawa, H., Tanabe, T., Yoshida, D., Hosseini, S.H.R., Katsuki, S., Akiyama, H.: Focusing system of burst electromagnetic waves for medical applications. IEEE Trans. Dielectr. Electr. Insul. **20**(4), 1321–1326 (2013)
62. Guo, F., Yao, C., Bajracharya, C., Polisetty, S., Schoenbach, K.H., Xiao, S.: Simulation of delivery of subnanosecond pulses to biological tissues with impulse radiating antenna. Bioelectromagnetics **35**, 145–159 (2013)
63. Xiao, S., Guo, F., Li, J., Hou, G.. Schoenbach, K.H.: Simulation of delivery of subnanosecond electric pulses into biological tissues. In: Proceedings of the 2012 IEEE International Power Modulator and High Voltage Conference, San Diego (2012)

Chapter 3
Special Electromagnetic Agents: From Cold Plasma to Pulsed Electromagnetic Radiation

Petr Lukes, Hidenori Akiyama, Chunqi Jiang, Andrea Doria, Gian Piero Gallerano, Alfonsina Ramundo-Orlando, Stefania Romeo, Maria Rosaria Scarfì, and Olga Zeni

Abstract Pulsed power is applied to gases and liquids to produce plasmas; these may interact with living organisms. Plasma, known as the fourth state of matter, consists of charged and neutral particles and is characterized by a quasi-neutral state. Two types of plasma exist, nonthermal and thermal equilibrium. Nonthermal plasma, the type most frequently employed in the bioelectric field, has a gas temperature much lower than its electron temperature. The biocidal agents which may be produced by nonthermal plasmas are heat, electric fields, ultraviolet radiation, shock waves, and reactive chemical species. Additionally, secondary biological effects can be induced through post-discharge reactions of chemical species produced by plasma in gases and liquid.

Interactions between pulsed electromagnetic fields and biological systems are studied in order to elucidate whether cellular systems are able to demodulate rapidly oscillating electric fields. Pulsed electromagnetic radiation comprises a central carrier frequency in the range from several GHz to about 1 THz and an electric field peak amplitude of tens of kV/m. High-power ultrawideband

P. Lukes (✉)
Institute of Plasma Physics, Czech Academy of Sciences, Prague, Czech Republic
e-mail: lukes@ipp.cas.cz

H. Akiyama
Kumamoto University, Kumamoto, Japan

C. Jiang
Old Dominion University, Norfolk, USA

A. Doria • G.P. Gallerano
ENEA, Fusion Physics Division, Frascati, Italy

A. Ramundo-Orlando
Institute of Translational Pharmacology of CNR, Roma, Italy

S. Romeo • M.R. Scarfì • O. Zeni
Institute for Electromagnetic Sensing of the Environment of CNR, Napoli, Italy

H. Akiyama, R. Heller (eds.), *Bioelectrics*, DOI 10.1007/978-4-431-56095-1_3

electromagnetic pulses (UWB-EMP) have recently been used for research in the bioelectric field to elucidate basic interactions with biological systems, to study induced biological effects, and to develop THz-based diagnostic technologies.

Keywords Nonthermal plasma • Ultrawideband electromagnetic pulse • Biocidal agents • Pulsed electromagnetic radiation • Secondary biological effects

3.1 Introduction

Hidenori Akiyama

Pulsed power is applied to gases and liquids to produce plasmas; these may interact with living organisms. Plasma, known as the fourth state of matter, consists of charged and neutral particles and is characterized by a quasi-neutral state. Two types of plasma exist, nonthermal and thermal equilibrium. The latter is used for applications such as nuclear fusion and welding. Nonthermal plasma, however, has a gas temperature much lower than its electron temperature and is produced primarily from gases at atmospheric pressure; such plasmas are known as nonthermal atmospheric-pressure plasmas (NTAPs). Plasma sources developed for application of NTAPs to biomedical and environmental fields include corona discharges, micro-hollow cathode discharges, dielectric barrier discharges, and plasma jets.

The biocidal agents which may be produced by nonthermal plasmas are heat, electric fields, ultraviolet radiation, shock waves, and reactive chemical species. Additionally, secondary biological effects can be induced in plasma-treated gases and liquid through post-discharge reactions of chemical species produced therein.

Interactions between pulsed electromagnetic fields and biological systems are studied in order to elucidate whether cellular systems are able to demodulate rapidly oscillating electric fields such as wideband pulses, with durations ranging from picoseconds to microseconds. The term "pulsed electromagnetic radiation" is used here to refer to high-power, radiated electromagnetic pulses, with central carrier frequency in the range from several GHz to about 1 THz and electric field peak amplitude of tens of kV/m. Radiation sources available for research will be detailed.

High-power ultrawideband electromagnetic pulses (UWB-EMP) may be utilized for such military applications as intentional electromagnetic interference, motor disablement, and improvised explosive device disruption; however, current bioelectric-related research mainly seeks to elucidate basic interactions with biological systems, the study of induced biological effects, and the development of THz-based diagnostic technologies.

3.2 Nonthermal Plasmas

Petr Lukes and Chunqi Jiang

Plasma is the fourth state of matter and characterized as a quasi-neutral entity that consists of charged and neutral particles (e.g., ions, electrons, and excited species). A plasma in strong ionization often exhibits a thermodynamic equilibrium of the plasma species where the electron temperature approaches the gas temperature, $T_e \approx T_g$. In this case the temperature describes the mean kinetic energy of the plasma species, and we have

$$E_{\mathrm{mean}} = 3/2KT_g \tag{3.1}$$

where K is Boltzmann constant. The velocities of the particles in a thermal equilibrium plasma (aka. thermal plasma) satisfy the Maxwellian distribution. On the contrary, a weakly ionized plasma sustained by an electric field usually does not have sufficient ionization degree that allows thermalization by collisions between electrons and heavy particles or reaches a thermodynamic equilibrium. The electron temperature $\mathrm{T_e}$, indicating the mean energy of electrons, needs to be much higher than the gas temperature T_g (i.e., $T_e >> T_g$), so that there is a sufficient number of energetic electrons to ionize atoms and/or molecules to form plasma. These weakly ionized plasmas are thereby nonequilibrium (aka nonthermal).

Recent development of nonthermal plasmas at atmospheric pressure is particularly interesting to the environmental and biomedical fields, although nonthermal plasmas can also be generated under lower or higher pressure. The biomedical applications of nonthermal atmospheric-pressure plasmas (NTAPs), in turn, have motivated the plasma source development itself.

3.2.1 Nonthermal Plasma Sources for Biomedical Applications

To date, there are many different types of NTAPs that have been applied for biomedical applications. According to the range of driving frequencies and electrode configurations, the NTAP discharges can be categorized into several basic types which are briefly described in further text.

3.2.1.1 Corona Discharges

Corona discharges occur when a sharp nonuniform electric field forms near one or both electrodes and the localized field near the electrode(s) is much stronger than in the rest of the gap [1]. Typical configurations of corona discharges are point-to-plane, point-to-point, and concentric rod-to-cylinder. For modern industrial applications, such as surface treatment and cleansing of gas and liquid exhaust streams, arrays of multiple points or large scales of concentric wire-to-cylinder are proposed and employed [2, 3]. However, the application of the continuous corona discharge is limited by very low currents, in order to prevent the arc formation, and hence a low rate of treatment of materials and exhaust streams. Pulsed corona discharges allow higher application voltage and peak power than continuous corona and thereby are more attractive in environmental and biomedical applications with their enhanced chemistry and increased energy efficiency.

3.2.1.2 Micro-hollow Cathode Discharges

Micro-hollow cathode discharges (MHCDs) [4] are nonthermal glow discharges capable of stable DC operation at atmospheric pressure. Perforated metal–dielectric–metal configurations and the submillimeter-range characteristic dimensions (i.e., the thickness and the aperture diameter) of the dielectric material enable high-pressure breakdown at relatively low voltages. These MHCD plasmas may serve as the cathode or the electron source for a larger-volume glow discharge, which allows more efficient processing and treatment in environmental and biomedical applications [5–7].

3.2.1.3 Dielectric Barrier Discharges

In a dielectric barrier discharge (DBD), dielectric layers are used to cover at least one of the electrodes. DBD discharges, also known as silent discharges, can be operated at a relatively wide-frequency range, i.e., from 50 Hz up to RF, and pulsed modes. The typical operation frequencies for DBDs are between 50 Hz and 500 kHz. By applying a sinusoidal voltage of sufficient amplitude to the electrodes, a large number of micro-discharges occur due to charge accumulation on the dielectric. The dielectric serves both as a discharge “current stopper” by limiting the amount of charge transported by a single micro-discharge and a discharge “transporter” by distributing the micro-discharges over the entire electrode surface [8]. Although the DBD was originally considered filamentary, relatively large volume, diffused atmospheric-pressure glow discharges can also be generated in a pulsed DBD in helium or mixture of helium and small trace of other gases [9, 10]. This resulted in biomedical applications of the diffused DBDs [11, 12]. In addition, one-atmosphere uniform glow discharges in air generated by a water-cooled DBD reactor powered with RF were also reported for room temperature sterilization of surfaces and fabrics [13].

3.2.1.4 Radio-Frequency (RF) Discharges

RF discharges are obtained when the gas is subjected to an oscillating electromagnetic field generated by inductively or capacitively coupling of the power. In an inductively coupled RF discharge, a high-frequency (typ. in the range of 0.1–100 MHz) current is passed through a solenoid coil that surrounds a dielectric tube filled with the gas to be studied. The most common industrial frequency is 13.56 MHz. Pulsed discharges are produced if a sufficiently strong current pulse is fed into the coil. High-pressure, inductively coupled RF plasmas have relatively high gas temperature (>7000 K) and have been typically used for material heating as equilibrium plasmas [1]. In capacitively coupled RF discharges, the systems of two electrodes behave as capacitors to the high-frequency voltage. The two electrodes may be bare and in direct contact with the discharge plasma or may be insulated by dielectric layers. Geometry of the electrodes may be as simple as two parallel plates or in more complex configurations such as coaxial wire-to-cylinders.

3.2.1.5 Microwave Discharges

Microwave-induced plasmas are mostly produced in waveguide structures or resonant cavities where the electromagnetic fields were excited at microwave frequencies between 300 MHz and 3 GHz [8]. The most common frequency is 2.45 GHz, which is used in microwave ovens. Large-volume nonequilibrium plasmas can be generated with the microwave-induced, surface-wave-sustained discharges [14], which provide a good option for applications including plasma–chemical processing and lasing.

3.2.1.6 Plasma Jets

Plasma jets have become one of the most attractive discharges for biomedical applications due to their advantage in extending the plasma to regions not confined by electrodes or dielectrics. Based on aforementioned NTAP plasma excitation schemes, the "internally" generated NTAP plasma plumes or jets are pushed outside the device by gases flowing at rates that are sufficient to "extend" the plumes externally without disturbing or distinguishing the discharge. DC, pulsed, AC, RF, and microwave plasma jets have all been reported and applied for environmental and biomedical applications. Due to a relatively large number of plasma jets that have been developed through the years, not all of them will be covered in this chapter. Instead only a few of them with representative or originative configurations and implications in biomedical applications are listed in the following. Additional examples of atmospheric plasma jets can be found in a recent review given by Laroussi et al. [15].

3.2.1.6.1 DC-Driven Plasma Jets

The so-called plasma brush, developed by Duan et al., was able to be ignited and sustained in both continuous and pulsed modes with different plasma gases such as argon or helium at atmospheric pressure [16]. A typical plasma brush electrode system consisted of a dielectric gas compartment and two metallic electrodes placed a certain distance apart inside the gas compartment [16]. At a typical DC power of 10.4 W, the gas temperature of the plasma was measured from 120 to 40 °C as the gas flow rate decreasing from 3500 to 1000 sccm [17]. Another reported DC operational atmospheric plasma jet was based on the afterglow of a micro-hollow cathode discharge [7]. The gas temperature of the plasma jet was above 200 °C near the device nozzle and dropped to below 55 °C as the distance from nozzle increased to 5 mm at a flow rate of 220 ml/min.

3.2.1.6.2 Pulsed Power-Driven Plasma Jets

Laroussi and Lu developed a pulsed plasma device that generated 5-cm long pencil-shaped plasma jets at atmospheric pressure [18]. The discharge system, based on a tubular DBD configuration consisted of two electrodes, each made of a thin copper ring attached to the surface of a centrally perforated alumina (Al2O3) disk. The electrodes were typically powered by up to 10 kV, 200 ns voltage pulses at a repetition rate up to 10 kHz. The authors reported a plume gas temperature of 290 K for the 5-cm-long atmospheric He plasma jet operating at its typical power conditions. Jiang et al. generated a several-centimeter-long He or He–O_2 plasma jet using a concentric tubular configuration-based device powered by 100 ns, 6 kV voltage pulses at typical repetition rate of 1–2 kHz [19]. The authors reported an average power of less than 1 W for the plasma source operated at its typical treatment condition. In addition, Lu et al. advanced the NTAP pulsed plasma jet with a single-electrode design in which up to 4-cm-long He plasma jet was generated between the high-voltage wire and surrounding room air [20]. The rotational and vibrational temperatures of the plasma were measured to be about 300 K and 2950 K, respectively.

3.2.1.6.3 AC-Driven Plasma Jets

Kedzierski et al. were among the earliest groups that reported an atmospheric-pressure helium plasma jet generated by a cylindrically symmetric DBD device, powered by sinusoidal high voltages (1–15 kV, 5–50 kHz) [21]. At a typical operation condition (AC voltage of 7 $kV_{p\text{-}p}$ at 13 kHz), the plasma jet consumed an electrical power of 4 W. Cheng et al. developed an AC-driven cold plasma jet generated by a concentric cylindrical configuration [22]. The high voltage was applied to the outer tube, separated from the inner ground tube by an air gap. The working gas (Ar) was flowing through the center ground tube. With sufficient gas flow (e.g., 50–2,500 l/h or 0.8–41.7 lpm), steady plasma jets were obtained when the device was powered by the AC voltage of 30–80 $kV_{p\text{-}p}$ at 6–20 kHz. The authors reported a plasma gas temperature of only 25–30 °C.

3.2.1.6.4 RF-Driven Plasma Jets

Janca et al. developed one of the early versions of RF jets, also named as a "HF plasma pencil," for surface processing [23]. The electrode configuration is based on a center hollow electrode surrounded with a pencil-shaped dielectric tube. The working fluids (gas or liquid or dispersed particles) were flowing through the center hollow tube electrode that was powered by RF (13.56 MHz) power ranging from 20 to 200 W or the RF voltage amplitude ranging from 100 to 1,000 V. The rotational temperature was measured to be 300 K to 550 K for varying power input (50–130 W). Forest et al. developed RF (at 27 MHz) plasma jets using two types of electrode configurations [24]: (1) one used a pair of ring electrodes wrapped around a quartz capillary tube through which the working gas flows and (2) the other consisted of center pin-type electrode, a surrounding glass capillary (as the gas flow channel), and a ground ring outside the capillary tube. The temperature was measured in the range from 40 to 100 °C depending on the gas flow (3–10 slpm) and the power (5–25 W) applied.

3.2.1.6.5 Microwave-Driven Plasma Jets

The development of microwave plasma jets/torches can be traced back to the 1960s. The recent work by Al-Shamma'a et al. reflected an interesting design for a 2.45-GHz waveguide-based applicator to generate a microwave plasma jet at atmospheric pressure [25]. The system consisted of a 1–6-kW magnetron power supply, a circulator, a water-cooled matched load, and the plasma applicator. Working gas could be Ar, He, or N_2. Although the primary applications of the microwave plasma jet were for high-temperature material processing, the authors also suggested other possible tasks by tailoring the jets. Later, Stonies et al. reported a microwave plasma torch (with length up to 4-mm long) generated by a small coaxial plasma source powered by 2-W microwave (at 2.45 GHz) power and with an argon flow of <70 ml/min [26].

3.2.2 *Biocidal Agents Produced by Nonthermal Plasmas*

Different types of nonequilibrium atmospheric-pressure plasmas (NTAPs) generated by electrical discharges in gases and liquids have been applied for biological and medical applications [27–40]. NTAPs have been shown to be effective in inactivation of wide range of microbes and have been also applied directly to the human body for wound healing [41, 42] and dental operations [43, 44]. Other direct applications of plasma to the human body that do not involve microbial disinfection include cancer treatment [45–49], lithotripsy [50–53], and electrosurgery [54, 55].

While this is a very broad scale of plasma applications, there are some common aspects with regard to basic chemical and physical mechanisms that may be involved in the biological effects induced by plasma. These are thermal, electric field, ultraviolet radiation, direct chemical reactions of neutral reactive species, and interactions of charged particles with living matter [33]. The contribution of these processes in particular biological effect induced by plasma is dependent on the type of discharge plasma, its energy, and the nature and the chemical composition of the surrounding environment. Chemical effects induced by the reactive oxygen species (ROS) are generally accepted to play the dominant role in the inactivation process of NTAPs (e.g., atomic oxygen, metastable oxygen molecules, ozone, and OH radicals) [34]. In addition to the ROS, reactive nitrogen species (RNS) produced by electrical discharge plasma under atmospheric conditions with suitable nitrogen sources such as with air or nitrogen carrier gases can participate in the inactivation process of NTAPs (e.g., nitric oxide, nitrogen dioxide radical) [56]. In the case of plasmas in gas/liquid environments, the presence of water adds more complexity to the system. Reactive species formed by discharges in the gas and at the gas–liquid interface can penetrate or dissolve into the liquid and initiate chemical and biocidal processes in the liquid [57]. When plasma is generated directly in the liquid phase, the physical processes such as high electric field, ultraviolet radiation, and shock waves may significantly contribute to the biocidal activity of the plasma in addition to the chemical effects (driven mainly by OH radicals and H_2O_2) [58–64]. There are also possible synergistic effects of the abovementioned processes and post-discharge chemical reactions in plasma-treated liquids (e.g., peroxynitrous acid) that may contribute to the biological effects induced by NATPs [56, 57, 65]. These effects are described in the following sections.

3.2.2.1 Heat

Thermal effects from the plasma depend strongly on the plasma energy. In low-energy (J/pulsed) corona-type discharges in and over water, small thermal effects are found as is the case of plasma jets, DBDs, or gliding arc discharges developed over the liquid. The development of NTAPs such as nanosecond- or microsecond-pulsed plasma jets with He as the carrier gas provides gas temperatures around 300 ± 50 K or below 34 °C [18, 19, 66] and reduces any potential heating effects during the microbial inactivation process. RF or other high-frequency dielectric barrier discharges reported heating target surface temperature ranged from 40 to 55 °C for direct plasma exposure for minutes [67–69]. These NTAPs of relatively low gas temperatures enabled surface treatment for heat-sensitive materials and implied that heat played a minor or negligible role for microorganism inactivation. However, in high-power kJ/pulse discharges, high-pressure radio-frequency, or microwave discharge-based plasmas, thermal effects can be an important process. Plasma generation in saline solutions for electrosurgery applications involves significant generation of thermal effects in addition to the chemical species formed [54, 55]. Local heating near the electrodes leads to the formation of vapor layers where plasma is generated [70–72]. The radical and ions formed in the plasma zone, coupled with possible thermal effects, lead to reactions with proteins such as collagen and tissue cutting and cauterization. Since the denaturing of tissue protein due to thermal necrosis initiates above a temperature of 43 °C [51], it is possible that local thermal effects as well as the chemical effects may participate in the tissue ablation, and clearly thermal effects can generally increase the rates of the chemical reactions.

3.2.2.2 UV Radiation

The effects of UV radiation on microorganisms have been studied since the late nineteenth century [30, 73]. It has been widely accepted that only light with wavelength below 300 nm was mostly effective on germicidal actions via nucleic acid (DNA/RNA) damage of microorganisms [74, 75]. Mainly the formation of pyrimidine dimers causes inhibition of cell replication and transcription and hence the prevention of bacteria or virus from multiplying. High UV absorbance of DNA was observed around 260 nm and below 230 nm, which explains the popular use of low-pressure mercury lamps (monochromatic UV radiation) and medium-pressure mercury lamps (polychromatic UV radiation) for a variety of UV-based disinfection applications.

The relative importance of ultraviolet radiation in the plasma-induced biological effects is strongly dependent on the type of electrical discharge (including its contact with liquid) and the energy input to the system. Ultraviolet radiation is one of the principal forms of energy, which is dissipated from high-current/high-voltage arc electrical discharges in water [60, 76–78]. For underwater plasmas initiated by exploding fine wires, it was shown that up to 28 % of the energy transferred into the plasma (1.5 kJ/pulse) is converted to UV radiation with a peak radiant power of 200 MW [79]. For the pulsed arc discharge generated in the plate–rod reactor, it was found that the UV light energy has a 3.2 % total energy injected into the reactor [80]. This radiation, especially vacuum ultraviolet (λ=75–185 nm), could cause photolysis resulting in the dissociation of water and oxygen molecules into the primary radicals, hydrogen and oxygen atoms, and hydroxyl radicals (Eqs. 3.2 and 3.3). However, the effect of VUV photolysis will be associated only with a very thin layer of liquid since strong absorption of the light with a wavelength smaller than 180 nm by water:

$$H_2O + h\nu \rightarrow H\bullet + OH\bullet \tag{3.2}$$

$$O_2 + h\nu \rightarrow O + O \tag{3.3}$$

UV light with a wavelength greater than 185 nm penetrates into the bulk of the solution and can inactivate microorganisms in water. UV light intensity generated by the pulsed arc discharge was estimated to be of the order 10^6 W/cm^2, and the UV radiation was determined as the main disinfection process in the discharge [60]. For pulsed streamer discharge in water with pulse energy of the order of 1 J/pulse, it was determined that the UV radiation (190–280 nm) could reach in dependence on the solution conductivity the intensity of the order 0.1–10 mW/cm^2 and that the UV radiation from the discharge contributes about 30 % to the overall inactivation of bacteria *E. coli* [64].

UV light emissions from atmospheric-pressure gas-phase plasmas are significantly lower compared to those produced by underwater plasma, although the amounts and intensities vary with type of reactor, gas composition, and other conditions [81]. UV emissions from air plasma discharges were reported mostly at UVA and UVB wavelengths with radiation intensity below 50 μW/cm^2 [33]; no significant emission was observed in the UVC spectrum. The addition of water vapor leads to characteristic emissions of OH• (at 284 and 309 nm) [82, 83]. The role of vacuum ultraviolet radiation in plasma-induced chemistry was considered in argon/oxygen plasma jets in contact with water [84, 85].

3.2.2.3 Shock Waves

In underwater discharges (especially pulsed electrohydraulic sparks/arcs with the pulse energy of the order kJ/pulse), a rather significant amount of discharged energy is also transformed into the formation of shock waves in water. An important role of shock waves in electrohydraulic treatment of water was demonstrated when used for decontamination of sludge, killing of bacteria, and removal of zebra mussels from the intake pipes of water treatment facilities [86]. Li et al. [62, 63] observed a mechanical rupture of intracellular gaseous vacuoles of cyanobacteria cells exposed to underwater streamer discharge, which was caused presumably by the physical effects of shock waves induced by the discharge. Underwater shock waves are also used in extracorporeal shock wave lithotripsy (ESWL) to noninvasively disintegrate kidney stones and gallstones [50–53]. The treatment involves focusing shock waves generated by an ESWL device (lithotripter) outside the patient's body to disintegrate the stone at a depth in tissue. In electrohydraulic lithotripters, the shock wave is generated by an underwater high-current spark discharge between a pair of electrodes placed at the focus of an ellipsoidal reflector. The targeted stone is located at the second focus of the ellipsoid and several hundreds to thousands of shock wave pulses are delivered to comminute the stone. Lithotripsy has been in clinical use worldwide for treating kidney stones and continues to be the favored method for uncomplicated, upper urinary tract calculi, even with the advent of percutaneous surgical methods. Shock waves are also used in other branches of medicine in orthopedics and traumatology, such as in the treatment of the nonunion of long bones, *plantar fasciitis*, *epicondylitis humeri radialis*, and calcific *tendinitis* of the shoulder, as well as in cardiac shock wave therapy. Shock waves have also been used to treat erectile dysfunction, Peyronie's disease, and chronic pelvic pain syndrome, in oncology, and to mediate cell transfection and bacterial transformation [87, 88]. Using micro-shock waves to facilitate particle delivery in gene therapy, food preservation and cancer treatment have also been studied, and new technologies for drug release and needleless vaccine delivery have been developed [89, 90]. Therefore, shock waves can cause various types of cell damage. Apart from immediate lethal effects such as fragmentation of cells, permeabilization of cell membrane, swollen mitochondria, altered vimentin structure, cytoplasmic cisternae, and nuclear changes have been described [91–96]. In vivo treatments have shown that shock waves induce necroses in tumors and can cause delay of tumor growth, which can be significantly potentiated using cytostatic drugs [97–101]. It has been demonstrated that enhanced cytotoxicity of anticancer drugs results from shock wave-induced transient increase of cell membrane permeability in which cavitation plays an important role [102–106]. Two shock waves, generated with an appropriately selected time delay between pulses (i.e., tandem shock waves), were shown to have considerably higher efficiency than traditional shock wave lithotripters and have been demonstrated to cause significant cytotoxic effects in tumor cells both in vitro and in vivo [49, 107, 108]. However, the optimal shock wave profile and pulse combination have yet to be established.

3.2.2.4 Reactive Chemical Species

The type and quantity of the reactive species formed by discharge plasma depend on the nature and the composition of the ambient gas and, in the case of liquid water contacting the plasma, the properties of the solution. The amount of energy delivered by the discharge, other parameters (e.g., utilization of a separator or a barrier), and the nature of the working electrodes can be important. Common to plasma systems interacting with living matter is the presence of a gas–liquid environment, in which plasma typically occurs in humid air and touches a wet surface (biofilms, cell tissue, skin). Among various chemical species produced by plasma at the gas–liquid environment, OH radical, atomic oxygen, ozone, and hydrogen peroxide are the main reactive oxygen species (ROS) generally accepted to play the dominant role in the biological effects of NTAPs [34]. Contribution of nitrogen-based reactive species (RNS) has to be also considered such as nitric oxide and its derivatives formed with water, including nitrites and peroxynitrites [56, 57, 65]. The ROS/RNS possess high oxidizing properties and are expected to cause oxidative attack of cell membrane as the primary target of cell or tissue injury by plasma in both dry and humid environments. Only relatively recently, the naturally developed mechanisms and strategies in aerobic organisms to modulate or control the generation of ROS and maintain redox (oxidation–reduction reaction) homeostasis intrigued "the forefront of biomedical research in view of the expanding knowledge on the roles" of these species [109, 110]. A review on ROS and RNS, produced by NTAPs, and for potential plasma-mediated applications was given by Graves [111]. For example, the degradation of bacterial outer structures by plasma species (i.e., OH radical) may involve attack on various functional groups and degrade specific cell membrane components such as the lipidic part of lipopolysaccharide or phospholipids and proteins. The membrane is being weakened, and the inner fluid may leak out, which prevents the bacteria from stimulating the anti-stress and repairing processes, or reactive species produced by plasma can pass through the membrane and cause damage inside the cell (e.g., hydrogen peroxide or peroxynitrite). In the following text, the main properties of these species in terms of their biological effects are described.

3.2.2.4.1 Hydroxyl Radical

The hydroxyl radical (OH•) is the major ROS produced in electrical discharge plasmas in water and gas/liquid environments. In gas-phase plasmas with humid gas, absolute ground state OH density was measured in a pulsed He plasma jet to be on the order of 10^{12} cm^{-3} near the nozzle using laser-induced fluorescence (LIF) [112]. Adding water vapor in the ambient air was able to increase the OH concentration up to one order of magnitude [112]. Another measurement using broadband UV absorption showed OH density to be in the range of 10^{13}–10^{14} cm^{-3} in an RF-powered, He–H_2O NTAP [113]. The OH radical has a high oxidizing power ($E^o = 2.85$ V), and it is the strongest oxidant that can exist in an aqueous environment. It reacts with most organic compounds with rates that approach diffusion-controlled limits. Its high reactivity causes a very short lifetime in the order of 100 μs in the gas phase and less than ns in aqueous solution, so that it may directly react with a target species only in its immediate surroundings. In the case of microbial cells, the primary target of the OH radical is the outer cell wall, including the cell membrane. The cell membrane, composed largely of organic compounds such as lipids, proteins, and polysaccharides, is susceptible to the OH radical attack. Lipids are the most vulnerable macromolecules of the cell membrane to oxidation. Lipid reactions with OH radicals proceed mainly via H abstraction from the unsaturated carbon bonds of fatty acids, which in the presence of oxygen cause lipid peroxidation [114, 115]. Similarly, OH radicals can damage membrane proteins by H abstraction from α-carbon of peptide bonds –CO–NH– between chain peptide-linked amino acids. OH radical attack leads then to peroxidation and backbone cleavage of proteins [116].

3.2.2.4.2 Ozone

Ozone is a powerful oxidant and it has the highest standard redox potential among conventional oxidants ($E^o = 2.07$ V). Ozone was detected by most NTAPs in contact with ambient air even without an addition of O_2 in the carrier gas (typ. noble gases). Ellerweg et al. quantified O_3 density in the effluent of a microscale atmospheric-pressure RF plasma jet either in ambient helium or air to be on the order of 10^{14}–10^{15} cm^{-3} using molecular beam mass spectroscopy

[117, 118]. Another quantitative measurement of ozone concentration in a 6-kV, 140-ns pulsed plasma jet via UV absorption spectroscopy showed a density in the order of 10^{15} cm^{-3} for less than 2-W average power [119]. For an RF He–O_2 plasma, 0.1 to 4.0×10^{15} cm^{-3} of O_3 was measured in the afterglow at a power density between 6.1 and 30.5 W/cm^3, a gas temperature of 100 ± 40 °C, and the O_2 concentration between 0.26 % and 2.6 % [120]. For the same plasma and power conditions, Jeong et al. also identified singlet-delta metastable oxygen, $O_2(^1\Delta_g)$, and singlet-sigma metastable oxygen, $O_2(^1\Sigma_g{}^+)$, to be on the order of 10^{16} cm^{-3} and 10^{15} cm^{-3}, respectively, using absolute visible or infrared optical emission spectroscopy [120]. Although ozone is formed by the gas-phase discharge, when a discharge is generated in close proximity to the water surface, ozone can transfer from the gas phase into the liquid and subsequently inactivate microorganisms or oxidize organic compounds in the water. The reactivity of ozone is, however, more selective compared to the reactivity of the OH radical. Ozone is less reactive with saturated and aliphatic hydrocarbons and its reactivity in water is strongly pH dependent. The direct oxidation with molecular ozone is of primary importance under acidic conditions and proceeds with organic compounds with specific functional groups in their molecules such as unsaturated and aromatic hydrocarbons with substituents such as hydroxyl, methyl, or amine groups. In water ozone is unstable and decomposes via a series of chain reactions to produce hydroxyl radicals (Eqs. 3.4 and 3.5):

$$O_3 + H_2O + OH^- \rightarrow OH\bullet + HO_2\bullet + O_2 \tag{3.4}$$

$$O_3 + H_2O_2 + OH^- \rightarrow OH\bullet + HO_2\bullet + O_2 + OH^- \tag{3.5}$$

The addition of hydrogen peroxide accelerates the decomposition of ozone and increases the hydroxyl radical concentration. This process, also referred to as the Peroxone process, occurs very slowly at low pH, but at pH values above 5, it is greatly accelerated. The process is further enhanced by the photochemical generation of hydroxyl radicals in the O_3/H_2O_2/UV process [121, 122].

3.2.2.4.3 Hydrogen Peroxide

Hydrogen peroxide is a very important species, which increases the collective oxidizing power of the plasma, especially in the case of underwater plasmas, in which hydrogen peroxide is the most abundant long-lived plasma–chemical product [123]. The antimicrobial properties of H_2O_2 are well recognized and a variety of applications have been developed. H_2O_2 is able to react in the gas phase, at the liquid surface, and in the bulk target solution with solutes since it is completely soluble in water and can readily transfer across cell membranes. Its action may take place both in direct exposure of the target to the discharge and in post-discharge conditions. The principal mechanism of the H_2O_2 cytotoxicity involves penetration into cells and the generation of hydroxyl radicals through interactions of H_2O_2 with intracellular transition metals (Cu^+ and/or Fe^{2+}) by the Fenton's reaction [124]:

$$Fe^{2+}/Cu^{+} + H_2O_2 \rightarrow Fe^{3+}/Cu^{2+} + OH\bullet + OH^{-} \tag{3.6}$$

In a bacterial cell, this process might be initiated with the reduction of ions Fe^{3+} to Fe^{2+} liberated from bacterial iron storage proteins, ferritin or bacterioferritin, by superoxide radical $\cdot O_2^{-}$. The released Fe^{2+} ions can then react with H_2O_2 outside the proteins or close to their surface to yield hydroxyl radicals that might cause oxidative damage in bacterial cells and stimulate, e.g., double-strand breaks of DNA [125]. Such mechanisms, however, require relatively large concentrations of hydrogen peroxide to penetrate into the cell since bacterial cells possess an adaptive response to low concentrations of H_2O_2, and they may protect themselves against its oxidant toxicity with defense proteins and enzymes, and they even produce H_2O_2[126–128]. Nevertheless, the protective mechanisms of bacterial cells against hydrogen peroxide might be strongly reduced under the influence of high electric field due to polarization and a subsequent increase of the cell membrane permeability [61, 129].

3.2.2.4.4 Peroxynitrite

The main RNS to be considered are nitric oxide NO, the primary species formed by an electrical discharge in air, and its derivatives formed with water, including nitrites NO_2^{-} and peroxynitrites $O{=}NOO^{-}$. Stoffels et al. investigated an RF $He–N_2–O_2$ plasma needle using mass spectrometry and found NO to be the dominant reaction product: the NO density reached up to 20 % of the total atmospheric-pressure density or on the order of 10^{18} cm^{-3} [130]. Interestingly, a recent study on air chemistry of NTAPs showed that there were two "regimes" in generation of the plasma reactive species for an AC-powered surface micro-discharge: the low-power, ozone-dominated mode and the high-power, NO_x-dominated mode [131]. That is, ozone production was dominant immediately after plasma ignition and for low-power density plasmas where the gas temperature is relatively low. With the increase of power density and for longer treatment times, production of NO and NO_2 started increasing and at the same time quenched ozone-producing reactions [131, 132], leading to the nitrogen oxide-dominated mode.

Peroxynitrite is produced by the reaction of nitric oxide and superoxide anion radicals (Eq. 3.7), by the reaction of the nitrite anion with hydrogen peroxide (Eq. 3.8) or by the reaction of NO_2 with an OH radical (Eq. 3.9) [133–137]:

$$O_2^{\bullet -} + NO\bullet \rightarrow O = N\text{-}OO^{-} \tag{3.7}$$

$$NO_2^{-} + H_2O_2 + H^{+} \rightarrow O = N - OO^{-} + H_3O^{+} \tag{3.8}$$

$$OH \bullet + NO_2 \bullet \rightarrow O = N - OO^- + H^+ \quad (3.9)$$

Among these three pathways, the second utilizing hydrogen peroxide (Eq. 3.8) is of particular importance in gas–liquid plasmas since a significant amount of NO_2^- and H_2O_2 is produced by discharges in contact with water or in humid atmospheres. Recently, the rate of formation of peroxynitrite in plasma-treated water was quantitatively determined through kinetic analysis of the post-discharge reaction between H_2O_2 and nitrite ions in air plasma-treated water [57] (see further text in Sect. 3.2.2.6).

The oxidant reactivity of peroxynitrite ($E^o = 2.05$ V) is highly pH dependent and both peroxynitrite anion ($ONOO^-$) and its protonated form peroxynitrous acid (ONOOH) can participate in direct one- and two-electron oxidation reactions. ONOOH oxidizes organic molecules directly or through H^+ or CO_2-catalyzed homolysis yielding nitrogen dioxide radical ($\cdot NO_2$), hydroxyl radical (OH•), or carbonate anion radical ($CO_3^{\cdot -}$). Only a few chemical groups directly react with peroxynitrite, which favors selective reactions with key moieties in proteins, such as thiols, iron/sulfur centers, and zinc fingers [134]. The half-life of peroxynitrite is short (~10–20 ms), but sufficient to cross cell membranes and diffuse quite far on a cellular scale (~ one to two cell diameters) [138] and allow significant interactions with most critical biomolecules [139]. This is a significant difference with reactivity and biocidal effects of hydroxyl radical which is so reactive that it can only diffuse about the diameter of a typical protein [134, 140, 141]. In contrast, peroxynitrite reacts slowly enough to react more selectively throughout the cell. Thus, peroxynitrite can have more subtle and specific actions on cells [137]. Koppenol [142] and Pryor and Squadrito [143] demonstrated that peroxynitrite was a key agent in oxidative stress and is involved in various diseases (e.g., HIV, Alzheimer's disease, arteriosclerosis, gut inflammation) [144–146].

3.2.2.5 Secondary Biocidal Agents in Plasma-Treated Liquids

In addition to the primary biological effects induced directly by plasma via chemical and physical processes described in previous sections, secondary biological effects can be induced in the plasma-treated liquid through the post-discharge reactions of chemical species produced by plasma in the liquid either directly or transferred from the gas-phase discharge plasma via a gas–liquid interface (e.g., hydrogen peroxide, ozone, nitrates, nitrites, peroxynitrites, etc.). Many of these chemical species, especially produced by air plasmas, are not stable in the liquid, and subsequent reactions can take place such as catalytic reactions (mainly of H_2O_2 with metal impurities sputtered from electrodes, e.g., Fe, Cu, W, Pt (Eq. 3.6)) [124, 147–149] or pH-dependent reactions (e.g., reaction between H_2O_2 and nitrites under acidic conditions (Eq. 3.8)) [57, 150] or decomposition of ozone under alkaline conditions (Eqs. 3.4 and 3.5) [121, 122]). Products of these reactions are giving rise to new transient species as OH•, $O_2^{\bullet -}$, NO•, and NO_2• radicals, and peroxynitrite, which have highly cytotoxic properties and cause prolonged antibacterial activity of plasma-treated solutions even several days after the solution's exposure to the discharge [151–159]. In addition to post-discharge chemical reactions of reactive species from plasma in liquid, the UV radiation from the plasma can induce photochemical reactions in water with hydrogen peroxide, dissolved ozone, nitrates, or nitrites to produce secondary OH•, O, NO•, and NO_2• radicals.

3.2.2.6 Post-discharge Chemical Processes in Plasma-Treated Liquids

Prolonged antibacterial activity of plasma-treated aqueous solutions was recently reported for electrical discharges generated directly in water and at gas–liquid environments [151–159]. This extended biological effect phenomenon, hereafter referred to as *plasma-activated water (PAW)* [151–154], can be generally defined as the chemical reactions that initiate or continue in plasma-treated water after the plasma discharge is switched off and in the absence of any external energy source, and they probably involve the presence of long-lived reactive species transferred from the discharge into the liquid. For solutions that were treated by air–liquid-phase plasmas, the antimicrobial properties of PAW were attributed to the synergetic effect of acidic conditions, nitrites, and peroxides through the cytotoxic activity of secondary reactive chemical species NO•, NO_2•, OH•, and ONOOH and possibly also contribution of ozone transferred from plasma into the liquid [57, 154, 156, 157, 160].

Nitrites and nitrates are formed in plasma-treated water through the dissolution of nitrogen oxides formed in air plasma by gas-phase reactions of dissociated N_2 and O_2. Along with the formation of NO_2^- and NO_3^- in the plasma-treated water, the dissolution of NO_x in water produces H^+ ions in water, which can be described by overall reactions (3.10) and (3.11) [161, 162]:

$$NO_{2(aq)} + NO_{2(aq)} + H_2O_{(l)} \rightarrow NO_2^- + NO_3^- + 2\,H^+ \tag{3.10}$$

$$NO_{(aq)} + NO_{2(aq)} + H_2O_{(l)} \rightarrow 2\,NO_2^- + 2\,H^+ \tag{3.11}$$

Nitrites are not stable under acidic conditions (pH <3.5). Nitrous acid HNO_2, which is in acid-base equilibrium ($pK_a = 3.3$) with nitrites (3.12), decomposes under acidic conditions into nitric oxide NO• and nitrogen dioxide NO_2• via reaction (3.13) or forms a nitrosonium ion via HNO_2 protonation via reaction (3.14). Nitrogen dioxide further undergoes hydrolysis in aqueous medium to yield nitrite ion as the final product via reaction (3.15) [161–164]. NO• and NO_2• may also react with dissolved oxygen to produce nitrite and nitrate ions according to the overall reactions (3.16) and (3.17), respectively [65, 156, 164]:

$$NO_2^- + H^+ \leftrightarrow HNO_2 \tag{3.12}$$

$$2\,HNO_2 \rightarrow NO\bullet + NO_2\bullet + H_2O \tag{3.13}$$

$$HNO_2 + H^+ \rightarrow H_2NO_2^+ \rightarrow NO^+ + H_2O \tag{3.14}$$

$$2\,NO_2\bullet + H_2O \rightarrow NO_3^- + NO_2^- + 2\,H^+ \tag{3.15}$$

$$4\,NO\bullet + O_2 + 2\,H_2O \rightarrow 4\,NO_2^- + 4\,H^+ \tag{3.16}$$

$$4\,NO_2\bullet + O_2 + 2\,H_2O \rightarrow 4\,NO_3^- + 4\,H^+ \tag{3.17}$$

Formed nitrogen radicals NO• and NO_2• possess strong cell toxic properties and are more likely to be the main causes of the cytotoxic effects of nitrites under acidic conditions (also known as *acidified nitrites*) [56, 157]. In addition, peroxynitrite is formed through the reaction of nitrites with hydrogen peroxide under acidic conditions (Eq. 3.8). Peroxynitrite can react either directly or, more often, indirectly through reactions of secondary OH• and NO_2• radicals, which are formed by the H^+ -catalyzed decomposition of peroxynitrite by reaction (3.18) at pH <6.8 [133–137]:

$$O{=}NOOH \leftrightarrow OH\bullet + NO_2\bullet \tag{3.18}$$

Post-discharge formation of peroxynitrite by the reaction between H_2O_2 and NO_2 can last in plasma-treated water for many hours, and peroxynitrite along with acidified nitrites was shown to be the key source of the prolonged antibacterial properties of plasma-treated water caused by the cytotoxic activity of secondary reactive chemical species NO•, NO_2•, and OH• [57]:

$$\begin{aligned} O_2^{\bullet-} + NO\bullet &\leftrightarrow O{=}NOO^- \overset{\pm H^+}{\leftrightarrow} O{=}NOOH \\ &\leftrightarrow \bullet OH + NO_2\bullet \end{aligned} \tag{3.19}$$

However, since the peroxysystem is reversible, intracellular cytotoxicity of the peroxynitrite system might be driven not only by its decomposition into OH• and NO_2• radicals (Eq. 3.18) but also through superoxide and nitric oxide radicals (Eq. 3.7), which might be formed through reversible dissociation of peroxynitrite, when pH becomes more alkaline (e.g., inside the cell) and thus more favorable for the reverse reaction of peroxynitrite (Eq. 3.19) [137].

3.2.2.7 Photochemical Processes in Plasma-Treated Liquids

It was shown in Sect. 3.2.2.2 that UV radiation from underwater plasma can significantly contribute to the inactivation of microorganisms in water. In addition, the UV radiation from the plasma can induce photochemical reactions in water with hydrogen peroxide, dissolved ozone, nitrites, or nitrates to produce secondary OH•, O, NO•, and NO_2• radicals.

The photolysis of ozone in the 200–280-nm region involves the light-induced homolysis of ozone and the subsequent production of OH radicals by the reaction of excited oxygen ($O(^1D)$) with water (Eqs. 3.20 and 3.21) [165]:

$$O_3 + h\nu \rightarrow O_2 + O(^1D) \quad (3.20)$$

$$O(^1D) + H_2O \rightarrow 2\,OH\bullet \quad (3.21)$$

The primary process of H_2O_2 photolysis in the 200–300-nm region is the dissociation of H_2O_2 to hydroxyl radicals with a quantum yield of two OH• radicals formed per quantum of radiation absorbed [166–168]:

$$H_2O_2 + h\nu \rightarrow 2\,OH\bullet \quad (3.22)$$

The OH• radicals thus formed enter a radical chain mechanism, in which the propagation cycle gives a high quantum yield of the photolysis of H_2O_2:

$$OH\bullet + H_2O_2 \rightarrow H_2O + HO_2\bullet \quad (3.23)$$

$$HO_2\bullet + H_2O_2 \rightarrow O_2 + H_2O + OH\bullet \quad (3.24)$$

This process has been demonstrated to occur in underwater plasma, especially in solutions of more conductive electrolytes. As the solution conductivity increased, the yield of H_2O_2 produced by the discharge decreased due to increasing photolysis of H_2O_2 accounting for up to 14 % of the total decomposition rate of H_2O_2 [64].

Nitrites and nitrates can also act as indirect photosensitizers to produce secondary oxidants such as the oxide atom and superoxide and hydroxyl radicals via the following mechanism [168]:

$$NO_2^- + h\nu \rightarrow NO\bullet + O\bullet^- \quad (3.25)$$

$$NO_3^- + h\nu \rightarrow NO_2\bullet + O\bullet^- \quad (3.26)$$

$$NO_3^- + h\nu \rightarrow NO_2^- + O \quad (3.27)$$

$$O + H_2O \rightarrow 2\,OH\bullet \quad (3.28)$$

$$O\bullet^- + H_2O \rightarrow OH\bullet + OH^- \quad (3.29)$$

$$2\,NO_2\bullet + H_2O_2 \rightarrow NO_2^- + NO_3^- + 2\,H^+ \quad (3.30)$$

UV(A) photolysis of nitrites was reported as the most likely reason for observed enhanced antimicrobial effect achieved in aqueous suspension by the synergistic interaction between UV(A) and ambient-condition air plasma treatment of *E. coli* [160].

3.3 Pulsed Electromagnetic Radiation

Andrea Doria, Gian Piero Gallerano, Alfonsina Ramundo-Orlando, Stefania Romeo, Maria Rosaria Scarfì, and Olga Zeni

3.3.1 Introduction

An intriguing issue in the study of the interactions between electromagnetic fields and biological systems is to understand whether cellular systems are able to demodulate rapidly oscillating electric fields, like wideband pulses with durations ranging from picoseconds to microseconds. Direct application of short and ultra-short, high-voltage electric pulses on cells and tissue, by means of electrodes, has been shown to elicit a variety of biological effects with cell membranes as primary site of interaction, while gaps in knowledge exist regarding how radiated electric fields with frequency components from the MHz to the THz interact with biological structures.

The term "pulsed electromagnetic radiation" will be used here to refer to high-power, radiated electromagnetic pulses, with central carrier frequency f_c in the range from several GHz to about 1 THz, pulse width (τ) from subnanoseconds to microseconds, and electric field peak amplitude E_p of tens of kV/m. In terms of bandwidth, these pulses can be narrowband at a specific frequency, as long as the inverse of pulse duration is smaller than the carrier frequency (τ f_c >> 1), or wideband when the pulse duration is so short that a burst of radiation is generated, with a bandwidth extending from virtually DC to radiofrequencies and even to the THz region.

We will also use definitions of the gigahertz and terahertz regions that extend each approximately two decades in frequency, namely, 0.1–10 GHz and 0.1–10 THz. In describing the available radiation sources, this should allow a better understanding of the effort in the extension of microwave electronics toward high frequencies on one side and the development of photonic devices from the optical region toward low frequencies on the other side.

Limited data on the effects of exposures to pulsed electromagnetic radiation on biological membranes are available in the literature, and they can be mainly extracted from papers dealing with high-power ultrawideband electromagnetic pulses (UWB-EMP) and THz radiation pulses.

UWB-EMP are pulsed electromagnetic radiations characterized by durations in the range of nanoseconds, rise and fall time of picoseconds, electric field amplitudes on the order of units to tens of kV/m, and very large bandwidth, ranging from DC to microwaves. High-power UWB-EMP are emitted by systems used for both military and civilian applications, such as for intentional electromagnetic interference, fast boat and car neutralization, and improvised explosive device disruption. Most of the literature dealing with the principles of penetration and propagation of such signals into biological matters, and with their biological effects, mainly addresses health issue aspects, with particular attention to the possible hazards to personnel involved in the use of such systems and to the possible exposure of general population [169]. These occupational safety issues have stimulated scientific curiosity toward the understanding of the basic interactions between such short-pulsed electromagnetic signals and living cells and tissues, along with the development of possible technological applications in medicine and biology.

THz waves are electromagnetic radiation in the frequency range from 100 GHz to 10 THz, occupying the region of the electromagnetic spectrum lying between the microwaves and the infrared. The applications of THz radiation in biology, biomedicine, and environmental studies have exponentially increased over the last decades due to the unprecedented development of many new types of THz sources, detectors, and components. Current research is mainly addressed to the understanding of basic interactions with biological systems, the study of induced biological effects, and the development of THz-based diagnostic technologies [170, 171]. The THz sources employed by different research groups worldwide very often show quite different characteristics, in terms of the radiation spectral content and its temporal structure, continuous wave or pulsed emissions, which may be crucial in eliciting specific responses from biological systems.

We provide, in this subchapter, a brief overview on the different sources of short radiated pulses together with the main literature results on bioeffects at membrane level, in order to highlight possible common features and differences when compared to the effects of short and ultrashort high-voltage electric pulses. As far as the effects of THz pulses are concerned, the main results have been obtained on artificial membranes by ENEA and CNR-IFT group and are reported in the last section.

3.3.2 Principles of Generation

In this section we will focus on the description of THz sources taking into account that the principles of generation for UWB-EMP have been covered in Chap. 2.

Although THz radiation was first observed about hundred years ago, the corresponding portion of the electromagnetic spectrum has been for long time considered a rather poorly explored region at the boundary between the microwaves and the infrared. This situation has changed during the past 20 years with the rapid development of coherent THz sources, such as solid-state oscillators, quantum cascade lasers, optically pumped solid-state devices, and novel free electron devices, which have in turn stimulated a wide variety of applications from material science to telecommunications and from biology to biomedicine.

THz sources include pulsed microwave generators, like klystrons, traveling-wave tubes, gyrotrons, and free electron lasers in the range from several GHz to about 10 THz. The spectral and temporal characteristics of the radiation pulses will be discussed, with an emphasis on high-peak power devices and on the peak electric field that can be reached by focusing the radiation on a small spot.

Regarding the pulse duration, we will limit ourselves to describing sources that can produce short pulses of radiation, typically from 10 ps to 10 ns. As to the output power, we will consider high-peak power radiated pulses that, when focused to a spot of about 1 cm^2, can generate a peak electric field in the range 10 kV/cm to 1 MV/cm.

We will not cover solid-state oscillators and amplifiers utilizing Gunn, IMPATT, and TUNNET diodes, which have been developed by several research groups [162], since they are essentially continuous wave (CW) devices that can provide an output power of about 100 mW at 100 GHz. The output power falls off as $1/f^2$ and then as $1/f^3$ as the frequency increases. Similarly we will omit quantum cascade lasers (QCLs), which lie at the high-frequency boundary of the THz region [173, 174].

A brief review of far-infrared gas laser and laser-driven THz emitters will be presented, which are the most widely used sources of THz radiation. We will also briefly discuss the physical principles of the generation of THz radiation from free electrons and the development of tabletop free electron laser (FEL) sources.

3.3.2.1 Far-Infrared Gas Lasers

Far-infrared (FIR) gas lasers are the oldest coherent sources developed in the THz region, their early development dating back to the 1960s [175]. They are optically pumped lasers, which use a CO_2 laser to excite the roto-vibrational levels of gas molecules at pressures in the mbar range. The most widely used gas is methanol, which provides a powerful (typically 100 mW) emission line at 118 μm. Gas lasers are line tunable in the range 0.3–5 THz (1,000–60 μm), although with limited power, and are commercially available. The technology of FIR lasers has seen modest development in recent years, as it has been the case of most gas lasers. However they still are ideal sources for specific applications in which continuous tunability is not a must. They can provide short pulses, if the CO_2 laser pump is operated in the so-called Q-switched mode. Subnanosecond pulsed can be achieved in this way [175].

3.3.2.2 Laser-Driven THz Emitters

Today's most widely used sources of pulsed THz radiation are laser-driven THz emitters based on frequency down-conversion from the optical region. Two main techniques have been developed to produce THz radiation. The first one is based on a short-pulse (femtosecond) Ti: sapphire laser [176] which illuminates the gap between closely spaced electrodes on a photoconductor (e.g., silicon on sapphire or GaAs) generating carriers, which are then accelerated by an applied bias field (100 V). The resulting current transient, which is generally coupled to an RF antenna through a strip line, radiates in a wideband at THz frequencies corresponding to the Fourier transform of the laser pulse time profile. The upper limit in frequency for these devices is given by the carrier recombination time in the semiconductor and by the bandwidth of the strip line.

A similar terahertz spectrum can be obtained by applying a sub-picosecond laser pulse to a crystal with a large second-order susceptibility like ZnTe [177]. Due to the nonlinear response of the crystal, photomixing occurs, producing a time-varying polarization, which in turn gives rise to THz emission. In this case higher frequencies can be reached due to the fast response of the crystal and to the absence of any strip line or conductor. Both techniques and the related electro-optical detection are at the basis of most THz imaging systems [178, 179].

Laser-driven solid-state emitters also include CW photomixers, in which offset-frequency-locked CW lasers are focused onto a photoconductor under bias. The laser-induced photocurrent is modulated at the laser difference frequency and is coupled to an antenna, which emits THz radiation [180]. In this case a narrowband continuous wave emission is obtained, which can be tuned over a fairly wide range by shifting the optical frequency of one of the two drive lasers.

The typical frequency range covered by laser-driven solid-state emitters is 0.2–2 THz or higher depending on the laser pulse parameters. Average power levels range from nanowatts to hundred microwatts and pulse energies are typically in the femtojoule to nanojoule range.

3.3.2.3 Free Electron-Based Sources

Historically, the first propagation of electromagnetic radiation was observed in experiments on the fields generated by moving charges. In 1887 R. Hertz demonstrated that the fields generated by spark gap closure could propagate in the free space and induce a current on a coil; this was the first evidence of an electromagnetic wave. Since then, many devices have been designed and constructed for the generation of radiation in the spectral region between 300 MHz and 300 GHz, based on the interaction of free electrons with "structures" of different natures. Free electron-based sources like klystrons, traveling-wave tubes (TWTs), backward wave oscillators (BWOs), and gyrotrons have been extensively studied since the mid of the past century to approach the high-frequency part of the microwave region. Gyrotrons are generally designed to reach high average power and are not described here since they deserve a review on their own [181].

The problem of energy transfer between a beam of relativistic electrons and a co-propagating wave can be easily addressed within the framework of classical electrodynamics [182]. It is known that the rate of energy exchange between flowing charges and a wave is given by the integration of the $\vec{J} \cdot \vec{E}$ product over the volume $V = \Sigma \cdot L$ of the region of space where the interaction occurs:

$$\Delta P = -\frac{1}{2}\int_V \Re(\vec{J}\cdot\vec{E})dV \tag{3.31}$$

Here $\vec{J}$ is the electron current density, which can be expressed as a sum over the distribution of electrons with individual velocity $\vec{v}_i$ and phase φ_i:

$$\vec{J} = 1/V\sum\nolimits_i e\vec{v}_i \exp[i\varphi_i] \tag{3.32}$$

$\vec{E}$ is the electric field of a wave of the form $\vec{E} = \vec{E}(x,y)\exp[i(\omega t - kz)]$ and the interaction volume V is defined by a mode cross-section Σ and an interaction length L along the direction of propagation z. It is clear from (3.31) that energy exchange can only occur between corresponding components of the electron current density and of the wave electric field. In general, the longitudinal component of the current density due to the drift motion of the electrons will not couple to an electromagnetic wave propagating in the free space, since the latter one has zero longitudinal electric field component. However, in a waveguide or in a suitable loaded structure, TM modes can be excited, which have a longitudinal component of the electric field.

Equations (3.31) and (3.32) tell us that the power exchanged between electrons and the radiated field increases significantly when the electrons, or a great number of them, have the same or similar phase φi. This happens if they are "bunched" on the scale of the wavelength of the electromagnetic field ($\lambda = 2\pi/\omega$). In order to bunch the electrons, we need to induce a velocity modulation to the beam and among the possible techniques, the most elementary one involves a structure, realized with metal components, that exploits, again, the interaction between a charged particle and an electromagnetic wave: this is a gridded gap as sketched in Fig. 3.1.

When the gridded gap is coupled to a resonant cavity, we can excite a specific frequency in the microwave spectral region, due to the energy exchange mechanism. The central emitted frequency has a bandwidth related to the quality factor of the resonant cavity.

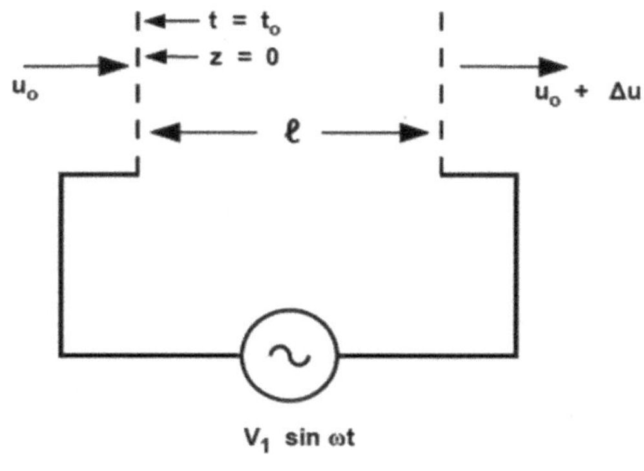

Fig. 3.1 The "gridded gap" circuit: an oscillating voltage is applied between two grids traversed by an electron beam

The simplest device exploiting the velocity modulation in a gridded gap is the klystron. It uses two coupled cavities as illustrated in the schematic layout of Fig. 3.2a.

The electron beam is generated by an electron gun and passes through the grids of the first cavity: the input cavity. In this case an RF signal is coupled into the cavity by means of a coaxial cable; the voltage developed across the cavity imparts the velocity modulation to the beam. After the first cavity, we find a drift space, where the velocity modulation is transformed into a spatial modulation or bunching.

A multi-cavity klystron and a klystron whose cavities are coupled by transmission lines are shown in Fig. 3.2b. A natural development of this scheme is to introduce a large number of cavities. This condition corresponds to have the electron beam that becomes only partially bunched between successive cavities. The result is that the gain per cavity is reduced with respect to the klystron, while, due to the increased number of cavities, the gain per unit length is increased. Moreover the transmission line loading has the effect to lower the values of the Q parameter of the cavities, increasing the bandwidth, without a corresponding power loss. A sketch of a multi-cavity device with a transmission line connecting them is reported in Fig. 3.2c: this device is a traveling-wave tube (TWT) amplifier. A TWT is designed in such a way that the backward-traveling-wave contributions coming from successive resonant cavities have canceling phases in order not to have energy propagating in the backward direction. Moreover, due to the necessity to maintain a phase relation between the electron beam and the electromagnetic wave, the TWT is a so-called slow-wave structure. A slow-wave structure is a periodic composition of elements in a guiding device in which the wave phase velocity ($|\mathbf{v}_p| = |\omega/\mathbf{k}|$ where k is the electromagnetic wave momentum) has a value lower than the vacuum speed of light c.

A periodic slow-wave structure is also used in the backward wave oscillator, which is discussed next. The BWO [183] is a slow-wave device where the electrons spiralize through a corrugated structure in an axial magnetic field interacting with the first spatial harmonic of the backward wave. Indeed, in this region of the dispersion relation, the phase velocity of the wave is positive and the group velocity is negative. BWOs are tabletop devices that can operate in the THz region at moderate power levels (1–100 mW). The technology of BWOs has been mostly developed in Russia and they are now commercially available through companies in the USA and Europe. BWOs operate with an accelerating potential in the range 1–10 kV and axial magnetic field of about 1 T. They can be tuned over tens of GHz by varying the accelerating potential. A number of different BWOs can be implemented in an integrated system to cover altogether a wide-frequency range extending from 30 GHz to 1.2 THz.

As the frequency is increased toward the THz region, klystrons, TWT, and BWO suffer from simple physical scaling problems and increasing metallic wall losses. The small size of the resonant cavities that compose the periodic structure has hindered the TWT from being developed at very short wavelengths, typically below

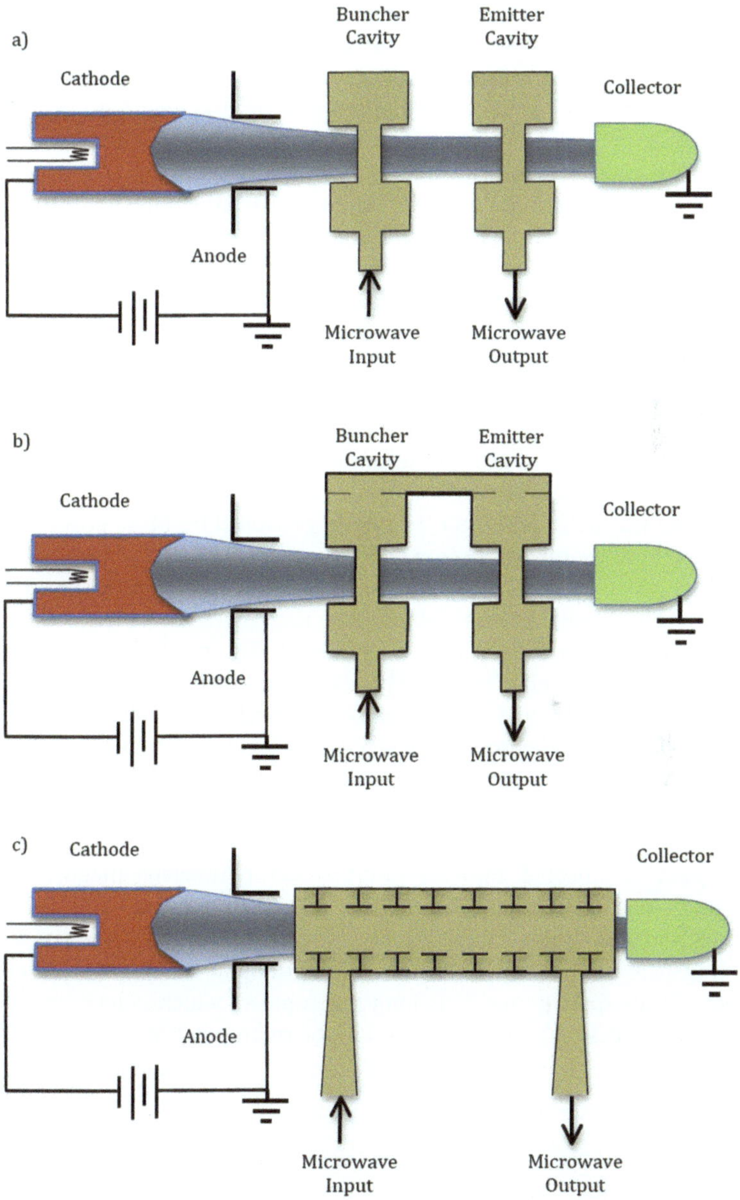

Fig. 3.2 Operating scheme of the Klystron and its derived devices: (**a**) two-cavity klystron amplifier; (**b**) klystron oscillator; (**c**) multi-cavity klystron or traveling-wave tube (TWT)

3 mm, i.e., frequencies higher than 100 GHz. The need for high electron current densities also becomes an issue.

The free electron-based radiation sources exhibit a time structure that is directly related to that of the electron beam generator, because the electromagnetic field generation mechanism is "quasi-instantaneous" and thus does not significantly affect the electron pulse profile. Generally speaking, the electron pulse duration, which can be as long as CW, and the repetition rate are related to the maximum power (voltage times current) available from the modulator and of the pulse-forming line. The electron guns used in commercial klystrons, for instance, have voltages of hundred kilovolts, pulse durations of few microseconds, and kilohertz repetition rates. Different electron guns have been developed in the past like, for instance, the field emission electron gun that has an accelerating voltage higher than 1 MV, currents in the range of 1 kA, and pulse durations of 10–30 ns. These could in principle be used as drivers of any free electron-based radiation source, leading to short-pulse operation (few nanoseconds).

To overcome the necessity of reducing the physical size of the source components as the frequency is increased, different schemes have been developed to let the electrons exchange momentum and allow photon emission. The most frequently used one is the magnetic undulator originally proposed by Motz [184] which was employed by Phillips in the ubitron [185] back in 1960 to generate mm-wave radiation and which led to the realization of the first free electron laser (FEL) in 1977 [186]. Other free electron devices are the Cherenkov FEL, based on the interaction with a dielectric-loaded waveguide [187], and the metal-grating FEL, based on the Smith–Purcell effect [188]. A summary of free electron-based generators of short pulses GHz and THz radiation is reported in Table 3.1.

Two free electron undulator-based sources of THz radiation have been developed at the ENEA Frascati Research Center, covering altogether the spectral range from 90 GHz to 0.7 THz. The ENEA Compact FEL utilizes a 2.3-MeV microtron to provide 13 ps electron bunches with four A peak current [189] (Fig. 3.3). The electron beam is injected into an eight-period permanent magnet undulator ($\lambda_u = 2.5$ cm), and the emitted radiation is stored in a hybrid resonator, which utilizes a WR42 waveguide for transverse confinement of the mode and wire grid electron transparent mirrors (ETMs) for the longitudinal confinement. Micropulses occur with 330-ps spacing in a 4-μs long macropulse, which is repeated at 10 Hz. Peak power in excess of 3 kW is obtained in the micropulse at 130 GHz. When the

Table 3.1 Free electron-based generators of short pulses GHz and THz radiation

Type	Frequency range	Electron beam type/voltage	Pulse duration	Peak power
Klystron	1–10 GHz	Diode or triode e-gun	1–5 μs[a]	5–50 MW
HV TWT	35 GHz	Diode or triode e-gun	50 ns	1 MW
Gyrotron	50–60 GHz	Diode or triode e-gun	50 ns	40–50 MW
HV BWO	0.1–1.4 THz	Diode or triode e-gun	20–100 ns	100 kW
Compact FEL	200–700 GHz	Electrostatic <1 MV	20–100 ns	1 MW
	300–3 THz	RF driven 1 MV–5 MV	10–100 ps	1 MW

[a]Typical pulse duration of commercially available devices

Fig. 3.3 View of the ENEA Compact FEL source

beam is focused to a spot size of about $0.5 \times 1\ \text{cm}^2$, a peak electric field greater than 2 kV/cm is obtained in the micropulse.

The second FEL source, named FEL-CATS (Compact Advanced THz Source), is a small-scale device that occupies an area of $0.5 \times 1 \times 2$ m, comparable to that of a standard optical table (Fig. 3.4). It exploits a high-efficiency generation scheme based on the mechanism of coherent spontaneous emission [190]. FEL-CATS utilizes a 2.5-MeV RF linac to generate the electron beam, which is injected into a linearly polarized magnetic undulator composed of 16 periods, each 2.5-cm long, with a peak magnetic field of 6,000 G. The electrons accelerated by the linac enter a second RF structure, called phase-matching device (PMD), placed between the linac and the undulator, which is controlled in phase and amplitude to correlate the electron distribution in energy as a function of time in the bunch. In this way the contributions to the total radiated field by individual electrons in the bunch are added in phase, leading to a manifold enhancement of the coherent emission in a single pass through the undulator and without the use of any optical cavity. Power levels up to several kilowatts have been measured over a pulse length of about 5 μs. The absence of resonators, and the use of a short length undulator, results in a broadband emission. Tunable operation was obtained in the spectral region between 0.4 and 0.7 THz (800 μm to 400 μm wavelength) with a relative line width of about 10 % FWHM [191].

A variety of biological systems have been studied with the ENEA Compact FEL in the frame of the European project THz-BRIDGE [192]. The peculiar temporal structure of the emitted radiation (Fig. 3.5) allows the investigation of the effects of high-peak power while maintaining a low-average power incident on the sample, typically few mW, thus avoiding heating effects.

Fig. 3.4 View of the ENEA FEL-CATS source

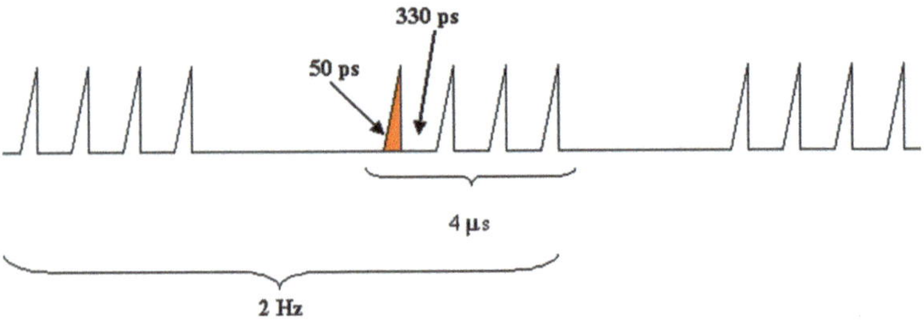

Fig. 3.5 Temporal structure of the ENEA Compact FEL

3.3.3 Biological Effects

Although a large number of experiments have been performed aimed at giving insight into the biological effects of electromagnetic fields [193, 194], the bioeffects of pulsed electromagnetic radiation have not been studied in as much detail as for continuous wave (CW) radiation. The available studies mainly address the health issue, giving emphasis on effects on carcinogenesis, teratogenesis, genotoxicity, behavior, and nervous and cardiovascular systems. The majority of studies have been carried out on laboratory animals, but cultured cells or yeasts have also been employed. Overall results do not seem to indicate any consistent effect, but are far to be conclusive since they are carried out under very different conditions in terms of biological models, electromagnetic conditions, and assessed parameters [169, 195].

A limited number of papers have been identified, which address the interactions between pulsed electromagnetic radiation and biological membranes. They cover the following radiation parameters ranges: electric field amplitude from 50 kV/m up to 15 MV/m, pulse duration from 150 ps up to 500 ns, pulse rise time from 70 ps up to 20 ns (not always indicated), and pulse repetition rate from 0.5 Hz up to 1 kHz. Such studies are presented in the following.

In five in vitro studies, permeabilization effect of plasma membrane was addressed either as evaluation of structural rearrangements or as uptake/release of drugs and small molecules.

In particular, atomic force microscopy scanning of Chinese hamster ovary (CHO) cells exposed to pulsed electromagnetic radiation revealed alteration in the structure of plasma membrane, which increased upon increasing either the electric field strength (from 50 to 400 kV/m) or the pulse number (from 10 to 200) [196]. It is worth noting that, in this paper, pulse parameters were not accurately reported.

Kolb and coworkers detected an increased intracellular calcium levels in platelets exposed to 150-ps long pulses at nonlethal electric field amplitudes (100 kV/cm) [197].

Jordan and coworkers employed two cell lines, the non-excitable African green monkey kidney cells (COS-1) and the excitable mouse neuroblastoma cells (N1E-115), to compare the effectiveness in electroporation of three distinct RF electromagnetic field modalities: (i) 25-kHz unipolar RF bursts, (ii) dc square pulses, and (iii) bipolar RF burst (0–1.5 kV/cm) at 13.56 MHz and 20 kHz. In both cell lines, the exposure modality at 25-kHz unipolar RF bursts resulted in killing fewer cells (trypan blue uptake) in comparison to exposure modality at dc square pulses, determining a measurable improvement in the electroporation yield (DAPI uptake). The authors also claimed that exposure to bipolar RF burst (0–1.5 kV/cm) at 13.56 MHz and/or 20 kHz was not able to cause electroporation; however, no data were presented in the manuscript supporting such a claim [198].

Further, the same authors observed an increased uptake of the commonly employed chemotherapy agent, bleomycin, in Jurkat cells. These cells were exposed, in an electroporation cuvette, to trains of 10–100-UWB pulses at 1-Hz pulse repetition rate and 50–200-kV/cm peak electric field strength, generated by a gigahertz transverse electromagnetic cell-like structure. As a matter of fact, 24 h after the exposure, UWB treatments gave rise to a significant increase in bleomycin-induced cell killing, compared to pulse-only or drug-only treatments [199].

Recently, Macaire and coworkers demonstrated an increased permeability in plasma membrane of murine melanoma cells (B16f10) exposed to pulsed electromagnetic radiation. In particular, two trains of 30,000 pulses, each 2.4-ns long, separated by 20-min interval at 100-Hz pulse repetition rate and 930-kV/m electric field amplitude, were able to evoke, in the absence of significant alteration of cell viability, the rapid release in the culture medium of the cytoplasmic adenosine triphosphate (ATP), a highly polar, energy-related, signal molecule [200].

In seven different papers, the blood–brain barrier (BBB) was used as biological model, either ex vivo from exposed rats or as BBB model coculture. One of these papers reported also results in the blood–testis barrier (BTB).

The BBB is a highly specialized structural, transport, and biochemical (enzymatic) barrier, which mainly consists of microvascular endothelial cells and overlying astrocytic foot processes. It regulates the entry of compounds and cells between blood and brain and, thus, has a fundamental role in brain homeostasis. It also forms a route of communication between circulating blood and underlying brain tissues. This barrier results from the selectivity of the tight junctions (TJs) between microvascular endothelial cells that restrict the passage of solutes. At the interface between the blood and the brain, endothelial cells are stitched together by these TJs, which are transmembrane proteins anchored into the endothelial cells by another protein complex. Transmembrane proteins of the TJ include occludin, claudins, and junctional adhesion molecule [201].

The BTB in mammalian testes is constituted almost exclusively by different specialized junctions between the Sertoli cells of the seminiferous tubule near the basement membrane, where they create a unique microenvironment for the development and maturation of meiotic and postmeiotic germ cells [202].

Different methods can be used to assess changes in the permeability of the blood barriers. Histological methods involve the preparation of brain sections after the experiment, which are then stained for the presence of the permeability marker in the brain extravascular space. The permeability marker is either naturally present in the blood or artificially injected during the experiment, and the most commonly employed are horseradish peroxidase (HRP), Evans blue (EB), lanthanum nitrate, and albumin. Physiological methods include comparison of the concentrations of the tracer and a reference molecule in the blood, brain tissue, or both [203].

The measurement of transendothelial electrical resistance (TEER) is also an accurate and sensitive indicator of the integrity of BBB. TEER represents the impedance of the barrier structure; thus a decrease in TEER reflects an increase in the permeability and a loss of barrier function.

Finally, the mechanisms underlying changes on permeability of BBB can also be addressed by studying the localization and the expression levels of the tight junction specialized proteins abovementioned.

Ding and coworkers studied changes in BBB permeability in Sprague–Dawley rat models after exposure to 200 and 400 pulses, 14-ns long, with 200-kV/m electric field amplitude, delivered at 1 Hz. The effect was detected by transmission electron microscopy and immunohistochemistry using lanthanum nitrate and endogenous albumin as vascular tracers, respectively. The increased BBB permeability resulted reversible and associated with altered localization and decreased expression levels of TJ proteins [204, 205]. In a series of subsequent studies, the same authors further characterized the permeabilization of BBB induced by electromagnetic pulses, in the same rat model, by determining (1) the role of protein kinase C signaling pathway [206]; (2) the expression levels of TJ proteins, in different organs of adult rats (i.e., the cerebellar cortex, hippocampus, heart, lung, and testis) [207]; and (3) the involvement of matrix metalloproteinases [208].

Zhou and coworkers demonstrated in an in vitro model, established by coculturing brain microvascular endothelial cells and astroglial cells isolated from rat brain, an increase of BBB permeability after exposure to 100, 200, and 400 pulses, 350-ns long, at 100 kV/m and 400 kV/m and 2-Hz repetition rate. In this case, changes in the permeability were assayed by measuring the TEER, the HRP transmission, and the levels of BBB TJ-related proteins [209].

In a further in vitro BBB model composed by the epithelial/endothelial-like human ECV304 cell line and primary rat cerebral astrocytes, increased BBB permeability was also demonstrated after exposure to pulsed electromagnetic radiation, 500-ns long, at a frequency of 2.856 GHz, with power density of 50 mW/cm^2, 500 Hz repetition rate, for a total treatment duration of 5 min. Also in this case, TEER and HRP transmission were assayed, and authors also demonstrated the involvement of TJ-related proteins [210].

One study reported increased permeability of the BTB in BALB/c mice subjected to 200 electromagnetic pulses, 350-ns long at 200-kV/m and 400-kV/m electric field amplitude, by measuring the expression of TJ-associated proteins [211].

Taken together, when considered over time, the effects on BBB and BTB were found to be reversible and provided certain evidences that plasma membrane structure represents a possible site of interaction of pulsed electromagnetic radiation with mammalian cells.

3.3.4 Experimental Results on THz-Pulsed Electromagnetic Radiation

As far as THz-pulsed electromagnetic radiation is concerned, the only investigations addressing direct effects on biological membranes were carried out on artificial membranes, such as liposomes [212]. Liposomes or lipid vesicles are polymolecular aggregates, formed in dilute aqueous solution on dispersion of individual bilayer-forming amphiphiles or of mixtures of geometrically complementary amphiphiles (cylindrical molecular shape). Liposomes are composed of one or more closed shells (lamellae with a thickness of usually 4–5 nm), entrapping a small volume of the aqueous solution in which the vesicles are formed [213]. Liposomes prepared from biologically relevant amphiphiles (e.g., phosphatidylcholine) are often used as biomimetic model system to study certain aspects of biological membranes, e.g., peptide–membrane interactions and membrane permeability. In particular, enzyme-containing liposome is a useful model to study even in small alterations in lipid bilayer permeability induced by external stimuli (i.e., physical or chemical), because the material exchange (i.e., substrate) between the bulk aqueous phase and the liposome aqueous interior is very low.

Carbonic anhydrase (CA)-containing liposomes (Fig. 3.6) are prepared by using a mixture of phosphatidylcholines, commonly present in natural membranes in

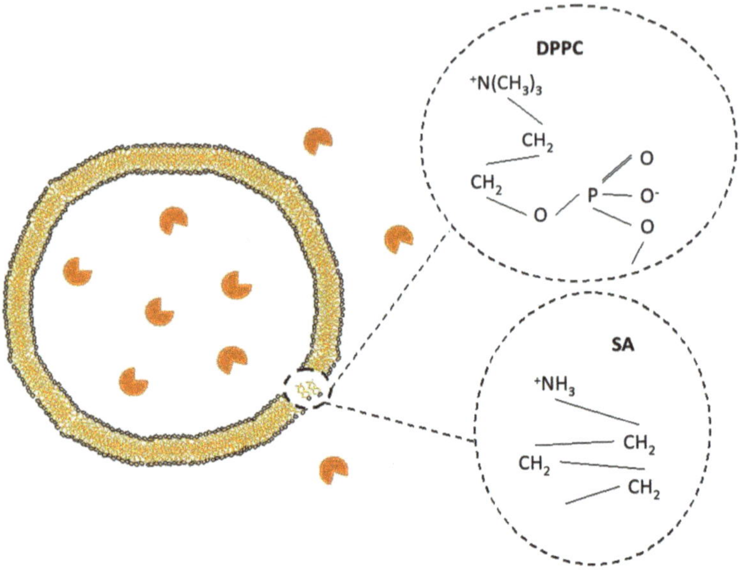

Fig. 3.6 Schematic representation of CA-loaded liposomes (*left*). The chemical structure of dipolar (DPPC) and unipolar (SA) portion of lipids is depicted (*right*). CA enzyme molecule partially located on the external surface of liposome is depicted

combination with cholesterol, to increase the stability of the lipid bilayers, and the charged lipids stearylamine added with the aim of preventing aggregation of the liposomes. Carbonic anhydrase-loaded liposomes have been well characterized, [214] and they are considered to retain typical permeation rates of solute across the lipid bilayer [215]. They also enable a direct comparison with the permeability function of the lipid bilayer in cell membranes.

The influence of 130-GHz radiation, at different peaks and average power levels, was studied on the above model by evaluating induced changes on the diffusion rate of the substrate p-nitrophenyl acetate (p-NPA) across the lipid bilayer. The influx of the p-NPA across the lipid membrane was followed by means of a spectrophotometer measurement of CA enzymatic activity. CA activity measurements were performed in TRIS–saline (pH = 7.55) using p-NPA as substrate [216]. The appearance of reaction product p-nitrophenolate anion (p-NP), at its peak absorbance at $\lambda = 400$ nm, was followed. The enzyme reaction rate, expressed as change in absorbance units at 400 nm/min (ΔA/min), was computed automatically as the slope of a linear fit to the experimental recorded curves.

The exposure system has been designed in a way that made it possible to perform the kinetic measurements of CA in real time during irradiation. For this purpose, the sample cell of the CARY-50 spectrophotometer was modified to directly deliver the 130-GHz radiations to one side of the cuvette containing the liposomes in a final volume of 1.5 ml (Fig. 3.7).

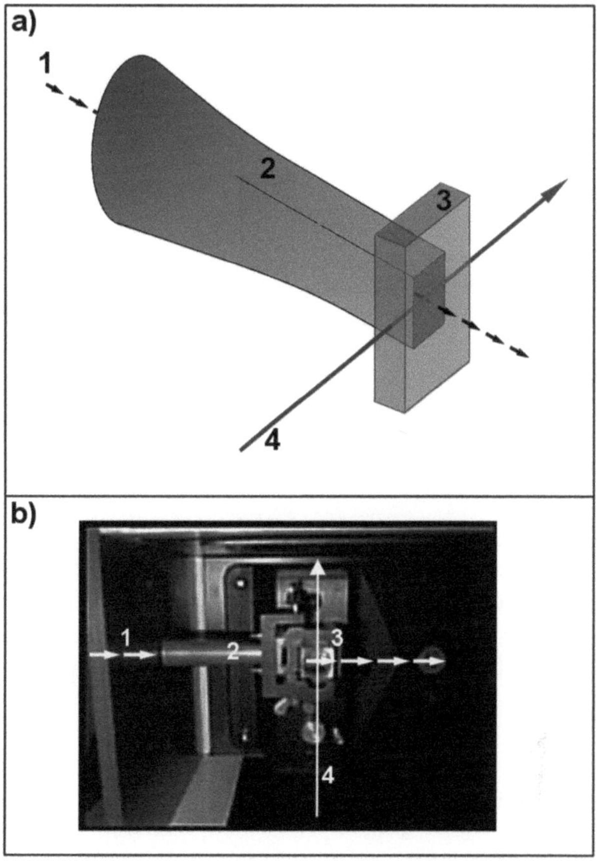

Fig. 3.7 Exposure system: (**a**) an electroformed copper horn is used to match the circular cross section of the THz beam to the rectangular cross section of the cuvette. (**b**) Internal view of the CARY-50 spectrophotometer modified to allow THz irradiation of the sample from the side of the cuvette. (*1*) THz beam, (*2*) electroformed copper horn, (*3*) cuvette containing the liposome preparation (width × height × thickness = 1 × 3 × 0.5 cm), (*4*) spectrophotometer probe beam

The CA-loaded liposomes were unilamellar and homogeneous in size (≈30 nm). They showed a typical basal enzymatic activity in the sham samples due to the presence of CA molecules partially located on the external surface of intact liposome (Fig. 3.8 white column). Overall twelve different liposome preparations were used for irradiation to 130 GHz, at different peaks and average power levels, as well as pulse modulated at low-frequency repetition rate of 5, 7, or 10 Hz. The analyses in both the frequency repetition rate domain and peak electric field domain showed a window effect of irradiation on liposome permeability. A significant enhancement of CA enzymatic activity, related to an increase of bilayer permeability, occurred when the pulse repetition rate was 7 Hz more than 5 Hz or 10 Hz. It was previously indicated that CA-loaded liposomes have a natural resonant frequency; they may respond selectively to 7-Hz frequency [217]. It is worth nothing that liposome bilayer is about 4-nm thick and that the peak electric field applied through the 130-GHz radiation was up to 2,700 V/cm, about two orders of magnitude lower than naturally developed across lipid bilayer. A clear role of the peak electric field in eliciting the observed effect resulted in a window effect around

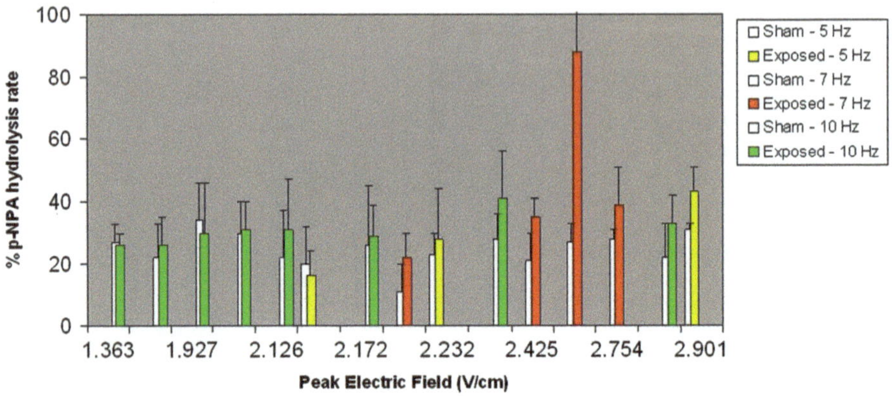

Fig. 3.8 Effects of pulsed 130-GHz radiation on permeability of CA-loaded liposome. The enzymatic hydrolysis rate increases as function of the calculated peak electric field is shown

2,580 V/cm (Fig. 3.8). In this case, by applying an electromagnetic pulse at this frequency, the liposome may store the combined energy of each pulse as some sort of vibration. This could enable it to rearrange phospholipids in the lipid bilayer leading to change of permeability, which would not have been possible from the energy of each pulse alone, but only at this resonant frequency.

3.4 Concluding Remarks

Although high-voltage electric pulses and ultrawideband pulsed electromagnetic radiation are generated and delivered to biological system in different modalities, they present some common specific properties, such as very high amplitude, a complex temporal signature, ultrashort duration, and very low-energy deposition in biological material [218]. As a matter of fact, in all cases cell membrane permeabilization occurred, consistently with the effect highlighted, in much more detail, in the case of high-voltage ultrashort electric pulses, on both cell membrane and intracellular targets. An intriguing question is whether biological systems are able to demodulate rapidly oscillating electromagnetic fields. A schematic representation of this issue is reported in Fig. 3.9, to help visualizing such a concept from a qualitative point of view. A nonlinear response of the biological system may filter out a low-pass envelope of the wave electric field, which resembles a short electric pulse comparable to those applied with electrodes. Short pulses of electromagnetic energy in the frequency range 30–3,000 GHz could then be used to investigate the basic interactions in cellular systems by comparing their effect to that of short electric pulses with same amplitude of the electric field and same pulse duration.

Whether the basic mechanisms of interaction are the same or not is still an open question. A possible unifying view comes out from studies of molecular dynamic

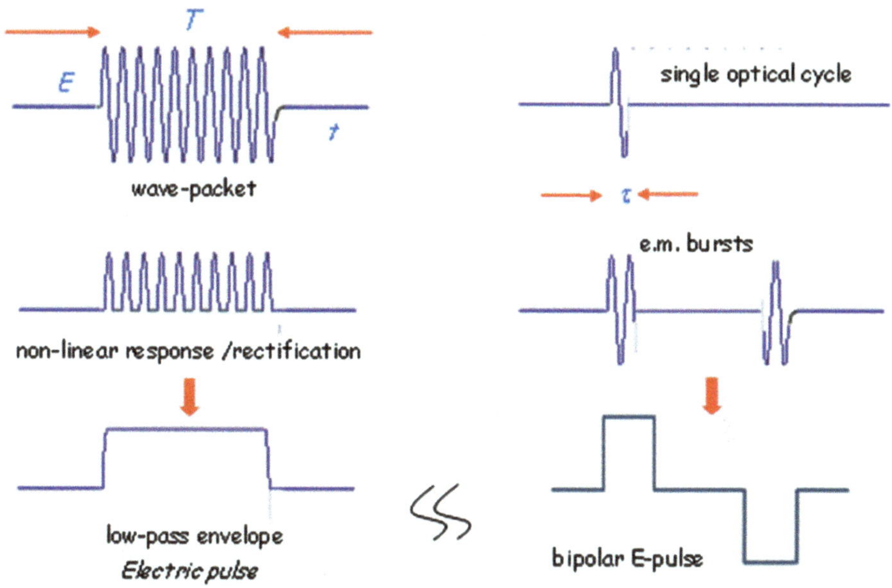

Fig. 3.9 Qualitative comparison of wave-packets and short electric pulses (T, duration of the wave packet; τ, optical cycle; E, amplitude of the wave electric field)

simulation, which identifies the lipid electropore, a nanometer-scale conductive pathway forming across the membrane structure, upon the electric stimulus application, as the primary event leading to plasma membrane permeabilization [219].

Only few studies address direct effects of pulsed electromagnetic radiation on biological membranes, the majority being related to health issues. For this reason, pulse parameters actually experienced by cell structures are often inaccurately described or even missing. Therefore more systematic investigations on this topic are needed both for elucidating basic interaction mechanisms and for the development of new minimal invasive diagnostic and therapeutic procedures. The latter are also fostered by the most recent technological advancements that have led to the availability of new radiation sources.

With this respect, the possibility of using pulsed electromagnetic radiation for reversible opening of blood–brain barrier is particularly attractive in view of the development of new therapeutic strategies for brain tumors and neurodegenerative disorders like Alzheimer's disease, Parkinson's disease, or different types of metabolic disorders.

Electrical discharge plasma in dry and humid gases as well as over and directly inside liquid water can also effectively inactivate a wide range of microorganisms over many orders of magnitudes by chemical and physical means. While our knowledge of many of the chemical pathways and the various physical factors that affect cellular structures are known, there remain many challenges to develop predictive tools and mathematical analysis to quantify these effects and to determine how much of a given inactivation is due to the specific factor. Beyond microbial inactivation, current and future efforts in biomedicine involve plasma interaction with eukaryotic cells and in particular with human cells from stem cells to cancer cells. Plasma interaction with such cells can certainly be destructive, as desired for cancer treatment, but perhaps new advances in plasma technology will enable more constructive alterations of cellular metabolism. These processes were not covered in this chapter; however, the research on biomedical applications of plasmas, also called *plasma medicine*, becomes a rapidly developing field during recent years. A wide range of studies in the field of plasma medicine include the treatment of skin diseases, wound sterilization, treatment of dental cavities, plasma stimulation of blood coagulation, skin cancer, prion, and protein and pyrogenic substance inactivation. Such developments clearly require the combination of new plasma technologies (e.g., new power supplies and methods to deliver and control the plasma) with advanced understanding of the biochemical and physiological responses.

References

1. Razer, Y.P.: Gas Discharge Physics. Springer-Verlag, LLC, New York (1997)
2. Vercammen, K.L.L., Berezin, A.A., Lox, F., Chang, J.-S.: Non-thermal plasma techniques for the reduction of volatile organic compounds in Air streamers: a critical review. J. Adv. Oxid. Technol. **2**, 17 (1997)
3. Akishev, Y.S., Deryugin, A.A., Kochetov, I.V., Napartovich, A.P., Trushkin, N.I.: Dc glow-discharge in air-flow at atmospheric-pressure in connection with waste gases treatment. J. Phys. D. Appl. Phys. **26**, 1630–1637 (1993)
4. Stark, R.H., Schoenbach, K.H.: Direct current glow discharges in atmospheric air. Appl. Phys. Lett. **74**, 3770–3772 (1999)
5. Jiang, C.Q., Mohamed, A.A.H., Stark, R.H., Yuan, J.H., Schoenbach, K.H.: Removal of volatile organic compounds in atmospheric pressure air by means of direct current glow discharges. IEEE Trans. Plasma Sci. **33**, 1416–1425 (2005)
6. Becker, K.H., Schoenbach, K.H., Eden, J.G.: Microplasmas and applications. J. Phys. D. Appl. Phys. **39**, R55–R70 (2006)
7. Kolb, J.F., Mohamed, A.A.H., Price, R.O., Swanson, R.J., Bowman, A., Chiavarini, R.L., Stacey, M., Schoenbach, K.H.: Cold atmospheric pressure air plasma jet for medical applications. Appl. Phys. Lett. **92**, 241501 (2008)
8. Eliasson, B., Kogelschatz, U.: Nonequilibrium volume plasma chemical-processing. IEEE Trans. Plasma Sci. **19**, 1063–1077 (1991)
9. Donohoe, K.G., Wydeven, T.: Plasma polymerization of ethylene in an atmospheric pressure-pulsed discharge. J. Appl. Polym. Sci. **23**, 2591–2601 (1979)
10. Kanazawa, S., Kogoma, M., Moriwaki, T., Okazaki, S.: Stable glow plasma at atmospheric-pressure. J. Phys. D. Appl. Phys. **21**, 838–840 (1988)
11. Laroussi, M.: Sterilization of contaminated matter with an atmospheric pressure plasma. IEEE Trans. Plasma Sci. **24**, 1188–1191 (1996)
12. Laroussi, M.: Nonthermal decontamination of biological media by atmospheric-pressure plasmas: review, analysis, and prospects. IEEE Trans. Plasma Sci. **30**, 1409–1415 (2002)
13. Kelly-Wintenberg, K., Montie, T.C., Brickman, C., Roth, J.R., Carr, A.K., Sorge, K., Wadsworth, L.C., Tsai, P.P.Y.: Room temperature sterilization of surfaces and fabrics with a one atmosphere uniform glow discharge plasma. J. Ind. Microbiol. Biot. **20**, 69–74 (1998)

14. Chan, C.M., Ko, T.M., Hiraoka, H.: Polymer surface modification by plasmas and photons. Surf. Sci. Rep. **24**, 3–54 (1996)
15. Laroussi, M., Akan, T.: Arc-free atmospheric pressure cold plasma jets: a review. Plasma Process. Polym. **4**, 777–788 (2007)
16. Duan, Y.X., Huang, C., Yu, Q.S.: Low-temperature direct current glow discharges at atmospheric pressure. IEEE Trans. Plasma Sci. **33**, 328–329 (2005)
17. Duan, Y.X., Huang, C., Yu, Q.S.: Cold plasma brush generated at atmospheric pressure. Rev. Sci. Instrum. **78**, 015104 (2007)
18. Laroussi, M., Lu, X.: Room-temperature atmospheric pressure plasma plume for biomedical applications. Appl. Phys. Lett. **87**, 113902–1 (2005)
19. Jiang, C.Q., Chen, M.T., Gorur, A., Schaudinn, C., Jaramillo, D.E., Costerton, J.W., Sedghizadeh, P.P., Vernier, P.T., Gundersen, M.A.: Nanosecond pulsed plasma dental probe. Plasma Process. Polym. **6**, 479–483 (2009)
20. Lu, X.P., Jiang, Z.H., Xiong, Q., Tang, Z.Y., Pan, Y.: A single electrode room-temperature plasma jet device for biomedical applications. Appl. Phys. Lett. **92**, 151504 (2008)
21. Kedzierski, J., Engemann, J., Teschke, M., Korzec, D.: Atmospheric pressure plasma jets for 2D and 3D materials processing. Sol. St. Phen. **107**, 119–123 (2005)
22. Cheng, C., Liu, P., Xu, L., Zhang, L.Y., Zhan, R.J., Zhang, W.R.: Development of a new atmospheric pressure cold plasma jet generator and application in sterilization. Chin. Phys. **15**, 1544–1548 (2006)
23. Janca, J., Klima, M., Slavicek, P., Zajickova, L.: HF plasma pencil – new source for plasma surface processing. Surf. Coat. Tech. **116**, 547–551 (1999)
24. Foest, R., Kindel, E., Ohl, A., Stieber, M., Weltmann, K.D.: Non-thermal atmospheric pressure discharges for surface modification. Plasma Phys. Contr. F. **47**, B525–B536 (2005)
25. Al-Shamma's, A.I., Wylie, S.R., Lucas, J., Pau, C.F.: Design and construction of a 2.45 GHz waveguide-based microwave plasma jet at atmospheric pressure for material processing. J. Phys. D. Appl. Phys. **34**, 2734–2741 (2001)
26. Stonies, R., Schermer, S., Voges, E., Broekaert, J.A.C.: A new small microwave plasma torch. Plasma Sources Sci. Technol. **13**, 604–611 (2004)
27. Laroussi, M.: Biological decontamination by nonthermal plasma. IEEE Trans. Plasma Sci. **28**, 184–188 (2000)
28. Sunka, P.: Pulse electrical discharges in water and their applications. Phys. Plasmas **8**, 2587–2594 (2001)
29. Montie, T.C., Kelly-Wintenberg, K., Roth, J.R.: An overview of research using the one atmosphere uniform glow discharge plasma (OAUGDP) for sterilization of surfaces and materials. IEEE Trans. Plasma Sci. **28**, 41–50 (2000)
30. Moisan, M., Barbeau, J., Crevier, M.C., Pelletier, J., Philip, N., Saoudi, B.: Plasma sterilization. Methods mechanisms. Pure Appl. Chem. **74**, 349–358 (2002)
31. Moisan, M., Barbeau, J., Moreau, S., Pelletier, J., Tabrizian, M., Yahia, L.H.: Low-temperature sterilization using gas plasmas: a review of the experiments and an analysis of the inactivation mechanisms. Int. J. Pharm. **226**, 1–21 (2001)
32. Odic, E., Goldman, A., Goldman, M., Delaveau, S., Le Hegarat, F.: Plasma sterilization technologies and processes. High Temp. Mater Process. **6**, 385–396 (2002)
33. Laroussi, M., Leipold, F.: Evaluation of the roles of reactive species, heat, and UV radiation in the inactivation of bacterial cells by air plasmas at atmospheric pressure. Int. J. Mass Spec. **233**, 81–86 (2004)
34. Gaunt, L.F., Beggs, C.B., Georghiou, G.E.: Bactericidal action of the reactive species produced by gas-discharge nonthermal plasma at atmospheric pressure: a review. IEEE Trans. Plasma Sci. **34**, 1257–1269 (2006)
35. Fridman, G., Friedman, G., Gutsol, A., Shekhter, A.B., Vasilets, V.N., Fridman, A.: Applied plasma medicine. Plasma Process. Polym. **5**, 503–533 (2008)
36. Stoffels, E., Sakiyama, Y., Graves, D.B.: Cold atmospheric plasma: charged species and their interactions with cells and tissues. IEEE Trans. Plasma Sci. **36**, 1441–1457 (2008)

37. Kong, M.G., Kroesen, G., Morfill, G., Nosenko, T., Shimizu, T., van Dijk, J., Zimmermann, J. L.: Plasma medicine: an introductory review. New J. Phys. **11**, 115012 (2009)
38. Weltmann, K.D., Kindel, E., von Woedtke, T., Hahnel, M., Stieber, M., Brandenburg, R.: Atmospheric-pressure plasma sources: prospective tools for plasma medicine. Pure Appl. Chem. **82**, 1223–1237 (2010)
39. Wende, K., Landsberg, K., Lindequist, U., Weltmann, K.D., von Woedtke, T.: Distinctive activity of a nonthermal atmospheric-pressure plasma jet on eukaryotic and prokaryotic cells in a cocultivation approach of keratinocytes and microorganisms. IEEE Trans. Plasma Sci. **38**, 2479–2485 (2010)
40. Weltmann, K.D., von Woedtke, T.: Basic requirements for plasma sources in medicine. Eur. Phys. J. Appl. Phys. **55**, 13807 (2011)
41. Daeschlein, G., von Woedtke, T., Kindel, E., Brandenburg, R., Weltmann, K.D., Junger, M.: Antibacterial activity of an atmospheric pressure plasma jet against relevant wound pathogens in vitro on a simulated wound environment. Plasma Process. Polym. **7**, 224–230 (2010)
42. Lloyd, G., Friedman, G., Jafri, S., Schultz, G., Friedman, A., Harding, K.: Gas Plasma: medical uses and developments in wound care. Plasma Process. Polym. **7**, 194–211 (2010)
43. Sladek, R.E.J., Stoffels, E., Walraven, R., Tielbeek, P.J.A., Koolhoven, R.A.: Plasma treatment of dental cavities: a feasibility study. IEEE Trans. Plasma Sci. **32**, 1540–1543 (2004)
44. Villeger, S., Ricard, A., Sixou, M.: Sterilization of dental bacteria in a N_2-O_2 microwaves post-discharge, at low pressure: influence of temperature. Eur. Phys. J. Appl. Phys. **26**, 203–208 (2004)
45. Sunka, P., Babicky, V., Clupek, M., Benes, J., Pouckova, P.: Localized damage of tissues induced by focused shock waves. IEEE Trans. Plasma Sci. **32**, 1609–1613 (2004)
46. Sensenig, R., Kalghatgi, S., Cerchar, E., Fridman, G., Shereshevsky, A., Torabi, B., Arjunan, K.P., Podolsky, E., Fridman, A., Friedman, G., Azizkhan-Clifford, J., Brooks, A.D.: Non-thermal plasma induces apoptosis in melanoma cells via production of intracellular reactive oxygen species. Ann. Biomed. Eng. **39**, 674–687 (2011)
47. Fridman, G., Shereshevsky, A., Jost, M.M., Brooks, A.D., Fridman, A., Gutsol, A., Vasilets, V., Friedman, G.: Floating electrode dielectric barrier discharge plasma in air promoting apoptotic behavior in melanoma skin cancer cell lines. Plasma Chem. Plasma Process. **27**, 163–176 (2007)
48. Hall, E.H., Schoenbach, K.H., Beebe, S.J.: Nanosecond pulsed electric fields induce apoptosis in p53-wildtype and p53-null HCT116 colon carcinoma cells. Apoptosis **12**, 1721–1731 (2007)
49. Lukes, P., Sunka, P., Hoffer, P., Stelmashuk, V., Pouckova, P., Zadinova, M., Zeman, J., Dibdiak, L., Kolarova, H., Tomankova, K., Binder, S., Benes, J.: Focused tandem shock waves in water and their potential application in cancer treatment. Shock Waves **24**, 51–57 (2014)
50. Coleman, A.J., Saunders, J.E.: A review of the physical properties and biological effects of the high amplitude acoustic fields used in extracorporeal lithotripsy. Ultrasonics **31**, 75–89 (1993)
51. Bailey, M.R., Khokhlova, V.A., Sapoznikov, O.A., Kargl, S.G., Crum, L.A.: Physical mechanisms of the therapeutic effect of ultrasound (a review). Acoust. Phys. **49**, 369–388 (2003)
52. Sunka, P., Babicky, V., Clupek, M., Fuciman, M., Lukes, P., Simek, M., Benes, J., Locke, B., Majcherova, Z.: Potential applications of pulse electrical discharges in water. Acta Phys. Slovaca **54**, 135–145 (2004)
53. Loske, A.M., Prieto, F.E., Fernandez, F., van Cauwelaert, J.: Tandem shock wave cavitation enhancement for extracorporeal lithotripsy. Phys. Med. Biol. **47**, 3945–3957 (2002)
54. Stalder, K.R., McMillen, D.F., Woloszko, J.: Electrosurgical plasmas. J. Phys. D. Appl. Phys. **38**, 1728–1738 (2005)
55. Stalder, K.R., Woloszko, J.: Some physics and chemistry of electrosurgical plasma discharges. Contrib. Plasma Phys. **47**, 64–71 (2007)

56. Graves, D.B.: The emerging role of reactive oxygen and nitrogen species in redox biology and some implications for plasma applications to medicine and biology. J. Phys. D. Appl. Phys. **45**, 263001 (2012)
57. Lukes, P., Dolezalova, E., Sisrova, I., Clupek, M.: Aqueous-phase chemistry and bactericidal effects from an air discharge plasma in contact with water: evidence for the formation of peroxynitrite through a pseudo-second-order post-discharge reaction of H_2O_2 and HNO_2. Plasma Sources Sci. Technol. **23**, 015019 (2014)
58. Edebo, L., Selin, I.: Effect of pressure shock wave and some electrical quantities in microbicidal effect of transient electric arcs in aqueous systems. J. Gen. Microbiol. **50**, 253–259 (1968)
59. Sato, M., Ohgiyama, T., Clements, J.S.: Formation of chemical species and their effects on microorganisms using a pulsed high-voltage discharge in water. IEEE Trans. Ind. Appl. **32**, 106–112 (1996)
60. Ching, W.K., Colussi, A.J., Sun, H.J., Nealson, K.H., Hoffmann, M.R.: Escherichia coli disinfection by electrohydraulic discharges. Environ. Sci. Technol. **35**, 4139–4144 (2001)
61. Abou-Ghazala, A., Katsuki, S., Schoenbach, K.H., Dobbs, F.C., Moreira, K.R.: Bacterial decontamination of water by means of pulsed-corona discharges. IEEE Trans. Plasma Sci. **30**, 1449–1453 (2002)
62. Li, Z., Sakai, S., Yamada, C., Wang, D., Chung, S., Lin, X., Namihira, T., Katsuki, S., Akiyama, H.: The effects of pulsed streamerlike discharge on cyanobacteria cells. IEEE Trans. Plasma Sci. **34**, 1719–1725 (2006)
63. Li, Z., Ohno, T., Sato, H., Sakugawa, T., Akiyama, H., Kunitomo, S., Sasaki, K., Ayukawa, M., Fujiwara, H.: A method of water-bloom prevention using underwater pulsed streamer discharge. J. Environ. Sci. Health A **43**, 1209–1214 (2008)
64. Lukes, P., Clupek, M., Babicky, V., Sunka, P.: Ultraviolet radiation from the pulsed corona discharge in water. Plasma Sources Sci. Technol. **17**, 024012 (2008)
65. Brisset, J.L., Hnatiuc, E.: Peroxynitrite: a re-examination of the chemical properties of non-thermal discharges burning in air over aqueous solutions. Plasma Chem. Plasma Process. **32**, 655–674 (2012)
66. Jiang, C., Carter, C.: Absolute atomic oxygen density measurements for nanosecond-pulsed atmospheric-pressure plasma jets using two-photon absorption laser-induced fluorescence spectroscopy. Plasma Sources Sci. Technol. **23**, 065006 (2014)
67. Deng, X.T., Shi, J.J., Chen, H.L., Kong, M.G.: Protein destruction by atmospheric pressure glow discharges. Appl. Phys. Lett. **90**, 013903 (2007)
68. Goree, J., Liu, B., Drake, D., Stoffels, E.: Killing of S-mutans bacteria using a plasma needle at atmospheric pressure. IEEE Trans. Plasma Sci. **34**, 1317–1324 (2006)
69. Weltmann, K.D., Brandenburg, R., von Woedtke, T., Ehlbeck, J., Foest, R., Stieber, M., Kindel, E.: Antimicrobial treatment of heat sensitive products by miniaturized atmospheric pressure plasma jets (APPJs). J. Phys. D. Appl. Phys. **41**, 194008 (2008)
70. Woloszko, J., Stalder, K.R., Brown, I.G.: Plasma characteristics of repetitively-pulsed electrical discharges in saline solutions used for surgical procedures. IEEE Trans. Plasma Sci. **30**, 1376–1383 (2002)
71. Schaper, L., Graham, W.G., Stalder, K.R.: Vapour layer formation by electrical discharges through electrically conducting liquids-modelling and experiment. Plasma Sources Sci. Technol. **20**, 034003 (2011)
72. Schaper, L., Stalder, K.R., Graham, W.G.: Plasma production in electrically conducting liquids. Plasma Sources Sci. Technol. **20**, 034004 (2011)
73. Rentschler, H.C., Nagy, R., Mouromseff, G.: Bactericidal effect of ultraviolet radiation. J. Bacteriol. **41**, 745–774 (1941)
74. Barnard, J.E., Morgan, H.D.R.: The physical factors in phototherapy. Br. Med. J. **2**, 1269–1271 (1903)
75. Hijnen, W.A., Beerendonk, E.F., Medema, G.J.: Inactivation credit of UV radiation for viruses, bacteria and protozoan (oo)cysts in water: a review. Water Res. **40**, 3–22 (2006)

76. Willberg, D.M., Lang, P.S., Hochemer, R.H., Kratel, A., Hoffmann, M.R.: Degradation of 4-chlorophenol, 3,4-dichloroaniline, and 2,4,6-trinitrotoluene in an electrohydraulic discharge reactor. Environ. Sci. Technol. **30**, 2526–2534 (1996)
77. Lang, P.S., Ching, W.-K., Willberg, D.M., Hoffmann, M.R.: Oxidative degradation of 2,4,6-trinitrotoluene by ozone in an electrohydraulic discharge reactor. Environ. Sci. Technol. **32**, 3142–3148 (1998)
78. Martin, E.A.: Experimental investigation of a high-energy density, high-pressure arc plasma. J. Appl. Phys. **31**, 255–267 (1960)
79. Robinson, J.W., Ham, M., Balaster, A.N.: Ultraviolet radiation from electrical discharges in water. J. Appl. Phys. **44**, 72–75 (1973)
80. Sun, B., Kunitomo, S., Igarashi, C.: Characteristics of ultraviolet light and radicals formed by pulsed discharge in water. J. Phys. D. Appl. Phys. **39**, 3814 (2006)
81. Lukes, P., Locke, B.R.: Plasmachemical oxidation processes in a hybrid gas-liquid electrical discharge reactor. J. Phys. D. Appl. Phys. **38**, 4074–4081 (2005)
82. Shuaibov, A.K., Shimon, L.L., Dashchenko, A.I., Shevera, I.V.: A water-vapor electric-discharge vacuum ultraviolet source. Tech. Phys. Lett. **27**, 642–643 (2001)
83. Shuaibov, A.K., Minya, A.J., Gomoki, Z.T., Shevera, I.V., Gritsak, R.V.: Vacuum-UV emitter using low-pressure discharge in helium-water vapor mixture. Tech. Phys. Lett. **37**, 126–127 (2011)
84. van Gils, C.A.J., Hofmann, S., Boekema, B., Brandenburg, R., Bruggeman, P.J.: Mechanisms of bacterial inactivation in the liquid phase induced by a remote RF cold atmospheric pressure plasma jet. J. Phys. D. Appl. Phys. **46**, 175203 (2013)
85. Reuter, S., Winter, J., Schmidt-Bleker, A., Schroeder, D., Lange, H., Knake, N., Schulz-von der Gathen, V., Weltmann, K.-D.: Atomic oxygen in a cold argon plasma jet: TALIF spectroscopy in ambient air with modelling and measurements of ambient species diffusion. Plasma Sources Sci. Technol. **21**, 024005 (2012)
86. Zastawny, H.Z., Romat, H., Karpel vel Leitner, N., Chang, J.S.: Pulsed arc discharges for water treatment and disinfection. In Electrostatics 2003, vol. 137, pp. 325–330. IOP Publishers, Bristol (2004)
87. Delius, M.: Extracorporeal shock waves: bioeffects and mechanisms of action. In: Srivastava, R.C., Leutloff, D., Takayama, K., Grönig, H. (eds.) Shock Focussing Effect in Medical Science and Sonoluminescence, pp. 211–226. Springer, Berlin (2003)
88. Loske, A.M., Campos-Guillen, J., Fernandez, F., Castano-Tostado, E.: Enhanced shock wave-assisted transformation of Escherichia coli. Ultrasound Med. Biol. **37**, 502–510 (2011)
89. Hosseini, S.H.R., Iwasaki, S., Sakugawa, T., Akiyama, H.: Characteristics of micro underwater shock waves produced by pulsed electric discharges for medical applications. J. Korean Phys. Soc. **59**, 3526–3530 (2011)
90. Hosseini, S.H.R., Menezes, V., Moosavi-Nejad, S., Ohki, T., Nakagawa, A., Tominaga, T., Takayama, K.: Development of shock wave assisted therapeutic devices and establishment of shock wave therapy. Minim. Invasive Ther. Allied Technol. **15**, 230–240 (2006)
91. Russo, P., Stephenson, R.A., Mies, C., Huryk, R., Heston, W.D., Melamed, M.R., Fair, W.R.: High energy shock waves suppress tumor growth in vitro and in vivo. J. Urol. **135**, 626–628 (1986)
92. Brauner, T., Brummer, F., Hulser, D.F.: Histopathology of shock wave treated tumor cell suspensions and multicell tumor spheroids. Ultrasound Med. Biol. **15**, 451–460 (1989)
93. Kohri, K., Uemura, T., Iguchi, M., Kurita, T.: Effect of high energy shock waves on tumor cells. Urol. Res. **18**, 101–105 (1990)
94. Clayman, R.V., Long, S., Marcus, M.: High-energy shock waves: in vitro effects. Am. J. Kidney Dis. **17**, 436–444 (1991)
95. Lifshitz, D.A., Williams, J.C., Sturtevant, B., Connors, B.A., Evan, A.P., McAteer, J.A.: Quantitation of shock wave cavitation damage in vitro. Ultrasound Med. Biol. **23**, 461–471 (1997)

96. Randazzo, R.F., Chaussy, C.G., Fuchs, G.J., Bhuta, S.M., Lovrekovich, H., de Kernion, J.B.: The in vitro and in vivo effects of extracorporeal shock waves on malignant cells. Urol. Res. **16**, 419–426 (1988)
97. Holmes, R.P., Yeaman, L.I., Li, W.J., Hart, L.J., Wallen, C.A., Woodruff, R.D., McCullough, D.L.: The combined effects of shock waves and cisplatin therapy on rat prostate tumors. J. Urol. **144**, 159–163 (1990)
98. Lee, K.E., Smith, P., Cockett, A.T.: Influence of high-energy shock waves and cisplatin on antitumor effect in murine bladder cancer. Urology **36**, 440–444 (1990)
99. Gambihler, S., Delius, M.: In vitro interaction of lithotripter shock waves and cytotoxic drugs. Br. J. Cancer **66**, 69–73 (1992)
100. Mastikhin, I.V., Teslenko, V.S., Nikolin, V.P., Kolosova, N.G., Gorchakov, V.N.: Tumor growth inhibition by combined action of shock waves and cytostatics. In: Loske, A.M. (ed.) New Trends in Shock Wave Applications to Medicine and Biotechnology, pp. 151–164. Research Signpost, Kerala (2010)
101. Kato, M., Ioritani, N., Suzuki, T., Kambe, M., Inaba, Y., Watanabe, R., Sasano, H., Orikasa, S.: Mechanism of anti-tumor effect of combination of bleomycin and shock waves. Jpn. J. Cancer Res. **91**, 1065–1072 (2000)
102. Delius, M., Denk, R., Berding, C., Liebich, H.G., Jordan, M., Brendel, W.: Biological effects of shock waves: cavitation by shock waves in piglet liver. Ultrasound Med. Biol. **16**, 467–472 (1990)
103. Huber, P., Debus, J., Peschke, P., Hahn, E.W., Lorenz, W.J.: In vivo detection of ultrasonically induced cavitation by a fibre-optic technique. Ultrasound Med. Biol. **20**, 811–825 (1994)
104. Coleman, A.J., Choi, M.J., Saunders, J.E.: Detection of acoustic emission from cavitation in tissue during clinical extracorporeal lithotripsy. Ultrasound Med. Biol. **22**, 1079–1087 (1996)
105. Cleveland, R.O., Sapozhnikov, O.A., Bailey, M.R., Crum, L.A.: A dual passive cavitation detector for localized detection of lithotripsy-induced cavitation in vitro. J. Acoust. Soc. Am. **107**, 1745–1758 (2000)
106. Zhong, P., Zhou, Y., Zhu, S.: Dynamics of bubble oscillation in constrained media and mechanisms of vessel rupture in SWL. Ultrasound Med. Biol. **27**, 119–134 (2001)
107. Lukes, P., Zeman, J., Horak, V., Hoffer, P., Pouckova, P., Holubova, M., Hosseini, S.H.R., Akiyama, H., Sunka, P., Benes, J.: In vivo effects of focused shock waves on tumor tissue visualized by fluorescence staining techniques. Bioelectrochemistry **103**, 103–110 (2015)
108. Lukes, P., Fernández, F., Gutiérrez-Aceves, J., Fernández, E., Alvarez, U.M., Sunka, P., Loske, A.M.: Tandem shock waves in medicine and biology: a review of potential applications and successes. Shock Waves **26**, 1–23 (2016)
109. Foyer, C.H., Noctor, G.: Redox regulation in photosynthetic organisms: signaling, acclimation, and practical implications. Antioxid. Redox Signal. **11**, 861–905 (2009)
110. Novo, E., Parola, M.: Redox mechanisms in hepatic chronic wound healing and fibrogenesis. Fibrogenesis Tissue Repair **1**, 5 (2008)
111. Graves, D.B.: The emerging role of reactive oxygen and nitrogen species in redox biology and some implications for plasma applications to medicine and biology. J. Phys. D. Appl. Phys. **45**, 263001 (2012)
112. Yonemori, S., Nakagawa, Y., Ono, R., Oda, T.: Measurement of OH density and air-helium mixture ratio in an atmospheric-pressure helium plasma jet. J. Phys. D. Appl. Phys. **45**, 225202 (2012)
113. Bruggeman, P., Cunge, G., Sadeghi, N.: Absolute OH density measurements by broadband UV absorption in diffuse atmospheric-pressure He-H_2O RF glow discharges. Plasma Sources Sci. Technol. **21**, 035019 (2012)
114. Marnett, L.J.: Lipid peroxidation – DNA damage by malondialdehyde. Mutat. Res. Fundam. Mol. Mech. Mugag. **424**, 83–95 (1999)
115. Machala, Z., Chladekova, L., Pelach, M.: Plasma agents in bio-decontamination by dc discharges in atmospheric air. J. Phys. D. Appl. Phys. **43**, 222001 (2010)

116. Davies, M.J. Protein oxidations: Concepts, mechanisms and new insights. Free radical school presentations. Society for Free Radical Biology and Medicine http://www.sfrbm.org/frs/Davies2003.pdf. 31th Oct 2011
117. Ellerweg, D., Benedikt, J., von Keudell, A., Knake, N., Schulz-von der Gathen, V.: Characterization of the effluent of a He/O_2 microscale atmospheric pressure plasma jet by quantitative molecular beam mass spectrometry. New J. Phys. **12**, 013021 (2010)
118. Ellerweg, D., von Keudell, A., Benedikt, J.: Unexpected O and O_3 production in the effluent of He/O_2 microplasma jets emanating into ambient air. Plasma Sources Sci. Technol. **21**, 034019 (2012)
119. Jiang, C., Puech, V., Magne, L., Jeanney, P.: Absolute ozone measurements for a low-energy pulsed plasma needle. In The 62nd Gaseous Electronics Conference, Saratoga Springs, NY, USA (2009)
120. Jeong, J.Y., Park, J., Henins, I., Babayan, S.E., Tu, V.J., Selwyn, G.S., Ding, G., Hicks, R.F.: Reaction chemistry in the afterglow of an oxygen-helium, atmospheric-pressure plasma. J. Phys. Chem. A **104**, 8027–8032 (2000)
121. Staehelin, J., Hoigne, J.: Decomposition of ozone in water – rate of initiation by hydroxide ions and hydrogen peroxide. Environ. Sci. Technol. **16**, 676–681 (1982)
122. Hoigne, J., Bader, H., Haag, W.R., Staehelin, J.: Rate constants of reactions of ozone with organic and inorganic compounds in water. 3. Inorganic compounds and radicals. Water Res. **19**, 993–1004 (1985)
123. Locke, B.R., Shih, K.Y.: Review of the methods to form hydrogen peroxide in electrical discharge plasma with liquid water. Plasma Sources Sci. Technol. **20**, 034006 (2011)
124. Walling, C.: Fentons reagent revisited. Accounts Chem. Res. **8**, 125–131 (1975)
125. Keyer, K., Gort, A.S., Imlay, J.A.: Superoxide and the production of oxidative DNA damage. J. Bacteriol. **177**, 6782–6790 (1995)
126. Pericone, C.D., Park, S., Imlay, J.A., Weiser, J.N.: Factors contributing to hydrogen peroxide resistance in Streptococcus pneumoniae include pyruvate oxidase (SpxB) and avoidance of the toxic effects of the Fenton reaction. J. Bacteriol. **185**, 6815–6825 (2003)
127. LeBlanc, J.J., Davidson, R.J., Hoffman, P.S.: Compensatory functions of two alkyl hydroperoxide reductases in the oxidative defense system of Legionella pneumophila. J. Bacteriol. **188**, 6235–6244 (2006)
128. Sabri, M., Leveille, S., Dozois, C.M.: A SitABCD homologue from an avian pathogenic Escherichia coli strain mediates transport of iron and manganese and resistance to hydrogen peroxide. Microbiology (UK) **152**, 745–758 (2006)
129. Lukes, P., Clupek, M., Babicky, V., Vykouk, T.: Bacterial inactivation by pulsed corona discharge in water. In 2007 IEEE Pulsed Power Conference, Albuquerque, NM, vol. 1–4, pp 320–323 (2007)
130. Stoffels, E., Gonzalvo, Y.A., Whitmore, T.D., Seymour, D.L., Rees, J.A.: A plasma needle generates nitric oxide. Plasma Sources Sci. Technol. **15**, 501–506 (2006)
131. Pavlovich, M.J., Clark, D.S., Graves, D.B.: Quantification of air plasma chemistry for surface disinfection. Plasma Sources Sci. Technol. **23**, 065036 (2014)
132. Eliasson, B., Kogelschatz, U.: Modeling and applications of silent discharge Plasmas. IEEE Trans. Plasma Sci. **19**, 309–323 (1991)
133. Beckman, J.S., Beckman, T.W., Chen, J., Marshall, P.A., Freeman, B.A.: Apparent hydroxyl radical production by peroxynitrite – implications for endothelial injury from nitric oxide and superoxide. Proc. Natl. Acad. Sci. U. S. A. **87**, 1620–1624 (1990)
134. Beckman, J.S., Koppenol, W.H.: Nitric oxide, superoxide, and peroxynitrite: the good, the bad, and the ugly. Am. J. Physiol. Cell Physiol. **271**, C1424–C1437 (1996)
135. Goldstein, S., Squadrito, G.L., Pryor, W.A., Czapski, G.: Direct and indirect oxidations by peroxynitrite, neither involving the hydroxyl radical. Free Rad. Biol. Med. **21**, 965–974 (1996)
136. Squadrito, G.L., Pryor, W.A.: Oxidative chemistry of nitric oxide: the roles of superoxide, peroxynitrite, and carbon dioxide. Free Rad. Biol. Med. **25**, 392–403 (1998)

137. Pacher, P., Beckman, J.S., Liaudet, L.: Nitric oxide and peroxynitrite in health and disease. Physiol. Rev. **87**, 315–424 (2007)
138. Denicola, A., Souza, J.M., Radi, R.: Diffusion of peroxynitrite across erythrocyte membranes. Proc. Natl. Acad. Sci. U. S. A. **1998**, 3566–3571 (1998)
139. Pryor, W.A., Squadrito, G.L.: The chemistry of peroxynitrite: a product from the reaction of nitric-oxide with superoxide. Am. J. Physiol. Lung Cell. Mol. Physiol. **268**, L699–L722 (1995)
140. Beckman, J.S.: Peroxynitrite versus hydroxyl radical: The role of nitric oxide in superoxide dependent cerebral injury. In: Chiueh, C.C., Gilbert, D.L., Colton, C.A. (eds.) The Neurobiology of NO and OH, Ann. New York Acad. Sci., pp. 69–75 (1994)
141. Hutchinson, F.: The distance that a radical formed by ionizing radiation can diffuse in a yeast cell. Radiat. Res. **7**, 473–483 (1957)
142. Koppenol, W.: The basic chemistry of nitrogen monoxide and peroxynitrite. Free Rad. Biol. Med. **25**, 385–391 (1998)
143. Pryor, W., Squadrito, G.: The chemistry of peroxynitrite: the good, the bad and the ugly. Am. J. Physiol. **268**, L699–L722 (1995)
144. Packer, L.: Nitric Oxide Part C: Biological and Antioxidant Activities. Section III – Peroxynitrite, pp. 279–381. Academic, San Diego (1999)
145. Laskin, D.L.: Oxidative/Nitrosative Stress and Disease. Ann. New York Acad. Sci. **1203**, Wiley, New York (2010)
146. Seago, N., Clark, D., Miller, M.: Role of inducible oxide synthase (NOS) and peroxynitrite in gut inflammation. Inflamm. Res. **44**, S153–S154 (1995)
147. Kirkpatrick, M.J., Locke, B.R.: Effects of platinum electrode on hydrogen, oxygen, and hydrogen peroxide formation in aqueous phase pulsed corona electrical discharge. Ind. Eng. Chem. Res. **45**, 2138–2142 (2006)
148. Sharma, A.K., Locke, B.R., Arce, P., Finney, W.C.: A preliminary-study of pulsed streamer corona discharge for the degradation of phenol in aqueous-solutions. Hazard. Waste Hazard. Mater. **10**, 209–219 (1993)
149. Lukes, P., Clupek, M., Babicky, V., Sisrova, I., Janda, V.: The catalytic role of tungsten electrode material in the plasmachemical activity of a pulsed corona discharge in water. Plasma Sources Sci. Technol. **20**, 034011 (2011)
150. Anbar, M., Taube, H.: Interaction of nitrous acid with hydrogen peroxide and with water. J. Am. Chem. Soc. **76**, 6243–6247 (1954)
151. Kamgang-Youbi, G., Herry, J.M., Bellon-Fontaine, M.N., Brisset, J.L., Doubla, A., Naitali, M.: Evidence of temporal postdischarge decontamination of bacteria by gliding electric discharges: application to *Hafnia alvei*. Appl. Environ. Microbiol. **73**, 4791–4796 (2007)
152. Kamgang-Youbi, G., Herry, J.M., Brisset, J.L., Bellon-Fontaine, M.N., Doubla, A., Naitali, M.: Impact on disinfection efficiency of cell load and of planktonic/adherent/detached state: case of *Hafnia alvei* inactivation by Plasma Activated Water. Appl. Microbiol. Biotechnol. **81**, 449–457 (2008)
153. Kamgang-Youbi, G., Herry, J.M., Meylheuc, T., Brisset, J.L., Bellon-Fontaine, M.N., Doubla, A., Naitali, M.: Microbial inactivation using plasma-activated water obtained by gliding electric discharges. Lett. Appl. Microbiol. **48**, 13–18 (2009)
154. Naitali, M., Kamgang-Youbi, G., Herry, J.M., Bellon-Fontaine, M.N., Brisset, J.L.: Combined effects of long-living chemical species during microbial inactivation using atmospheric plasma-treated water. Appl. Environ. Microbiol. **76**, 7662–7664 (2010)
155. Oehmigen, K., Hahnel, M., Brandenburg, R., Wilke, C., Weltmann, K.D., von Woedtke, T.: The role of acidification for antimicrobial activity of atmospheric pressure plasma in liquids. Plasma Process. Polym. **7**, 250–257 (2010)
156. Oehmigen, K., Winter, J., Hahnel, M., Wilke, C., Brandenburg, R., Weltmann, K.D., von Woedtke, T.: Estimation of possible mechanisms of Escherichia coli inactivation by plasma treated sodium chloride solution. Plasma Process. Polym. **8**, 904–913 (2011)
157. Traylor, M.J., Pavlovich, M.J., Karim, S., Hait, P., Sakiyama, Y., Clark, D.S., Graves, D.B.: Long-term antibacterial efficacy of air plasma-activated water. J. Phys. D. Appl. Phys. **44**, 472001 (2011)

158. Ercan, U.K., Wang, H., Ji, H.F., Fridman, G., Brooks, A.D., Joshi, S.G.: Nonequilibrium plasma-activated antimicrobial solutions are broad-spectrum and retain their efficacies for extended period of time. Plasma Process. Polym. **10**, 544–555 (2013)
159. Brisset, J.L., Moussa, D., Doubla, A., Hnatiuc, E., Hnatiuc, B., Kamgang-Youbi, G., Herry, J. M., Naitali, M., Bellon-Fontaine, M.N.: Chemical reactivity of discharges and temporal post-discharges in plasma treatment of aqueous media: Examples of gliding discharge treated solutions. Ind. Eng. Chem. Res. **47**, 5761–5781 (2008)
160. Pavlovich, M.J., Chang, H.W., Sakiyama, Y., Clark, D.S., Graves, D.B.: Ozone correlates with antibacterial effects from indirect air dielectric barrier discharge treatment of water. J. Phys. D. Appl. Phys. **46**, 145202 (2013)
161. Seinfeld, J.H., Pandis, S.N.: Atmospheric Chemistry and Physics: From Air Pollution to Climate Change, 2nd edn. Wiley, Hoboken (2006)
162. Park, J.Y., Lee, Y.N.: Solubility and decomposition kinetics of nitrous acid in aqueous solution. J. Phys. Chem. **92**, 6294–6302 (1988)
163. Naitali, M., Herry, J.M., Hnatiuc, E., Kamgang, G., Brisset, J.L.: Kinetics and bacterial inactivation induced by peroxynitrite in electric discharges in air. Plasma Chem. Plasma Process. **32**, 675–692 (2012)
164. Katsumura, Y.: NO_2 and NO_3 radicals in radiolysis of nitric acid solutions. In: Alfassi, Z.B. (ed.) The Chemistry of Free Radicals: N-Centered Radicals, pp. 393–412. Wiley, Chichester (1998)
165. Peyton, G.R., Glaze, W.H.: Destruction of pollutants in water with ozone in combination with ultraviolet-radiation. 3. Photolysis of aqueous ozone. Environ. Sci. Technol. **22**, 761–767 (1988)
166. Hunt, J.P., Taube, H.: The photochemical decomposition of hydrogen peroxide – Quantum yields, tracer and fractionation effects. J. Am. Chem. Soc. **74**, 5999–6002 (1952)
167. Baxendale, J.H., Wilson, J.A.: The photolysis of hydrogen peroxide at high light intensities. Trans. Farad. Soc. **53**, 344–356 (1957)
168. Tarr, M.A.: Chemical Degradation Methods for Wastes and Pollutants: Environmental and Industrial Applications. Marcel Dekker, New York (2003)
169. Schunck, T., Bieth, F., Pinguet, S., Delmote, P.: Penetration and propagation into biological matter and biological effects of high-power ultra-wideband pulses: a review. Electromagn. Biol. Med. **30**, 1–18 (2014)
170. Ramundo-Orlando, A., Gallerano, G.P.: Terahertz radiation effects and biological applications. J. Infrared Milli. Terahz. Waves **30**(12), 1308–1318 (2009)
171. Son, J.H.: Principle and applications of terahertz molecular imaging. Nanotechnology **24**, 214001 (2013)
172. Eisele, H.: Recent advances in the performance of InP Gunn devices and GaAs TUNNETT diodes for the 100–300 GHz frequency range and above. IEEE Trans. Microwave Theory Tech. **48**(4), 626–631 (2000)
173. Williams, B.S., Kumar, S., Callebaut, H., Hu, Q., Reno, J.L.: Terahertz quantum cascade laser operating up to 137 K. Appl. Phys. Lett. **83**, 5142–5144 (2003)
174. Williams, B.S., Kumar, S., Hu, Q., Reno, J.L.: Resonant-phonon terahertz quantum-cascade laser operating at 2.1 THz (λ= 141 μm). Electr. Lett. **40**(7), 431–432 (2004)
175. Crocker, A., Gebbie, H.A., Kimmitt, M.F., Mathias, I.E.S.: Stimulated emission in the far infrared. Nature **201**, 250–251 (1964)
176. Auston, D.H., Cheung, K.P., Valdmanis, J.A., Kleinman, D.A.: Cherenkov radiation from femtosecond optical pulses in electro-optic media. Phys. Rev. Lett. **53**, 1555–1558 (1984)
177. Fattinger, C., Grischkowsky, D.: Terahertz beams. Appl. Phys. Lett. **54**, 490–492 (1989)
178. Jiang, Z., Zhang, X.C.: Terahertz imaging via electrooptic effect. IEEE. Trans. Microwave Theory Tech. **47**(12), 2644–2650 (1999)
179. Wu, Q., Hewitt, T.D., Zhang, X.C.: Two-dimensional electro-optic imaging of THz beams. Appl. Phys. Lett. **69**, 1026–1028 (1996)
180. Brown, E.R., McIntosh, K.A., Nichols, K.B., Dennis, C.L.: Photomixing up to 3.8 THz in low-temperature-grown GaAs. Appl. Phys. Lett. **66**, 285–287 (1995)

181. Thumm, M.: State-of-the-Art of High Power Gyro-Devices and Free Electron Masers. KIT Scientific Reports 7641, KIT Scientific Publishing, Karlsruhe, Germany (2012)
182. Gold, S.H., Nusinovich, G.S.: Review of high power microwave source research. Rev. Sci. Instrum. **68**, 3945–3974 (1997)
183. Staprans, A., McCune, E., Ruetz, J.: High-power linear-beam tubes. Proc. IEEE **61**(3), 299–330 (1973)
184. Motz, H.: Applications of the radiation from fast electron beams. J. Appl. Phys. **22**(5), 527–535 (1951)
185. Phillips, R.M.: History of the ubitron. Nucl. Instr. Meth. Phys. Res. **A272**, 1–9 (1988)
186. Deacon, D.A.G., Elias, L.R., Madey, J.M.J., Ramian, G.J., Schwettman, H.A., Smith, T.I.: First operation of a free-electron laser. Phys. Rev. Lett. **38**, 892–893 (1977)
187. Ciocci, F., Doria, A., Gallerano, G.P., Giabbai, I., Kimmitt, M.F., Messina, G., Renieri, A., Walsh, J.E.: Observation of coherent millimeter and submillimeter emission from a microtron-driven Cherenkov free-electron laser. Phys. Rev. Lett. **66**, 699–702 (1991)
188. Doucas, G., Kimmitt, M.F., Doria, A., Gallerano, G.P., Giovenale, E., Messina, G., Andrews, H.L., Brownell, J.H.: Determination of longitudinal bunch shape by means of coherent Smith-Purcell radiation. Phys. Rev. ST Accel. Beams **5**(7), 072802 (2002)
189. Gallerano, G.P., Doria, A., Giovenale, E., Renieri, A.: Compact free electron lasers: from cerenkov to waveguide FELs. Infrared Phys. Tech. **40**, 161–174 (1999)
190. Doria, A., Gallerano, G.P., Giovenale, E., Letardi, S., Messina, G., Ronsivalle, C.: Enhancement of coherent emission by energy-phase correlation in a bunched electron beam. Phys. Rev. Lett. **80**, 2841–2844 (1998)
191. Doria, A., Gallerano, G.P., Giovenale, E., Messina, G., Spassovsky, I.: Enhanced coherent emission of THz radiation by energy-phase correlation in a bunched electron beam. Phys. Rev. Lett. **93**, 264801 (2004)
192. Doria, A., Gallerano, G.P., Giovenale, E., Messina, G., Lai, A., Ramundo-Orlando, A., Sposato, V., D'Arienzo, M., Perrotta, A., Romanò, M., Sarti, M., Scarfì, M.R., Spassovsky, I., Zeni, O.: THz radiation studies on biological systems at the ENEA FEL Facility. Infrared Phys. **45**, 339–347 (2004)
193. Scientific Committee on Emerging and Newly Identified Health Risks (SCENIHR) Opinion on Potential health effects of exposure to electromagnetic fields (EMF), available at http://ec.europa.eu/health/scientific_committees/emerging/docs/scenihr_o_041.pdf
194. Singh, S., Kapoor, N.: Health Implications of electromagnetic fields, mechanisms of action, and research needs. Adv. Biol. 198609 (2014)
195. Seaman, R.L.: Effects of exposure of animals to ultra-wideband pulses. Health Phys. **92**(6), 629–634 (2007)
196. Zeng, L., Zou, C., Zhang, J., Wang, J., Miao, X., Ren, D., Li, Y., Guo, G.: Perforate on CHO cell membranes induced by electromagnetic pulses irradiation observed by atomic force microscopy. Afr. J. Biotechnol. **8**(12), 2796–2781 (2009)
197. Kolb, J.F., Xiao, S., Camp, J.T., Migliaccio, M., Bajracharya, C., Schoenbach, K.H.: Sub-nanosecond electrical pulses for medical therapies and imaging, Proceedings of the 4th European Conference on Antennas and Propagation (EuCAP), Barcelona, Spain 12–16 April 2010
198. Jordan, D.W., Gilgenbach, R.M., Uhler, M.D., Gates, L.H., Lau, Y.Y.: Effect of pulsed, high-power radiofrequency radiation on electroporation of mammalian cells. IEEE. Trans. Plasma Sci. **32**(4), 1573–1578 (2004)
199. Jordan, D.W., Uhler, M.D., Gilgenbach, R.M., Lau, Y.Y.: Enhancement of cancer chemotherapy in vitro by intense ultrawideband electric field pulses. J. Appl. Phys. **99**, 094701 (2006)
200. Macaire, S., Catrain, A., Tortel, S., Joly, J.C., Girard, S., Bonnet, P., Vian, A.: Radiated ultrashort high-power electromagnetic pulses induce ATP release in B16F10 murine melanoma cells. JEMAA **7**(3), 66–74 (2015)
201. Stamatovic, S.M., Keep, R.F., Andjelkovic, A.V.: Brain endothelial cell-cell junctions: how to "Open" the blood brain barrier. Curr. Neuropharmacol. **6**, 179–192 (2008)

202. Cheng, C.Y., Mruk, D.D.: The blood-testis barrier and its implications for male contraception. Pharmacol. Rev. **64**(1), 16–64 (2012)
203. Stam, R.: Electromagnetic fields and the blood–brain barrier. Brain Res. Rev. **65**, 80–97 (2010)
204. Ding, G.R., Li, K.C., Wang, X.W., Zhou, Y.C., Qiu, L.B., Tan, J., Xu, S.L., Guo, G.Z.: Effects of exposure to electromagnetic pulse on the permeability of brain micro vascular in rats. Biochem. Environ. Sci. **22**(3), 255–258 (2009)
205. Ding, G.R., Qiu, L.B., Wang, X.W., Li, K.C., Zhou, Y.C., Zhou, Y., Zhang, J., Zhou, J.X., Li, Y.R., Guo, G.Z.: EMP-induced alterations of tight junction protein expression and disruption of the blood–brain barrier. Toxicol. Lett. **196**, 154–160 (2010)
206. Qiu, L.B., Ding, G.R., Li, K.C., Wang, X.W., Zhou, Y., Zhou, Y.C., Li, Y.R., Guo, G.Z.: The role of protein kinase C in the opening of blood-brain barrier induced by electromagnetic pulse. Toxicology **273**(1–3), 29–34 (2010)
207. Qiu, L.B., Chen, C., Ding, G., Zhou, Y., Zhang, M.: The effects of electromagnetic pulse on the protein levels of tight junction associated-proteins in the cerebral cortex, hippocampus, heart, lung, and testis of rats. Biomed. Environ. Sci. **24**(4), 438–444 (2011)
208. Qiu, L.B., Zhou, Y., Wang, Q., Yang, L.L., Liu, H.Q., Xu, S.L., Qi, Y.H., Ding, G.R., Guo, G. Z.: Synthetic gelatinases inhibitor attenuates electromagnetic pulse-induced blood-brain barrier disruption by inhibiting gelatinases-mediated ZO-1 degradation in rats. Toxicology **285**(1–2), 31–38 (2011)
209. Zhou, J.X., Ding, G.R., Zhang, J., Zhou, Y.C., Zhang, Y.J., Guo, G.Z.: Detrimental effect of electromagnetic pulse exposure on permeability of in vitro blood-brain-barrier model. Biomed. Environ. Sci. **26**(2), 128–137 (2013)
210. Wang, X.W., Ding, G.R., Shi, C.H., Zeng, L.H., Liu, J.Y., Li, J., Zhao, T., Chen, Y.B., Guo, G.Z.: Mechanisms involved in the blood-testis barrier increased permeability induced by EMP. Toxicology **276**(1), 58–63 (2010)
211. Wang, L.F., Li, X., Gao, Y.B., Wang, S.M., Zhao, L., Dong, J., Yao, B.W., Xu, X.P., Chang, G.M., Zhou, H.M., Hu, X.J., Peng, R.Y.: Activation of VEGF/Flk-1-ERK pathway induced blood-brain barrier injury after microwave exposure. Mol. Neurobiol. **52**(1), 478–491 (2014)
212. Ramundo-Orlando, A., Stano, P., Gallerano, G.P., Doria, A., Giovenale, E., Messina, G., D'Arienzo, M., Spassovsky, I.: Permeability changes of cationic liposomes loading carbonic anhydrase induced by 130 GHz pulsed radiation. Bioelectromagnetics **28**(8), 587–671 (2007)
213. Walde, P. Preparation of vesicles (liposomes). In: Nalwa HS, editor. Encyclopedia of Nanoscience and Nanotechnology. vol. X, pp. 1–37. American Scientific Publishers (2003)
214. Ramundo-Orlando, A., Morbiducci, U., Mossa, G., d'Inzeo, G.: Effect of low-frequency, low-amplitude magnetic fields on the permeability of cationic liposomes entrapping carbonic anhydrase. I. Evidence for charged lipid involvement. Bioelectromagnetics **21**, 491–498 (2000)
215. Cevc, G. editor. Solute transport across bilayers. In: Phospholipids Handbook, pp. 639–661. Marcel Dekker, New York (1993)
216. Pocker, Y., Stone, J.T.: The catalytic versatility of erythrocyte carbonic anhydrase. Kinetic studies of the enzyme-catalyzed hydrolysis of p-nitrophenyl acetate. Biochemistry **6**, 668–678 (1967)
217. Ramundo-Orlando, A., Mattia, F., Palombo, A., d'Inzeo, G.: Effect of low-frequency, low-amplitude magnetic fields on the permeability of cationic liposomes entrapping carbonic anhydrase. II. No evidence for surface enzyme involvement. Bioelectromagnetics **21**, 499–507 (2000)
218. Gallerano,G.P., Doria, A., Giovenale,E., Messina, G., Spassovsky, I., Ramundo-Orlando, A., Scarfì, M.R., Romeo,S., Zeni, O.: Electric vs. electromagnetic ultrashort pulses: Search for a unifying view, Millimeter and Terahertz Waves, Congress IRMMW-THz (2011)
219. Vernier, P.T., Levine, Z.A., Ho, M.C., Xiao, S., Semenov, I., Pakhomov, A.G.: Picosecond and Terahertz perturbation of interfacial water and electropermeabilization of biological membranes. J. Membr. Biol. **248**(5), 837–847 (2015)

Chapter 4
Biological Responses

Ken-ichi Yano, Lea Rems, Tadej Kotnik, Damijan Miklavčič, James C. Weaver, Kyle C. Smith, Reuben S. Son, Thiruvallur R. Gowrishankar, P. Thomas Vernier, Zachary A. Levine, Marie-Pierre Rols, Justin Teissie, Lluis M. Mir, Andrei G. Pakhomov, Peter Nick, Wolfgang Frey, David A. Dean, Keiko Morotomi-Yano, Robert E. Neal II, Suyashree Bhonsle, Rafael V. Davalos, and Stephen J. Beebe

Abstract Cells are the structural and functional unit of all living organisms and exhibit fundamental properties of life. Cells are surrounded by the cell membrane and subdivided into various compartments. Pulsed electric fields (PEFs) exert profound effects on cells by interacting with the cell membrane and other cellular components. This chapter describes the biological effects of PEF at cellular and subcellular levels. First, this chapter begins with the overview of cell exposure to PEF from a biophysical point of view. Second, the interaction of PEF with biological membranes, membrane pore formation, and their physiological significance is described from multifaceted standpoints. Next, this chapter explains subcellular events induced by PEF, including the effect on cytoskeleton and signal transduction.

K.-i. Yano (✉)
Kumamoto University, Kumamoto, Japan
e-mail: yanoken@kumamoto-u.ac.jp

L. Rems • T. Kotnik • D. Miklavčič
Faculty of Electrical Engineering, University of Ljubljana, Ljubljana, Slovenia

J.C. Weaver • K.C. Smith • R.S. Son • T.R. Gowrishankar
Harvard-MIT Division of Health Sciences and Technology, Massachusetts Institute of Technology, Cambridge, MA, USA

P.T. Vernier • A.G. Pakhomov • S.J. Beebe
Frank Reidy Research Center for Bioelectrics, Old Dominion University, Norfolk, VA, USA

Z.A. Levine
Department of Physics, University of California Santa Barbara, Santa Barbara, CA, USA

M.-P. Rols • J. Teissie
Institute of Pharmacology and Structural Biology, CNRS and University of Toulouse, Toulouse, France

L.M. Mir
Vectorology and Antitumor Therapies, UMR 8203, CNRS, Univ. Paris-Sud, Université Paris-Saclay, Gustave Roussy, Villejuif, France

H. Akiyama, R. Heller (eds.), *Bioelectrics*, DOI 10.1007/978-4-431-56095-1_4

Lastly, detailed description on irreversible electroporation and cell death by PEF is provided. The topics covered in this chapter serve as the basis for the applications of PEF in medicine, environmental science, and food and biomass processing.

Keywords Electroporation • Electropermeabilization • Computational model • Cell membrane • Cellular effect • Cell death

4.1 Biophysical Aspects of Cell Exposure to Electric Pulses

Lea Rems, Tadej Kotnik, and Damijan Miklavčič

Abstract Exposure of a cell to an external electric field results in induced transmembrane voltage ($\Delta\Psi_m$) which superimposes onto the membrane resting potential. If the absolute value of $\Delta\Psi_m$ is high enough and present long enough, an increased transmembrane transport of ions as well as charged and neutral molecules can be observed, i.e., the cell membrane electroporates. Theoretical models predict that $\Delta\Psi_m$ (related to increased electric field in the membrane) reduces the energetic barrier for formation of small pores in the membrane lipid bilayer and allows pore expansion and stabilization. Although the process of formation and dynamics of each pore is stochastic, on the scale of cells and tissues, the effects of membrane electroporation only become detectable at $\Delta\Psi_m$ exceeding a certain "critical" value. This critical value has been reported in the range from few hundreds of millivolts to about 1 V; though, it was found that it depends on experimental conditions and cell type. Since $\Delta\Psi_m$ appears to be the driving mechanism of electroporation, it is crucial to know its time and spatial distribution during application of an electric pulse. In this chapter we thus review three different approaches for determining $\Delta\Psi_m$: analytical, numerical, and experimental. Based on combined measurements of $\Delta\Psi_m$ and transmembrane molecular transport, we also demonstrate that transmembrane transport is indeed confined to the membrane regions, where $\Delta\Psi_m$ exceeds a certain critical value.

P. Nick
Botanical Institute, Karlsruhe Institute of Technology, Karlsruhe, Germany

W. Frey
Institute for Pulsed Power and Microwave Technology, Karlsruhe Institute of Technology, Eggenstein-Leopoldshafen, Germany

D.A. Dean
Department of Pediatrics, University of Rochester, Rochester, NY, USA

K. Morotomi-Yano
Institute of Pulsed Power Science, Kumamoto University, Kumamoto, Japan

R.E. Neal II
AngioDynamics Inc., Queensbury, NY, USA

S. Bhonsle • R.V. Davalos
Biomedical Engineering, Virginia Tech, Blacksburg, VA, USA

4.1.1 Cell in the Electric Field

From the electrical point of view, a cell can roughly be described as an electrolyte-resembling solution (the cytoplasm) surrounded by a thin layer of dielectric material (the cell membrane) and immersed into another electrolyte-like solution (the extracellular medium). The cell can be thus considered as a closed capacitor. This (electrically) heterogeneous structure allows electric fields to act particularly on the cell membrane and modify the transmembrane voltage. Modification of the transmembrane voltage can then result in a variety of profound biochemical and physiological responses in biological cells, such as cell migration, modulation of cell growth, triggering of action potentials in excitable cells, and membrane electroporation [1–3].

In practically every cell, there is already an endogenous transmembrane voltage present in physiological conditions, which is termed the *resting potential* (or *resting voltage*). The resting potential arises from a charge imbalance on two sides of the cell membrane and is maintained by a system of ion channels and pumps in the membrane. Typically, the interior of the cell is slightly more negative than its exterior, and the resting potential reaches up to several -10 mV (depending on the cell type), when measured from the inside toward the outside of the cell. When exposed to an external electric field (such as by placing the cell between two electrodes connected to a voltage pulse generator), an additional component termed the *induced transmembrane voltage* and denoted by $\Delta\Psi_m$ is superimposed onto the resting potential. Unlike the resting potential, $\Delta\Psi_m$ is present only for the duration of the exposure to the electric field, is proportional to the electric field strength, and varies with the position on the membrane.

Let us first consider the general physical picture of what happens when the cell is placed between two electrodes, and we change the electric potential on one of the electrodes (i.e., a step change in the voltage between the electrodes). At first instance, an electric field is established between the electrodes. The electric field then acts to rotate (polarize) water and other dipolar molecules, e.g., lipid head groups and proteins. The relaxation time of water molecules in pure water is about 8 ps at room temperature, the relaxation time of water molecules bound to the membrane interface is on the order of 100 ps, lipid head groups on the order of 1 ns, and proteins up to the order of 100 ns [4–6]. Particularly the polarization of dipoles at the water–membrane interface partially contributes to an increase in $\Delta\Psi_m$ [7]. The other contribution comes from redistribution of charged ions and molecules in the extracellular and intracellular solutions by the influence of the electrophoretic force, which begins to charge the membrane. How fast this process is, depends on the cell size and the conductivity of the solutions (the mobility and concentration of charged particles in the solution). For cells in medium with physiological conductivity, it takes up to few microseconds to fully charge the cell membrane; however, in low conductive extracellular medium, this process can also take tens of microseconds [8]. As long as the cell membrane is charging and its transmembrane voltage is changing, the interior of the cell is exposed to the electric field and the membranes of the intracellular organelles are charged as well. In the first tens of nanoseconds, the voltage induced on an organelle membrane is comparable to the voltage induced on the cell membrane. However, the more the cell membrane is charged, the lower is the electric field inside the cell. Due to lowering of the intracellular electric field, the voltage on the organelle membranes starts to drop. After a certain time, the cell membrane is completely charged and reaches a steady-state value of $\Delta\Psi_m$. The cell membrane at this point acts as a shield, preventing its interior to be exposed to the electric field.

This physical picture can be mathematically described in terms of the partial differential equation:

$$\nabla\left[\left(\sigma+\varepsilon\frac{\partial}{\partial t}\right)\nabla\Psi(x,y,z,t)\right]=0 \tag{4.1}$$

with Ψ denoting the electric potential, σ denoting the electrical conductivity, and ε denoting the dielectric permittivity. If we replace the partial derivative $\partial/\partial t$ with its complex equivalent s, the equation can also be solved in the complex frequency domain for sinusoidal electric fields or arbitrary pulse shape using Laplace transform [9, 10]. This approach is general regardless of the size, shape, and arrangement of the cells, but is only valid as long as the finite speed of electromagnetic wave propagation can be neglected, i.e., we assume that the electric field is established between the electrodes instantly. This equation needs to be solved in every region that can be attributed a certain electrical conductivity σ and electric permittivity ε (Fig. 4.1a). The $\Delta\Psi_m$ can then be calculated as the difference between the electric potentials on each side of the region, representing the membrane.

If at any point $\Delta\Psi_m$ gets high enough to induce sufficient number of conductive pores, the pores allow enhanced ionic transport through the membrane which increases the membrane conductivity by several orders of magnitude. This in turn partially discharges the membrane and thereby reduces $\Delta\Psi_m$ during the pulse [11, 12]. In such case the membrane conductivity has to be considered as a function of $\Delta\Psi_m$ [13], which will be in more detail described in Sect. 4.2. In this chapter, however, we will only consider the case of $\Delta\Psi_m$ for nonporated membrane, where the membrane conductivity can be considered constant.

4.1.2 Determination of the Induced Transmembrane Voltage

For cells with regular shapes (spheres, spheroids, cylinders) that are sufficiently far apart (in dilute suspensions), Eq. (4.1) and thereby the time dependence and spatial distribution of the $\Delta\Psi_m$ can be derived *analytically* [15]. In the case of irregular shapes, cells close to each other (in dense suspensions, cell clusters, tissues), or when including nonlinear equations describing pore formation and pore dynamics accompanied by changes in the membrane conductance, the analytical approach is no longer possible. In such case, Eq. (4.1) has to be solved *numerically*. Another

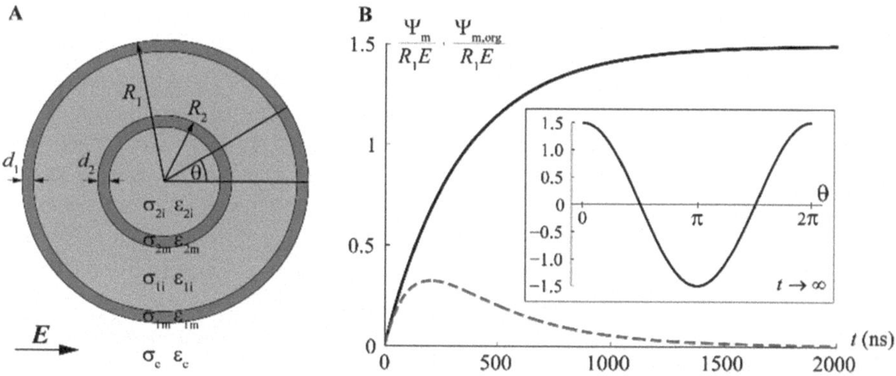

Fig. 4.1 (**a**) Model of a spherical cell with a concentric spherical organelle. The model consists of five regions, each characterized by an electrical conductivity (σ, in S/m) and a dielectric permittivity (ε, in As/Vm). Subscript index "e" corresponds to the extracellular solution, index "1" to the cell and index "2" to the organelle. (**b**) The time course of $\Delta\Psi_m$ (*solid*) and $\Delta\Psi_{m,org}$ (*dashed*), normalized by the electric field strength E and the cell radius R_1, at the point on the membrane, where the $\Delta\Psi_m$ and $\Delta\Psi_{m,org}$ reach the highest value ($\theta = 0$). Calculation was obtained based on analytical approach for solving (4.1) (see Sect. 4.1.2.1) for a spherical cell with radius $R_1 = 10$ μm containing a concentrically positioned spherical organelle with radius $R_2 = 3$ μm in medium with physiological conductivity. Inset shows the normalized $\Delta\Psi_m$ along the cell membrane in steady state, when the charging of the cell membrane is completed (Adapted from Ref. [14])

way for determining the $\Delta\Psi_m$ is by *experimental measurements* with voltage-sensitive fluorescence dyes. In the following subsections, we describe these three methods in more detail.

4.1.2.1 Analytical Derivation

We will begin with a model of an isolated spherical cell placed in a homogeneous electric field. Although biological cells are not perfect spheres, in theoretical treatments they are often considered as such. Particularly for cells in suspensions, this is a reasonable approximation.

If an electric pulse is long enough so that $\Delta\Psi_m$ reaches steady state, the time derivatives in (4.1) are zero, and the partial differential equation simplifies to the Laplace equation: $\Delta\Psi(x, y, z) = 0$. This equation can be solved in a particular coordinate system: for a spherical cell, in spherical coordinates by applying physically realistic boundary conditions (finiteness of Ψ, continuity of Ψ and its derivatives, and asymptotic vanishing of the cell's effect on Ψ with increasing distance from the cell). The detailed derivation of the spatial distribution of $\Delta\Psi_m$ can be found in [15]; we will only outline the results here.

As a first approximation, we can consider that the cell membrane has zero conductivity and that the intra- and extracellular solutions are perfect conductors. Moreover, since the cell membrane acts as a shield, the interior of the cell can also be neglected. This leads to a well-known steady-state Schwan equation:

$$\Delta\Psi_{\mathrm{m}} = 1.5\,ER\cos\theta \tag{4.2}$$

The $\Delta\Psi_{\mathrm{m}}$ is proportional to the electric field strength E and the cell radius R and varies with the cosine of the angle θ between the direction normal to the membrane surface and the direction of the applied electric field (Fig. 4.1, inset in B). The highest absolute value of $\Delta\Psi_{\mathrm{m}}$ is thus established at the points, which are closest to the electrodes (often referred to as the "poles" of the cell). This approximation is well suited for cells, which are placed in a solution of physiological conductivity (~1 S/m); however, for media with much lower conductivity (particularly 0.01 S/m and less [8]), the factor 1.5 reduces considerably. Moreover, when pulses are few microseconds long or less, the time dependence of $\Delta\Psi_{\mathrm{m}}$ of the membrane charging should also be taken into account. The more general expression describing $\Delta\Psi_{\mathrm{m}}$, which also takes into account the membrane conductivity, is the first-order Schwan equation:

$$\Delta\Psi_{\mathrm{m}} = fER\cos\Theta\left(1 - e^{-t/\tau_{\mathrm{m}}}\right) \tag{4.3a}$$

$$f = \frac{3\sigma_e\left[3dR^2\sigma_i + \left(3d^2R - d^3\right)(\sigma_m - \sigma_i)\right]}{2R^3(\sigma_m + 2\sigma_e)\left(\sigma_m + \frac{1}{2}\sigma_i\right) - 2(R-d)^3(\sigma_e - \sigma_m)(\sigma_i - \sigma_m)} \tag{4.3b}$$

$$\tau_m = \frac{R\varepsilon_m}{d\frac{2\sigma_e\sigma_i}{2\sigma_e+\sigma_i} + R\sigma_m} \tag{4.3c}$$

Note that both the factor f and the time constant of membrane charging τ_m depend on the properties of the extracellular medium as well as the electrical and geometrical properties of the cell.

If the pulse length is further reduced below 1 μs, the dielectric permittivity of the aqueous media surrounding the cell needs to be taken into account, and expression (4.3) expands into the second-order Schwan equation [10]. This also applies for sinusoidal electric fields with frequencies above 1 MHz.

In the submicrosecond range, particularly for pulses that are on the order of 10 ns, the transmembrane voltage on intracellular organelles becomes comparable to the transmembrane voltage on the cell membrane (Fig. 4.1) [14]. For a spherical organelle positioned concentrically inside a spherical cell, the $\Delta\Psi_{\mathrm{m}}$ (here denoted $\Delta\Psi_{\mathrm{m,org}}$) can also be derived analytically. Such representation of a cell is referred to as the "double-shell model" and is also often used for determining electrical properties of cells using dielectric spectroscopy [6, 16]. The analytical expression is rather complicated and lengthy and will not be presented here, but can be found in [14]. Let us just note that both the $\Delta\Psi_{\mathrm{m}}$ and $\Delta\Psi_{\mathrm{m,org}}$ are proportional to the electric field strength and vary with cosine of θ, but both are also dependent on the electrical properties of the extracellular medium, as well as geometrical and electrical properties of the cell and the organelle. For cells that have a large nucleus, the nucleus can considerably affect the $\Delta\Psi_{\mathrm{m}}$ during the charging phase, so a double-shell model needs to be used also for correctly determining $\Delta\Psi_{\mathrm{m}}$ on the cell membrane, even if $\Delta\Psi_m$ on the nuclear membrane is not of interest [17]. The $\Delta\Psi_{m,org}$ is also roughly proportional to the size of the organelle, so a higher electric field is required to induce a similar $\Delta\Psi_{m,org}$ on a smaller organelle [18].

In the nanosecond range, $\Delta\Psi_m$ and $\Delta\Psi_{m,org}$ become comparable; therefore, nanosecond pulses which result in sufficiently high electric field can be used to electroporate both the cell membrane and membranes of intracellular organelles. This corroborates with experiments [19–22]. An example is shown in Fig. 4.2, where multiple 60-ns, 50-kV/cm pulses were used to electroporate membranes of postendocytotic vesicles in mouse melanoma cells B16-F1 [22]. Electroporation of vesicles (detected by release of fluorescent dye Lucifer yellow) was also accompanied by plasma membrane electroporation, detected by uptake of propidium ions.

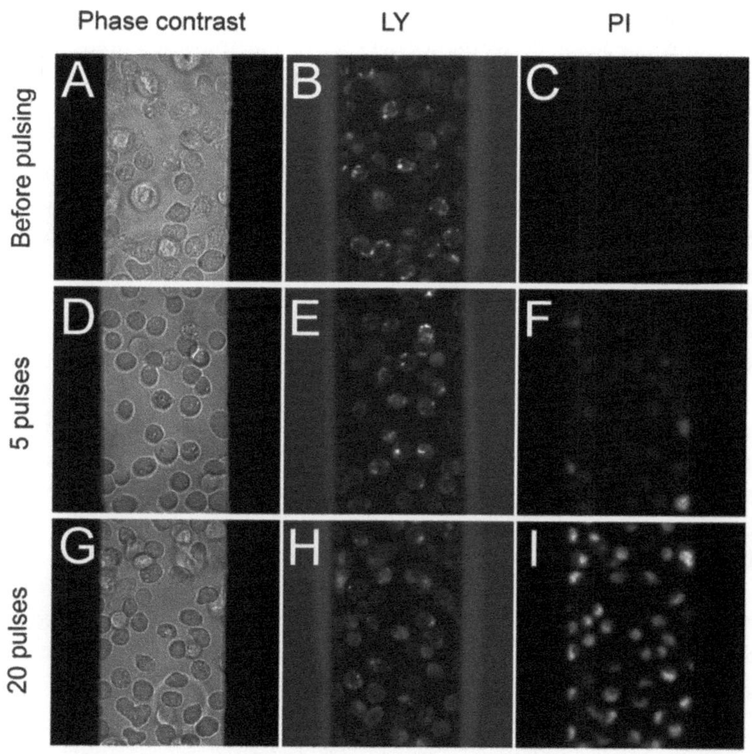

Fig. 4.2 Electroporation of postendocytotic vesicles in mouse melanoma cells B16-F1 cells after applying multiple 60 ns, 50 kV/cm at 1 kHz repetition frequency. Vesicles were loaded with membrane-impermeable dye Lucifer yellow (LY), and the extracellular medium contained membrane-impermeable dye Propidium Iodide (PI). Phase contrast, LY, and PI fluorescence images of cells before pulsing (**a–c**), 10 min after applying five pulses (**d–f**) or 20 pulses (**g–i**). When no pulses are applied, vesicles appear in LY fluorescence image (**b**) as distinct bright dots. The cell membrane is intact as can be seen by the absence of fluorescence in PI fluorescence image (**c**). After application of five pulses, only electroporation of the cell membrane is detected by an increase in PI fluorescence inside the cells (**f**). After 20 pulses, electroporation of vesicles is also detected by release of LY from the vesicles, which can be seen as decrease in fluorescence of the vesicles accompanied by increased LY fluorescence in the cytosol (**h**) (From Ref. [22])

4.1.2.2 Numerical Computation

Numerically, (4.1) can be solved, e.g., with finite difference method [23], finite volume method [24], finite element method [25], distributed equivalent circuit representation [26], or a transport lattice approach [27]. Here, we present the finite element method, which is well suited for handling curved boundaries and has been used by our group to compute the steady-state $\Delta\Psi_m$ of irregularly shaped cells [25, 28]. Finite element computation of $\Delta\Psi_m$ is generally performed in four steps. First, the three-dimensional geometry of the cell(s) of interest is constructed. Second, the continuous geometry is "meshed" into discrete, usually tetrahedral elements, and the partial differential Eq. 4.1 describing the electric potential is transformed by the finite element method into a matrix equation of algebraic expressions. Third, the matrix equation is solved, either directly or iteratively (until reaching adequate convergence). Finally, the transmembrane potential is extracted from the computed spatial distribution of the electric potential as the difference between the electric potential on each side of the membrane. The three-dimensional model of an irregularly shaped cell can be constructed from a sequence of cross-sectional images of the cell under consideration, based on microscopic images of the cell membrane, stained with a fluorescent membrane dye, such as di-8-ANEPPS (Fig. 4.3). This model can be imported into a suitable software package (e.g., Comsol Multiphysics), which discretizes the model into finite elements and solves (4.1) by the finite element method.

Explicit representation of the cell membrane in the model can nevertheless lead to problems. The meshing of the cell membrane, which is over 1000-fold thinner than the dimensions of a typical cell, requires the model to consist of an extremely large number of small finite elements. This consequently requires a large amount of computer memory and long computational time to solve the matrix equation. However, unless the spatial distribution of the electric potential inside the membrane is of interest, this can elegantly be avoided by representing the membrane with a boundary condition describing the transmembrane current density J_n [25]. As far as intracellular and extracellular potential is concerned, the effect of the membrane with thickness d, electrical conductivity σ_m, and electric permittivity ε_m is equivalent to the effect of an interface with thickness 0, surface electrical conductivity σ_m/d, and surface electric permittivity ε_m/d:

$$J_n = \frac{\sigma_m}{d}(\Psi_{\text{int}} - \Psi_{\text{ext}}) + \frac{\varepsilon_m}{d}\frac{\partial}{\partial t}(\Psi_{\text{int}} - \Psi_{\text{ext}}) \tag{4.4}$$

The first term on the right-hand side represents the conductive and the second term the capacitive component of the electric current flowing through the membrane. By assuming that σ_m is a function of $\Delta\Psi_m$, this approach can be extended further, e.g., to simulate the time course of electroporation [28].

In addition to irregular cells, numerical computation can be used for determination of $\Delta\Psi_m$ in cells in dense suspensions. When cells are close to each other, the local field around each cell is affected by other cells, and the spatial distribution of $\Delta\Psi_m$ starts to deviate significantly from that given by (4.2) and (4.3). As the volume

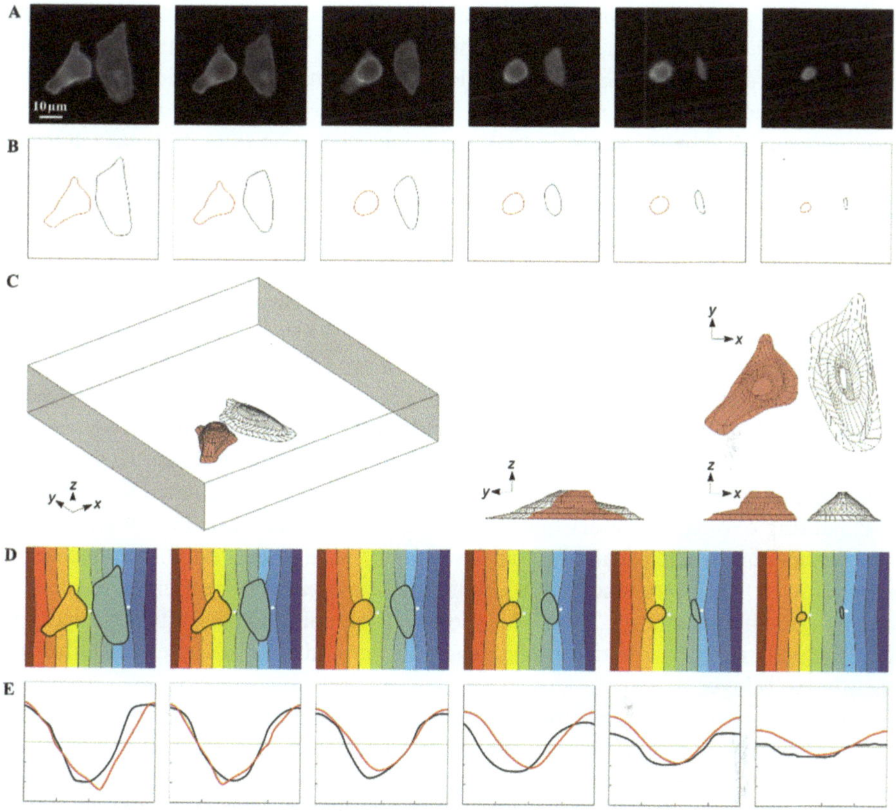

Fig. 4.3 Numerical computation of $\Delta\Psi_m$: (**a**) fluorescence cross-section images of two cells, (**b**) contours of their cross sections, (**c**) three-dimensional model of the cells constructed from the contours, (**d**) computed distribution of electric potential, (**e**) computed $\Delta\Psi_m$ (*red*: left cell, *black*: right cell) (From Ref. [25])

fraction occupied by the cells increases beyond 10 % and approaches 50 %, the factor 1.5 gradually decreases toward 1, and the distribution also starts to diverge from the ideal cosine shape [29, 30]. Due to the lower $\Delta\Psi_m$, the efficiency of electroporation with the same pulse parameters is typically lower in dense suspensions than in dilute ones [31].

4.1.2.3 Experimental Measurement

An alternative to both analytical and numerical determination of $\Delta\Psi_m$ is the experimental approaches. These include measurements with microelectrodes and with potentiometric fluorescent dyes. Microelectrodes (either conventional or patch clamp) were used in pioneering measurements of the action potential propagation [32, 33] and were preferred for their simple use and high temporal resolution. Nevertheless, the invasive nature of microelectrodes, their low spatial resolution, and physical presence, which distorts the external electric field, are considerable shortcomings. In contrast, measurements by means of potentiometric dyes are noninvasive, offer higher spatial resolution and do not distort the field and by that the $\Delta\Psi_m$, and allow the measurement to be performed on a number of cells simultaneously. Potentiometric dyes such as di-8-ANEPPS [34–36], RH292 [11, 12], and ANNINE-6 [37, 38] have thus become the preferred tool in experimental studies and measurements of $\Delta\Psi_m$, experimental studies of voltage-gated membrane channels, and monitoring of nerve and muscle cell activity. These dyes incorporate into the lipid bilayer of the cell membrane, where they start to fluoresce, with their fluorescence spectra being dependent on the amplitude of the transmembrane voltage; the relative change in fluorescence of these dyes is linearly dependent on the transmembrane voltage in a certain range of voltages. With a suitable experimental setup incorporating a pulse laser, a fast and sensitive camera, and a system for synchronizing the acquisition with the field exposure, these dyes enable monitoring of the time variation of $\Delta\Psi_m$ with a resolution of microseconds, and in the case of ANNINE-6, down to nanoseconds. The latter dye was used to determine $\Delta\Psi_m$ during exposure to electroporating 60-ns long pulses [37, 38]. However, the measurements revealed that $\Delta\Psi_m$ on the cathodic side of the membrane is considerably lower that on the anodic side, which was attributed to reorientation of lipid head groups by the electric field. This reorientation could locally affect the electric field, which is felt by the dye molecules incorporated into the outer leaflet of the membrane bilayer.

The measured voltage can then be compared to theoretical prediction, either by analytical derivation (e.g., the case of spherical cells) or by numerical computation for irregularly shaped cells. In both cases, one obtains a very good agreement, which confirms the applicability of (4.1) for determining the transmembrane voltage (Fig. 4.4), at least in the microsecond and millisecond range and below electroporation threshold, i.e., while membrane conductivity can be considered constant. Figure 4.4c compares $\Delta\Psi_m$ measured with di-8-ANEPPS during a nonporative 50-ms pulse to numerically predicted steady-state $\Delta\Psi_m$.

4.1.3 Correlation Between Induced Transmembrane Voltage and Molecular Transport

To monitor transmembrane molecular transport and thereby determine the electroporated regions of the membrane, cells are exposed to the electroporating electric field in the presence of an otherwise membrane–impermeant fluorescent dye such as propidium iodide (PI). The fluorescence of PI increases by several orders of magnitude when the dye is bound to nucleic acids, due to which the localized entry of the dye through electroporated regions of the membrane into the cell is detected as a local increase in fluorescence. The correlation between the $\Delta\Psi_m$ and the electroporation-mediated transport across the membrane can be demonstrated particularly clearly by combining potentiometric measurements and

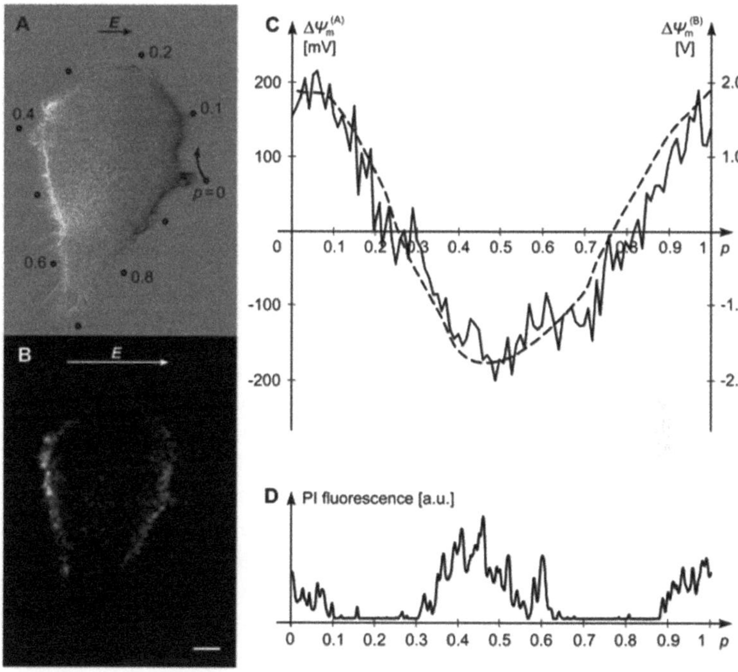

Fig. 4.4 The induced transmembrane voltage ($\Delta\Psi_m$) and electroporation of an irregularly shaped Chinese hamster ovary (CHO) cell: (**a**) changes in fluorescence of di-8-ANEPPS reflecting $\Delta\Psi_m$, with dark regions corresponding to membrane depolarization and bright regions corresponding to membrane hyperpolarization; (**b**) fluorescence of propidium iodide (PI), reflecting transport of PI across the membrane; (**c**) $\Delta\Psi_m$ along the path shown in (**a**) as measured (*black solid*) and as predicted by numerical computation (*gray dashed*); (**d**) fluorescence of PI along the path shown in (**a**) (Adapted from Ref. [45])

monitoring of transmembrane transport on the same cell. An example is shown in Fig. 4.4. These experimental results confirm the theoretical prediction that the highest values of $\Delta\Psi_m$ are found in the membrane regions facing the electrodes (the "poles" of the cell) and show that electroporation-mediated transport is detected in these same regions, i.e., the areas for which the absolute value of $\Delta\Psi_m$ is above a certain critical value. This critical value is though a very rough estimate. It has been reported in the range from few hundreds of millivolts to about 1 V; however, it was found that it depends on experimental conditions and cell type and can vary even among cells of the same type [39]. Furthermore, the threshold of uptake detection depends on the sensitivity of the imaging system and generally on the dye being used for detection of molecular transport [40–43]. Due to stochastic nature of pore formation, higher $\Delta\Psi_m$ is also predicted to be required for formation of a similar number of pores, and hence detectable electroporation, if the pulses are shorter [44]. Albeit the critical value of $\Delta\Psi_m$ cannot be considered as a general predictor of detectable electroporation, the dependence of membrane electroporation and transmembrane molecular transport on sufficiently high $\Delta\Psi_m$ clearly demonstrates that molecular transport is closely correlated to $\Delta\Psi_m$.

4.2 Continuum Modeling for Bioelectrics

James C. Weaver, Kyle C. Smith, Reuben S. Son, and Thiruvallur R. Gowrishankar

Abstract A model is a quantitative hypothesis. For bioelectrics a model predicts what will happen for particular electrical conditions. Here we consider isolated, single cells with a single, outer membrane (plasma membrane or PM). Related models with multiple membranes treat cells with organelles or multiple cells close together (tissue). We use computational models, which can be both precise and complex, while still allowing quantitative testing. Existing models include multiple interactions and parameters and accept a wide range of applied field waveforms as inputs. Complex computational models are increasingly key to science and engineering. Here we present bioelectric modeling based on nonequilibrium processes. Physically, fields drive movement and/or reorientation of entities with permanent or induced charge. Biologically, living cells exist far from equilibrium, driven by metabolism. We recognize both sources of nonequilibrium processes as a basis for bioelectric models. Living cells involve many active processes. Some take place near or within membranes, which amplify externally applied electric fields by responding with larger membrane field changes. These amplified membrane fields couple to cell functions such as voltage-gated channels and the transient pores of electroporation (EP). For bioelectrics the electrical response (transmembrane voltages, local fields) and chemical changes (transport) are of interest. Many early models are passive, with electrical properties remaining fixed when a field is applied. We describe a meshed transport network method, focusing on EP, which causes changing membrane properties. These models' responses can be estimated for fields with strengths of 0–100 kV/cm and durations from ~1 ns to ~1 s, for both idealized and digitized experimental pulse waveforms.

4.2.1 Model Construction

Passive models with analytical expressions for cell responses are well established [10, 48]. Models with irregular membrane geometries and complex physiologic behavior are also established for normal physiology where bioelectrics is unimportant. Here we focus on models with large, spatially distributed membrane conductance changes by EP, using a general method. EP is the generally accepted mechanistic hypothesis for rapid (nanosecond) increases in local membrane conductance by field pulses [46, 47]. EP accounts for key experimental behaviors (e.g., flattening of the transmembrane voltage profile) and transport of dissolved (free) ions and molecules through the outer (plasma) and inner (organelle) membranes. EP is fundamentally nonthermal: there is Joule heating, but temperature rise itself causes insignificant pore creation.

4.2.2 Modeling Based on Nonequilibrium Transport

Models based on transport recognize that (1) fields drive charged species transport, e.g., electrodiffusion of small ions and large molecules, and (2) living cells operate far from equilibrium, e.g., metabolically driven resting potential sources. Basing models on transport inherently allows inclusion of both gentle and steep gradients in space and time. Gradients in electric potential define fields; gradients in solute concentrations (ions and molecules) drive electrodiffusion. Ubiquitous small ions control electrical behavior, while electrodiffusion of larger ions and molecules occurs within bulk media and through dynamic pore populations [46]. Meshing constrained by membrane geometry establishes transport networks [46]. Transport between nodes is governed by established equations, with solutions obtained by Kirchhoff's laws [49]. Here we discuss basic features of our modeling methods. Details and basic equations are publicly available [46, 47].

4.2.3 Simplest Cell Model

For basic concepts we use a cell model with a single curved membrane (cylindrical geometry, Fig. 4.5a, d) [46, 47]. In addition to being the simplest case, it is a starting point for multiple membrane models (Fig. 4.5b, c, e, f). In our models "transport" is general, including sources and sinks for transported free solutes. Even "movement" of pore radius (r_p) within a changing landscape where pores expand and contract is considered transport, mathematically analogous to electrodiffusion of solutes, described by the Smoluchowski equation (SE) [46, 50]. Here we focus on the simplest model (Fig. 4.5a, d).

4.2.4 Pore Models

Models of nanometer-scale pores include cylindrical, toroidal, and trapezoidal holes. Trapezoidal geometry reflects some cylindrical and toroidal features (Fig. 4.6a, b) and compares favorably with molecular dynamic (MD) pore geometries (Sect. 4.1). The trapezoidal model has some of the focusing fields within the flared openings, with the full focusing fields having well-known "spreading resistance" or "access resistance" [46, 47]. Focusing fields guide free charged solutes (ions and molecules, approximated as cylindrical) toward and away from pore interior regions. Within heterogeneous focusing field regions, solutes move by electrodiffusion toward and away from membrane pore interiors. Inside the restrictive, cylindrical pore interior (half the membrane thickness), electrodiffusion is more complicated.

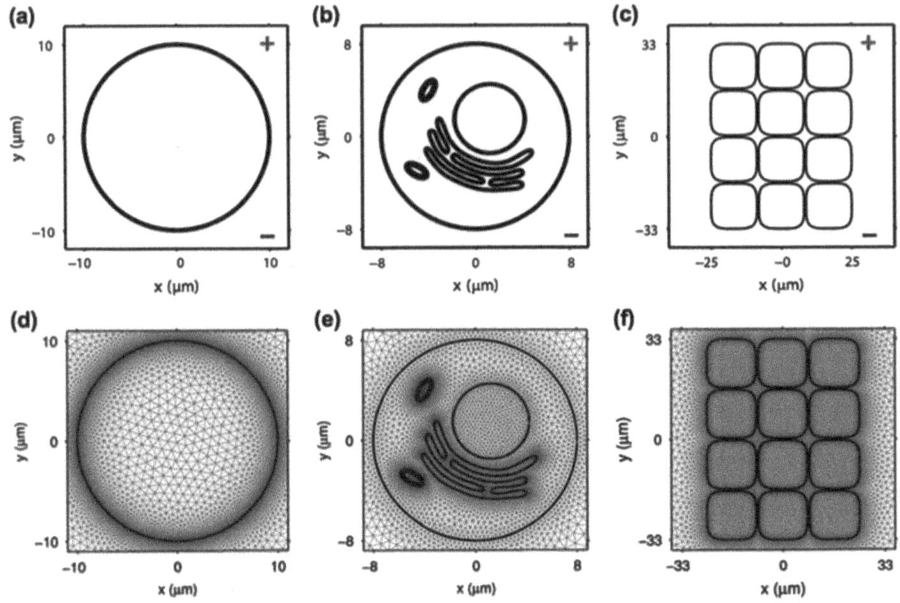

Fig. 4.5 Geometry and meshed transport network for cell models. (**a**) PM with the simplest, single curvature [46, 47]. (**b**) Multiple intracellular membranes, for example, two mitochondria (*ovals*) with mitochondrial outer and inner membranes (MOM and MIM, respectively), a very large endoplasmic reticulum membrane (*convoluted*) and a large double-membrane nuclear envelope (*circular*). As depicted here, a single curve represents either a single or double membrane, but the models have separate membranes, with different electrical properties when appropriate [46]. (**c**) Multiple cell model for studying basic EP-mediated solute transport in tissue. (**d–f**) Corresponding meshed transport networks for (**b–c**)

The mesh interconnects all nodes, with most in bulk aqueous media (Fig. 4.5). The most complex interactions occur in the membrane. Transmembrane node pairs contain local models for fixed (passive) components of the conductive and capacitive membrane properties, a metabolically driven resting potential source, and also a complete nonlinear, hysteretic membrane EP model with (1) an absolute rate equation for pore creation at r^*, (2) the Smoluchowski equation (SE) for pore expansion/contraction within the changing (U_m-dependent) energy landscape (Fig. 4.6c), and (3) pore destruction. The many interconnected interactions yield a cell system model.

The model includes K, the partition factor, which depends on pore size (radius, r_p) and electrical charge (charge number, z_s). Transport also depends on H, the hindrance factor, which is most restrictive for the pore interior. H depends on the relative size of the solute (radius, r_s) and pore radius (r_p) [46]. Electrodiffusion of each solute (e.g., propidium, calcein) is computed separately [47].

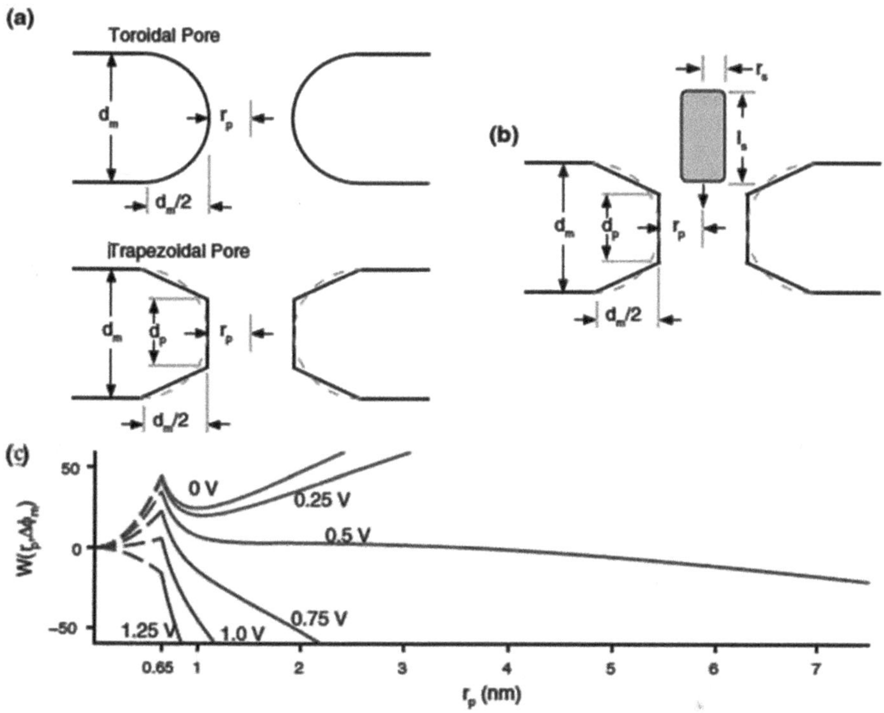

Fig. 4.6 Pore and solute geometries and energy landscape [46, 47]. (**a**) Toroidal and trapezoidal pores. (**b**) Trapezoidal pore with approaching cylindrical solute. (**c**) Pore energy landscapes for transmembrane voltage $U_m = 0$ to 1.25 V

4.2.5 Pore Energy Landscape and Pore Behavior

When a pore is created, it appears in the landscape, W, at r^* (0.65 nm, Fig. 4.6c). The Smoluchowski equation and landscape together govern the resulting "motion" of r_p, with pore expansion equivalent to r_p increasing and for pore contraction r_p decreasing. During a pulse W at each of the many transmembrane node pairs changes, so that pore populations (pore-size histograms) at different membrane sites evolve and eventually vanish postpulse. Pore populations interact through the aqueous portion of the mesh, so that the model's behavior is a system response. The landscape is usually shown for the case with a single minimum size pore ($r_{min} \approx$ 1 nm) present. The landscape is populated by stochastic pore creation at r^*, estimated by an absolute rate equation with a highly nonlinear dependence on U_m. After a typical pulse, the membrane is depolarized ($U_m = 0$). The landscape is then depopulated as pores diffuse to reach a small pore radius r_d slightly below r^* (a model feature) [46]. Due to the landscape's downward slope away from r_d at elevated U_m, a negligible fraction of pores reach r_d during a pulse.

4.2.6 Active and Passive Versions of Cell Models

To understand U_m-mediated effects, we need to know $U_m(t)$ over the PM. Early models use fixed (passive) electrical properties [10, 48]. But experiments [51] showed vividly that the local membrane conductance increases dramatically in some regions of a cell membrane for large field pulses, those now associated with electroporation. This is an active response. For the same membrane geometry, both passive and active versions can be created, with the only difference that electroporation is "knocked out" (omitted) in passive models, leaving only fixed local conductance and capacitance (dielectric) membrane values (Supplementary Information of [47]).

Figure 4.7a is early ($t = 1$ μs) in the long pulse. It exhibits the traditional cosinusoidal passive (dot-dash curve) response, with "approximate amplification" $(R_E)_{max} \approx 2300$, but for the active (solid) curve $(R_E)_{max} \approx 1800$, and at the poles only ~50, the site of appreciable R_E flattening [51]. (b) As in (a) but at 99 μs, end of the short pulse flat peak value. Here the passive case has $(R_E)_{max} \approx 4000$, but the active case has smaller $(R_E)_{max}$ within porated regions, owing to significant expansion of some pores, which increases membrane conductance that diminishes E_m. Both (a) and (b) show voltage division dominated by membrane conductance, which due to poration is much larger for (b) than (a). (c) Early (1 ns) in short pulse with passive and active curves indistinguishable, as only a few pores are created, with insignificant membrane conductance increase. $(R_E)_{max}$ is only ~10, due to dielectric voltage division (displacement current >> conductance density) [13] (Fig. 4.7c). (d) Late (9 ns) in the short pulse, the passive and active responses are indistinguishable except near the poles, where enough pores exist that conduction currents exceed displacement currents. Here $(R_E)_{max} \approx 100$ for the passive case and ≈ 90 for the active case.

Figure 4.7 illustrates important constraints. Comparison of the long pulse (conventional EP) and short pulse (surpa-EP by nsPEF) reveals the increasingly important contribution of dielectric effects as pulse durations are shortened. R_E becomes

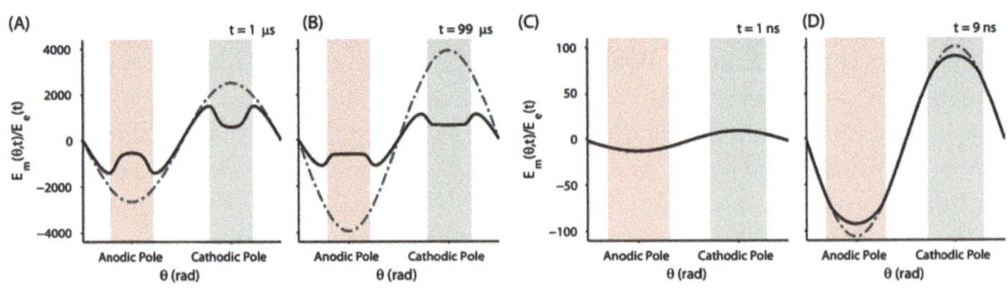

Fig. 4.7 Amplification gain factor, E_m/E_e, is shown as a function of angle at different times. (**a, b**) Show the response to a 1.5 kV/cm, 100 μs pulse at the start and end of the pulse maximum. (**c, d**) Show the response to a 40-kV/cm, 10-ns pulse at the start and end of the pulse maximum. The *dash-dotted curve* represents the passive membrane amplification response, in which the dynamic electroporation model has been "knocked out." In (**c**), the two curves are indistinguishable because the onset of electroporation has not occurred [47]. *Light red* and *gray* regions show porated regions

relatively small, and an increasingly large creation rate is needed to compensate for a shorter pore creation time. Both small R_E and short pulse duration tend to decrease pore creation. Nevertheless nsPEF is very effective in porating multiple cell membranes. As field strength increases progressively, smaller organelle membranes are porated, such as the MOM and MIM of mitochondria are permeabilized and then depolarized after the pulse.

4.2.7 Electrical and Poration Behavior

EP is characterized by first rapid and then slow changes. Figure 4.8d, j shows pore number, $N(t)$ for the entire membrane. Figure 4.8a, g shows $U_m(t)$, averaged over quadrants of the membrane. Although the total pore number, $N(t)$, increases for widely used pulses, the pore-size distributions (dynamic pore populations) change during the pulse. Importantly, $U_m(t)$ and $N(t)$ are interrelated, each affecting the other. This complexity arises from rapid electrical interactions within the cell and almost equally rapid changes in membrane properties (pore creation/expansion). This results in spatially distributed, conducting pores of changing sizes, which interact electrically through the aqueous media within the model [46, 47].

Figure 4.8a shows $U_m \approx$ first rising by membrane charging for the long pulse. As U_m approaches ~1 V, a burst of pore creation occurs, creating a large membrane conductance, and U_m falls in ~1 μs to ~0.5 V. This phenomenon is termed reversible electrical breakdown (REB). Figure 4.8g has simpler behavior for the short pulse, with a peak U_m due simply to the pulse maximum ending at 9 ns. For both pulses (Fig. 4.8a, g), pore creation and destruction are well separated on the pulse timescales. Creation rates are finite, so a nonzero time is needed to create enough pores to cause a detectable EP effect. This means there are no general critical values of U_m. For typical pulses longer than a few microseconds, two pore populations evolve during a pulse. One has small pores, while the other has dynamic large pores. At the pulse end, large pores rapidly contract, leaving a narrow distribution of small pores, and these slowly (seconds) vanish stochastically [46, 47].

4.2.8 Solute Transport Rates and Cumulative Amounts

Figure 4.9 shows illustrate influx rates and cumulative amounts (molecule numbers) for transport of two fluorescent probes, calcein and propidium, for the long pulse. This provides estimates of the relative importance of free solute transport during and after a pulse. These show the rates of influx, both during and after the pulse, with postpulse diffusive transport estimated for five 4 s pore lifetimes (20 s), for two versions of the basic model.

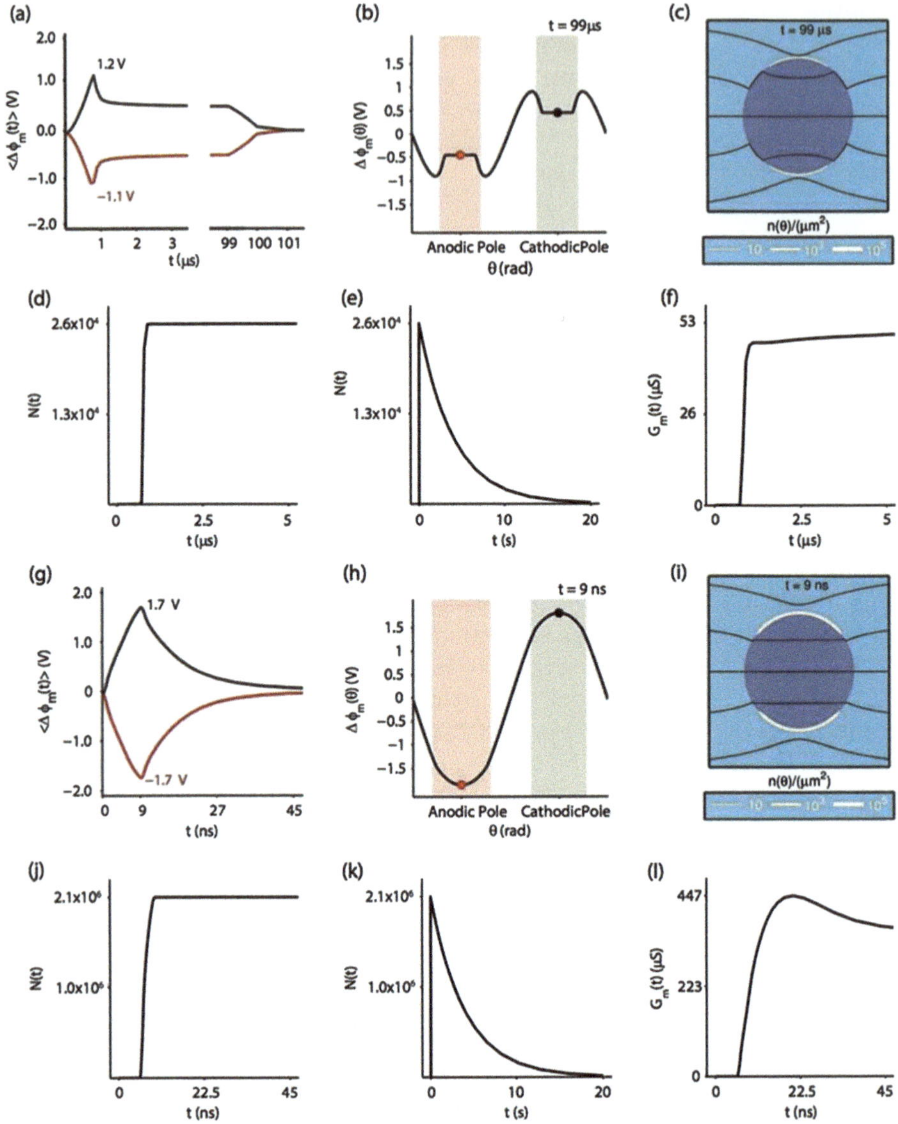

Fig. 4.8 Illustrative EP responses for a "long pulse" (1.5 kV/cm, 100 µs; **a–f**) and a "short pulse" (40 kV/cm, 10 ns; **g–l**). Average transmembrane voltage, $< \Delta\varphi_m(t) >$, is spatially averaged over the polar quadrants. Angular transmembrane voltage, $\Delta\varphi_m(\theta)$, at the end of the pulse maximum shows the spatial heterogeneity in the electrical response across the membrane. Light red and black shaded regions represent electroporated regions of the membrane. Similarly, pore density, $n(\theta)$, at the end of pulse maximum shows the spatial extent of electroporation. The total number of pores, $N(t)$, rises after REB and decreases exponentially after the pulse in accordance with pore lifetime, $\tau_p = 4$ s. Similarly, the total membrane conductance, $G_m(t)$, increases with pore creation and expansion [47], with short pulse creating ~10× larger membrane conductance

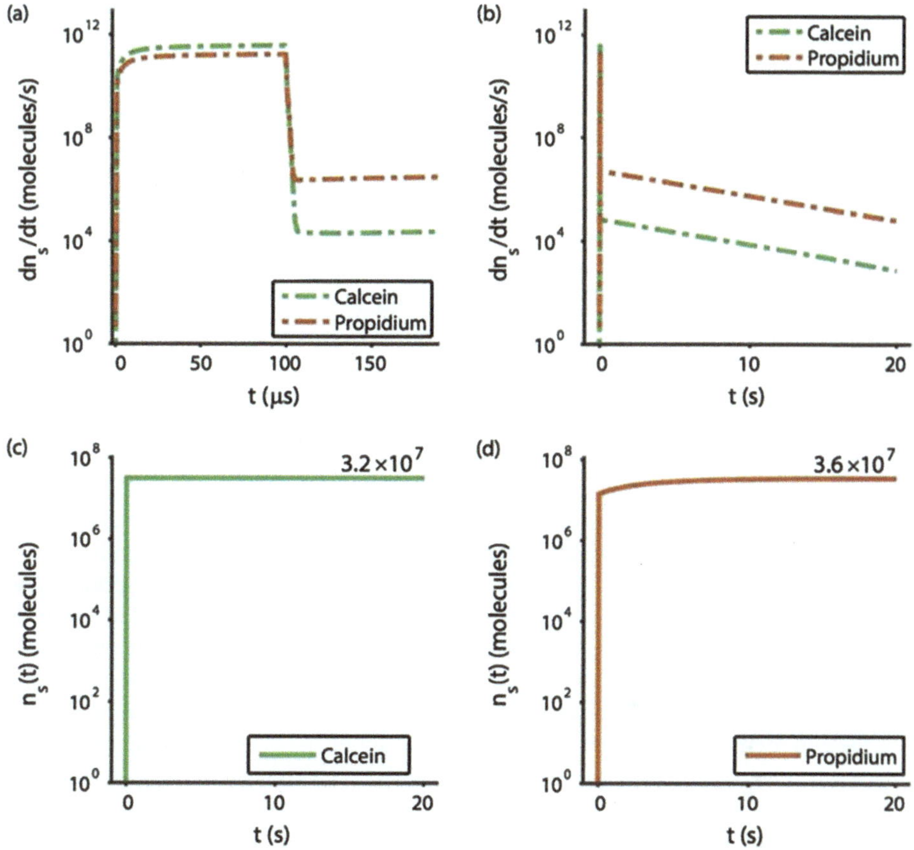

Fig. 4.9 Electrodiffusive transport rates of propidium and calcein, dn_s/dt, into the cell for a 1.5-kV/cm, 100 μs pulse on two different timescales (**a**, **b**) [47]. Cumulative solute influx, $n_s(t)$ (**c**, **d**), shows the total amount of solute inside the cell as a function of time. During the 1.5-kV/cm, 100 μs pulse, 3.2×10^7 molecules of calcein and 1.5×10^7 molecules of propidium are transported into the cell. After the 100 μs pulse, 2.8×10^5 molecules of calcein and 2.1×10^7 molecules of propidium are transported into the cell

4.3 Molecular Models of Lipid Electropores

P. Thomas Vernier and Zachary A. Levine

4.3.1 Molecular Models of Electroporation Before Molecular Dynamics

Early studies of electroporation (electropermeabilization) combined experimental observations of artificial and living cell membranes in porating electric fields with analytical models of the transformation of an impermeant barrier to a permeabilized membrane. Electroporation at first was understood primarily phenomenologically, as an increase in membrane conductance caused by the application of external electric fields. This change in the electrical properties of the membrane was associated with an increase in the permeability of the membrane to ions and small molecules, but there were at the time no experimental methods capable of revealing the membrane structural modifications that might be involved. Thus the first models of electropermeabilization were based on selected physical properties of lipid bilayers combined with the understanding of cell membrane structure that was available at the time, summarized in the fluid mosaic model [52], and did not include a description of the molecular rearrangements of membrane constituents that must be involved. The phenomenon was clear, but the mechanism was poorly understood.

Even the earliest models of electropermeabilization, however, did not entirely ignore the molecular composition and organization of cell membranes and lipid bilayers. The stochastic membrane defects whose formation is favored when the transmembrane potential is increased were defined, almost from the beginning, not simply as "defects" or as sites of "breakdown", but as populations of membrane-spanning pores, with probabilities of creation and annihilation, subject individually and in composite to expanding and contracting forces. Theoretical models and experimental observations were consistent with pore diameters on the order of 1 nm, comparable to the area occupied by the individual phospholipid molecules making up the lipid bilayer [53]. Placing the electric field-facilitated evolution of a population of transient aqueous pores on the framework of the fluid mosaic model of the cell membrane implied nanometer-scale dimensions for individual pores, but the molecular architecture of the pores themselves was not known.

Not surprisingly, given the absence of observational evidence and the likelihood that none would be forthcoming soon, this lack of knowledge did not prevent theoreticians (and some experimentalists) from picturing lipid electropores in molecular detail, at least in schematic form. Diagrams ("cartoons") of porated membranes, showing the individual phospholipids in the bilayer with the arrangement of the head and tail portions of the molecules emphasized, appear in some of the earliest electropermeabilization papers [54–57]. These drawings demonstrate the conceptualized structures of hydrophobic and hydrophilic pores, and they help to visualize how the head group dipole might participate in pore formation.

Because these schematic views do not include water and because they do not quantitatively represent how head group dipoles respond to the electric fields that are present when pores form, they fostered assumptions that led to incomplete descriptions of the roles of water and head group dipoles in lipid electropore initiation and construction [7, 58, 59]. With no direct observational methods on the horizon, the next step toward a more complete understanding of the mechanisms of electroporation was a physics-based, atomically detailed model of the lipid electropore, through molecular dynamic simulations.

4.3.2 Molecular Dynamic Models of Lipid Bilayers

The term molecular dynamics is commonly applied to two broad categories of molecular models. One kind of molecular dynamics—quantum mechanical (QM)—incorporates quantum mechanical considerations into the electronic interactions, both intramolecular and intermolecular, among the constituent atoms of the system. This approach is necessary when simulating chemical reactions, including electron transfer, or any system in which the conformation or any other property of a molecule is dependent on electronic energy levels or electron density distributions.

Quantum electronic molecular dynamics is computationally demanding, however, which limits simulations in most cases to a relatively small number of atoms and relatively short timescales. Representing pore formation in a lipid bilayer requires the inclusion of hundreds of thousands of atoms and hundreds of nanoseconds, system sizes that are at present out of reach.

The second kind of molecular dynamics—molecular mechanical (MM)—does not explicitly represent electronic states. Instead, MM tracks the mechanical and electrostatic interactions of atoms described by utilizing classical potential functions (with coefficients that are determined by "force fields," in the jargon of molecular dynamics). This greatly reduces the computational complexity per atom, enabling molecular simulations of systems large enough to represent lipid electropore formation.

The mathematics that permit this depend on several assumptions. For instance, the Born–Oppenheimer approximation allows us to calculate nuclear (atomic) trajectories without taking electron distributions into account. Atoms are treated as point masses with fractional charges, classically distributed to mimic the true electronic structure. Bonded interactions are described by three functions: axial (stretching), angular (planar), and torsional (twisting). The functional forms of these functions are harmonic, necessitating the use of spring constants across multiple degrees of freedom. Nonbonded interactions are tabulated over longer ranges (electrostatic) and shorter ranges (van der Waals). Each of these functions are optimized empirically, atom type by atom type, to produce spatially dependent potential energy expressions that map as closely as possible to the experimentally known behavior of the molecules represented in a given system.

A third variant of molecular dynamics, which we may expect to see more in the future, is a hybrid of the quantum and molecular mechanical approaches (QM/MM). In QM/MM, a critical portion of a molecule (e.g., the active site of an enzyme) is described quantum mechanically and then mapped onto a set of molecular mechanical coordinates across the entire system [60].

Within the framework of the molecular mechanical approach, which has proven to be both useful and practical for studies of phospholipid bilayers in electric fields, several levels of detail and definition are possible. One must judge whether the increased precision or accuracy justifies the longer simulation times or the larger number of computing cores needed for a given simulation. For example, instead of "all-atom" force fields, which represent each individual atom in the system, one may elect to use a "united-atom" model, in which small groups of atoms are treated as single entities. The $-CH_3$ and $-CH_2$ groups in a hydrocarbon chain, including those in the "tail" of a membrane phospholipid, behave very similarly in united-atom and "all-atom" simulations, but the computational demands of united-atom models are considerably smaller. For lipid bilayers, at least, the united-atom approach has proven effective [61, 62].

Some groups have moved even further in this direction, with coarse-grained models that extend the united-atom method to larger atom groups. This has the benefit of further savings in computational costs, but it remains to be seen whether coarse-grained simulations of electroporation are accurate enough to be quantitatively predictive or to capture the details of pore creation and annihilation, which seem to be critically dependent on the interactions of individual water molecules and phospholipid atoms. It seems likely that atomic-scale resolution is necessary, not only to represent the molecule-by-molecule construction of membrane-spanning bridges of intruding water that marks the pore initiation process and the closely coordinated hydration of the phospholipid head groups that follow the water into the membrane interior but also to model the interactions of ions and other solutes with pore walls in simulations of electrophoretic and diffusive transport through electropores.

Additional simulation details for phospholipid bilayer systems that must be taken into account include the specific models chosen for water and inorganic ions and for other constituents such as amino acids, sugars, mononucleotides, and larger molecules. More accurate models, such as those including polarizable bonds (e.g., for water), may be essential for accurate predictions of pore creation time and pore lifetime and for better representations of pore geometries in the presence of different salts at varying ionic strengths and interfacial lipid and water properties at the pore mouth and along the pore walls. Increasing access to more computing power will make the implementation of these options more efficient and standardized within the molecular dynamic community.

Although great insights and improved understanding have come from molecular dynamic simulations of lipid bilayers, it is important to keep in mind the limitations of these models, which constrain what they can tell us about living cell membranes in electric fields. The patch of membrane represented is very small, about one ten-millionth of the area of a typical cell. Even areas of a few hundred square nanometers, where multiple pores might form simultaneously and which would be large enough so that the diverse composition of a cell membrane (multiple lipid types, membrane attachments and associations with intracellular and extracellular structures, membrane proteins) might begin to be represented, are barely within reach. Simulation times are similarly limited by practically available computing power, to less than a microsecond for typical computing clusters. Finally, as mentioned above, most of the systems simulated today in electroporation studies contain only a few lipid types (often only one) and usually no other membrane constituents, the most glaring missing component being membrane proteins.

4.3.3 Molecular Dynamic Simulations of Lipid Bilayers in Electric Fields

To simulate electroporation, we first assemble a membrane. Most molecular dynamic simulations of electroporation in the literature use homogenous, fully hydrated phospholipid bilayers, for reasons discussed above. A system might contain, for example, palmitoyloleoylphosphatidylcholine (POPC, a common phospholipid found in many cell membranes), water, and nothing else (Fig. 4.10). Although this is far from the complexity of a living cell membrane and the

Fig. 4.10 Water bridge in POPC bilayer formed after application of an external electric field. Water molecules are *red* (oxygen) and *white*. Note the alignment of the water dipoles in the electric field in the low-permittivity environment of the membrane interior. The *gold spheres* are the acyl oxygens on the glycerol backbone of POPC. The phospholipid hydrocarbon tails are not shown. The simulated volume is approximately $7 \times 7 \times 10$ nm

surrounding medium, much has been learned from experimental observations of electroporation of artificial membranes of similar composition, and this body of knowledge can be used to validate the molecular models. With accumulated experience and increasing computing power, molecular simulations can iteratively become more complex and more realistic.

More and more, as molecular simulation tools are used to probe the process of electroporation more deeply, it becomes necessary to move beyond homogenous (single-phospholipid) bilayers in water; and consistently to include other lipids like phosphatidylserines, cardiolipins, sphingolipids, glycolipids, and cholesterol, and to represent a more physiological aqueous medium containing sodium, potassium, chloride, calcium, and other ions.

After the construction and stabilization of the membrane, a transmembrane electric potential must be applied. This can be done either by applying an external electric field or by creating an imbalance in the distribution of free charges on either side of the membrane. (These two methods can also be combined.)

The first method, which is simpler and is more often used, is implemented by globally applying a force vector $q\mathbf{E}$ to each charged atom in the system, in the direction of a specified external electric field $\mathbf{E}$. This force vector is summed with the other force vectors acting locally on each atom at each simulation step, which updates the position and velocity of each atom in the system. The field can be applied with the same magnitude at each simulation step or the magnitude can be time varied, for example, to approximate the effect of a finite rise time for the electrical stimulus.

It is important to note that the "applied" field is only one component of the electric force experienced by any given charge in the system. In a typical phospholipid bilayer simulation, the water dipoles and the phospholipid head group dipoles reorient in a globally applied electric field (which is the field that would be felt in a vacuum at that point). This produces a counter "polarization field", so that the net electric field experienced by a particular atom or atom group, which is the sum of the vectors of the external field, the fields from nearby charges, and all of the dipole fields at that point, is considerably less than the value of the externally applied field alone.

In the second method, a charge imbalance is imposed across the membrane by placing an excess of positive ions on one side and an excess of negative ions on the other. Because molecular dynamic simulations use periodic boundary conditions to minimize edge effects in small simulation volumes, it is not possible to create an ion imbalance in a system containing a single bilayer with water on each side, since the solvent on the two sides of the bilayer is actually connected (the simulation box "wraps around" in the x-, y-, and z-directions). One way to work around this is to place two parallel bilayers in the system, so that there is an inner volume of water that is confined between the two bilayers in the direction normal to the bilayers. This inner compartment is isolated from the outer compartment, which is located on the other side of the two bilayers, as long as the membranes remain intact. Another approach is to create a simulation with a single bilayer with a stack of water covering the bilayer on each side, followed by a slab of vacuum (or air) covering the water layer on each side. The vacuum is continuous in the z-direction and "wraps around" the water slabs, but the water on each side of the bilayer is isolated by the vacuum from the water on the other side.

4.3.4 Lipid Electropore Formation in Molecular Dynamic Simulations

Initial reports of molecular simulations of the electroporation of lipid membranes [63–65] were followed by an extension of the early results to a variety of lipid types and simulation conditions and to detailed molecular descriptions of pore creation and annihilation [59]. These processes are similar for many types of amphiphilic phospholipid bilayers, for heterogeneous and asymmetric bilayers containing anionic phospholipids, and even for "membranes" composed of octane [66] or vacuum [67]. The common sequence of events for all of these systems is the intrusion of interfacial water into the low-permittivity environment of the membrane interior and the stabilization of membrane-spanning water columns by the applied electric field [68]. For phospholipid bilayers the construction of these water bridges is accompanied by the energy-minimizing reorganization of the head groups along the field-stabilized water columns to form the wall of a lipid pore.

Although the construction of an electropore, once initiated, is deterministic and driven by the electric field, the first step—the appearance of the first stack of two or three water molecules climbing away from the interface into the membrane interior—is a random event. Increasing the transmembrane potential increases the probability that a pore will form at a certain place within a certain time [59]. The statistical nature of pore initiation at the molecular scale is consistent with the stochastic pore model that is the basis of popular continuum models of electroporation, bolstering confidence that the two perspectives will someday be unified.

With model membrane systems that are more complex and more representative of living cell membranes, we will be able to see how this simple, field-driven, water-directed mechanism of lipid electropore formation is modified by a variety of factors: ions in the medium, cholesterol, regions of different lipid compositions (lipid rafts, sites of peroxidation), membrane protein–lipid boundaries, permeabilizing peptides, DMSO, detergents, cytoskeletal attachments and other mechanical constraints on lipid motion, and many others. In this way we can move from molecular-scale reconstructions of the membrane in an applied electric field to a better understanding of electropermeabilization in living cells and tissues.

4.4 Electropermeabilization

Marie-Pierre Rols, Justin Teissie, and Lluis M. Mir

Electropermeabilization using μs and ms pulses has also been termed "classical electroporation," as opposed to the nsPEF effects achieved using ns pulsed electric fields. This subchapter reviews what is known about the electropermeabilization processes of cells and tissues submitted to μs and ms electric pulses. It aims to describe the basic aspects at the membrane level (the modulation of the transmembrane potential, the way to conduct experiments) and summarizes its short- and long-term consequences at the membrane and cell levels.

4.4.1 Resting and Electro-induced Transmembrane Potentials

As already described in Sects. 4.1.1 "Cell in the Electric Field" and 4.1.2 "Induced Transmembrane Voltage," the key nonthermal effect of electric field on cells is a position-dependent change in the resting transmembrane potential difference $\Delta\Psi_o$ of the plasma membrane [69–71]. The electrically induced potential difference $\Delta\Psi_E$, which is defined as the difference between the potentials inside (Ψ_{in}) and outside (Ψ_{out}) the target cell, at a point M on the cell surface, is given by

$$\Delta\Psi_E(t) = \Psi_{in} - \Psi_{out} = g(\lambda) rE \cos(\theta_M) \left\lfloor 1 - e^{\frac{-t}{\tau}} \right\rfloor \quad (4.5)$$

where t is the time after the onset of the electric pulse; g is a function that depends on the conductivities λ of the cytoplasm, the plasma membrane, and the extracellular medium; r is the radius of cell; E is the field strength; θ_M is the angle between the normal to the plasma membrane at the position M and the direction of the field; and τ is the membrane-charging time (≈1 μs in the case of eukaryotic cells in suspension in a physiologically conductive medium) [72].

For isolated spherical cells, in a conductive buffer a few microseconds after the beginning of the pulse, a steady-state potential difference is present:

$$\Delta\Psi_E = 1.5\ rE\cos\ \theta \quad (4.6)$$

The field-induced potential difference is added to the resting transmembrane potential difference:

$$\Delta\Psi = \Delta\Psi_0 + \Delta\Psi_E \quad (4.7)$$

Being dependent on the angular parameter θ, the field effect is position dependent on the cell surface. Membrane electropermeabilization is observed when the transmembrane potential reaches a critical value (between 0.2 and 0.4 V) [73]. The transmembrane potential difference of a cell exposed to an electric field is a critical parameter for successful cell permeabilization, whatever the size of cell, its shape, and its orientation. It defines the sites (location, size) where molecule uptake can take place [74–76].

4.4.2 Basics Aspects of Electropermeabilization

4.4.2.1 How to Conduct Experiments (Determination of the Experimental Protocol According to Cell Characteristics: Choice of Electrodes, of Cuvette or Petri Dish, of Pulsing Medium, of Temperature, etc.)

In most experiments, square wave electric pulses generators are used. They allow the independent control of the amplitude of the electric field and the duration of the pulse [77]. The electric pulses are delivered through a set of electrodes connected to the generator. Like the electric pulse parameters, the choice and the placement of electrodes have to be carefully selected considering the characteristics of the cells, both in the case of in vitro 2D and 3D cell cultures or of tissues in vivo [78].

The biodistribution of the local electric field is dependent on the electrode geometry and placement. The most commonly used applicators are plate, contact, and needle electrodes [79]. The plate electrodes are mostly used for attached cells grown on Petri dishes, for cells in suspension and for skin, muscles, and other superficial tissues. The tissue must be pinched between the electrodes to obtain an optimized distribution of the field effect [80]. With the contact electrodes, the penetration depth of an effective electric field is rather small and depends on the interelectrode distance. In contrast to plate electrodes, needle electrodes are invasive and have to be inserted throughout the tissue. Regardless of the kind of electrodes, the local electric field is highest around the electrode for needle and contact, whereas for plate electrodes, it decreases very rapidly outside of the electrode gap. Thus, if the cells (or the tissues) to be treated are grown on a surface that is larger than the interelectrode distance, the entire surface (or tissue) will have to be treated by successive displacements of the electrodes to cover the entire surface by the repetitive application of electric pulses [81]. Because of the structural heterogeneity of cell cultures grown at high density, especially in the case of 3D cell cultures and tissues, the electric field is difficult to determine. So the applied voltage (V) to distance between the electrode (d) ratio has to be used to describe the pulse amplitude instead of an electric field value. To this end, both V and d must be reported.

While tissues can be "simply" permeabilized by the direct application of the electric pulses through contact or needle electrodes, permeabilization of cells in culture can be conducted on different ways. The bottom of the Petri dish serves as an electropulsation chamber. For cells in suspension, special cuvettes can be purchased but "homemade" chambers can be designed: the cells can even hold by surface tension forces between two plate electrodes if the distance between the plates is 1 or 2 mm. Alternatively, the plate electrodes can be placed directly on the bottom of the Petri dish to form a sort of open cuvette.

Another parameter that can be adjusted is the temperature. Experiments are usually performed at room temperature, but incubating the cells on ice before pulsation or during the minutes following can improve the uptake of molecules. However, for preserving cell viability and depending on the goal of the experiment, the tip is to place the cells immediately at 37 °C after the pulses [82]. In another respect, to limit the Joule effect that strongly depends on the buffer conductivity, classical culture medium can be replaced by a low ionic, iso-osmotic buffer particularly in the case of cells grown on Petri dishes. The composition of this medium is generally a 10-mM phosphate buffer pH 7.4, 250-mM sucrose (to keep the isotonicity), and 1-mM $MgCl_2$ (for preserving the cell physiology) [83].

4.4.2.2 Limits on Detection Due to the Sensitivity of the Assay Method

The first experiments to detect membrane permeabilization were performed on cell populations by measuring conductivity changes or radioactive and/or small-molecule uptake (such as sugars, trypan blue, Lucifer yellow) [84–87]. The use of fluorescent dyes allows detecting membrane permeabilization by a more convenient way. Visualization can be performed at the single-cell level using a fluorescent light microscope, while flow cytometry permits to quantify the uptake of molecules on a large number of cells [88, 89]. Quantification can be subject to artifacts as the sensitivity of the detection method highly depends on the fluorescent dye (size, fluorescent quantum yield) and on the cell autofluorescence [86]. Another very sensitive method constitutes the exposure of the cells in the presence of a toxic compound such as Ca^{2+} or bleomycin, a non-permeant cytotoxic drug at very low external concentrations [90]. Indeed, the uptake of 500 molecules is sufficient to kill the cells in culture. Interestingly, quantitative comparisons can be performed through the determination of cell viability using precise methods like the cloning efficiency approach. Consequently, the statement that a cell has been successfully permeabilized or not requires the delivery of additional information including the type of molecule used in the assay and the detection method employed (Fig. 4.11).

4.4.2.3 Effect of the Electric Field Parameters (Pulse Amplitude, Pulse Shape, Pulse Duration, Number, Polarity, and Repetition Frequency)

Permeabilization occurs only on the areas of the plasma membrane where the membrane potential difference has been brought above its critical value ($\Delta\Psi c$, close to 200–300 mV for ms pulses independently of the cell type). Membrane permeabilization is therefore controlled by the electric field strength. This means

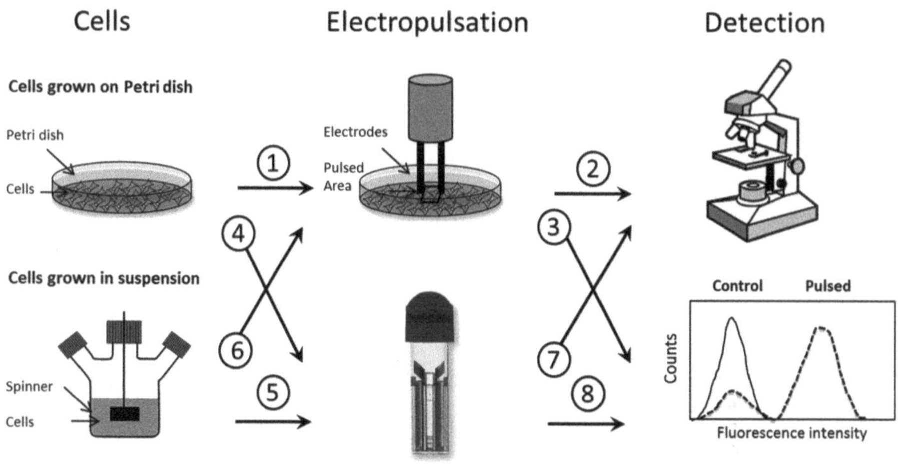

Fig. 4.11 Tips for your experiments. Cells grown in Petri dish can be directly electropermeabilized on the Petri dish by using plated electrodes (*1*). Detection and quantification of the permeabilized cells can be performed by direct counting under microscope (*2*) or by flow cytometry after careful trypsinization of the pulsed area (*3*). They can also be electropulsed in suspension on cuvettes after trypsinization (*4*). Cells grown in suspension can be electropermeabilized on cuvettes (*5*) or on "homemade" pulsation chamber by using plated electrodes (*6*). Detection of the permeabilized cells can be performed under microscope (*7*) or by flow cytometry (*8*)

that a field intensity E larger than a critical value, Ep, needed to induce $\Delta\Psi c$ must be applied. Ep is dependent on the size of the target cells (Eqs. 4.5 and 4.6, above). Large cells are more sensitive than the small ones. Electric field values have to be adapted to each cell line in order to avoid affecting their viability. The field strength triggers permeabilization: when E > Ep, it controls the area of the cell surface, which is affected. From Eqs. 4.5 and 4.6, it is clear that for field intensities close to Ep, permeabilization is only present for θ values close to 0 or π. Under that condition, only the localized parts of the membrane surface facing the electrodes are affected. However, within these permeabilized cell regions, the local level of permeabilization is controlled by the number and the duration of the electric pulses. So, membrane permeabilization only occurs for electric field values E higher than the threshold value Ep, whatever the number and duration of electric pulses. Increasing E, above Ep, leads to increase in the extent of membrane area where permeabilization takes place, and in that specific area, the extent efficacy of permeabilization is determined by the number and duration of electric pulses [75, 86].

As it will be described in Chap. 5, "Electrotransfer Mechanism for Drug and Gene and Delivery," permeabilization can be improved by changing the polarity of the electric pulses and the frequency.

4.4.3 Electropermeabilization, a Fast, Transient, and Localized Process

4.4.3.1 Available Direct Methods for Cell Electropermeabilization Measurement: Optical Imaging, Atomic Force Microscopy, Nuclear Magnetic Resonance, Electron Microscopy, and Coherent Anti-Stokes Raman Scattering

These methods are not based on the transport of substance across the electropermeabilized membranes. Video microscopy allows visualization of the process at the single-cell level. In addition to fluorescence microscopy, phase contrast can be used to detect cell swelling or bleb formation that can appear following cell electropermeabilization or to detect lipid loss inducing several membrane alterations (pore, tubule, and vesicle formation) as observed on giant unilamellar vesicles [91]. The need of data recorded directly at the single-cell level without any staining or preparation is of great interest. In this context, the effects of membrane destabilization resulting in permeabilization to small molecules can be addressed by performing direct quantitative biophysical measurements using atomic force microscopy (AFM) [92, 93] and nuclear magnetic resonance (NMR) [94, 95] or phase-contrast and coherent anti-Stokes Raman scattering (CARS) microscopy as relevant optical approaches to gain insight into membrane changes [96]. The resolution of these techniques is quite poor compared to electron microscopy where ultrastructural modifications of membranes and cell organelle can be detected (but EM requires fixation of the cells).

4.4.3.2 Visualization of the Uptake of Molecules and Its Dependence to the Molecule Properties

Membrane electropermeabilization is a fast and localized process. Propidium iodide, a fluorescent non-permeant molecule, can be used as a probe for small molecules ($\leq$4 kDa). Its uptake in the cytoplasm is a fast process that can be detected during and after the application of electric pulses. In less than 1 min, it appears at the nucleus level [41, 89]. PI does not allow for quantitative measurements as its fluorescence depends on its molecular environment. As reported by Silve and Mir, there is an inverse relationship between the size and number of molecules internalized using identical cells and pulse parameters [97].

As described in Sect. 4.4.2.2, fluorescent dyes allow visualizing membrane permeabilization on a convenient way. Visualization can be performed at the single-cell level under fluorescent microscopes but can be subject to artifacts as the sensitivity of the detection depends on the molecule properties (size, charge) and the way the experiments are conducted (microscope adjustment).

4.4.3.3 Direct and Localized Access to the Cytoplasm

Exchange across the pulsed cell membrane is not homogeneous on the whole-cell membrane. It occurs on the two caps at the cell surface facing the electrodes. Uptake is asymmetrical. Using PI (which has a direct access to the cytoplasm after the cell membrane permeabilization), it has been shown that uptake is more pronounced at the anode-facing side of cells than at the cathode one, i.e., in the hyperpolarized area than in the depolarized one, in agreement with above theoretical considerations and Eqs. 4.5 and 4.6 [98]. As will be described in Chap. 5 "Electrotransfer Mechanism for Drug and Gene and Delivery," large molecules such as plasmid DNA are internalized in the cells by more complex mechanisms.

4.4.3.4 Resealing (Assays, Temperature, Energy, Cytoskeleton)

Once electropermeabilized, cells can stay permeable after switching off the pulse generator. Lifetime of permeabilization can be assayed by adding fluorescent dyes at various times following the pulses. If the cell membrane is still permeable, then the cell will be fluorescent. One key actor of that phenomenon is the cytoskeleton [99]. Lipid vesicles can be permeabilized but resealing occurs in less than 1 s. Cells pretreated with drugs that affect the cytoskeleton can be permeabilized by electric pulses but have a very fast resealing rates. Resealing varies from a few seconds (when cells are put at 37 °C just after pulsation) to several hours (when cells are maintained on ice) according to the experimental conditions (temperature and pulse parameters). However, one has to take into account that viability can be affected since ATP release will occur [100].

4.4.3.5 Short- and Long-Term Membrane Permeabilization and Cell Alterations (Lipids and Proteins and Cytoskeleton)

Mechanisms at the molecular level are detected after the pulse application. These data could be correlated with the transient reorientation of the phospholipid head group reported on mammalian cells and lipid bilayers by ^{31}P NMR as well as a field pulse correlated phosphatidylcholine flip/flop restricted to the permeabilized regions [101]. Transient phosphatidylcholine flip/flop in nucleated mammalian cells appears as a direct effect of electric fields and not a secondary one as reported in red blood cells.

Changes in living cell membrane properties due to EP process can be assessed by AFM in real time. AFM can be used for the quantification of the destabilization process in terms of elasticity. Stiffness data gives new insight of the permeabilization process. Data obtained on fixed cells give access to information at a specific moment of the EP phenomenon. Topological images of membrane on living cells at high resolution can be recorded and give direct visual information about EP effect. Moreover, AFM senses an effect of EP that is shorter than plasma membrane permeabilization and that is linked to cortical actin destabilized by electric pulses [92].

Altogether, these short- or long-term membrane alterations will have direct consequences on molecule uptake as will be described in the next chapter.

4.4.4 Cells Responses: From 1D (Single-Cell Analysis) to 3D Studies (Using 3D Cell Aggregates)

4.4.4.1 Single Cells: Endocytosis-Like Processes Are Present

Electric pulses alter the plasma membrane permeability leading to the uptake of molecules that can enter the cells. The membrane structures driving the permeability are thought to result from lipid mismatches (aqueous conductive defects), allowing small molecules to have a direct access to the cytoplasm. Although these membrane defects are present for several minutes and are associated to its passage, DNA has not a direct access and accumulates at the membrane as will be described in chapter 5. DNA internalization thus occurs after membrane resealing and involves the passage of large-sized DNA aggregates [102]. Endocytosis-like process takes place as it is the case for many viral or chemical vectors. Macropinocytosis was earlier reported for the delivery of proteins, when added during the minutes following pulse delivery. One has to notice that electric pulses, applied under conditions that do not lead to "classical" permeabilization, can enhance natural endocytosis pathways [103, 104].

4.4.4.2 Monolayers: A Fusogenic State Is Triggered

One direct consequence of membrane permeabilization, which may occur when cells are into contact, as it is the case of cells growing in Petri dishes up to confluence or artificially brought into contact, is membrane fusion [105, 106]. Fusion can also be obtained when cells submitted to electric pulses are brought into contact by centrifugation or filtration. This fusogenic state can also be observed in vivo in the liver [107].

4.4.4.3 Spheroids: The Use of 3D Models of Cell Culture as Relevant Tools to Access In Vivo Electropermeabilization Constraints (Mainly Cell Density and Contact)

Even if the high majority of studies underlying molecule transfer by electric fields have been performed on 2D cell culture in Petri dish or in cells cultured in suspension, 3D multicellular spheroids represent a nice, relevant, cheap, easy-to-handle in vitro model to gain understanding of in vivo tissue electropermeabilization and electrogene delivery processes [108] and optimization of electromediated drug delivery protocols [109]. Small-molecule uptake is present but spatially heterogeneous within a 3D multicellular spheroid after electroporation, with a progressive decrease from peripheral to interior cells [110]. This can be due to the increased diffusional lag time required for solutes to penetrate homogeneously throughout the extracellular space within spheroid before applying electric field, to limited extracellular reservoir of molecules surrounding densely packed cells, to local perturbation of the electric field due to heterogeneous high-density multicellular environment electrical properties, and finally to a position-dependent variation in cell size as reported in Sect. 4.1.2.2 "Numerical Computation."

Moreover, in the case of large molecules as plasmid DNA, spheroids allow showing that electrophoresis, and not tissue deformation or electroosmosis, is the driving force for interstitial transport. At the opposite of cells in 2D cultures, only cells on one side of the outer leaflet expressed the reporter gene [111]. This low expression is in fair agreement with in vivo experiments on tumors [112]. Close contacts between cells and extracellular matrix may act as a physical barrier that limits/prevents uniform DNA distribution and explain the absence of gene expression in the inner region of spheroid [113]. The limited access of plasmid DNA to central region of spheroid remains a significant barrier to efficient gene delivery in tissue. In addition, three-dimensional reconstructed human connective tissue model could become a useful tool to study skin electrotransfer mechanisms and would help improve electrogene therapy approaches such as the systemic delivery of therapeutic proteins and DNA vaccination [114].

4.5 Basic Properties of Nanoelectropores and Their Impact on Cell Function

Andrei G. Pakhomov

4.5.1 Nanoelectropore Definition

High-amplitude electric pulses weaken the barrier function of the cell membrane, making it more "leaky." As a result, solutes that were unable or had a limited ability to cross the membrane move more freely down their electrochemical gradients. For example, the concentration of free Ca^{2+} inside biological cells is roughly 10^5 times lower than in the outside medium, so opening of membrane pores permeable to Ca^{2+}

will cause its transmembrane flow from outside into the cytosol. The negative charge inside the cell will assist the flow of positively charged Ca^{2+}.

Likewise, electropermeabilized membranes may allow the transmembrane flow of anions and neutral solutes, small molecules (such as ATP and dyes like YO-PRO-1), and water. The size of pores puts a limit on which species can cross the membrane; therefore testing different species may be the first step to estimate the size of pores. In the first approximation, the size of pores formed by electroporation depends on the electric pulse amplitude and duration. Once pores are created, the continued application of external voltage leads to pore expansion. That is why "long" (micro- and millisecond duration) pulses may open pores permeable to larger solutes than pores opened by shorter nanosecond-duration electric pulses (nsEP). The diameter of pores formed by the conventional electroporation using micro- and millisecond pulses is estimated to fall within 1–100-nm range [115–117]. The principal primary effect of nsEP, as demonstrated by both in silico models [13, 118–120] and live cell measurements, is the formation of small pores having the diameter on the order of 1–1.5 nm [43, 121–123]. Consequently, the term "nanoelectroporation" has established a dual meaning as (1) application of nanosecond-duration pulses for electroporation and other bioeffects and (2) a treatment that causes preferential formation of the smallest membrane pores (also called "nanopores" or "nanoelectropores"). The size-based definition is more common and will be used below in this chapter. With that said, application of high-voltage nsEP is perhaps the most common way to create nanometer-sized pores, so the two definitions of "nanoelectropore" are not necessarily contradictory.

At present, there is no consensus as to whether nanoelectropores are discrete membrane structures with long lifetime; or they are transient structures which randomly emerge and expire in membranes disrupted by electric pulses; or the term "nanopores" is just a convenient way to characterize the selective leakiness of electropermeabilized membranes to certain solutes. Although the exact nature of electropermeabilization of membranes by nsEP is not fully understood, the concept of the formation of long-lived nanopores is consistent with most if not all available experimental data.

4.5.2 How to Make Nanoelectropores in Live Cells

As mentioned above, exposure of living cells to high-voltage nsEP is a common way to open long-lived nanopores, but other approaches may be employed as well. In principle, nanopores may be the first stage of larger pore formation by conventional electroporation [115, 124], and stable nanometer-sized pores were also reported to form by gradual shrinking of larger electropores [115–117]. Some chemical and physical factors cause membrane perturbations and bioeffects similar to nsEP, which could potentially be explained by nanopore formation [125–127]. Finally, prolonged holding of cell membrane at marginally permeabilizing transmembrane potentials (~200 mV) using whole-cell patch clamp increases the membrane conductance by opening nanometer-sized pores [128]. The sameness of these pores to those opened by nsEP has yet to be proven, but there are no reasons to suspect that they are different.

While nanopores are not a unique effect of nsEP, stimulation by nsEP is a convenient way to create nanopores to study their properties and impact on cells. With longer electric pulses ("conventional electroporation"), it is problematic to identify nanopores and explore their properties in heterogeneous pore populations. The time interval until larger pores shrink is variable and uncertain, and profound consequences of large pore opening can mask nanopore effects. On the contrary, nsEP forms a relatively uniform population of stable nanopores, thereby offering an opportunity to explore nanopore properties and their impact on cells. At the same time, anticipated findings with nsEP-opened nanopores will likely broadly apply to nanopores created by longer pulses and nonelectrical stimuli as well.

4.5.3 Detection of Nanoelectropores and Their Properties

Contrary to larger pores formed by the conventional electroporation, nanopores exert complex behaviors that are traditionally thought to be unique for protein ion channels [122, 129]. Nanopores are inward rectifying, voltage and current sensitive, and ion selective; they can remain open for minutes while oscillating between electrically silent and open states. Nanopores appear adequately equipped to replace or complement a number of transmembrane ion transport functions that are now ascribed to classic ion channels. Consequently, the lasting disruption of the cell membrane barrier function impacts a wide spectrum of living cell functions.

The detection of nanopores is inherently related to their specific properties, which is why they are considered together in this chapter. We will consider nanopore detection by (1) differential uptake of fluorescent markers (dyes, reporter ions, or molecules), (2) changes of the cell volume due to water uptake or loss, and (3) direct measuring of membrane current using patch clamp.

4.5.3.1 Fluorescent Dye Uptake Techniques

The cell membrane integrity can be routinely tested by measuring the transmembrane flow of substances which do not penetrate through the intact membrane. DNA stains like propidium iodide and YO-PRO-1 are essentially nonfluorescent in extracellular media, but show bright fluorescence upon binding to nucleic acids (DNA in particular) inside the cell. The intact cell membrane is essentially impermeable to these dyes, and adding them to the external medium does not change the fluorescence of cells. However, if the membrane barrier function is compromised, these dyes can enter the cell and bind to DNA and RNA, which is detected by an increase in cell fluorescence (Fig. 4.12).

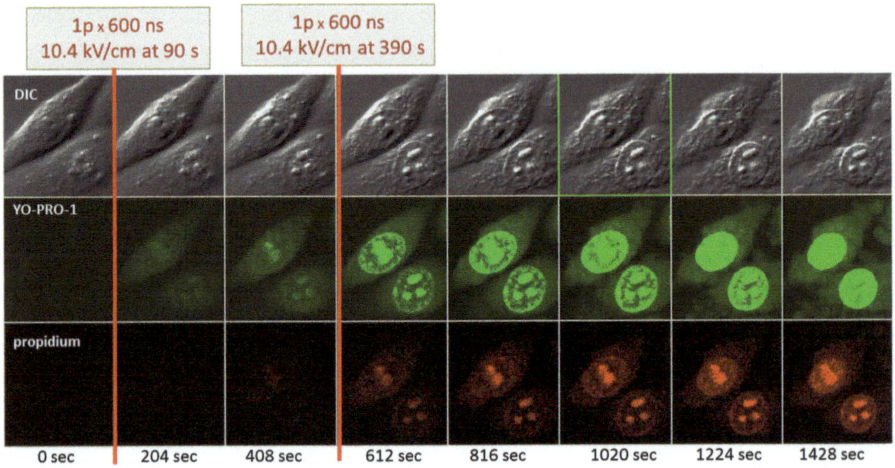

Fig. 4.12 Detection of membrane nanoporation by the entry of fluorescent dyes YO-PRO-1 (*center row*) and propidium (*bottom*). Shown are sequential images of CHO cell images subjected to two treatments with nsEP, at 90 and 390 s into the experiment (a single 600-ns pulse at 10.4 kV/cm) . The *top row* shows the cells as seen with differential interference contrast (DIC) optics. The time from the start of the experiment is indicated under the images. Note that the entry of YO-PRO-1 after the first nsEP treatment was not accompanied by any detectable entry of propidium

(Of note, traditionally most publications abbreviate propidium iodide to "PI" and discuss the PI uptake by electroporated cells. However, it should be taken into account that in water, PI dissociates into propidium cation and iodide anion; it is the propidium cation that penetrates the membrane and causes fluorescence changes, not the entire PI molecule. This difference is particularly important when talking about the pore size. Therefore, below we will use a more accurate term "propidium uptake" (Pr uptake). The YO-PRO-1 dye actually is supplied and added to solutions as an iodide salt (which also dissociates), so referring to YO-PRO-1 uptake is correct.)

Pr fluorescence is commonly used to detect dead cells and cells with membrane disrupted by conventional electroporation. However, the entry of Pr through nanopores is relatively modest, so when fluorescence detectors are tuned to the detection of dead cells, the Pr entry through nanopores can remain undetected. In Fig. 4.12 (bottom), there is no detectable Pr uptake after a single 600-ns pulse at 10.4 kV/cm, whereas the application of the second pulse triggers the uptake. In early studies with nsEP, this peculiarity was erroneously interpreted as a lack of damage to the cell membrane by nsEP. More detailed later studies suggested that it is the small size of membrane pores that limits Pr uptake, whereas the fact of membrane permeabilization can be readily demonstrated by other methods, such YO-PRO-1 uptake. The molecule of YO-PRO-1 is smaller and its passage through nanopores is not restricted to such a degree as for Pr. In Fig. 4.12 (center row), the entry of YO-PRO-1 is obvious even after the first nsEP treatment.

Even more sensitive detection of nanoporation is based on the uptake of still smaller solutes, such as Ca^{2+} and Tl^{+} ions [123, 130, 131]. For Tl^{+} uptake detection, cells are preloaded with a Tl^{+}-sensitive dye FluxOR® (Invitrogen, Eugene, OR) and placed in a medium with Tl^{+} added (of note, Tl^{+} precipitates with Cl^{-}, so the medium should be properly formulated to exclude Cl^{-} or minimize its concentration). Since Tl^{+} cation is not found in living cells in any appreciable amount, the fluorescent method of detection of Tl^{+} entry is highly sensitive and always reflects Tl^{+} entry through the plasma membrane (in contrast to Ca^{2+} detection, since Ca^{2+} can come into the cytosol from intracellular depots). Intact lipid membrane is impermeable to Tl^{+}, but this ion can potentially enter the cell via "classic" voltage-gated K^{+}-selective ion channels and various nonspecific cation channels. The differentiation between Tl^{+} entry through endogenous channels and through nsEP-opened nanopores may be challenging and requires careful selection of the cell type and/or pharmacological inhibition of the channels.

Most of living cells at rest have about 100 nM of free Ca^{2+} in the cytosol, which is about 10,000-fold less than in extracellular fluids. Thanks to this naturally high transmembrane Ca^{2+} gradient, opening of nanopores quickly leads to an increase in the intracellular Ca^{2+} concentration, which can be detected by various Ca^{2+}-sensitive fluorescent dyes. Ca^{2+} detection methods are among the best developed, most affordable, and most reliable. A broad selection of Ca^{2+}-sensitive dyes is available, including dyes like Fura-2 for fast quantitative measurement of Ca^{2+} concentration by ratiometric imaging. Detection of nanopores by Ca^{2+} uptake is almost as sensitive as with Tl^{+}, but more caveats need to be taken into account. First, the resting level of Ca^{2+} in the cytosol can transiently increase into the millimolar range when cells are activated. Second, Ca^{2+} can come into cells through voltage-gated Ca^{2+} channels and many types of other channels. Third, Ca^{2+} can enter cytosol not only from the outside but also from intracellular stores, such as the endoplasmic reticulum (ER). Finally, cells express multiple mechanisms that can either amplify weak Ca^{2+} signals by opening Ca^{2+} channels in both the plasma membrane and the ER or reduce and terminate Ca^{2+} increase by actively pumping Ca^{2+} out of the cytosol. While it is tempting to use ratiometric Ca^{2+} imaging to quantify its inflow through the nanoporated cell membrane (in order to measure the nanopore transport), the impact of biological factors may be difficult to rule out. This is the reason why Ca^{2+} imaging is more often employed to study the downstream consequences of nanopore formation rather than for quantitation of the transport.

The dye and ion uptake methods described above are only indicative of the pore size and, as a rule, are more useful for comparison of different treatment conditions rather than for accurate pore-size measurements. Smaller pores will restrict the flow of larger molecules to a greater extent than the flow of smaller ones. For example, if a 100-ns pulse triggers high uptake of Tl^{+} (a small solute) and little uptake of Pr (a relatively large solute), whereas a 10-ms pulse triggers high uptake of both Tl^{+} and Pr, such data indicate that pores opened by the 10-ms pulse are larger. Under certain conditions, the ratio of the entry of two solutes of different molecular size can be used for a more definitive estimation of pore size [43].

4.5.3.2 Changes of the Cell Volume Due to Water Uptake or Loss

Typically, permeabilization of cell membrane leads to cell swelling. The mechanism of cell volume changes is reasonably well understood and results from Donnan-type colloid osmotic pressure [116, 122, 132]. It should be emphasized that the colloid osmotic force depends on the molecular size of solutes rather than on the overall osmolarity of the medium.

The mechanism of the colloid osmotic effect is explained in Fig. 4.13. The simplified scenario presented in this figure disregards any active volume control mechanisms [132, 133] and considers only three types of solutes which differ in the molecular size. In the top row (before poration), the extracellular buffer contains only small molecules, whereas inside the cell, there are both large and small molecules. The osmolality inside and outside the cell is made the same (300 mOsm/kg), which results solely from small solutes outside and from both small and large solutes inside (250 and 50 mOsm/kg, respectively). We assume that electroporation (right panel) makes the membrane permeable to the small solutes but not to the large ones. Then, the small solutes enter the cell up to the same concentration as outside and their osmolality inside reaches 300 mOsm/kg (the cell volume is assumed to be negligible compared to the outside solution volume). Large solutes remain trapped in the cell, still contributing their 50 mOsm/kg. Hence the total osmolality inside will be 350 mOsm/kg (more than outside), causing water entry and cell swelling. In real life, even a small increase in osmolality is offset by water uptake and continual swelling (so that the maximum osmolality stays just above 300 mOsm/kg).

Alternatively, if we initially have only large (pore-impermeable) solutes outside the cell, the electroporation leads to the loss of small intracellular solutes which is not compensated by any uptake (Fig. 4.13, middle row). The osmolality within the cell drops to 50 mOsm/kg (from large solutes only), and water leaves the cell, causing its shrinking. Finally, an extracellular solute which is marginally permeable through the pores (comes in very slowly, e.g., because it can utilize just a small fraction of pores) may cause biphasic volume changes, namely, initial shrinking followed by swelling (bottom row).

Thus, the direction of volume changes (shrinking or swelling) after the electroporation depends on whether the outside solute(s) can or cannot enter the cell through the pores. One can put molecules of different sizes into the outside solution to test if they inhibit swelling in electroporated cells. When a solute blocks swelling (and converts it into shrinking), it indicates that pores are smaller than the molecule of this solute.

This approach was employed for measuring nanopore size, by using differently sized neutral sugars and different polyethylene glycols to "calibrate" the pore size by blocking swelling [121]. This method was remarkably sensitive, e.g., it readily distinguished between the effects of adonitol and mannitol, the molecules which have the same cross section and differ just by a single carbon alcohol group. This study established that cell volume changes after the exposure to either 60- or 600-ns

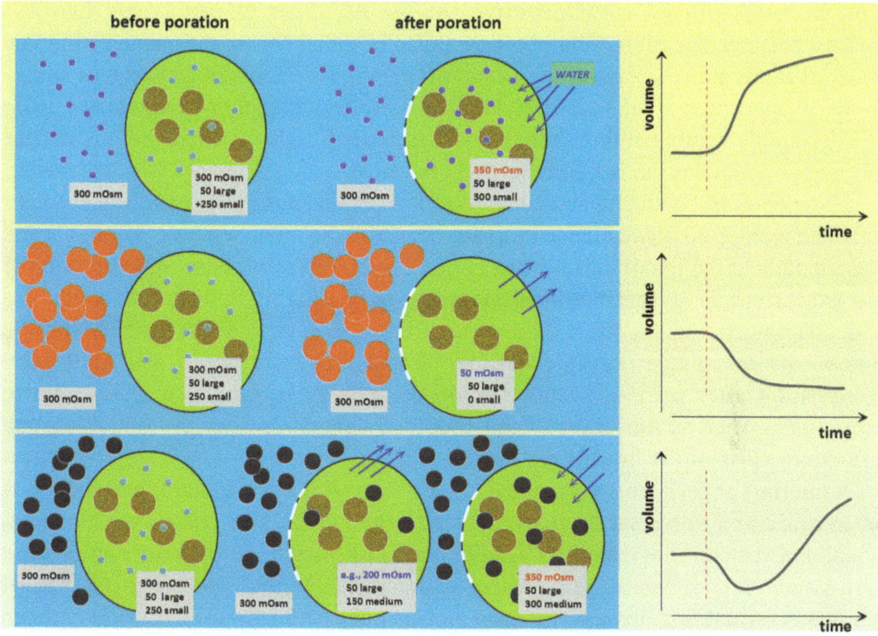

Fig. 4.13 A simplified schematic explaining the colloid osmotic mechanism of cell swelling or shrinking after membrane permeabilization. Smaller, intermediate, and larger solutes are depicted by different size circles. Three rows represent three different scenarios depending on the size of solutes in the extracellular medium. The scenarios are with a small solute which easily passes the electroporated membrane (*top*), with a large solute that cannot pass the electroporated membrane (*center*), and with an intermediate-sized solute with slow or limited pore permeability (*bottom*). The initial osmolality inside and outside the cell is the same for all scenarios (300 mOsm/kg). The intracellular osmolality (before poration) is produced by large, pore-impermeable solutes (50 mOsm/kg) and small, pore-permeable solutes (250 mOsm/kg). In the bottom row, the cell is shown before poration and early and late after the poration. See text for more details

pulses are determined by pores smaller than 0.9 nm in diameter, which is arguably the most accurate to date assessment on nanopore size. However, there is a small fraction of Pr-permeable (larger) pores, which would allow Pr entry, but have little impact on cell volume. The study also suggested that the pore diameter may increase with more intense nsEP treatments.

4.5.3.3 Direct Measuring of Membrane Current Using Patch Clamp

Once the cell membrane is permeabilized, its electrical conductance increases. This change is most reliably and quantitatively measured using the whole-cell configuration of the patch clamp method [134]. To date, electrophysiological measurements proved more sensitive for the detection of nanopores than any fluorescent dye detection. With a single pulse of either 60- or 600-ns duration and using different pulse voltages, the extent of membrane permeabilization was shown to be proportional to the absorbed dose, with the threshold at about 10 mJ/g [135]. Sample heating at this dose is less than 0.01 °C, which shows that membrane electropermeabilization is a nonthermal effect.

The electrical conductance of nanopores is dependent on the membrane potential. The maximum conductance increase occurs at the most negative potentials and corresponds to the maximum inward current. Such differential enhancement of the inward current, or inward rectification, proved to be a hallmark feature of nanopores and distinguishes them from larger, propidium-permeable pores (which show a linear current–voltage dependence). We found that cells which display inward rectification after nsEP exposure remain completely or largely impermeable to propidium, whereas the onset of the propidium uptake marks the reduction and subsequent loss of the inward rectification [129].

Numerous experiments performed to date suggest that the inward rectification is an intrinsic and universal property of nanopores, which also points to their functional and structural asymmetry. The induction of inward rectification can be used as a nanopore "signature," i.e., as an indirect but reliable and sensitive sign of nanopore formation. However, it may not be easy to distinguish nanopores from classic ion channels which may also display inward-rectifying properties (e.g., [136, 137]).

Other remarkable features of nanopores include their extended lifetime (minutes), the sensitivity to the electric current, and ion selectivity [122, 138]. Among various pharmacological agents that block ion channels and were tested for blocking nanopores, only lanthanide ions (Gd^{3+} and La^{3+}) inhibited nanopore currents, probably by restructuring of the lipid bilayer [122, 139, 140].

4.5.4 Impact of Nanopore Formation on Cell Physiology

Long-lasting disruption of the plasma membrane barrier function expectedly affects many if not all physiological processes in living cells. Therefore, it is usually not productive to explore the impact of nanopores on a certain intracellular structure in isolation; instead, one should look at mechanisms and pathways which result in biochemical, physiological, and morphological effects.

In simple terms, bioeffects of nanopore formation can be divided into physiological and pathological. Physiological effects usually are associated with (1) depolarization of the membrane and activation of voltage-gated Na^+ and Ca^{2+} channels and/or (2) Ca^{2+} entry into the cytosol through nanopores. In both cases, nanopore formation leads to an increase of the cytosolic Ca^{2+} level.

This increase is perhaps the most consequential effect of nanopore formation. The resting Ca^{2+} concentration in most cells is held tightly at about 100 nM, and its changes (increases) serve as a versatile and universal signal for activation of Ca^{2+}-dependent cascades. Depending on the cell type and physiology, as well as on the amplitude and timing of Ca^{2+} transients, they may lead to such diverse effects as cell differentiation or division, cytoskeleton rearrangements, endo- and exocytosis, synthesis and release of neuromediators, activation of immune cells, and apoptotic or necrotic cell death [141]. Ca^{2+} entry through nanopores mimics Ca^{2+} signaling that would normally originate from the activation of membrane receptors and channels and may lead to similar downstream effects. Among the first downstream effects are the amplification of the Ca^{2+} signal by calcium-induced calcium release (CICR) [130, 131] and activation of the phosphoinositide signaling cascade [142, 143]. Nanopore formation by nsEP stimulation may be a unique approach for nonchemical activation of Ca^{2+} signaling in various types of cells, including cells that have no voltage-sensitive membrane channels. It has been proposed earlier to employ nsEP for heart pacing [128] and stimulation of catecholamine release [130, 138], but better understanding of nsEP effects may open the way for numerous other applications.

However, if nanopores are too numerous, flooding of cells with Ca^{2+} may become the primary cause of cell death. Ca^{2+} signaling is critically involved, in many ways, in both the initiation and effectuation of the cell death (see [144] for a review and other chapters of this book). Damaged cells develop necrotic blebbing, swelling, cytoplasm granulation, destruction of the cytoskeleton, shrinkage of the nuclei (pyknosis), and other pathological signs. While nanoporation can lead to either necrotic or apoptotic cell death [145–150], the increased Ca^{2+} facilitates the early necrosis and thereby decreases the cell population that could potentially become apoptotic. Recently we demonstrated that Ca^{2+}-mediated necrosis results from a delayed, abrupt, irreversible, and osmotically independent expansion of pores in the cell membrane [148].

Even with lower levels of ambient Ca^{2+}, the disruption of the membrane barrier function by nanopore formation may culminate in cell death. In this case, the primary cause of necrosis is the persistent plasma membrane permeabilization to small solutes (<1 nm). As described above, it results in the osmotic imbalance, water uptake, cell swelling, and cell membrane rupture. When the uptake of water is blocked by an iso-osmotic addition of a pore-impermeable solute such as sucrose [121], cells are rescued from the necrotic death, but nonetheless die later on by apoptosis [149]. However, the cause of the apoptosis in cells rescued from the necrosis is not fully understood. The cell death mechanisms and pathways resulting from nanopore formation are discussed in more details in other chapters of this book. For a more detailed analysis of nanopore properties, impact on cell physiology, and their hypothetical structure, please see a specialized review [122].

4.6 The Cytoskeleton as Target: Electromanipulation of Sensing at the Plant Cell Membrane

Peter Nick and Wolfgang Frey

Electromanipulation by pulsed electric fields can induce very specific and sometimes surprising, cellular effects. In this context, the current chapter wants to transport two main ideas: (1) The cellular effect is not determined by physics alone, but decisively shaped by cell biology. (2) The cellular effect is intimately linked with the cytoskeleton, and the specific differences in the electrically induced responses of animal and plant cells have to be seen in the context of qualitative differences in cytoskeletal functions. This chapter will therefore first survey the structural features of the plant cytoskeleton and then explore the different cellular functions conveyed by these plant-specific cytoskeletal arrays. Whereas the cytoskeleton in animal cells has been mostly discussed with respect to its structural and architectural role, the plant cytoskeleton seems to be rather of a sensory nature. It is this sensory function that renders the plant cytoskeleton a versatile and rewarding target for bioelectrical manipulation. Since plant cells are immobile and stabilized by a cell wall, the architectural function of the cytoskeleton has become partially obsolete. However, the tensegral organization of the cytoskeleton cannot only be used to confer structural organization; it can simultaneously be used as integrative sensory structure. This sensory function of the cytoskeleton has become dominant in plant cells and provides the conceptual framework to understand the effect of electromanipulation on the membrane-associated cytoskeleton, and the physiological functions conveyed by this membrane–cytoskeletal signaling hub with focus on volume control and programmed cell death/apoptosis. We propose a role for a membrane-associated, highly dynamic population of actin as a sensor for membrane integrity and a primary target for electromanipulation.

4.6.1 *The Cytoskeleton as Central Switch for Plant Cells*

Similar to their animal counterparts, plant cells can be specifically manipulated by bioelectrical treatments. However, the response of plant cells shows certain characteristics that can only be understood, when the plant-specific peculiarities of cellular structure and function are considered.

Although it is important to describe the biophysics behind these effects, a mere reduction of the phenomena to physics will fail to explain most of the biological effects. Cells are not passive targets of physical manipulation, but they actively *respond* to this physical manipulation by specific cellular events that are integrated into a biological context. This subchapter is motivated by the idea that electric manipulation will be *actively* perceived as *signal* that conveys *information* and therefore will elicit in the receiver cell an *active response*. When we understand the *biological context* of this *active response*, we will be able to design smarter and more specific strategies to evoke the cellular effects we desire to achieve. This subchapter will attempt to illustrate this mission using electromanipulation of the plant cytoskeleton as a proof of concept.

Plant cells move only rarely, and therefore this central topic of animal development does not play a role in plants. On the cellular level, plants respond to signals (such as electromanipulation) by three phenomena: (1) directional cell expansion, (2) directional cell division, and (3) patterned cell differentiation. All three phenomena are intimately linked with plant-specific arrays of the cytoskeleton.

4.6.2 Players of the Plant Cytoskeleton

In contrast to their animal counterparts, plant cells harbor only two of the cytoskeletal systems: microtubules and actin filaments. Despite intensive efforts, the third cytoskeletal system, the intermediate filaments, could never be convincingly demonstrated in plants, neither on the structural nor on the molecular level. However, microtubules and actin filaments are complemented by a structurally complex cell wall that is linked with the cytoskeleton through transmembrane connections into a tensegral network that is often termed *cell wall cytoskeletal continuum* [151].

4.6.2.1 Microtubular Arrays

During interphase, plant cells display a peculiar array of parallel microtubule bundles subtending the plasma membrane and usually oriented perpendicular with the axis of preferential cell expansion (Fig. 4.14a). These so-called *cortical microtubules* control the movement of cellulose-synthesizing enzyme complexes in the plasma membrane and thus the direction of cellulose deposition. Since cortical microtubules can reorient in response to various stimuli, they represent a tool, by which plant cells can regulate the directionality of expansion [152].

When a cell prepares for mitosis, cortical microtubules are complemented by additional arrays that are not seen during interphase. As the first event heralding the ensuing mitosis, the nucleus moves into the cell center which is nothing else than the site where later the new cross wall will be laid down. This nuclear movement is linked with a *radial microtubule* array that emanates from the nuclear surface and merges with the cortical cytoskeleton (Fig. 4.14b), tethering the nucleus to its new position. This nuclear movement sets the pace for the actual mitotic division—in tobacco cells, where the movement is delayed—due to overexpression of a plant-specific microtubule motor, cell division is delayed as well [153]. Once the nucleus has reached its final position in the cell center, the *cortical microtubules* suddenly disappear and are replaced by a broad band of microtubules around the cell equator (Fig. 4.14c). This so-called *preprophase band* predicts site and plane of the new cell plate that will be formed much later, once mitosis has been completed. The

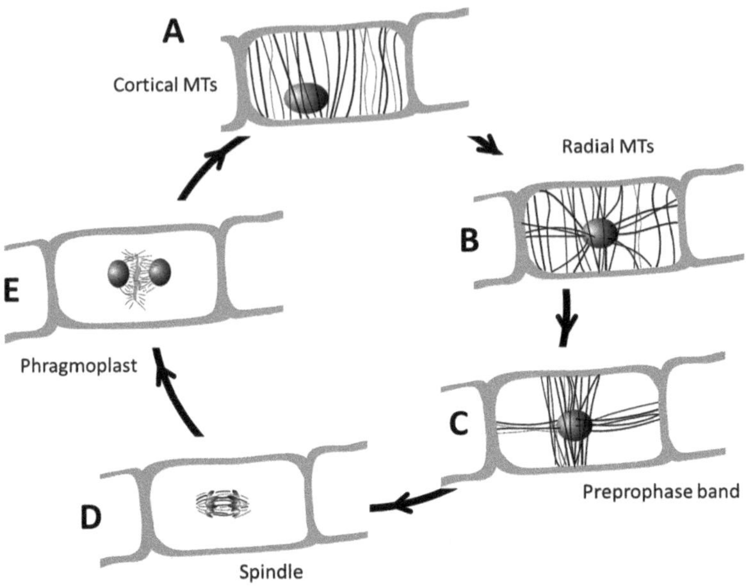

Fig. 4.14 Dynamic organization of plant microtubules during interphase and the cell cycle

preprophase band is necessary and sufficient for the correct organization of the cell plate [154]. The division *spindle* is always laid down orthogonal with the *preprophase band* [155] such that the *spindle* equator is located in the plane of the *preprophase band* (Fig. 4.14d). As soon as the chromosomes have separated, a new array of microtubules, the *phragmoplast*, appears at the site that had already been marked by the preprophase band (Fig. 4.14e). The phragmoplast controls the transport of vesicles to the periphery of the growing cell plate and consists of a double ring of interdigitating microtubules that increases in diameter with growing size of the cell plate and acts as a track for plant-specific minus-end-directed kinesins [156].

4.6.2.2 Actin Arrays

The dynamic complexity of microtubular organization is mirrored and complemented by a complex structure of the actin cytoskeleton.

During interphase, three distinct arrays can be discriminated (Fig. 4.15a): an array of transvacuolar actin cables (1), connecting a highly dynamic network of short actin filaments underneath the plasma membrane (2), and a perinuclear basket of actin bundles (3).

Unlike microtubules, the overall setup of the transvacuolar actin cables does not change much during the cell cycle, except for the fact that these cables are reoriented into a structure resembling a Maltesian cross during the premitotic nuclear migration (Fig. 4.15b). This so-called *phragmosome* seems to be interconnected with the premitotic *radial microtubules*. In contrast to the

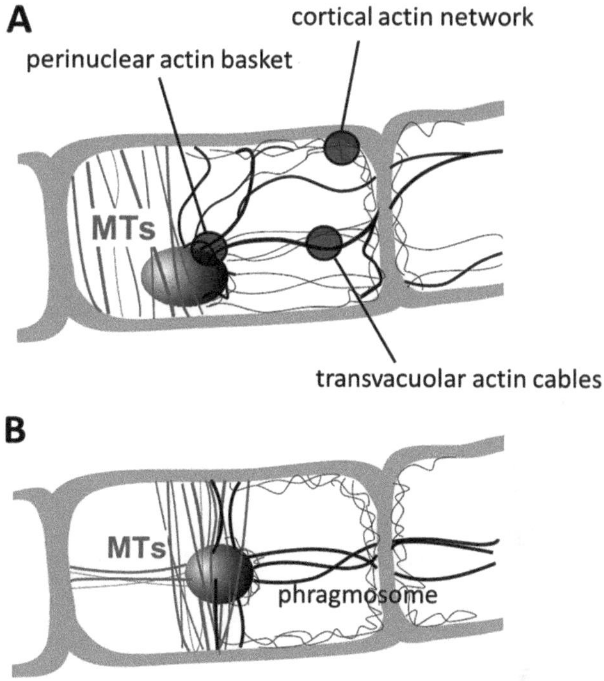

Fig. 4.15 Organization of plant actin filaments during interphase and the cell cycle

microtubular *preprophase band*, the phragmosome persists during meta- and anaphase and seems to participate in the organization of new microtubules during the formation of the phragmoplast [156]. During interphase, these transvacuolar cables are oriented parallel with the axis of cell expansion [157], and the rigidity of these transvacuolar strands as well as the degree of their bundling is under control of signals including plant hormones [158], kinase cascades [159], or light [160].

The transvacuolar cables are complemented by a fine network of highly dynamic microfilaments subtending the plasma membrane. This network had first been overlooked, because it could be rendered visible only by specific pretreatment with protein cross-linkers [161] or upon very mild fixation [160]. This peripheral meshwork was found to be linked with auxin-triggered cell elongation [162, 163], but only recently the advances of GFP-based live cell imaging made it possible to visualize this network in a consistent way such that it could be followed through the development or the response to signals [164, 165].

The perinuclear actin basket had first been discovered in dividing cells [166] and overlaps with *radial microtubules*. The actin basket and radial microtubules are cross-linked by a plant-specific minus-end-directed kinesin motor [153, 156]. The function of this actin basket is to move the nucleus to the division site. By super-resolution microscopy, we were able to see the three-dimensional topology during nuclear movement [167] and found, surprisingly, that the nucleus is not pulled by actin, but squeezed in a peristaltic manner.

4.6.2.3 Molecular Players of the Cytoskeleton

The qualitatively different organizations and functions of the plant cytoskeleton are achieved on the base of surprisingly similar building blocks: The major components tubulin and actin are almost identical between plants and animals (for a review on tubulins, see [168]; for a review on actins, see [169]). Although both proteins are encoded by gene families, whose members are differentially expressed depending on tissue, developmental stage, or signals (reviewed in [170] for actin and in [171] for tubulin), there seems to be little, if any, functional diversification of these isotypes. For instance, tubulin from animals can readily co-assemble with plant tubulin in vitro as well as in vivo. Upon microinjection into plant cells, it participates in the dynamic reactions of the host cytoskeleton in the same way as the endogenous tubulin (cell division [172, 173], cell expansion [174, 175]. From this, it can be concluded that the factors responsible for the specific organization of the plant cytoskeleton are extrinsic to actin and tubulin themselves. In fact, there is considerable diversification between plants and animals with respect to actin-binding proteins (for a review see [176]), microtubule-binding proteins (for a review see [177]), kinesin motors (for a review see [178]), and myosin motors (for a review see [179]).

4.6.2.4 The Cell Wall Cytoskeletal Continuum

In animal cells, a continuum between the cytoskeleton and the extracellular matrix is central for mechanosensing (reviewed in [180]) and employs membrane-spanning integrins that bind on their one side to proteins of the extracellular matrix containing Arg-Gly-Asp (RGD) motives and on the other side to the actin cytoskeleton (for a review see [181]). Plants seem to lack integrin homologues, but there is evidence for cytoskeletal reorganization after treatment with RGD peptides [182–184]. As a molecular basis for the plant cytoskeleton–plasma membrane–cell wall continuum, cell wall-associated kinases, arabinogalactan proteins, pectins, and cellulose synthases are discussed (for a review see [185]). The rich, but more or less circumstantial evidence for such transmembrane interactions of the cell wall and cytoskeleton has been assembled into a model of a so-called *plasmalemmal reticulum* as a third element of the plant cytoskeleton [151]. This *plasmalemmal reticulum* is probably a tensile structure and seems to participate in the control of cellulose deposition. Moreover, it was proposed to represent a manifestation of the plant version of lipid rafts.

4.6.3 The Plant Cytoskeleton as Membrane Sensor

Plants have evolved the peculiar cytoskeletal arrays described in the previous section to address specific functions. It is important to consider these functions, because they set the logical framework to understand the specific effects of electromanipulation. For all of these functions, a system property of the cytoskeleton is relevant: *tensegrity*. Microtubules are fairly stiff structures, which, with respect to their mechanic properties, can be compared to delicate glass fibers and therefore they are able to transduce compression forces [186]. In contrast, actin filaments are more flexible (mechanically comparable to silk) and therefore able to transduce traction forces. The concept of *tensegrity* became popular by the fundamental architectural work of Richard Buckminster Fuller (1895–1983). He accomplished the highest structural stability by combining a minimum of stiff and tensile elements. In fact, it is cytoskeletal *tensegrity* that shapes animal cells (for a review see [187, 188]). However, this architectural function of the cytoskeleton became mostly obsolete in plant cells due to the presence of a cell wall which is built as a composite structure with elongate load-absorbing elements (cellulose microfibers) embedded in an amorphous matrix of hemicellulose pectins and proteins.

Cellular architecture employs the tensegral principle to reach maximal mechanical stability and, simultaneously, flexibility on the basis of parsimonious use of resources and load-bearing elements. In addition, it can adapt continuously to the ever-changing conditions of growing and developing cells. This requires efficient sensing of forces and strains followed by appropriate reorganization of the tensegral building blocks. Thus, the tensegral cytoskeleton is not only a device to provide mechanical stability. It must also sense patterns of stresses and strains. This mechanical stimulation feeds back to the organization of the cytoskeleton in such a way that a stable minimum of mechanical energy is reached and continuously adjusted. It is this hidden sensory function of the tensegral cytoskeleton that dominates over architectural roles in the walled plant cells that are under continuous turgor pressure and use this pressure for regulated expansion. During plant evolution, the interphasic plant cytoskeleton was therefore shaped by selective pressures toward optimized sensing and integration at the plasma membrane. It is this sensory nature of the cytoskeleton that becomes also relevant for the response of plant cells to electromanipulation.

4.6.3.1 Cortical Microtubules and Cell Wall Texture

The parallel bundles of *cortical microtubules* are usually oriented perpendicular to the axis of preferential cell expansion. The *cortical microtubules* were found to define the biophysical properties of the yielding cell wall and thus the geometry of expansion. In cylindrical cells, where isotropic action of turgor pressure is predicted to produce only half of the strain in the longitudinal direction relative to the transverse direction, a transverse orientation of *cellulose microfibrils* maintains the lateral reinforcement needed to drive elongation [189]. In fact, filamentous structures acting as "reinforcement mechanism" to counteract lateral expansion were predicted merely from these biophysical considerations and termed "microtubules" [190]. This prediction stimulated an intense search for such structures and 1 year later led to the discovery of "microtubules" by electron microscopy [191]. The classical model assumes that *cortical microtubules* control the orientation in which newly synthesized *cellulose microfibrils* will be laid down (reviewed in [192, 193]).

Cortical microtubules can change their orientation in response to a broad range of signals, both exogenous and endogenous, and thus allow tuning of plant morphogenesis with the challenges of the environment. Signal-dependent reorientation of microtubules will cause altered deposition of *cellulose microfibrils*, a mechanism that allows adjusting the direction in which the cell wall yields to the turgor pressure exerted by the expanding protoplast and eventually alters the proportionality of cell expansion in response to the stimulus. The cellulose-synthesizing enzyme complexes are integrated into the membrane by fusion of exocytotic vesicles and are thought to move within the fluid membrane leaving a "trace" of crystallizing cellulose. The movement of the enzyme complex will determine cellulose orientation and thus the anisotropy of the cell wall. It is the direction of this movement where cortical microtubules interfere with the mechanical anisotropy of the expanding cell wall.

The direct contact between *cortical microtubules* and newly emerging *cellulose microfibrils* has been demonstrated by electron microscopy, but is also supported by a wealth of data, where signal responses of cell expansion were preceded by a corresponding reorientation of *cortical microtubules*. As to be expected from a microtubule-based mechanism for cellulose orientation, elimination of *cortical microtubules* by inhibitors produces a progressive loss of ordered cellulose texture. The resulting loss of axiality causes lateral swelling and bulbous growth. The striking parallelism between *cortical microtubules* and newly deposited *cellulose microfibrils* led to a "monorail" model proposing that motor proteins move along *cortical microtubules* and pull cellulose synthetases [194]. A concurrent "guardrail" model, where the crystallizing cellulose pushes the synthetase complex within microtubule-dependent protrusions, has been recently discarded, based on direct molecular evidence: Mutants with reduced cell wall integrity were affected in the microtubule-severing protein katanin [195] or in kinesin-related proteins [196]. Moreover, fluorescently tagged cellulose synthases were shown to move in tracks adjacent to the subtending cortical microtubules [197], and a cellulose synthase (CSI1) binds directly to microtubules [198].

This means that *cortical microtubules* are physically linked to proteins that span the plasma membrane and might transduce conformational changes (such as those caused by electromanipulation) directly upon the microtubular cytoskeleton.

4.6.3.2 Cortical Microtubules Act as Sensors for Abiotic Stress

As a central element of their adaptive strategy for survival, plants must integrate the signaling evoked by different stress factors into a balanced and appropriate response. Many of these stress factors can be sensed through alterations of membrane tensions: These not only include evident situations such as direct mechanic load or osmotic forces (that in the case of ionic stress are complemented by electrostatic forces) but also stimuli reporting on other environmental factors, such as gravity, wounding, wind, touch, or pathogen attack, as well as factors modulating membrane fluidity, such as cold and heat (for a review, see [199]).

As the tensegral cytoskeleton is connected with the cell wall through integrative linkers, the mechanical strains produced by cellulose microfibrils can align cortical microtubules, thus closing a self-referring circuit between the cell wall and cytoskeleton in the growing plant cell. The cell expansion reinforced in a direction perpendicular to the orientation of microtubules and microfibrils will generate forces parallel with the major strain axis [200]. These forces will then relay back through the plasma membrane upon cortical microtubules that are aligned in relation to these strains. As individual microtubules mutually compete for tubulin heterodimers and as the number of microfibrils is limited by the quantity of cellulose synthase rosettes, this regulatory circuit should follow the rules of a reaction–diffusion system [201] and therefore be capable of self-organization and patterning.

Microtubules might sense mechanical stress themselves, because during growth, they build up considerable mechanical tension [202]. This tension has to be compensated by specific proteins complexing the growing end (so-called +TIP proteins). Any mechanic stimulus at the tip will impair this compensation and the accumulated tension will be discharged as catastrophic outward bending of protofilaments. This innate mechanosensitivity makes microtubules ideal signal amplifiers in concert with other mechanosensors. In fact, microtubules were identified as interactors of stretch-activated ion·channels from genetic studies in the worm *Caenorhabditis* [203] and pharmacological studies suggest that mechanosensitive calcium channels are also active in plants [204, 205]. Moreover, osmotic challenge of membranes causes assembly of cortical microtubules [206, 207]. This will stimulate phospholipase D, whose product phosphatidic acid-dependent activates a membrane-bound NADPH oxidase, producing an oxidative burst as a trigger for plant adaptation to drought stress [208].

Cortical microtubules have therefore emerged as important components of a signaling hub at the membrane of plant cells that is able to sense and process different stress signals into different and specific signatures [209].

4.6.3.3 Cortical Actin Filaments Act as Sensors of Membrane Integrity

Plant actin is essential for intracellular transport of various cargoes such as peroxisomes [210], chloroplasts [211], mitochondria [212], or Golgi vesicles [213]. Also, the active transport of plasmid DNA after electrotransfer is accomplished by the actin cytoskeleton [214]. Transvacuolar actin cables structure transvacuolar cytoplasmic strands [215] and drive the premitotic migration of the plant nucleus [153, 156]. The function of the dynamic actin network subtending the plasma membrane has remained enigmatic, however. The cortical actin net seems to be physically interconnected with the distal ends of the transvacuolar cables that on their proximal ends are linked with the perinuclear actin basket. The bundling of cortical actin filaments can be modulated by signals such as the plant hormone auxin [160, 163], and this will feed back on the cellular sensitivity to auxin [216], leading to the concept of an actin–auxin oscillator [217]:

Auxin regulates actin organization [165] by controlling actin dynamics through auxin-dependent differential membrane association of actin depolymerization factor 2 [167]. Actin, in turn, regulates auxin transport [218] constituting a self-referring oscillating circuit. This circuit oscillates with a period of around 20 min and underlies the regulation of cell expansion and cell division by auxin. Any disturbance of membrane integrity will impact on these oscillations and result in a dissociation of actin from the membrane and a contraction of the transvacuolar cables toward the nucleus.

This remodeling of actin is usually followed by programmed cell death. Similar to apoptosis of animal cells, programmed cell death in plants fulfills important functions for genetic integrity and plant survival. In particular, it is a central feature of immunity against so-called biotrophic pathogens (for a review, see [219]). This specialized group of pathogens does not kill the host cell, but invades them after manipulating basal immunity and reprogramming the host cell to deliver nutrients, thus turning its victim into a kind of defenseless zombie. The most efficient strategy to encounter these pathogens is to activate programmed cell death. Thus, the infected cell will commit suicide for the sake of its healthy neighbors. Prior to its final sacrifice, the reprogrammed cell activates the accumulation of antimicrobial toxins in the vacuole. The controlled breakdown of the vacuole will then execute suicidal cell death, but at the same time kill the invader in a kind of "kamikaze" strategy.

Elicitors or pharmaceutical compounds that can trigger programmed cell death cause a rapid breakdown of the dynamic meshwork of cortical actin filaments followed by a contraction of transvacuolar cables [220–222]. This phenomenon is not confined to plant immunity, but is found across eukaryotic cells in general (for reviews see [223, 224]) and for plant cells in particular [225, 226].

Cortical actin filaments have therefore emerged as part of an oscillatory sensor that can sense membrane integrity. Any impact on membrane integrity will cause a phase shift of the actin–auxin oscillations and result in actin remodeling, which is then transduced into programmed cell death. Since actin filaments are linked with the unknown plant functional analogues of animal integrins, they should be able to respond to conformational protein changes at the membrane. From these considerations it is predicted that electromanipulation, even at energies that are not sufficient to cause irreversible pores, should impact on this actin-based switch between life and death.

4.6.4 The Plant Cytoskeleton Responds to nsPEFs

Methods shape concepts—the availability of fluorescent proteins allowing for imaging of specific organelles in living cells has revealed the seemingly static plant cells as highly dynamic systems. The term "cytoskeleton" was coined at a time when actin filaments and microtubules were accessible only through electron microscopy in ultrathin sections of fixed material. The concept of a static "cellular skeleton" has been replaced meanwhile by the model of a dynamic cytoskeletal equilibrium as evident, when fluorescent markers are followed after local bleaching (so-called fluorescence recovery after photo bleaching, FRAP) or when local changes of fluorescent color are followed, a novel approach made possible by the use of photoconvertible fluorescent proteins [227]. The fluorescent protein technology in combination with advanced fluorescence microscopy provided the tools to get insight into the dynamic changes of the plant cytoskeleton and the plant endomembrane system in response to electromanipulation. A panel of transgenic lines of the cellular plant model tobacco BY-2 (reviewed in [228]) expressing either GFP fusions of plant tubulin, the actin-binding domain 2 of plant fimbrin, or the retention motif for the endoplasmic reticulum, HDEL, in fusion with GFP, made it possible to follow the response of cytoskeleton and endomembrane system in living cells. These reporters were chosen, because they do not constrain the functionality of the cytoskeleton (in contrast to fluorescent phalloidin). A second prerequisite was the construction of a device that allowed administering nanosecond pulsed electric fields (nsPEFs) directly under the microscope, such that the cellular responses could be followed already during the first minutes after challenge. In fact, this approach revealed that the membrane-associated cytoskeleton responds swiftly and dramatically to electromanipulation [229].

Already a single pulse of 33 $kV.cm^{-1}$ at a duration of 10 ns was sufficient to disorder and disassemble cortical microtubules such that the fluorescently tagged tubulin used for visualization diffused into the mesh-like cortical cytoplasmic strands. This response had already initiated in the first minute after the pulse and was fully manifest at 3 min after pulsing. At the same time, the nucleus lost its ellipsoidal shape and was rounded up, which was a few minutes later followed by blebbing of the nuclear envelope. The response of actin was even swifter: cortical actin meshwork was rapidly depleted, while the transvacuolar cables contracted toward the nucleus within the first minute after the pulse. Similar to actin, the endoplasmic reticulum is subdivided into a highly dynamic submembrane meshwork of sheets that are connected by strands, whereas the transvacuolar ER strands and the nuclear envelope appear to be relatively static. Nevertheless, a treatment with nsPEFs not only degraded the cortical ER, but produced a disintegration of the nuclear envelope followed by invasion of the GFP signal into the karyoplasm, from where it remained excluded in non-challenged control cells. This breakdown of the nuclear envelope occurred around 3–4 min after pulsing, which parallels the blebbing of the perinuclear microtubules.

These rapid responses of cortical microtubules and especially cortical actin were followed by somewhat slower changes in the behavior of the plasma membrane. To probe for potential disintegration of the plasma membrane, we measured the uptake of trypan blue, a dye that cannot permeate the membrane of living cells and is therefore classically used for viability assays. A variation of the electric dose by increasing the number of pulses showed that uptake of the dye became detectable from two pulses and was saturated at ten pulses [230].

A detailed time course of trypan blue permeability using five pulses revealed that the penetration of the dye initiated from the tip of the cells, whereas it was slower at the lateral flanks suggesting a strong impact of membrane curvature. This is to be expected from geometrical considerations of membrane charging: in a sphere, the local membrane voltage induced by a field depends on the radius, the field strength, and the cosines of the angle with the field vector [76].

Although the leakage of the dye became detectable already from the first minute after the five pulses, it required 6 min to become fully expressed [230]. In contrast, the cytoskeletal effects described above were already seen for a single pulse and earlier, which means that the response of the cytoskeleton was more sensitive and more rapid as compared to the permeabilization of the plasma membrane.

In the search for other cellular responses to nsPEFs, we had to decrease the stringency to a range, where cells still were able to survive for a sustained period. A mapping of prolonged viability over electrical energy density revealed that the cellular response followed an all-or-none pattern: below a certain threshold, all cells remained viable; above this threshold, all cells died [231]. For a treatment above the threshold, we observed a stratification of cytoplasmic strands, a block of cytoplasmic streaming, and a loss of nuclear positioning within 3–5 h after the treatment. The auxin efflux carrier PIN that was strictly localized to the plasma membrane in unpulsed controls partially detached from the membrane, such that its localization became diffuse. Since this loss of membrane localization of PIN should impair auxin flow through the cell file, this should also impinge on the synchronization of cell division within a file, a phenomenon, which can be scored as a reduction in the frequency of a diagnostic peak for hexacellular files and an increase of quadricellular files [232]. In fact, this very specific readout was observed following nsPEF treatment, indicative of an impaired auxin flow, corroborating the observed delocalization of the PIN protein. It should be mentioned that both, this division synchrony and the localization of the PIN protein, depend on actin filaments (reviewed in [217]).

In the next step, a subthreshold treatment of 20 pulses of 25-ns duration and 10 $kV.cm^{-1}$ was used to detect more subtle changes of cellular physiology. In fact, two phenomena could be observed: A delay of mitotic activity by 1 day and a delay of premitotic nuclear migration by 1 day.

The (rapid) response of the submembranous cytoskeleton [229–231] is thus followed by (slower) increases of membrane permeability, impaired localization of auxin efflux carriers, delayed premitotic nuclear migration, and delayed onset of mitosis, and all these responses are intimately linked with the cytoskeleton. Although these correlations are consistent with a model, where the response of the membrane-associated cytoskeleton might be the cause of all these slower cellular responses to nsPEFs, they remain correlations and are no proof for a causal link. The question of causality will be investigated in the next section.

4.6.5 Plant Actin Controls the Response to nsPEFs

To state that a phenomenon A (nsPEF response of the actin cytoskeleton) is causative for a different phenomenon B (membrane permeability and cell death), three conditions have to be met: (1) A must precede B in time, (2) A must be necessary for B, and (3) A must be sufficient for B.

The first condition has already been discussed in the previous section: the cytoskeletal responses, especially the response of actin filaments, are observed earlier than the leakage of trypan blue into the cell [229, 230].

To test the second condition, it is necessary to manipulate the actin cytoskeleton before challenging the cell by nsPEFs. In addition to pharmacological manipulation, it is possible to use genetic engineering. Both strategies have been employed to probe for changes in the response to nsPEFs: In fact, a mild pretreatment by the actin-stabilizing drug phalloidin was able to efficiently suppress the penetration of trypan blue, indicative of a stabilization of membrane integrity [229]. In an alternative approach, actin-binding proteins that confer different degrees of actin stabilization were constitutively overexpressed under the control of a strong viral promoter (cauliflower mosaic virus 35S) in tobacco BY-2 and then trypan blue uptake recorded over increasing pulse numbers [230]. These curves were dampened correlated with the degree of actin stabilization—whereas the Lifeact probe (an - actin-binding peptide currently used as state-of-the-art marker for actin, because it is reported not to interfere with actin dynamics) yielded a dose–response curve that was identical to that recorded for non-transformed cells, the FABD2 marker (conferring a very mild stabilization of actin), already caused conspicuous dampening of the amplitude and a shift toward higher pulse numbers. The mouse talin probe, producing the strongest stabilization of actin, suppressed trypan blue uptake almost completely. To drive the assay to the highest possible stringency, we generated a transgenic line, where the actin-stabilizing LIM domain is under control of a glucocorticoid-inducible promoter. By addition of the artificial glucocorticoid dexamethasone, actin can be stabilized and compared to a non-treated aliquot of the same cell line as negative control, such that any effect of potential genetic differences can be ruled out. Doing so, the induced cell line showed a strongly reduced uptake of trypan blue even for the highest pulse numbers tested. Thus, the second condition holds true as well: actin dynamics are necessary for membrane leakage. Although it cannot be excluded that there is also some impact of the transvacuolar actin cables and/or the perinuclear actin cage, this impact is estimated of minor importance, because these structures are already quite stable even in non-treated or non-transformed cells, whereas it is the dynamics of the cortical actin meshwork that are affected most.

To test the third condition, it is necessary to cause depletion of cortical actin filaments and contraction of transvacuolar cables in the absence of nsPEFs and to verify whether this will result in cell death. This experiment has become possible by using specific bacterial elicitors that produce a rapid actin response that down to the details of subcellular structure and timing matches the actin responses observed after nsPEFs. In fact, treatment of cell cultures of grapevine with the elicitor HrpN from the phytopathogenic bacterium *Erwinia amylovora* [220], of tobacco BY-2 with the elicitor HrpZ from *Pseudomonas syringae* [222], or simply with the plant defense compound resveratrol [221] was able to trigger rapid breakdown of cortical actin and contraction of transvacuolar actin cables, and these responses were followed by cell death. Thus, to induce a breakdown of cortical actin is sufficient to trigger cell death in the absence of nsPEF challenge.

Thus, all three conditions have been experimentally verified and found to be valid. It is therefore feasible to assume a causal link between the actin response at the plasma membrane and subsequent cellular responses (membrane leakage and cell death).

However, the conclusion of a dynamic actin population underneath the plasma membrane as a primary target of electromanipulation through nsPEFs is prone to raise some controversy: Whereas longer pulses with lower field strength are thought to cause reversible pore formation in the plasma membrane and can induce membrane passage of macromolecules, shorter pulses with higher field strength are discussed to cause mainly intracellular electromanipulation [233, 234]. The reason for this prediction is that the charging of the cell membrane should be slower than the penetrance of the electric field into the cell interior. Thus, intracellular membranes or organelles should be targeted before the energy can be dissipated into the plasma membrane [19, 235–237]. In fact, rupture of intracellular granules or vacuoles without detectable electroporation to the outer membrane [20, 237] or calcium release [234, 238] has been reported for mammalian cells after exposure to nsPEFs. However, rapid changes of plasma membrane permeability explained by the formation of nanopores [236, 239, 240] even for pulse durations of 10–60 ns [121, 241, 242] challenge this idea even for animal cells. In addition, the distribution of electric fields is definitely different in a cell, where, in addition to the plasma membrane, a second extensive membrane system, the tonoplast lining the vacuole, is present.

Rather than following theoretical considerations, we addressed this topic experimentally. A good deal of the discrepancy is caused by the fact that experiments conducted in different systems under different conditions and with different scopes have been compared. It was therefore relevant to address this question in a single experimental system, where the same molecular target is localized either in the cell interior or near the plasma membrane and where different cellular functions can be assigned to these two subpopulations of the target molecule. Through our investigation of the plant cytoskeleton, we attained access to such a candidate molecule: KCH, a plant-specific member of the kinesin-14 family, links microtubules and actin filaments [243] occurring in two subsets that confer two different cellular functions [156]—one subset is linked with the perinuclear actin basket and important for premitotic nuclear migration and mitosis [153]. The second subset is not linked to actin filaments and moves slowly with cortical microtubules and thus plays a role for cell expansion. To address the influence of KCH, we simply used a tobacco BY-2 cell line overexpressing this protein in fusion with GFP [231]:

As a cellular readout for the nuclear function of KCH, nuclear movement heralding the ensuing cell division was quantified [153] and found to be delayed by overexpression of KCH, as well as by a mild nsPEF treatment. A combination of both factors acted synergistically. Although this might indicate a site of action at the nucleus, i.e., in the cell interior, it should be kept in mind that the nucleus is tethered, through a radial network of actin filaments and microtubules, to the cell wall by means of transmembrane proteins [151]. A similar delay of nuclear migration might therefore be produced, when the connections of this radial network to the plasma membrane were disrupted by nsPEFs. Thus, the analysis of nuclear migration alone is not sufficient to decide on the site of action.

We therefore exploited cell expansion as a second cellular function that is unequivocally linked to the cortical microtubules subtending the plasma membrane. In fact, we observed that overexpression of KCH promoted cell expansion, and that this promotion was stimulated by nsPEFs in the KCH overexpressor but not in the non-transformed wild type, and that cell expansion responds more sensitively as compared to cell division. In other words, there is a strict synergy between KCH overexpression and nsPEFs with respect to a function that is clearly located in the cytoskeleton adjacent to the plasma membrane. In contrast to the hypothetical nsPEF target in the cell interior, the experimentally verified target site at the plasma membrane can explain all observations of this study supporting the cytoskeletal signaling hub at the membrane as primary target for nsPEFs in plant cells.

Although a functional model of this hub is still far from conceived, it is already possible to distil from our data and those of others a structural model as a conceptual framework to understand the cellular effect of electromanipulation.

Although the cortical cytoskeleton has been observed to be very close to the membrane, it is not clear, whether there is a direct connection or whether the link is rather indirect through third molecular partners. The resolution of light microscopy is limited to about 250 nm in xy and to about 500 nm in z, even for advanced confocal microscopy (for a review see [244]). However, it is possible to break the resolution barrier in z-direction by means of total internal reflection fluorescence (TIRF) microscopy. This novel technique uses an evanescent wave from a totally reflected excitation beam. This wave can only penetrate 50–100 nm into the specimen, such that the fluorescence observed under these conditions must come from the very periphery of the cell. For walled plant cells, this technique is not applicable, because the cell wall is by far thicker than the range for the evanescent wave. Several reports that had claimed TIRF imaging for plant cells have later been shown to be misled by an optical artifact called variable-angle epifluorescence (VAEF), which allows to view the uppermost few μm of a specimen without optical bleedthrough from the deeper layer [245]. In contrast, true TIRF records only the uppermost 50 nm and is, due to the curved topology of the specimen, confined to a narrow field of observation. To see whether the cytoskeleton is localized in close proximity to the plasma membrane, protoplasts were generated from cells expressing either the actin marker ATFABD or the microtubule marker AtTuB6 in fusion with GFP and viewed by TIRF microscopy. We observed that actin filaments were organized in starlike plaques, from where finer filaments emanated. The crossings of filaments appeared as punctate structures. The pattern for microtubules appeared similar, but the starlike centers were much finer and the setup appeared to be more delicate as compared to the actin filaments. The distance between these structures and the membrane is estimated to be in the range of a single microtubule; otherwise they would not be imaged by TIRF. The field of observation was small (around 5 μm in diameter), which would be expected from a penetration depth of 50 nm and which is clear evidence that the recorded structures were imaged by true TIRF and not by VAEF. We think that the punctate centers of the starlike cytoskeletal structures are the structural manifestation of the signaling hub.

This signaling hub seems to be embedded into a complex topology of the plasma membrane that deviates from the straight surface underlying most theoretical models of membrane charging. This topology depends on the cytoskeleton. There are two lines of evidence for this idea.

Stabilization of the submembranous cytoskeleton caused an increase in the apparent thickness of the cell membrane. Since the elementary membrane cannot be resolved by light microscopy, these changes of apparent thickness must be caused by membrane topologies, leading to a model of tubulovesicular membrane folds or invaginations that increase membrane surface and might be structurally maintained by actin filaments [230].

Plant protoplasts can swell within seconds in response to hypoosmotic shock without bursting. Since intensive expansion ability of plasma membranes is confined to ~2 % [246], there must be membrane material released from internal stores during hypoosmotic swelling which is difficult to be reconciled with a model of a straight membrane surface.

To get more insight into the functional relevance of actin for membrane integrity, we used the regulatory volume control in protoplasts as model and used hydraulic conductivity (L_p) as readout [247]. This readout could be derived from the time course of protoplast swelling and was then manipulated by different factors to identify molecular components interfering with this response. We found that chelation of calcium, inhibition of calcium channels, or manipulation of membrane fluidity by benzyl alcohol did not significantly alter L_p. In contrast, direct manipulation of the cytoskeleton via specific compounds either destabilizing or stabilizing actin filaments (latrunculin B, phalloidin) or microtubules (oryzalin, Taxol) modulated conductivity. In addition, the bacterial elicitor harpin or modulation of phospholipase D was effective.

We further used optochemical engineering of actin using a caged form of the phytohormone auxin to break the symmetry of actin organization [248] in a single protoplast and found that the transcellular gradient of actin rigidity resulted in a localized deformation of cell shape (nota bene: under isotonic conditions) indicative of a locally increased L_p.

Whereas auxin-triggered actin dynamics were promoting the swelling response, bundling of actin blocked expansion, probably through mechanic tethering of the microtubule constrained swelling, leading to a model where a dynamic population of microtubules impedes the integration of membrane material into the expanding membrane. This microtubule population could be controlled by specific pharmacological activation of phospholipase D, a central component of a membrane–cytoskeleton signaling hub. This should produce two consequences:

1. The activation of phospholipase D would produce phosphatidic acid, an important activator of the membrane-bound NADPH oxidase RboH culminating in an apoplastic oxidative burst (for a review see [209]) that provides a central input for plant stress signaling including activation of programmed cell death.
2. The detachment of the cytoskeleton from the membrane invaginations would release additional membrane material for resealing of nanopores and for release of mechanical tensions within the plasma membrane.

It should be emphasized that this cytoskeletal membrane hub is subject to numerous positive and negative feedback regulations. For instance, the apoplastic reactive oxygen species will penetrate through aquaporins into the cortical cytoplasm and interfere with the bundling state of actin creating a signal, where, due to cytoskeletal tensegrity, the actin will contract toward the nucleus, which seems to be an important trigger to activate programmed cell death. In contrast, the integration of vesicles into the resealing membrane will remove PIP2, a downstream product of phospholipase D sequestering actin depolymerization factor 2 [249], such that actin dynamics will be reinstalled and the dynamic filaments will relink to the membrane preventing actin contraction. Whether a cell challenged by electromanipulation will rescue membrane integrity by resealing or whether it will be doomed to programmed cell death leading to a long-term loss of membrane integrity does not only depend on the physics of membrane charging, but is crucially decided by the relative timing of these antagonistic feedback loops.

4.6.6 Consequences for Current Imaginations on nsPEF Interactions with Cells

The results discussed in previous sections demonstrate that the plant cytoskeleton has a major impact on cellular responses upon nsPEF administration. In particular, in the short pulse range, below 100 ns, the cytoskeleton bears a strong influence on membrane permeabilization. This fact might have been overseen in the past, because in the long pulse exposition regime, the effects of pore formation dominate and nowadays still are considered as a basic principle of membrane permeabilization.

Despite the existing differences in cytoskeletal structure between mammalian and plant cells, a role of the cytoskeleton for the response to pulsed electric fields is also supported by experimental results obtained from mammalian cells. Consistently, it is reported that cortical actin filaments disintegrate, and microtubular structures depolymerize after PEF treatment [250–252] within some minutes and recover later in a time frame of an hour [252]. Already back in the 1990s, it was shown that the cytoskeleton is strongly involved in membrane resealing after permeabilization by pulsed electric fields [99, 253]. Treatment with compounds acting on the cytoskeleton prolonged resealing times from minutes to hours. PEF-induced disintegration of cortical cytoskeleton structures is reported to decrease the Young's modulus of the cell boundary by more than a factor of 2 [250, 254], underlining the importance of the tensegral properties of the cytoskeleton for mechanical stabilization and shaping of mammalian cells.

The findings on the plant cytoskeleton provide unambiguous evidence for a physical link between the cortical actin cytoskeleton and the plasma membrane at distinct points. Actin filaments emanate in a starlike manner from these interconnections. It also could be demonstrated that nsPEF-induced dissolution of the cortical actin meshwork provokes permeability of the plasma membrane for larger molecules, whereas chemical stabilization of the actin meshwork maintained membrane integrity when challenging the plant cells by the same pulse treatment [229, 230]. This demonstrates that membrane permeabilization is not solely caused by the formation of pores in the PL bilayer, which undoubtedly is one of the primary effects of membrane permeabilization, but also can be evoked via destabilization of the submembranous cytoskeleton, which obviously feeds back to membrane integrity. In addition, it could be demonstrated that cytoskeleton plays an active and necessary role in membrane permeabilization of plant cells.

Besides membrane site-associated responses, nsPEFs affect the cell cycle as evident from a delay of premitotic nuclear positioning. Also intercellular auxin transport is disturbed.

The causal sequence for the different cellular responses and the cytoskeletal response remains to be elucidated. Whether dissolution of the cytoskeleton is the cause of osmotic swelling as strongly suggested by the current work or whether it is the consequence of osmotic swelling as proposed for mammalian cells [255] has to be clarified for each system by functional analysis (including time-course studies). Also the biophysical trigger for cytoskeletal dissolution is pending so far. Forces due to dipole alignment of actin or tubulin monomers along the direction of the electric field direction appear to be feasible to evoke filament disruption. Alternatively, a complete loss of phospholipid ordered structure [101] around the cytoskeleton–membrane links in consequence of the pulsed electric field might also cause the subsequent loss of cytoskeletal structure.

Irrespective of the underlying biophysical mechanisms, the sensorial properties of cytoskeleton and in particular on the sensory feedback of stimulation by pulsed electric field stimulation will open promising new applications for electromanipulation of plant cells by pulsed electric fields.

4.7 Intracellular Trafficking of Plasmid and siRNA After Electroporation

David A. Dean

The fate of nucleic acids once translocated across the plasma membrane depends on the nucleic acids themselves. For example, while plasmids must make their way to the nucleus in order to be expressed, siRNA, miRNA, shRNAs, mRNA, and most RNaseH-dependent and independent DNAs and RNAs (such as DNAzymes or hammerhead ribozymes) function within the cytoplasm so they do not need to traffic to the nucleus. Although the biophysical and biological environment of the cytoplasm impacts movement of all types of nucleic acids, the greatest amount of information has been gathered for the movement of plasmids, and as such, unless noted, most of the following discussion will be directed at the productive and nonproductive movement of plasmids throughout the cell.

4.7.1 A Possible Role for Actin

Once the plasma membrane has been destabilized by electroporation allowing for nucleic acid entry, both DNA and RNAs are confronted by a number of barriers that must be overcome for their function, all of which are independent of the electric field (Fig. 4.16). Studies with fluorescently labeled plasmids and fluorescently labeled RNAs show that the translocated nucleic acids remain near the inner surface of the plasma membrane for a period of time, prior to movement toward the interior of the cell [102, 214]. Using fluorescence labeling and fluorescent protein techniques, it was shown that DNA and actin co-localized to the same spots at the permeabilized membrane facing the anode immediately following electroporation [214]. Moreover, when the actin network was destabilized with latrunculin B, reduced levels of DNA were present at the spots, suggesting that the actin may play a role in the internalization of the DNA itself. Cortical actin forms a meshwork beneath the plasma membrane to provide structure to the cell and to link the intracellular and extracellular environment for signaling. It is likely that this network acts to temporarily "trap" the molecules after they have crossed the plasma membrane. Similar findings have been reported for entry of a number of viral particles [256, 257]. It is currently not known how the DNA escapes this region

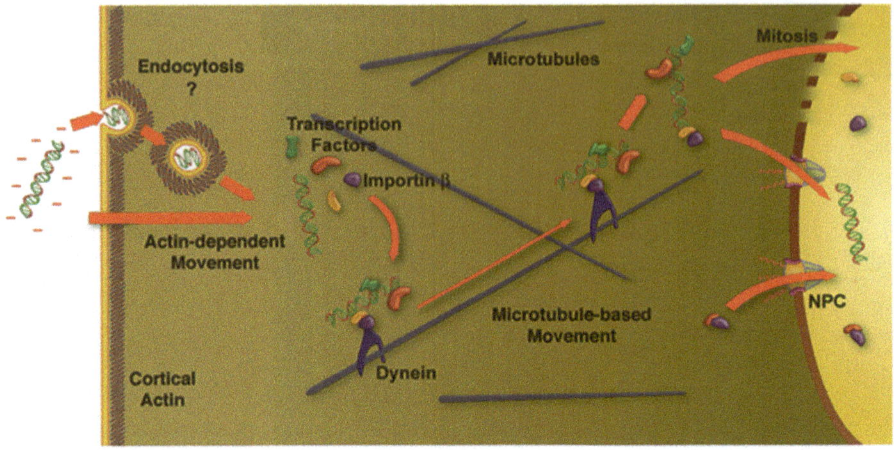

Fig. 4.16 Post-electroporation intracellular trafficking of plasmids. Following electropermeabilization of the membrane, plasmids may enter the cell by either endocytosis and/or direct entry into the cytosol at which point they must traverse the cortical actin layer, perhaps using actin-based movement [102, 214]. Once free in the cytoplasm, plasmids are rapidly complexed by a number of DNA-binding proteins in the cytoplasm which in turn bind to other proteins to form large protein–DNA complexes [259]. Transcription factors bound to the DNA interact with importinβ and other proteins that in turn link the complex to dynein for movement along microtubules to the nucleus where it falls apart at the nuclear periphery [264]. Nuclear entry is then mediated by importinβ in a sequence- and importin-dependent manner through the nuclear pore complex (NPC) in nondividing cells or independent of importins and any DNA sequence requirement during mitosis and the associated dissolution of the nuclear envelope (*top*)

or whether it is by directed movement or diffusion, but the nonuniform nature of the cortical actin meshwork may allow for localized diffusion of the nucleic acids at distinct sites or at "holes" out of this region of the cell and toward the interior.

Apart from the possible role of cortical actin in the internalization and trafficking of plasmids, several studies have also implicated a role for actin filaments in the cytoplasmic trafficking of plasmids in addition to that of the microtubule network [102, 214]. When actin dynamics were perturbed with drugs, both stabilization and destabilization of the networks resulted in slight decreases in the percent of fluorescently labeled plasmids showing active transport or the total distance traveled by the particles using single-molecule particle tracking [102, 214]. In contrast to these findings, earlier work showed that disruption of the actin cytoskeleton resulted in greatly enhanced diffusion of large DNA fragments within the cytoplasm of microinjected cells [258]. Proteomic studies from our laboratory have found that several actin-based motors (myosin 1B, 1C, and 9) are found in the protein–DNA complexes at early times after electroporation (15 min) along with a number of different microtubule-based motors [259], supporting a possible role for actin-based movement of DNA particles, at least at times between entry of the DNA into the cytosol and its binding to microtubules.

4.7.2 Microtubule-Based Movement of Plasmids

The cell cytoplasm has been described as early as the 1940s as resembling a reversible gel–sol system that is relatively stiff and does not allow for a great deal of free diffusion of large molecules [260]. It is a complex system composed of multiple cytoskeletal elements, including actin microfilaments, microtubules, and intermediate filaments, all of which are organized into a complex, crowded latticework that is constantly remodeling in response to a variety of internal and external stimuli. It has been shown that both double-stranded DNA fragments greater than 1000–2000 bp and macromolecule-sized solutes display almost no passive diffusion and are largely immobile in the cytoplasm of HeLa cells and fibroblasts [261, 262]. Thus, in order to traffic to the nucleus, plasmids must use directed movement, and it has been demonstrated that microtubules and their associated motor proteins are used to do so [102, 259, 263–267].

Several studies have shown that an intact microtubule network and motor proteins such as dynein are necessary for transfected naked DNA and some viruses to traverse the cytoplasm and reach the nucleus [102, 266, 268, 269]. When the microtubule network was disassembled by treatment with nocodazole, movement of DNA toward the nucleus was largely abolished [102, 266]. By contrast, modulation of the actin cytoskeleton had minor effects [266, 268]. Not only was movement of labeled DNA inhibited, but gene expression of reporter genes on the plasmids was also blocked [266, 268]. Inhibition of dynein, the major microtubule-based motor protein driving movement toward the nucleus, resulted in the same loss of plasmid movement and expression [102, 266]. Biochemical analysis has confirmed these findings. In one set of experiments, a spin-down assay in which plasmid DNA, cell extracts, and stabilized microtubules were incubated and then centrifuged to separate polymerized microtubules and any proteins or DNA interacting with them from the reaction was used to demonstrate that DNA interacted with microtubules [264]. In these studies, DNA interacted with microtubules only when cytoplasmic extracts were provided as a source of adapter proteins. By testing the ability of plasmids carrying different genes, transcriptional control elements, and DNA sequences to bind to the microtubules in this assay, it was found that plasmids interacted with microtubules in a sequence-specific manner. While the bacterial plasmid pBR322 failed to interact with microtubules in this spin-down assay, as did a number of plasmids containing various viral or cellular promoters, plasmids carrying the CMV promoter bound to the microtubules in the presence of cytoplasmic extracts [264]. Sequence analysis revealed that the only DNA element common to microtubule-binding promoters was the binding site for the cAMP response element-binding protein (CREB), and when a single CREB site was placed into pBR322, it interacted with microtubules. When the movement of individually fluorescently labeled plasmids was followed by particle tracking, this sequence specificity of movement was recapitulated in living cells: while pBR322 shows only marginal diffusive movement limited to small regions of the cytoplasm near its site of cytosolic entry, plasmids carrying CREB-binding sites or one of several other eukaryotic promoters and enhancers showed active and directed movements at much faster rates [102, 259, 263, 264]. Real-time particle tracking of quantum dot-labeled plasmids has shown that the DNA moves along microtubules with kinetics and dynamics that are in line with those seen for microtubule-based movement of organelles, virus particles, and proteins [102, 259, 263, 264].

As the spin-down assays suggest, plasmids do not interact with microtubules directly, but require adapter proteins in the cytoplasm to act as intermediaries between the DNA and the microtubules. Once the DNA has entered the cytoplasm and moved beyond the cortical actin domain, it binds to a number of sequence-specific and nonspecific DNA-binding proteins as well as a number of other proteins to form large protein–DNA complexes [259, 270, 271]. The formation of these complexes is rapid: within 15 min of electroporation in adherent cells, plasmids can be detected to physically interact with transcription factors, including CREB, as well as a number of other proteins [264]. These additional proteins include a number of microtubule accessory proteins and the motors dynein and kinesin, as well as tubulin itself.

4.7.3 Composition of the Protein–DNA Complexes

Using a proteomic approach, the nature of cytoplasmic plasmid–protein complexes has been investigated by several groups. Two approaches have been taken, both of which have produced similar and complementary data. The first approach has relied on in vitro formation of protein–DNA complexes by either incubating free plasmids with cytosolic extracts and then purifying the plasmids with a pull-down assay prior to analysis by mass spectrometry [270] or by incubating cytoplasmic extracts with plasmids that have been immobilized on a chromatography support and then identifying proteins eluted from the columns by mass spectrometry [271]. A second approach has been to cross-link plasmids with proteins inside of living cells at various times after electroporation and then isolate the plasmid–protein complexes for proteomic analysis [259]. Results from the cell-free studies identified upward of 200 proteins present in the DNA complexes, but the proteins identified were likely those that bound most stringently to the DNA or within the higher-order structure. The precipitation studies in living cells were able to identify many more proteins (>500) and showed that these complexes were dynamic and showed changing composition over time. This is likely more representative of the real situation in transfected cells.

In addition to several transcription factors as seen in previous studies, a number of proteins that fell into specific categories were identified and shown to bind specifically to plasmids that displayed active movement through the cytoplasm, but not to DNAs that showed very little movement (e.g., pBR322). At all time points, more unique proteins were bound to transcriptionally active plasmids that move (e.g., DNAs containing the CMV promoter) than pBR322. For example, at the 30-min time point, 324 unique proteins were identified in CMV promoter-containing plasmids but only 60 in pBR322 samples [259]. Specific proteins bound to moving plasmids (but not pBR322) included microtubule-based motor proteins (e.g., kinesin and dynein), proteins involved in protein nuclear import (e.g., importinβ-1, importin7, importin4, importinα, transportin, RAN, and several RAN-binding proteins), a number of hnRNP- and mRNA-binding proteins, heat shock proteins, chaperones, and transcription factors. Many of these were also identified in the cell-free binding studies [270, 271]. The significance of several of the proteins involved in protein nuclear localization and plasmid trafficking was determined by the monitoring movement of microinjected plasmids via live cell particle tracking in cells following protein knockdown by siRNA or through the use of specific inhibitors. Knockdown of importin7 had no effect on trafficking or nuclear import (see below), but knockdown of importinβ-1 inhibited trafficking and nuclear import down to the level of pBR322, a plasmid that neither traffics on microtubules nor is imported into the nucleus [259]. By contrast, knockdown of importinα had no effect on cytoplasmic movement but inhibited nuclear import by 50 %, suggesting that the different proteins involved in protein nuclear localization play distinct roles in intracellular DNA movement.

4.7.4 Nuclear Entry of Plasmids

Following trafficking of plasmids to the interior of the cell, they must enter the nucleus for productive gene expression to take place. As for microtubule-based movement, it is the proteins that bind to the plasmids that mediate entry into the nucleus in nondividing cells following electroporation or any transfection method. When plasmids carry certain DNA sequences termed DNA nuclear-targeting sequences (DTS), they form complexes with specific nuclear localization signal (NLS)-containing proteins that in turn bind to importins for entry of the complex through the nuclear pore complex into the nucleus (Fig. 4.17) [272–279]. The most common proteins that bind to these sequences are transcription factors (such as

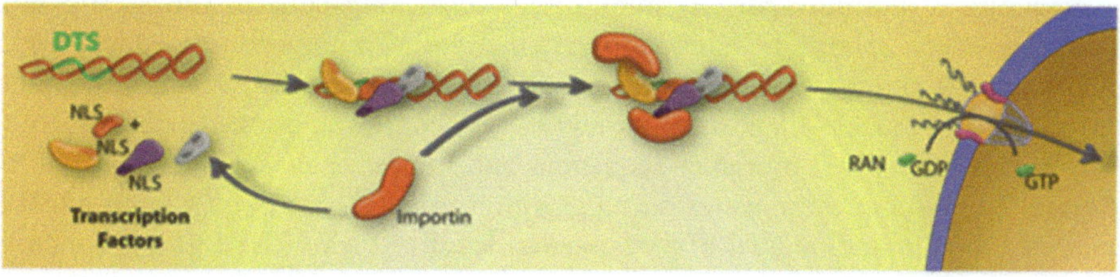

Fig. 4.17 Model for DNA nuclear import in nondividing cells. If plasmids containing sequences that act as scaffolds for transcription factors and other DNA-binding proteins (termed DTS, or DNA nuclear-targeting sequences) are deposited into the cytoplasm during transfection, they can form complexes with these proteins, thereby attaching NLSs to the DNA. Some, but not all, of these NLSs may be in a conformation able to interact with importins for transport of the DNA–protein complex into the nucleus through nuclear pores

CREB) which are translated and many times retained in the cytoplasm in order to regulate their function [280]. Under normal circumstances, a typical transcription factor would be transported into the nucleus after synthesis or upon receiving an activation signal (e.g., TNFα-induced NF-kB translocation), bind to its DNA target sequence present in various promoters, and activate or repress transcription. However, if a plasmid containing the transcription factor binding site is present in the cytoplasm, the cytoplasmic transcription factor may bind to this site before nuclear import. The NLS import machinery (importins) will then bind to the DNA-bound transcription factors and translocate the DNA–protein complex into the nucleus [281, 282]. While it could be assumed that any eukaryotic promoter would act as a DTS, this appears not to be the case, since many strong viral and cellular promoters and enhancers have no DTS activity [272, 273, 277]. Alternatively, any plasmid, regardless of sequence, has the ability to enter into the nucleus during mitosis, once it reaches the area of the nuclear periphery. Either way, these trafficking events through the cytoplasm and into the nucleus are not unique to electroporation, but are those seen for all nonviral methods for gene delivery.

4.7.5 Plasmid Movement Within the Nucleus

Although it is well accepted that DNA must enter the nucleus in order for gene expression to occur during gene transfer, what happens to the DNA once inside the nucleus has not been extensively investigated. Numerous strategies to increase efficiency of gene transfer and transcription have been developed, but almost all have assumed that the end goal of DNA trafficking is to reach the nucleus and that transcription is independent of any other nuclear function. A number of studies using transfected and microinjected cells show that expressing plasmids do indeed display discreet staining patterns, suggesting movement and/or localization of the DNA once inside the nucleus [272, 273, 283–286]. This suggests that gene expression and subnuclear localization of transfected plasmids may be linked. Indeed, several reports have found that the transcriptional capability and transcribed sequence of a plasmid alters its intranuclear trafficking patterns [287–290]. For example, when plasmids carrying no eukaryotic promoter sequences were microinjected into the nuclei, they remained diffuse and evenly spread throughout the nucleus for at least 4 h postinjection [287]. By contrast, plasmids carrying RNA polymerase (RNAP) II promoter sequences and transcribable mRNA localized to discreet foci within the nucleus starting within minutes and by 4 h were mostly localized to a limited number of areas that co-localized with RNAP II and splicing machinery [287]. Similarly, when plasmids carrying an RNAP I promoter and rRNA sequences were injected into the nucleus, they localized to nucleoli with nucleolar transcription and processing factors [287, 288].

When cells were treated with transcription inhibitors, both RNAP I and RNAP II promoter plasmids failed to redistribute [287, 288]. Similarly when the RNAP II TATA box was mutated or the rDNA sequences were removed from the RNAP I plasmid such that neither could produce RNA, the plasmids also failed to redistribute throughout the nucleus [288]. These results suggest that transcription is needed for the large-scale redistribution of plasmids throughout the nucleus, although how is not yet understood. It is possible that once in the nucleus, low-level transcription initiates from the plasmids and the nascent RNA chain, while still bound to the plasmid via RNAP, may interact with modifying and splicing enzymes and other nuclear proteins that mediate active movement of the DNA to transcription centers that engage in high-level expression. Further experimentation is needed to understand these dynamics and their implications.

4.8 Intracellular Signaling Pathways Activated by Nanosecond Pulsed Electric Fields

Ken-ichi Yano and Keiko Morotomi-Yano

Nanosecond pulsed electric fields (nsPEFs) are increasingly regarded as a novel means for life sciences. One of the prominent effects of nsPEFs is efficient induction of cell death, which has attracted considerable interest for their applications in cancer therapy. Recent studies have demonstrated that human cells respond to nsPEFs through activation of various intracellular events. These intracellular responses can even be induced by weak nsPEFs that are insufficient to induce cell death or visible morphological changes under the microscope. Exposure to nsPEFs elicits sequential protein phosphorylation involved in signaling pathways and leads to the induction of downstream events such as gene expression and suppression of protein synthesis. This section offers an overview of the current knowledge of the intracellular signaling pathways activated by nsPEFs. Effects of nsPEFs on cytoskeleton and mechanisms for cell death are described in other Sects. 4.6 and 4.10.

4.8.1 General Principles of Intracellular Signal Transduction Activated by External Stimuli

Cells have elaborate mechanisms to respond to physical and chemical stimuli from their outer environment. Cells can detect such external stimuli and rapidly transmit them by evoking sequential changes in cellular proteins, mainly via protein phosphorylation, which is a covalent addition of a phosphate group to a specific residue in a target protein. Phosphorylation alters protein conformation and can either increase or decrease the activity of the protein. Protein phosphorylation is mediated by protein kinases, and the catalytic activity of a protein kinase itself is commonly modulated by phosphorylation by other protein kinases. A cascade of protein

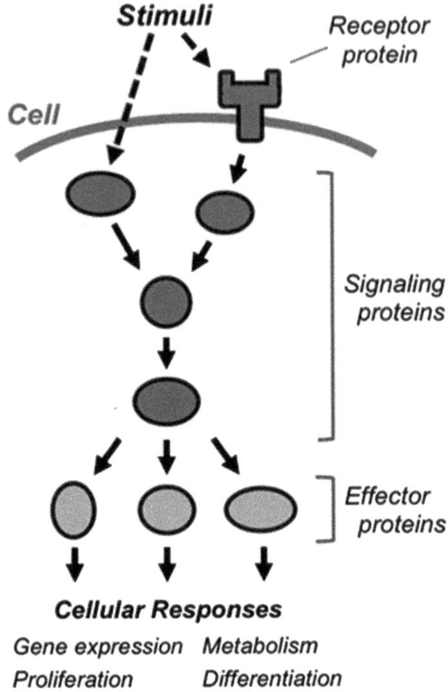

Fig. 4.18 Simplified diagram of signal transduction in human cells. External stimuli are perceived by the cell via surface receptor proteins or intracellular proteins and are transduced to downstream signaling proteins, many of which are protein kinases. Protein kinases in the signaling pathway phosphorylate downstream effector proteins that in turn affect various cellular activities

phosphorylation by protein kinases forms an intracellular signaling pathway. An intracellular signaling pathway can be activated by an external stimulus, relaying and amplifying it by phosphorylation of multiple downstream target proteins in the cell (Fig. 4.18). The effect of protein kinases is opposed by the activity of protein phosphatases that catalyze the dephosphorylation of proteins. When dephosphorylated, proteins return to their original conformation and functional states. The combined activities of protein kinases and protein phosphatases ensure the tight control of intracellular signaling pathways in response to external stimuli [291].

Activation of an intracellular signaling pathway results in the phosphorylation of downstream effector proteins that are involved in various cellular activities such as metabolism, proliferation, differentiation, and motility (Fig. 4.18). Thus, intracellular signaling pathways serve as a means for the control of various cellular activities. Furthermore, many transcription factors are located downstream of signaling pathways, and their functions are positively or negatively regulated via protein phosphorylation. Hence, the activation of a signaling pathway often leads to altered gene expression and consequent increase or decrease in specific proteins, which ultimately affect the nature and behavior of the cell. For these reasons, intracellular signal pathways play critical roles in proper responses to external stimuli, and their aberrant control is frequently associated with cellular dysfunctions such as malignant transformation [291].

4.8.2 *Activation of Intracellular Signaling Pathways by nsPEFs*

4.8.2.1 MAPK Pathways

The mitogen-activated protein kinase (MAPK) pathways are intracellular signaling pathways that are activated by extracellular stimuli and transduce them into cellular responses [292]. A typical MAPK pathway consists of three sequentially acting protein kinases, namely, MAPK, MAPK kinase (MAPKK), and MAPK kinase kinase (MAPKKK). Each MAPK pathway is named after its MAPK component. Activation of the MAPK pathway is initiated by the activation of MAPKKK that is triggered by an external stimulus. Activated MAPKKK phosphorylates its downstream MAPKK, and MAPKK in turn phosphorylates its downstream MAPK. This MAPKKK–MAPKK–MAPK module is well conserved among eukaryotes. Human cells have multiple MAPKs, and each MAPK functions in a distinct pathway and plays its physiological role. Among the multiple MAPK pathways in humans, three MAPK pathways are best characterized so far. The extracellular signal–regulated kinase (ERK) pathway is activated mainly in response to mitogens and growth factors. The c-jun N-terminal kinase (JNK) and p38 pathways are activated by environmental stress and inflammatory cytokines [292].

Activated MAPKs affect various intracellular processes by phosphorylating many downstream effector proteins, which include transcription factors. Phosphorylation of transcription factors by the activated MAPKs rapidly induces expression of a limited set of genes without de novo protein synthesis. These genes are called immediate-early genes and include *c-fos*, *c-jun*, and *Egr1*, which also encode transcription factors and further regulate the expression of many other genes. Thus, the activation of the MAPK pathways elicits complex changes in intracellular processes by changing protein activities and gene expression [292].

At the cellular level, the outcome of the MAPK activation is dependent on the cellular context, because different cell types contain different amounts of the MAPK pathway components. For example, relative activities of protein kinases and their counteracting phosphatases determine the magnitude and duration of the activation of the MAPK pathways. The amounts of downstream targets for a certain member of MAPKs vary among different cell types. Thus, a single external stimulus yields different outcomes in different cell types. Under physiological conditions, the MAPK pathways play critical roles in the control of proliferation, differentiation, migration, and apoptosis, depending on the cellular context. Furthermore, dysregulation of the MAPK pathways is frequently implicated in inflammation, obesity, malignant transformation, and tumor invasion.

Effects of nsPEFs on the MAPK pathways were demonstrated by experiments using the HeLa S3 cell line, which is one of the most common cell lines used in molecular biology [293]. The cells were exposed to shots of nsPEFs and subjected to Western blot analyses of major protein components in three representative MAPK pathways, JNK, p38, and ERK. The obtained results are summarized in Fig. 4.19. Application of nsPEFs induces phosphorylation of JNK, p38, and ERK in different temporal patterns. Their upstream MAPKKs (MEK1/2, MKK3/6, and MKK4) are also phosphorylated after exposure to nsPEFs. Furthermore, multiple downstream factors in the MAPK pathways, including MSK1, Hsp27, ATF2, p90RSK, and c-Jun, are phosphorylated after nsPEF exposure. In addition to protein phosphorylation, expression of the immediate-early genes of the MAPK pathways is transiently activated. The expressions of *Egr1*, *c-jun*, and *c-fos* are increased by more than tenfold, whereas *c-myc* expression is not significantly affected in nsPEF-exposed cells. Specific inhibitors for the JNK, p38, and ERK pathways suppress the phosphorylation of distinct downstream proteins and gene expression, which validate the functional connection between an individual MAPK pathway and specific downstream events in nsPEF-exposed cells [293]. In these experiments, HeLa S3 cells were used as a model to understand the key aspects of MAPK activation by nsPEFs. As described above, different cell types have different amounts of components of the MAPK pathways, and the outcome at the cellular levels is determined depending on the cellular context. The observation of the nsPEF-induced MAPK activation raises the possibility that nsPEFs can be utilized to control MAPK-related cellular activities such as proliferation and differentiation.

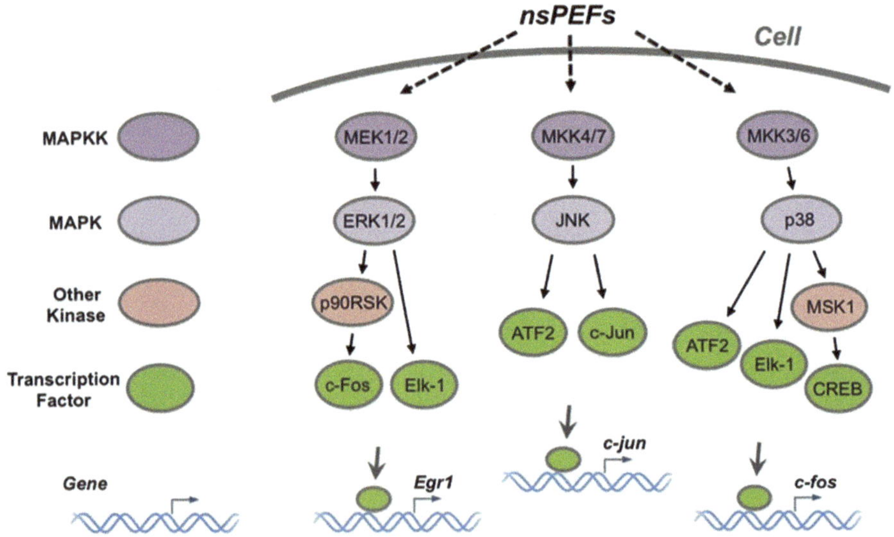

Fig. 4.19 Activation of the MAPK pathways by nsPEFs. Human cells possess multiple MAPK pathways that are differently activated by external stimuli. MAPKs phosphorylate various effector proteins, including transcription factors. Exposure of the cell to nsPEFs induces the phosphorylation of the proteins indicated in the figure. The three genes shown in the figure are transcriptionally activated after nsPEF exposure

4.8.2.2 AMPK Pathway

AMP-activated protein kinase (AMPK) plays a critical role in energy homeostasis [294]. AMPK is activated in response to decreased cellular energy that is represented by decreased ATP and increased AMP concentrations. Once activated, AMPK phosphorylates multiple metabolic enzymes and their regulators to modulate their activities, contributing to the restoration of energy balance. In human cells, two protein kinases are involved in the phosphorylation and consequent activation of AMPK (Fig. 4.20). LKB1 is a primary AMPK kinase and phosphorylates the AMP-bound form of AMPK that arises under low-energy conditions. Because LKB1 has cellular functions as a tumor suppressor, its gene is frequently lost in cells derived from malignant tumors such as HeLa S3 cells. Ca^{2+}/calmodulin-dependent protein kinase kinase (CaMKK) has a potential role as an alternative AMPK kinase and acts on the AMPK phosphorylation in LKB1-deficient cells. CaMKK requires increased intracellular Ca^{2+} for its enzymatic activity, whereas LKB1 is a Ca^{2+}-independent protein kinase.

Exposure of cultured human cells to nsPEFs activates AMPK in a cell type-dependent manner (Fig. 4.20) [295]. In Jurkat cells, LKB1 is a primary AMPK kinase, and nsPEFs induce AMPK activation. On the other hand, HeLa S3 cells lack functional LKB1, and CaMKK serves as an AMPK kinase. Accordingly, a CaMKK inhibitor suppresses AMPK activation by nsPEFs in HeLa S3 cells, but not in Jurkat

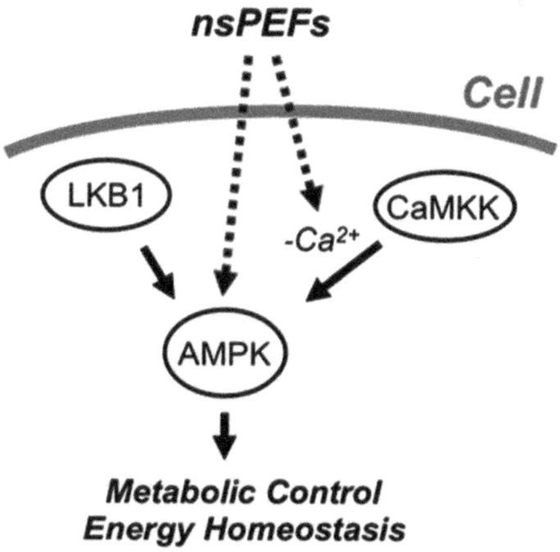

Fig. 4.20 Activation of the AMPK pathway by nsPEFs. AMPK is a critical regulator of cellular homeostasis and is activated via phosphorylation by two protein kinases. LKB1 is a major AMPK kinase. CaMKK is an alternative AMPK kinase that requires increased cytoplasmic Ca^{2+} for its enzymatic activity. Exposure of the cell to nsPEFs induces AMPK phosphorylation by LKB1. In LKB1-deficient cells, such as HeLa S3 cells, AMPK phosphorylation by nsPEFs is mediated by CaMKK and requires the presence of extracellular Ca^{2+}

cells. In both cell lines, AMPK phosphorylation quickly occurs within 1 min after nsPEF exposure.

Exposure to nsPEFs is known to increase cytosolic Ca^{2+} concentrations, mainly due to the influx of extracellular Ca^{2+}. The Ca^{2+} influx differently affects nsPEF-induced AMPK activation in Jurkat and HeLa S3 cells [295]. In Jurkat cells, AMPK activation is quickly induced by nsPEFs, irrespective of the presence or absence of extracellular Ca^{2+}. In HeLa S3 cells, the absence of extracellular Ca^{2+} causes a substantial reduction in the nsPEF-induced AMPK activation, indicating that Ca^{2+} influx is critical for the CaMKK-mediated AMPK activation in nsPEF-exposed HeLa S3 cells. These observations provide an example of a causal relationship between the cellular context and the Ca^{2+} dependency of nsPEF-induced intracellular responses.

4.8.2.3 Phosphoinositide Signaling Pathway

In addition to proteins, small molecules play important roles in intracellular signaling [291]. A small amount of phosphatidylinositol 4,5-bisphosphate (PIP2) exists in the inner layer of the plasma membrane. The hydrolysis of PIP2 yields two compounds, inositol 1,4,5-triphosphate (IP3) and diacylglycerol (DAG), both of which function as potent second messengers. IP3, which is water soluble, diffuses into the cytoplasm and binds to IP3 receptors in the ER to release Ca^{2+} from the ER. DAG remains tethered in the inner surface of the plasma membrane and activates protein kinase C (PKC) family members, many of which are Ca^{2+}-dependent kinases and further phosphorylate multiple proteins to induce various cellular responses (Fig. 4.21). PIP2 hydrolysis is catalyzed by phospholipase C (PLC), and the enzymatic activity of PLC is generally stimulated by external signals, such as hormones, via the activation of G-protein coupled receptors [291]. PIP2 binds to the inner layer of the plasma membrane. Exposure of the cell to nsPEFs yields two hydrolysis products of PIP2, namely, IP3 and DAG. IP3 elicits Ca^{2+} release from the ER, while DAG activates PKC.

Effects of nsPEFs on phosphoinositide signaling were investigated by using two sensor proteins for DAG and IP3, respectively [142, 143]. One sensor protein consists of a green fluorescent protein (GFP) fused to the PLCδ PH domain, which binds to both PIP2 and IP3. When exogenously expressed in cultured mammalian cells, this protein localizes in the plasma membrane and, after nsPEF exposure, rapidly diffuses into the cytoplasm, demonstrating the generation of IP3 in nsPEF-exposed cells. Another sensor protein contains GFP fused to the C1 domain of PKCγ, which binds to DAG. This sensor protein exists in the cytoplasm and translocates to the plasma membrane after nsPEF exposure, indicating DAG generation by nsPEFs. These observations clearly demonstrate that nsPEFs cause

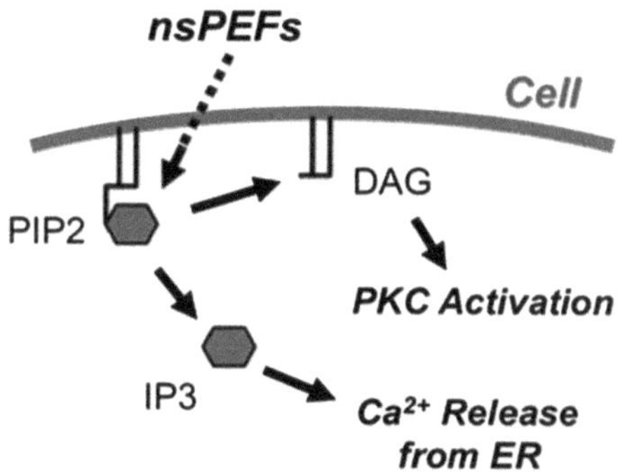

Fig. 4.21 Activation of the phosphoinositide signaling pathway by nsPEFs. PIP2 binds to the inner layer of the plasma membrane. Exposure of the cell to nsPEFs yields two hydrolysis products of PIP2, namely, IP3 and DAG. IP3 elicits Ca^{2+} release from the ER, while DAG activates PKC

increased levels of key molecules in the phosphoinositide signaling (Fig. 4.21). Future studies on the detailed mechanism for nsPEF-induced PIP2 hydrolysis as well as the effects on downstream events will provide important information for a better understanding of nsPEF actions.

4.8.3 Stress Response

4.8.3.1 Overview of Stress Responses

Cells often encounter various adverse circumstances and deleterious stimuli such as nutrient limitation, hypoxia, UV irradiation, heating, virus infection, and toxic agents. To withstand these conditions, cells activate multiple homeostatic mechanisms that are collectively called stress responses. Activation of stress responses frequently leads to transient suppression of general protein synthesis [297]. Because protein synthesis consumes large amounts of amino acids and energy, its transient suppression saves resources and serves as a protective mechanism to endure adverse conditions. The rate of protein synthesis is primarily regulated at the step of translation initiation, and the key event in the stress-induced translational suppression is the phosphorylation of the α-subunit of eukaryotic translation initiation factor 2 (eIF2α). This phosphorylation hinders the assembly of a protein complex required for translation initiation and thus results in the suppression of general protein synthesis.

The stress response involving eIF2α-phosphorylation is evolutionarily conserved among eukaryotic organisms from yeast to humans. In mammals, including humans, eIF2α-phosphorylation is mediated by four protein kinases that are activated in response to distinct forms of cellular stress (Fig. 4.22) [297]. Activation of PERK is induced by misfolded proteins in the ER. GCN2 is activated by amino acid deprivation and UV irradiation. PKR activation is triggered by double-stranded RNA produced during virus infection. HRI is activated by heme deprivation, heat shock, and oxidative stress in erythroid cells. At least one of these protein kinases is activated under adverse conditions, and eIF2α-phosphorylation acts as a

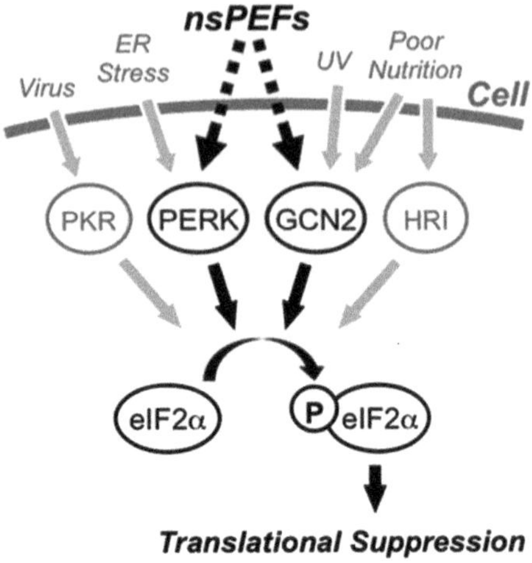

Fig. 4.22 Activation of stress responses by nsPEFs. Mammalian cells, including human cells, have four stress-responsive eIF2α-kinases, which are activated by different external stimuli. Among the four eIF2α-kinases, PERK and GCN2 are activated by nsPEFs and phosphorylate eIF2α, which further lead to translational suppression

convergence point for intracellular signaling activated by different forms of cellular stress. Furthermore, cells have a mechanism to recover from the translational suppression mediated by the phosphorylated eIF2α. GADD34 is a stress-inducible gene, and its transcript is translated by an alternative mechanism that is active even in the presence of the phosphorylated eIF2α. GADD34 protein forms a complex with protein phosphatase 1 and dephosphorylates eIF2α, leading to the restoration of protein synthesis [297].

4.8.3.2 Stress Responses Induced by nsPEFs

A recent study demonstrated that nsPEFs elicit stress responses in human and mouse cells [296]. Induction of stress responses requires relatively intense nsPEF conditions that cause growth retardation but not cell death. Under these conditions, eIF2α-phosphorylation is rapidly induced within a minute. In accordance with the induction of eIF2α-phosphorylation by nsPEFs, the rate of general protein synthesis is acutely decreased after nsPEF exposure. In addition to eIF2α-phosphorylation, the exposure to nsPEFs causes 4E-BP1 dephosphorylation, which is known to be a part of an eIF2-independent alternative mechanism for translational suppression [296]. The signaling pathway leading to 4E-BP1 dephosphorylation in nsPEF-exposed cells remains to be fully understood.

Concomitant with eIF2α-phosphorylation, PERK and GCN2, which are eIF2-α-kinases, are activated in nsPEF-exposed cells (Fig. 4.22). In double-knockout mouse embryonic fibroblasts that lack both *PERK* and *GCN2* genes, eIF2-α-phosphorylation by nsPEFs is significantly decreased, indicating that both kinases contribute to the nsPEF-induced eIF2α-phosphorylation. Single knockout of either *PERK* or *GCN2* has marginal effects on the nsPEF-induced eIF2-α-phosphorylation, suggesting the presence of functional compensation between PERK and GCN2 in nsPEF-induced stress responses [296].

PERK is known to be activated by ER stress, which is caused by unfolded proteins existing in the ER. GCN2 plays a critical role in the UV-induced stress response. The activation of PERK and GCN2 by nsPEFs raised the question whether nsPEFs activate the ER stress- and UV-induced pathways or whether nsPEFs act as a novel stress distinct from these cellular stresses. To clarify this point, activation of stress-specific gene expression was analyzed. ER stress generally leads to increased expression of specific genes, including *CHOP* and *BiP*. UV irradiation causes changes in gene expression, such as *GADD45β*. Increased expression of these stress-inducible genes is indicative of the activation of downstream events in the ER stress- and UV-induced pathways. However, quantitative RT-PCR analysis showed no increase in the expression of these genes after nsPEF exposure. Thus, despite the activation of PERK and GCN2, nsPEFs do not activate the canonical signaling pathways for ER stress and UV-induced stress and appear to act as a novel form of cellular stress [296].

4.8.4 Future Perspective

Recent studies have demonstrated that human cells respond to nsPEFs by activating multiple intracellular pathways, but important questions remain to be answered. First, the most upstream event(s) triggered by nsPEFs in each pathway is largely unclear, especially in the activation of MAPK pathways. Regarding the stress responses, nsPEFs seem to directly affect PERK and GCN2 molecules because there is no protein upstream of these eIF2α-kinases in their signaling pathways. However, molecular details of direct effects of nsPEFs on the PERK and GCN2 molecules should be understood. Second, how Ca^{2+} influx caused by nsPEFs is converted to intracellular signaling events remains unclear. Calmodulin (CaM) is a ubiquitous protein and universally serves as an intracellular Ca^{2+} receptor among eukaryotes. Ca^{2+}-bound CaM can activate many proteins, including Ca^{2+}/CaM kinases that further transduce Ca^{2+} signals. Similar to many Ca^{2+}-dependent cellular responses, Ca^{2+} signaling elicited by nsPEFs appears to be mediated by CaM and its associated proteins, which will be experimentally confirmed in future. Finally, contributions of activated intracellular signaling events to cell survival and death should be understood. For example, stress responses are generally protective reactions, but their excess activation promotes cell death. MAPK pathways are known to be differently involved in the initiation of cell death in a cell type-dependent manner. A better understanding of the physiological significance of nsPEF-induced intracellular signaling events will provide a mechanistic basis for the medical application of nsPEFs.

4.9 Cell- and Tissue-Level Response to Irreversible Electroporation: Implications for Treatment Planning and Outcome

Robert E. Neal II, Suyashree Bhonsle, and Rafael V. Davalos

The manipulation of cellular transmembrane potential and induction of the electroporation response for irreversible electroporation (IRE) and other electroporation-based therapies (EBTs) can be correlated to electric field exposure, as well as several secondary pulse parameters. However, there are a number of considerations that must be identified and accounted for when designing and implementing IRE therapies in vivo. These relate both to the identity and nature of the cells within an electroporated region, as well as the structural environment of the tissue containing the cells. This section of Chap. 4 identifies some of the complex environments, as well as cell- and tissue-level responses that can alter electric field distribution and the concurrent localized response to IRE electric pulses.

4.9.1 Specific Pulse Protocol Considerations for IRE Therapy

While much of this chapter focuses on general aspects of cellular electroporation from the molecular to cellular scale, this section considers aspects of particular importance to IRE. A key consideration in this regard relates to the overall energy of a pulse protocol and its ability to result in effective cell death regions. Reversible electroporation therapies, such as electrogene transfer and electrochemotherapy, aim to maintain an overall cell exposure energy to induce sufficient electroporation to facilitate transport of their targeted macromolecule without killing the cell, which results in typical protocols utilizing approximately eight pulses delivered at rates between 1 Hz and 5 kHz [298–300]. However, IRE pulse parameters are designed to kill the cells in a tissue region while confining thermal effects to those below which would induce morbidity and damage to the sensitive structures within and around the IRE ablation zone.

IRE feasibility for therapeutic targeted ablation first showed clinically relevant lesions were attainable with basic protocols while maintaining thermal effects below those known to cause thermal damage to the tissue [301, 302]. Subsequent studies further showed that guided adjustment of pulse parameters could manipulate the thermal implications from IRE therapy as well as reduce the effective electric field threshold required to ensure cell death. Such means for altering cumulative energy of a pulse protocol to change the effective lethal electric field threshold, thermal effects, and overall tumor response includes the strength of the pulse, use of additional pulses, and manipulation of pulse length [304–306]. Further evidence suggests a nonlinear response for the lethality of a pulse protocol that does not require additional energy, whereby altering the timing of electric pulses with a modulated pulse delivery regime may increase ablation zone and improves tumor response [303, 306, 307].

The thermal effects during multi-pulse IRE protocols have been determined experimentally and via numerical modeling. The goal of these studies was to delineate IRE cell death from the thermal damage that occurs when tissues are exposed to temperatures higher than their physiological norm for extended periods of time. It has been suggested that when the period of exposure is long, thermal damage can occur at temperatures as low as 43 °C [308, 309]. Although 50 °C is typically chosen, what is considered instantaneous thermal damage occurs as high as 83.6 °C (prostate) or 74.7 °C (liver) [310]. In order to assess tissue temperature, numerical models employ a modified Pennes' bioheat equation that includes the effects of blood perfusion and metabolism and an additional Joule's heating term to account for resistive heating [311–313], as defined in the equation:

$$\nabla \cdot (k\nabla T) + w_b c_b (T_a - T) + q''' + \sigma|\nabla\varphi|^2 = \rho c_p \frac{\partial T}{\partial t} \tag{4.8}$$

where k is the thermal conductivity of the tissue, T is the temperature, ρ_b is the blood density, w_b is the blood perfusion rate, c_b is the heat capacity of the blood, T_a is the arterial temperature, q''' is the metabolic heat generation, ρ is the tissue density, c_p is the heat capacity of the tissue and $\sigma|\nabla\varphi|^2$ is the heat generated due to the electric field, where σ is the electrical conductivity of the tissue and σ is the electric potential.

In regard to determining the outcome from elevated temperature effects, recent models utilize an Arrhenius-type analysis to assess thermal damage from the temperature distribution [314–317]. Numerical models for IRE that predict temperature changes have been validated from actual data and can therefore serve as a reference for appropriate treatment parameter selection to minimize thermal effects while retaining sufficient IRE exposure [318, 319]. Experimental work performed that physically recorded measured temperatures during IRE treatments shows effective ablations without significant thermal effects [314, 320]. Importantly, in vivo and clinical studies show that IRE's nonthermal mechanism of death helps preserve surrounding critical structures [321, 322].

For larger tumors or while dealing with tissues with higher electric field thresholds for ablation, stronger pulse protocols may be necessary to ensure complete tumor coverage. This correlates with higher voltages, larger pulse widths, pulse numbers, and insertions. In such cases, particular care must be taken to avoid significant thermal damage. Such approaches include reducing the pulse delivery rate or delivering small sets of pulses in a "phased" delivery approach around all pairs, allowing more time for conductive tissue cooling between each particular pair of electrodes [306, 307]. Additionally, proactive thermal countermeasures have been suggested, including using actively cooled electrodes, cooling patient baseline temperature, or cooling the region surrounding the target organ with hydrodissection.

4.9.2 Heterogeneous Systems: Static and Dynamic Conductivity Environments and Their Effect on IRE Ablation Zone Distribution and Therapy Outcome

After considering manipulation and determining appropriate secondary pulse parameters in IRE and EBT protocols, IRE ablation zones can be correlated with an effective lethal electric field threshold, which is unique to the tissue and selection of secondary parameters. As a result, electric field distributions, and their accurate prediction and understanding, play an important role in designing and implementing successful IRE therapies. The electric field relates to the governing equation:

$$\nabla \cdot (\sigma \ \nabla\Phi) = 0 \tag{4.9}$$

where Φ indicates the electric potential and σ represents the electrical conductivity. While electrode geometries and applied voltages will partially affect the electric field that a volume of tissue is exposed to, the electrical conductivity will also alter the electric field distribution. Thus, when heterogeneous distributions of electrical conductivity exist in the targeted tissue, both within and between different tissue types, electric field exposure will be different from an isotropic distribution, thus altering ablation zone shape and volume.

4.9.2.1 Effects of Heterogeneous Conductivity

Early numerical investigation into the effect of heterogeneous tissue distributions showed that when the electrodes delivering the electric pulses are placed around a volume of higher electrical conductivity, the voltage drop within that region is reduced, and this region is thus subjected to a reduced electric field [323], while the opposite was found for a lower conductivity central region. Many varieties of tumor, a common ablation target for IRE therapy, contain a relatively high cellular density and higher conductivity than surrounding tissues with more extracellular constituents, such as the connective and fatty tissue components implicated adjacent to breast cancer tumors. However, it was shown that by placing the energy delivery electrodes within the outer boundary of the more conductive region, the effect of decreased electric field distribution within a more conductive region is eliminated, and thus effective treatment of more conductive tumors is possible while maintaining thermal effects below those which induce patient morbidity [312].

4.9.2.2 Heterogeneous Environments In Vivo

Because heterogeneous electrical conductivity distributions have been shown to significantly affect electric field distributions and treatment zones, it is important to understand and consider environments where these effects may be most relevant. There are two conditions where heterogeneous conductivities are relevant at the tissue level. These include inherent heterogeneities relating to tissue structure and composition and induced heterogeneities that result from manipulation of the targeted region environment.

4.9.2.2.1 Inherent, Static Conductivity Heterogeneities

Relevant inherent tissue heterogeneities include highly localized effects, such as those due to the presence of diverse interstitial constituents, including macro- and microvasculature, connective tissue, general cellularity, cellular morphology, and discrete functional organ units, such as the proximal convoluted tubules or glomerular capsules within the kidney. It is typically regarded as overly cumbersome to consider all of the subcellular to cellular-level variability within a targeted environment, and thus effective bulk tissue conductivity is used as a metric to approximate the overall conductivity for a particular tissue type. This bulk electrical conductivity variation for different tissue types may more readily be discretized and regarded separately in predictive EBT simulations. Certain tissues may contain stochastic or organized aspects that will have varying extents of an effect on electric field distributions, including the presence of randomly distributed calcifications, and structural tissue orientations that exhibit anisotropic effects to electric field distribution. While anisotropy will occur for several tissues, the strongest effect is clearly documented in muscle tissue, where the high conductivity along muscle fibers compared to perpendicular to muscle fibers imparts a strong effect on electric field distribution and ablation shape [324–326].

In addition to inherent variability in tissue conductivity distributions, there are numerous potential intervention-related changes to conductivity that may occur due to the presence of implanted materials or secondary therapy demands. Metallic objects will behave as high-conductivity spots within the tissue, which can serve as conduits for electric current, altering electric field distributions. In addition, greater electrical conductivity results in increased Joule heating from the electric pulses and will serve as localized regions of increased thermal effects. This effect correlates with the intervention type and tissue. It was shown that the presence of small metallic objects, such as brachytherapy seeds, may not induce significant alterations to bulk electrical conductivity, electric field distribution, and thermal effects [327]. However, the relative effect to the heterogeneous environment will increase with the total relative amount of metallic object, and caution should remain warranted for larger metallic objects in the vicinity of EBTs [324], such as stents, biomechanical correction hardware, or pacemakers. Nonmetallic implants and devices will serve as low-conductivity regions in the tissue, overall likely evoking less significant alterations of EBT outcomes overall. Finally, intervention-related alterations to tissue properties may include the infusion of conductivity-altering fluids, such as saline or low-conductivity buffers. These solutions may be intentionally infused into the targeted or adjacent tissue or may occur as secondary means to delivering IRE therapy, such as hydrodissection. The alteration of osmolarity and conductivity of tissue from infused fluids should be carefully considered for their effect on EBTs. For example, high-conductivity saline within the targeted region will substantially increase bulk tissue conductivity, resulting in higher electric current and thermal effects.

4.9.2.2.2 Therapy-Induced Dynamic Electrical Conductivity Considerations

While inherent tissue heterogeneities will have a varying cumulative effect on electric field distributions and affected EBT volumes, it is also critical to consider dynamic changes to tissue properties that occur in response to the pulsed electric fields. This includes increases to electrical conductivity due to temperature rise from Joule heating. Further, it includes changes from electroporation-induced effects, where electroporated cells in a region no longer serve as dielectrics in the tissue, permitting improved electrolyte mobility within the environment and thus increased conductivity (Fig. 4.23). Such effects are pronounced, nonlinear, and highly dependent on local electric field intensity.

Thermal effects on mammalian tissue electrical conductivity are well documented and typically result in increases of 1–3 % per degree Celsius (known as temperature coefficient) [328]. Values of temperature coefficients for different tissues are given in Table 4.1. In an ex vivo study, it was shown for a typical clinical IRE pulse protocol that nearly all of the cumulative inter-pulse rise in electrical

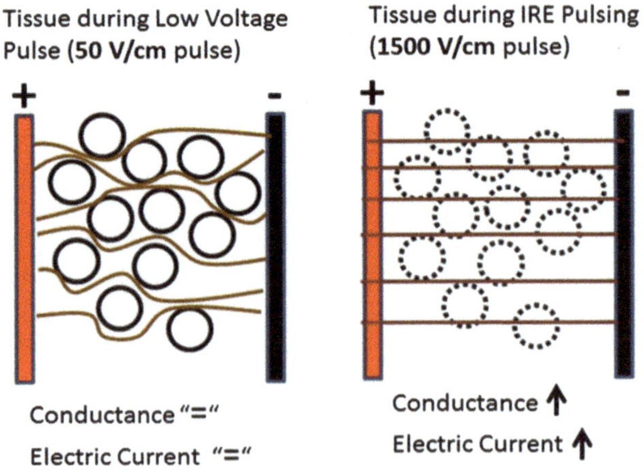

Fig. 4.23 Behavior of electric current pathway through cells in tissue without (*left*) and with (*right*) electroporation occurrence

Table 4.1 Change in electrical conductivity of soft tissue with temperature

Tissue	Temperature coefficient $\left(\frac{\Delta\sigma}{\sigma}\right)\left(\Delta T^{-1}\right)100\ \%°C^{-1}$	Reference
Brain, cow, pig	3.2	[328, 330]
Kidney, cow, pig	1.7	[328]
Liver, cow, pig	1.5	[328]
Pancreas, cow, pig	1.4	[328]
Spleen, cow, pig	1.0	[328]
Breast tumor, rat	1.45	[331]

conductivity during pulse delivery may be accounted for by considering thermal effects [329]. For pulse protocols with modest temperature rise, thermally induced increases in electrical conductivity are relatively mild, and the dominant consideration is the electroporation-induced effects on electrical conductivity.

Electroporation-induced dynamic tissue conductivity plays a significant role in redistribution of electrical conductivity and thus the electric field and ablation zone. When cells become electroporated, their capacity to behave as a dielectric is compromised, thus facilitating improved electric current flow through cellular regions relative to non-electroporated tissues. It has been shown in several studies that tissue typically exhibits increases in effective bulk tissue conductance of two to five times and even up to 1000 times depending on the tissue type and protocol strength [332–334]. A study using ex vivo porcine renal cortex core samples pulsed with IRE using plate electrodes showed that the increase in electrical conductivity is directly correlated with intensity of electric field exposure, with the increase plateauing at approximately 2000 V/cm [329]. Such electroporation-induced increases in conductivity can be regarded as increasing from a baseline conductivity with no electroporation, σ_0, up to a maximum, σ_{max}. The maximum conductivity, σ_{max}, occurs when the cellular membranes no longer restrict the extent of interstitial electrolyte mobility. It was found that the increase in bulk tissue conductivity was best approximated with an asymmetrical sigmoid Gompertz function, calculated as

$$\sigma_G(|E|) = \sigma_0 + (\sigma_{max} - \sigma_0) \cdot \exp[-A \cdot \exp(-B \cdot E)] \qquad (4.10)$$

where A and B are unitless coefficients that vary with pulse length, t(s). For a 100-μs long pulse, it was found $A = 3.053$ and $B = 0.00233$ [329]. The shape and behavior of this function can be found in Fig. 4.24.

The ex vivo study determined an equivalent circuit model representing cells and tissue undergoing electroporation can include a variable electroporation-based resistor, which varies with extent of electric field exposure. Additional examination of the equivalent circuit model suggests that the σ_{max} value for any particular tissue may be approximated by equating complete dielectric breakdown of all cell membranes in electroporated tissue to that of the membrane dielectric breakdown encountered when subjected to *β-dispersion* AC frequencies, which represent the range of frequencies where interfacial polarization of the lipid bilayer occurs.

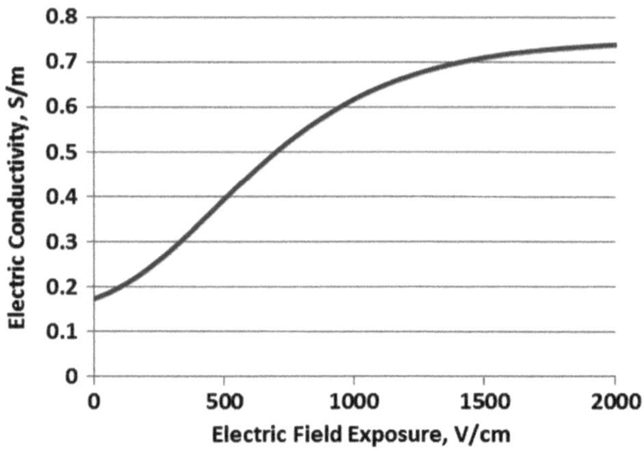

Fig. 4.24 Chart of $\sigma_G(|E|)$ function

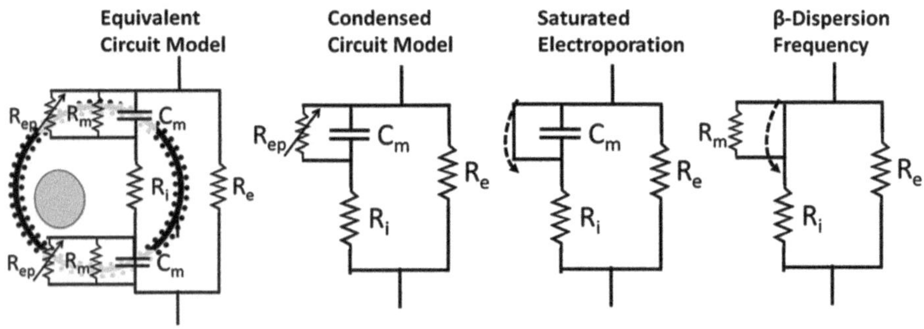

Fig. 4.25 Equivalent circuit model and response to varying electrical parameters. The lipid bilayer for a cell in suspension behaves as a membrane resistance, R_m; a variable electroporation-related resistance, R_{ep}; and a capacitor, C_m, in parallel relative to the extracellular resistance, R_e, and intracellular resistance, R_i. The high membrane resistance can be ignored, resulting in a condensed circuit model. When the electric field is of sufficient strength to induce a saturated amount of pores, the variable resistor behaves like a current shunt. Equivalently, when the frequency of an AC signal is in the β-dispersion range (200–500 MHz), the capacitive element behaves as a current shunt

In this range of frequency, the reactance of the membrane capacitance behaves as a short circuit to membrane resistance, thus providing an upper threshold for effective bulk tissue conductivity with fully saturated electroporation of the cells (Fig. 4.25).

4.9.2.2.3 Confirmation of Dynamic Conductivity Improved Numerical Simulation Accuracy

While it is well determined in the literature that electroporation-induced conductivity changes occur in tissue in response to electroporation pulses, it is further detailed that accounting for these effects in numerical simulations of electric field distributions results in improved predictions of EBT procedure outcomes. Consideration of dynamic electrical conductivity from electroporation increases the tissue conductance and thus simulated electric current, matching experimental findings. It also preferentially grows the lesion height dimension perpendicular to electrode pair orientations relative to the width dimension between the electrodes. This change in shape offers a more accurate depiction of true lesion shapes found from in vivo IRE ablations. For instance, it was shown using potato tuber that electric current and affected volume shape were better approximated with numerical simulations that used dynamic conductivity [335]. It was also shown that dynamic electrical conductivity simulations resulted in improved predictions of electric current than static models in animal tissue [336]. In an additional study, the effect of numerical models with dynamic conductivity using the asymmetrical Gompertz function was compared with one using linear dynamic and static electrical conductivity. It was shown using in vivo canine renal ablations with clinically relevant pulse protocols that the Gompertz function produced the best approximations of lesion shape and electric current, while linear dynamic conductivity simulations also offered markedly improved correlations to ablation shape and electric current relative to static conductivity models [337].

4.9.3 Cell- and Tissue-Specific Susceptibility to IRE Electric Pulses

While extensive evidence exists indicating that alterations to secondary pulse parameters (pulse length, pulse number, modulated pulse delivery) may manipulate the effective electric field threshold required to induce IRE cell death, it should also be considered that different cell varieties and the structural arrangement of different parenchymal and pathologic tissues will exhibit unique susceptibilities to IRE electric pulses.

4.9.3.1 Cell-Specific Susceptibility

In regard to cell-level differences in IRE pulse susceptibility, heterogeneous cell and functional unit distributions have shown distinct lethal electric field thresholds. One consideration is the size of cells in a targeted region, since calculations show that larger cells will experience larger transmembrane voltage change for a given electric field exposure, and thus increased likelihood of electroporation induction [338]. In addition, a veterinary clinical study showed complete destruction of all tumor cells in a targeted region, while the immediately adjacent muscle cells were able to recover from the electroporation pulses [339].

Furthermore, two in vivo renal ablation studies showed a penumbra ablation zone of varying cell deaths outside the region of complete cell ablation, but proximal to the region where no IRE cell death was observed [337, 340]. This penumbra region contrasts with the sub-millimeter demarcation between completely dead and completely viable cell regions found in liver ablations [302]. This difference may be because liver tissue is a relatively isotropic and homogeneous distribution of functional units composed primarily of hepatocytes. The penumbra transition zone of heterogeneous cell death distributions in kidney indicated viability may directly relate to the tissue functional unit variety, where viability was found in progressively lower electric fields for blood vessels and glomeruli, distal convoluted tubules, and proximal convoluted tubules. Interestingly, the transition of viability with structure variety seemed to correlate with metabolic demands of the cells in each structure, suggesting that more metabolically active cells are less resilient to the electroporation-induced cellular stresses. Such a behavior may have implications in tumors, which contain neoplastic cells that are very metabolically active.

4.9.3.2 Organ-Specific Susceptibility

While different cells within an organ appear to show a variation in their effective lethal electric field threshold for a given pulse parameter algorithm, several studies have characterized the unique effective thresholds for different organs by simulating electric field distributions relative to in vivo ablation studies (Table 4.2). In [318], brain lesions were found to correlate with an electric field of 502 V/cm for 90 pulses, each 50-μs long, delivered at 4 pulses per second. This value is relatively similar to a threshold of 575 V/cm determined for renal cortex from a protocol of 100 pulses, each 100-μs long delivered at a rate of 1 pulse per second and simulated

Table 4.2 Electric field thresholds of cell death determined for different parameters and tissue types

Organ	Parameters	Electrical conductivity model (static vs. dynamic)	Average electric field threshold of cell death (V/cm)	Reference
Brain, dog	50-μs, 90 pulses, 4 Hz	Dynamic	502	[319]
Kidney, dog	100-μs, 100 pulses, 1 Hz	Dynamic	575	[337]
Liver, rabbit	100-μs, 8 pulses, 1 Hz	Static	637	[341]
Liver, rat	20 ms, 1 pulse	Static	400	[302]
Pancreas, pig	70-μs–100-μs, 70–90 pulses	Static	>650	[342]
Prostate, human, dog	70-μs–100-μs, 90–100 pulses, 1 Hz	Dynamic	1072	[322]
Mammary tumor, rat	100 μs, 100 pulses, 0.33 Hz	Static	1000	[325]

using the Gompertz sigmoid dynamic conductivity [337]. This is also similar to the threshold of 637 V/cm in the liver tissue using static conductivity simulations for a protocol of 8 pulses, each 100-μs long delivered at a rate of 1 Hz [341], since consideration of dynamic electrical conductivity facilitates reduced calibrated electric field thresholds. Similarly, pancreas simulation calibrations found an effective electric field threshold of below 650 V/cm [342]. A notable exception is the prostate, where an electric field threshold of 1072 V/cm was determined from healthy prostate ablations in canines 6 h post-IRE and human lesions 3–4 weeks post-IRE using protocols of 70–100 μs with 90–100 pulses delivered, delivered at an ECG synchronous rate [322].

When exhibited, variability in IRE threshold may depend on the cellular constituents within an organ and their unique metabolic demands. Further, it may result from differences in interstitial components in the tissue altering electrical conductivity and electric field distributions. For instance, tissues that are highly calcified, contain extensive fibrous components, or exhibit glandular organization with low-conductivity surrounding membranes may induce significant voltage drop across these low-conductivity regions, thus reducing electric field exposure to the cells.

4.9.4 Conclusion

Cellular response to pulsed electric fields includes the development of reversible and irreversible electroporation defects in the cell membrane. The type and extent of these effects will vary with secondary pulse parameters that alter the total effective strength of an electroporation pulse protocol. Several cellular- and tissue-level aspects must be considered in relation to accurately predicting outcomes for electroporation therapies, including cell variety and distribution, organ, tissue properties, and the explicit pulse parameters used. When all secondary parameters are reasonably consistent, the effect can be directly correlated to electric field exposure. Thus, electric field distribution is often used as a surrogate to predict and identify affected regions of tissue in EBTs. When planning and implementing therapeutic IRE ablation procedures and other EBTs, it is vital to consider the effects of complex environments, including heterogeneous systems, which will alter the conductivity and electric field distribution in the tissue. These aspects include inherent conductivity heterogeneities between different tissue types, the composition of tissue constituents, and any prior intervention-related implants or manipulations of tissue. In addition, electroporation procedure delivery will also alter tissue properties, via thermal effects as well as electroporation-induced changes to tissue conductivity. Consideration of these dynamic effects and general tissue heterogeneities are important for creating accurate predictive models and determining optimally effective electroporation procedure protocols.

4.10 Nanosecond Pulsed Electric Field-Induced Cell Death Responses and Mechanisms

Stephen J. Beebe

4.10.1 Introduction

Life is not possible without death; they go hand in hand, like a coin with two different sides. Cell death occurs by a number of different mechanisms during development and throughout life. Very generally, cell death can be divided into two general categories. Accidental cell death (ACD), caused by severe physical, chemical, or mechanical insults that result from immediate structural breakdown, is not compatible with life and cannot be prevented by genetic or pharmacologic intervention of any kind. This was often called necrosis, but as will be seen, this term for ACD is no longer adequate. The term necrosis was used by Virchow in 1858, when histological stains and microscopy were not available, to mean an advanced stage of tissue breakdown, similar to what we would now call gangrene [343]. In contrast, regulated cell death (RCD), be it apoptosis or necrosis, can be triggered by exogenous stimuli and is genetically coded molecular mechanisms and can be pharmacologically and genetically modulated. These regulated processes occur during microenvironmental distresses, pre- and postembryonic development, tissue homeostasis, and immune responses [344]. Events within these RCD mechanisms are those that are part of completely genetically coded physiological programs such as (post)-embryonic development or the protection of tissue homeostasis. These are referred to as programmed cell death (PCD) mechanisms. To be clear, these instances of PCD are by definition regulated, but RCD mechanisms are not necessarily PCD ones [345].

Kerr and colleagues recognized a stereotyped form of cell death and called it apoptosis taken from the Greek meaning "falling off" as autumn leaves from trees. It was strictly characterized morphologically by cytoplasmic shrinkage, chromatin condensation eventually involving the entire nucleus, nuclear fragmentation, cell blebbing or a "boiling-like" process, and formation of apoptotic bodies [346]. In *Caenorhabditis elegans*, apoptosis was the first form of PCD to be characterized and was found in mammals to be highly conserved, but during evolution had evolved much more complex mechanisms at each step along the pathway. This indicates the development of redundancy and specialization of apoptotic mechanisms in higher organisms [347, 348]. Evidence now indicates that cell death as necrosis and apoptosis is an oversimplification and that there are multiple RCD programs that overlap, but are mutually exclusive with apoptosis [347]. Within the cellular morphological phenotypes described by Kerr and colleagues are diverse functional, biochemical, and immunological processes. Furthermore, morphological and biochemical markers are not necessarily linked [349].

Programmed cell death is well characterized during embryonic development as apoptosis, which plays important roles in shaping organisms. As examples, the tadpole loses cells in its tail by apoptosis as it morphs into a frog. Hands initially appear as tiny buds where apoptosis takes place in interdigital cells or "webs" between fingers in order to separate them from each other. Likewise, the lens of the eye is crafted by cell proliferation, migration, and processes that use the same regulators of those in apoptosis [350]. Apoptosis also provides a mechanism for the safe disposal of potentially destructive inflammatory cells such as neutrophils that are phagocytized by local macrophages when they are no longer needed. There are approximately 2.5×10^{10} neutrophils in the human body at any one time that live several hours to a few days, so the turnover of these potentially dangerous cells is significant. During development and homeostasis such as maintenance of neutrophil numbers, PCD occurs with conservation of plasma membrane integrity and without an immune response. This prevents the release of proinflammatory mediators and protects surrounding cells and tissues [351]. However, it is now clear that apoptosis does not always take place in the absence of inflammation or the absence of an immune response. This will be reiterated later.

A Nomenclature Committee on Cell Death (NCCD) was formulated to describe and evaluate distinct modalities of cell death, make recommendation on their definition, facilitate communication among scientists, and accelerate the pace of discovery [349]. That the 2012 committee formulated several previous rounds of recommendations [352–354] and published again in 2014 [344] indicates the need for a flexible, yet consistent process for nomenclature as our understanding of cell death mechanisms progresses. The 2012 NCCD described 13 regulated cell death mechanisms. Since even a brief description of these RCD mechanisms is beyond the scope of this chapter, the focus here will be on RCDs that are most common and especially those that have been described in response to electric fields. These will include caspase-dependent intrinsic and extrinsic apoptosis by death receptors and dependence receptors, caspase-independent intrinsic apoptosis, necroptosis, and parthanatos (PARP-mediated RCD) as well as roles for autophagy in cell death. Other cell death mechanisms can be reviewed in [349].

Regardless of external or internal stresses that induce RCD, cell responses are tightly regulated and coordinated. Generally, responses to threatening stimuli are aimed to avoid or remove the stimulus, initiate repair mechanisms, and reestablish homeostasis [355]. If these stress-adaptive approaches fail, cells initiate RCD. In this transition, RCD-inhibitory signals terminate and are replaced by RCD-promoting signals. Alternatively, these two signaling mechanisms coexist as RCD-inhibitory signals dissipate and RCD-promoting signals increase until they become predominant [345].

Before cell death mechanisms and their specific components are introduced, it may be useful to describe one of the first nomenclature systems used; it will be seen in some earlier literature and presenting them will introduce some important distinctions about RCD and what we have learned about them over time. One of the earliest cell death classifications included type 1 cell death as apoptosis, showing the morphological features described by Kerr et al. [346]; type II cell death as autophagy (self-eating), featuring cytoplasmic vacuolization for recycling cellular organelles; and type III cell death as necrosis, exhibiting neither apoptotic nor autophagic characteristics [356]. In practice, morphological characterizations of apoptosis are less valuable than initially appreciated. First, this classification system considered apoptosis as the only RCD mechanism known and necrotic cell death was defined as an ACD type, which is now known to be programmed. In addition, autophagy as a cell death mechanism has been considerably debated, but it is now considered as a cell death subroutine as opposed to a cell death mechanism on its own. Autophagy is a highly conserved, genetically programmed, integrated stress response whereby cells form intracellular membrane vesicles that engulf and degrade cytoplasmic organelles providing a survival advantage as cells undergo nutrient deprivation and other cellular stresses [357]. It is now also known that apoptosis and autophagy are not mutually exclusive pathways, but share some molecular regulators and can act in synergy or counter to each other [358]. Finally, pharmacological and genetic manipulations of cell death mechanisms can shift morphological markers from one morphologically defined cell death classification to another. Thus, *classification of cell death based on morphology has generally been abandoned in favor of classifications based on quantifiable biochemical parameters*. The NCCD defined molecular subroutines that characterize specific cell death mechanisms as well as pharmacological/genetic interventions that can be used to discriminate among them [349].

4.10.2 Regulated Cell Death (RCD) by Caspase-Dependent Mechanisms

Caspase 3-dependent cell death is generally considered to be apoptotic in nature and most generally exhibits classic apoptotic morphology defined by Kerr and colleagues [346]. This can occur by intrinsic apoptosis, which is initiated by intracellular stresses, or extrinsic apoptosis, which is initiated by extracellular death receptor signaling (FAS/CD95, tumor necrosis factor α (TNFα), and TNF-related apoptosis-inducing ligand, TRAIL) or dependence receptor signaling [netrin receptors such as patched, deleted in colorectal carcinoma (DCC) and UNC5A-D]. Increasing death receptor signaling increases cell death. In contrast, dependence receptors are only lethal when concentrations of their cognate ligands fall below a threshold level. Extrinsic apoptosis by dependence receptors expresses caspase 3 and caspase 9 through patched and DCC receptors or caspase 3 and activated protein phosphatase 2A (PP2A) and death-associated protein kinase 1 (DAPK1) through UNC5B receptors. Anoikis (ancient Greek for "the state of being homeless") exhibits caspase 3 activation, but is executed by the molecular mechanisms for intrinsic apoptosis as a response from cells that have lost cell-to-matrix interactions and β-1-integrin signaling and also express other distinguishing markers including downregulation of EGFR and inhibition of ERK1 signaling.

As mentioned earlier, genetic studies in *C. elegans* defines apoptosis as an evolutionarily conserved PCD pathway that is as fundamental to life as proliferation and differentiation. Studies in *C. elegans* provide the basis for understanding PCD in higher organisms that are referred to as apoptosis [359]. *C. elegans* is an excellent model for these studies because the history of every cell in the organism has been detailed. The male exhibits 1179 somatic nuclei and during development 148 undergo PCD. By using methods of genetics, developmental biology, molecular biology, and biochemistry, 13 genes have been identified that regulated PCD in *C. elegans*. All of these genes have several homologues in mammals. Mutations in many of these mammalian gene homologues cause diseases in humans [359]. This enhances the importance of understanding apoptosis mechanisms in humans.

The general strategy for regulation of PCD by apoptosis occurs when an antagonist (EGL-1, BH3-only protein) relieves inhibition of a repressor (CED-9, BCL-2/xl) that blocks a proapoptotic protein (BAX, BAK); in other words inhibition of an inhibitor induces activation (Fig. 4.26). This is analogous to releasing the cell death "breaks". In *C. elegans* this occurs when an apoptotic stimulus upregulates the transcription factor *egl-1*, a proapoptotic mammalian homologue BCL-2 homologue 3, called a BH3-only protein. EGL-1 binds to antiapoptotic CED-9, a mammalian BCL-2 homologue, which releases inhibition that CED-9 exerts on the adaptor protein CED-4, a mammalian homologue of apoptosis peptidase-activating factor 1 (APAF-1). This allows proapoptotic CED-4 to bind the cysteine protease CED-3, a mammalian caspase homologue. CED-3 cleaves multiple substrates and executes cell death.

In mammalian cells, the process is more evolved and complex in several ways, yet fundamentally the same. During evolution, genes were duplicated and selected to function in more complex multicellular organisms based on their environmental needs. A major outcome of gene duplication is specialization of function. This replication and complexity provides selective mechanisms to regulate apoptosis [347]. In mammals, there are several isoforms for all of the *C. elegans* apoptosis-related proteins at every step. These include three classes of BCL-2 family proteins that regulate outer mitochondrial membrane (OMM) integrity. Two classes include proapoptotic proteins BAX and BAK and antiapoptotic proteins BCL-2, BCLxl, BCLw, MCL1, and A1. Antiapoptotic proteins protect the integrity of the OMM by binding to proapoptotic proteins, which prevents them from permeabilizing the OMM that leads to caspase activation and apoptosis. The third class of BCL-2 family proteins includes proapoptotic BH3 only proteins that are subdivided into two groups based on their interactions with the other two classes of BCL-2 family members. Direct activators can bind to and inhibit antiapoptotic BCL-2 proteins as well as directly activate proapoptotic proteins (indicated in dotted line in Fig. 4.26). These include BID, BIM, and maybe PUMA. The other subgroup of proteins is sensitizers or derepressors that cannot directly activate the proapoptotic group, but bind to and inhibit the antiapoptotic proteins. These BH3-only proteins include

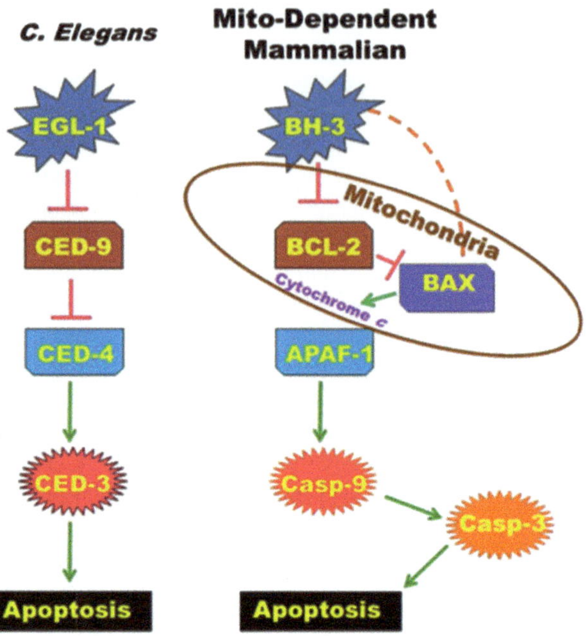

Fig. 4.26 A comparison of the primordial apoptosis pathways in C elegans (*left*) and more complex eukaryote/mammalian mitochondria-mediated apoptosis (*right*)

BAD, BIK, BMF, NOXA, and maybe PUMA. Some BH3-only proteins are further regulated by phosphorylation (BAD, BIK, BIM, BMF). Mammalian cells also express multiple caspase isozymes including initiator (caspase 8, 9, 10) and executioner (caspase 3, 6, 7) caspases as well as inflammatory caspases (caspase 1, 4, 5 in humans; caspase 1, 11, 12 in mice).

In mammalian cells, extrinsic apoptosis is induced by death receptor ligands or agonistic antibodies binding to their cognate death receptors, shown here with FAS ligand (FAS-L) and its receptor (FAS-R) (Fig. 4.27). After FAS ligand binding, intracellular signals are initiated by recruiting and anchoring FAS-associated protein with death domain FADD and caspase 8 to the cytoplasmic side of the transmembrane death receptor forming the death-induced signaling complex or DISC, which activates initiator caspase 8. Caspase 8 can now take either of two different pathways to activate executioner caspases depending on the cell type [360]. In type I cells such as thymocytes, the mitochondria-mediated intrinsic pathway is not used and caspase 8 directly activates caspase 3 (or other executioner caspases). In these cells there is ample formation of the DISC and enough caspase 8 is activated to directly activate caspase 3. This cascade cannot be inhibited by antiapoptotic proteins from the B-cell CLL/lymphoma 2 (BCL-2) family. An explanation for the efficient activation of caspase 3 in type I cells is the presence of FAS in membrane microdomains rich in cholesterol and glycosphingolipids called lipid rafts [361]. It is possible that lipid rafts form of a platform that allows efficient formation of the DISC and caspase 8 activation.

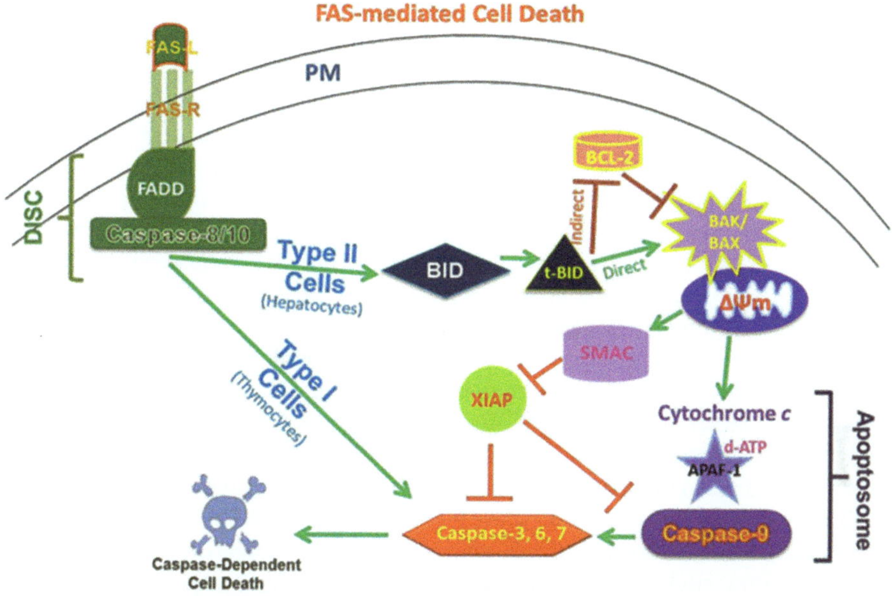

Fig. 4.27 FAS-mediated cell death in mammal. Cell death in higher organisms includes extrinsic apoptosis that is independent of mitochondria (type I cells) and signal amplification through mitochondrial mechanisms involved in intrinsic apoptosis (type II cells). *Green arrows* indicate activation. *Red* perpendicular lines indicate inhibitions. See the text for presentation and discussion

In contrast, for type II cells such as hepatocytes and pancreatic β-cells, activation of the DISC is insufficient to produce enough caspase 8 to activate caspase 3 and it is essential that the apoptotic signal is amplified through the mitochondria-mediated intrinsic pathway by caspase 8-mediated cleavage and activation of BH3-only protein BH3-interacting domain death agonist (BID) to form truncated BID or t-BID, which leads to cytochrome c release. These steps lead to creation of proteolipid pores that permeabilize outer mitochondrial membrane integrity by pore-forming proapoptotic proteins BCL-2-associated X protein (BAX) and BCL-2-agonist/killer (BAK). This cascade can be inhibited the by antiapoptotic BCL-2 family. Pore-forming activities of BAX and BAK are physically inhibited by interaction with antiapoptotic proteins BCL-2, BCL-2-like 1 (BCLxl), and myeloid cell leukemia (MCL1). In turn, the interactions of pro- and antiapoptotic proteins are under control of BH3-only proteins such as BCL-2-binding protein (PUMA), BCL-2-like 11 (BIM), and BID. Pore-forming proapoptotic proteins lead to the release of cytochrome c from mitochondria. Cytochrome c binds with APAF-1, deoxy-ATP, and caspase 9 to form the apoptosome to activate initiator caspase 9. Active caspase 9 then activates executioner caspase 3 to execute apoptosis and cell death [362]. Also released from mitochondria is second mitochondria activator of caspase (SMAC) (also called DIABLO), which facilitates apoptosis by inhibiting X chromosome-linked inhibitor of apoptosis (XIAP) and other IAPs that block activation of caspase 3, 6, 7, and 9. Another explanation for the difference between type I and type II cells is levels of XIAP in FAS death receptor-stimulated cells, such that type II cells require amplification through t-BID to overcome XIAP inhibition by way of the mitochondrial pathway for caspase activation, while type I cells do not [363]. t-BID functions to amplify this response by directly activating BAK/BAX by oligomerization and/or indirectly activating it by relieving BCL-2 mediated inhibition of BAK/BAX. Figure 4.27 shows three activation mechanisms for caspases: direct cleavage as in caspase 8 and caspase 9 cleavage of caspase 3, induced proximity activation of caspase 8 at the DISC, and holoenzyme formation for activating caspase 9 at the apoptosome. Caspases carry out specialized function and distinct roles based on subcellular localization and protein–protein interaction with substrate specificity playing a lesser important role. This division of labor permits multicellular organisms to sense and differentially respond to diverse environmental stimuli [347]. For example, caspase 8 is localized near and binds to the cytoplasmic side of death receptors and upon death receptor ligand binding with FADD to form the death-induced signaling complex (DISC) initiating extrinsic apoptosis. In contrast, caspase 9 binds to APAF-1 to form the apoptosome with cytochrome c initiating intrinsic apoptosis. Intrinsic apoptosis is activated by a number of intracellular stresses such as DNA or organelle damage, Ca^{2+} overload, hypoxia, oxidative stress, intracellular pathogens, increases in unfolded proteins, and excitotoxicity in neurons, among others. All of these diverse stimuli are connected to mitochondrial mechanisms. So, intrinsic apoptosis is initiated through mitochondria-mediated events that include cytochrome c release, APAF1-dependent activation of caspase 9 and caspase 3. Classification of initiator and executioner caspases is important to differentiate initiation of RCD (in this case apoptosis) and its actual execution [345, 364, 365]. In response to these same apoptosis initiators, caspase-independent processes also occur when apoptosis-inducing factor (AIF) and endonuclease G are released from mitochondria and translocated to the nucleus to degrade DNA.

Other complexities in RCD occur because FADD has multiple functions through its death effector domain (DED) and its death domain (DD) [366]. In addition to forming the DISC using its DD to bind to the cytoplasmic domains of the FAS receptor and using its DED to bind pro-caspase 8/10, FADD can also form complex II, which does not include the FAS receptor, but includes FADD, pro-caspase 8, and c-FLIP, which appear to amplify caspase activation. c-FLIP binding to the FADD DED signals proliferation and survival. FADD also forms complexes in TNF-R1 signaling that promotes either survival or apoptosis. The DD of FADD also binds Atg5 that promotes autophagy and RIP1 that induces necroptosis, another type of RCD (discussed later). Thus, the FADD DED and DD domains regulate many essential cell processes at the crossroads of survival and death by apoptosis or programmed necrosis.

Until recently, necrosis was considered accidental cell death with morphological characteristics that were neither defined as apoptosis nor autophagy, called type III cell death. As indicated earlier, apoptosis is not the only mechanism of cell suicide or RCD. It is now known that necrosis can also be regulated and occurs without caspase activation. Programmed necrosis is distinctly different than apoptosis and may operate as a backup system when apoptosis mechanisms collapse such as with viral infections that express caspase inhibitors or when apoptosis programs are inhibited in cancer. Necrosis exhibits morphological characteristics and has been defined in several pathological conditions. As opposed to apoptosis, in regulated necrosis, cells and organelles swell, nuclei remain intact, the nuclear membrane dilates, chromatin condenses into small, irregular blotches, DNA is randomly degraded, and cells do not fragment to form apoptotic bodies.

Regulated necrosis is not one signaling cascade, but appears to be interactions among multiple molecular events from more than one signaling pathway. However, it is difficult to define and characterize these mechanisms and pathways without knowing the initiators, propagators, and executioners in the processes. Analyzing the final characteristics does not does not indicate which pathways were executed. Nevertheless, significant progress has been made to delineate some of these events and has demonstrated their importance in both physiological and pathological processes. Programmed necrosis is responsible for negative selection of lymphocytes, termination of immune responses, regulation of bone growth, ovulation, cellular turnover in the intestine, or post-lactational regression of the mammary gland. It also has pathological correlates in neurodegeneration, excitotoxicity, stroke and ischemia–reperfusion, infection, and oxidative stress [367, 368]. However, there is more to learn about these mechanisms and in time they will be as well defined as apoptosis mechanisms and pathways.

4.10.3 RCD by Necroptosis

One subroutine of programmed necrosis that is well characterized through death receptors is necroptosis [344, 345, 347, 369]. Necroptosis can be induced by death receptors (FAS, TNFR, TRAIL) and pathogen recognition receptors (PRRs) on their plasma membranes that respond to molecules with pathogen-associated molecular patterns (PAMPs) such as viral or bacterial nucleotides, lipopolysaccharides, and lipoproteins that induce inflammation. Biochemical markers for necroptosis are best categorized from responses produced by tumor necrosis factor receptor (TNFR) when caspases are inhibited. Complexes in response to TNFR can induce survival, apoptosis, or necroptosis. Upon TNF ligand binding, an intracellular assembly called complex I is formed at the cytoplasmic side of the receptor by TNFR-associated death domain (TRADD), receptor-interacting protein kinase 1 (RIP1), cellular inhibitor of apoptosis protein (cIAP), and TNFR-associated factor 2 (TRAF2) and TRAF5. If cIAP-mediated ubiquitylation of RIP1 predominates over de-ubiquitylation by cylindromatosis, ubiquitinated RIP1 recruits transforming growth factor β–activated kinase (TAK1) and TAK1-binding protein 2 (TAB2) and TAB3 and Nf-kB is activated to promote inflammation and survival. If de-ubiquitylation predominates, cell death is induced by one of two different assembles called complex II. One complex II induces apoptosis by activating caspase-8 in the DISC. The other is similar to that shown in Fig. 4.27 for the DISC, except TRADD, RIPK1 and RIPK3 are present with FADD and caspase 8 cleaves and inactivates RIP1 and RIP3 to induce caspase-dependent apoptosis. If caspase 8 is inhibited, RIP1 and RIP3 become auto- or cross-phosphorylated (or by an unidentified kinase) and induce necroptosis. Necroptosis can be considered to be RIP3 dependent. This assembled complex II is called the necrosome, and it activates several bioenergetic changes, elevates reactive oxygen and nitrogen species, produces ceramide, induces lipid peroxidation, opens the mitochondria permeability pore complex, elevates intracellular calcium, and causes DNA damage and PARP-1 activation, ATP and NAD depletion, activation of calpains and cathepsins, and feedforward signaling loops that cause an energy crisis and cell death. The switch that modulates the apoptosis/necroptosis cell death scenarios is regulated by initiator caspase 8 and the ubiquitylation system. Necroptosis is pathologically relevant in defense of pathological organisms and conditions causing excessive loss of cell viability including ischemia–reperfusion injury, chronic neurodegenerative diseases, acute neurotoxicity, sepsis, and pancreatitis [345].

4.10.4 RCD by Parthanatos (PARP-1-Mediated RCD)

It has been debated whether there is one core program for regulated necrotic cell death or if there are distinct individual programs. Recent studies suggest that the necroptosis program described above is actually two separate pathways that can be activated independently. In this model, PARP-mediated cell death appears to occur with excessive DNA damage and overactivation of poly (ADP ribose) polymerase-1 (PARP-1). While PARP-1 is a molecular sensor of DNA strand breaks under normal conditions and is cleaved and inactivated by caspase 3 during apoptosis, hyperactivation leads to oversynthesize branched, long poly (ADP ribose) (PAR) chains attached to glutamate or aspartate residues of acceptor proteins. This leads to genotoxicity and mitochondriotoxicity causing irreversible loss of $\Delta\Psi_m$ and AIF release from mitochondria. This leads to excessive hydrolysis of NAD+ into nicotinamide and PAR and ATP hydrolysis causing a dramatic energy crisis and caspase-independent cell death. PAR is a common marker for PARP-mediated cell necrosis, which is also called parthanatos. H_2O_2 and DNA-alkylating agents induce cell death by PARP-1. RIP1 and TRAF appear to be downstream mediators required for JNK activation that leads to activation of calpains and cathepsins and compromise outer mitochondrial membrane integrity and release of AIF. It is likely that the ROS generated by necroptosis causes DNA damage and activation of PARP-1 and downstream processes that are characterized as PARP-mediated programmed necrosis. Thus the PARP-mediated pathway can be considered part of the TNF necroptosis pathway, but can function independently of it.

PARP-1-dependent cell death induces inflammation in cells in the nervous, cardiovascular, and immune systems. Inhibition of necroptosis by blocking RIP1 and inhibiting PARP-1 lowered ischemia–reperfusion damage after stroke. PARP-1 inhibitors have also been effective in treating tissue injury in response to septic and hemorrhagic shock, as well as acute lung inflammation, peritonitis, and acute forms of cardiomyopathies and heart failure. Inhibiting PARP-1 may also be effective in autoimmune diseases such as type I diabetes and rheumatoid arthritis [369].

4.10.5 Nanosecond Pulsed Electric Fields Activate RCD

Some initial studies investigating nsPEF effects on mammalian cells were basically extensions of studies using pulse power to decontaminate or kill bacteria [370]. Given that nsPEFs could kill bacteria, it was likely that they could induce cell death in cancer cells, perhaps by RCD. In the mid-to-late 1990s, mechanisms for regulated or programmed cell death, commonly referred to as apoptosis, were being defined at a rapid pace and disease-causing mutations in apoptotic proteins had been defined. At this time, apoptosis was the only known form of RCD. Thus, nsPEF-treated cells and tumors were analyzed for apoptotic characteristics in cells in vitro and tumor tissue in vivo. Essentially all studies with nsPEF demonstrated RCD mechanisms because cell death was delayed and ACD was not observed or not reported. Also of interest was defining cellular targets that were sensitive to nsPEFs. Generally, these included the plasma membrane, nucleus, DNA, mitochondria, and cytoskeleton.

4.10.6 In Vitro Studies Showing nsPEF-Induced RCD

Caspase activation and other markers for apoptosis have been shown in vitro in human Jurkat and HL60 cells [147, 233, 238, 371], Jurkat and rat glioma cells [372], HCT-116 cells with and without p53 [373], B16f10 cells [374], E4 squamous cell carcinoma cells [375], and several hepatocellular carcinoma cell lines [376], among others.

NsPEF-induced alterations in asymmetrically located plasma membrane lipids have also been used as an apoptosis marker, but there are some special considerations using this as a cell death marker. Phosphatidylcholine and sphingomyelin are predominantly located in the outer (luminal) leaflet. Phosphatidylethanolamine and phosphatidylinositol reside mainly in the inner (cytoplasmic) leaflet, while phosphatidylserine (PS) is located almost exclusively in the inner leaflet [377]. Externalization of PS to the outer leaflet of the plasma membrane is generally considered to be an apoptosis marker. Although externalization of PS can be context dependent, it does require caspase-mediated cleavage and inactivation of cell flippase(s), which transports PS from the outer to the inner leaflet [378]. It was shown that nsPEF-induced PS externalization was at least partially caspase dependent [390]. Factors released from mitochondria have also been implicated in the activation of lipid scramblases resulting in bidirectional lipid scrambling that disrupts the asymmetric distribution of PS [377]. Finally, the transmembrane protein Xkr8 promotes caspase-dependent PS exposure during apoptosis, probably acting at a late step in PS exposure, possibly in phospholipid scrambling [379]. This is sufficient to serve as an "eat-me" signal and engulfment by macrophages. However, electric fields can "pull" PS by electrophoretic migration from the inner leaflet to the outer leaflet of the plasma membrane through nanometer-sized electropores induced by pulses as short as 3 ns [380]. Furthermore, these cells can be phagocytized as apoptotic mimicry [381]. In this context, PS externalization is not an apoptosis marker and is reversible. Also parthanatos, which will be presented later, also shows externalization of PS [382]. Thus, using PS externalization as an apoptotic marker in response to nsPEFs requires caution and should only be used at times significantly after pulse application and supported by other apoptosis markers. For example, the demonstration that caspases are activated and that nsPEF-induced PS externalization is attenuated by the caspase inhibitor z-VAD-fmk satisfies this criterion for using PS externalization as an apoptosis marker [390].

Detection of DNA fragments using terminal deoxyribonucleotidyl transferase (TDT)-mediated dUTP-digoxigenin nick end labeling (TUNEL) assay has often been used as an indicator of apoptosis. However, a number of studies indicate that TUNEL-positive cells can occur in other forms of nonapoptotic cell death. For example, TUNEL-positive cells have been identified in a number of pathological conditions that cause necrosis [383–385]. The TUNEL assay does not discriminate among apoptosis, necrosis, or autolytic cell death [386] and therefore should not be used alone as an apoptotic marker. In general, multiple markers should be used to identify cell death by apoptosis.

Other nsPEF-induced indicators apoptosis include morphological changes such as cell shrinkage, membrane blebbing and DNA fragmentation, but more recent data indicate that these may not be specific to apoptosis [344]. Other markers such as changes in BCL-2 family proteins can be suggestive of apoptosis, but are not indicative. In addition to caspase activation, PARP cleavage and cytochrome c release into the cytoplasm have been used as a valid apoptosis marker.

After the initial studies by Beebe [233, 238] and Vernier [372] and colleagues, several other studies provided additional evidence for nsPEF-induced apoptosis; however, based on other observations, it appeared that that other RCD mechanisms were likely present. In $p53^{-/-}$ and $p53^{+/+}$ human HCT-116 colon carcinoma cells treated with lethal, square wave nsPEFs (50 pulses, 1 Hz, 60 ns 60 kV/cm or 5–30 pulses, 1 Hz, 60 kV/cm) [373], cell shrinkage was observed in cells with the presence of active caspases occurring before increases in BAX and cytochrome c release. Caspase activation could have occurred without mitochondrial involvement, which would implicate the extrinsic pathway like that in type I cells. Since no major differences in cell death mechanisms were observed in cells that did or did not express p53, cell death by nsPEF does not require p53.

Murine B16f10 melanoma cells treated with nsPEF (1–10 pulses, 1 Hz, 300 ns, 12–60 kV/cm) also expressed some apoptosis indicators in an electric field-dependent manner, cell shrinkage, membrane blebbing, and active caspases, but other apoptosis markers were absent [374]. Increasing numbers of cells exhibited loss of $\Delta\Psi_m$ and increases in ADP/ATP ratio were observed, which would cause mitochondria to swell, rupture, and cytochrome c, AIF, or SMAC release, which were not observed. Interestingly, dying cells did not exhibit PS externalization. Degradation of the actin cytoskeleton indicated the cells were undergoing disassembly. While apoptosis appeared to be present, it is likely that other RCD mechanisms were operating.

In E4 squamous cell carcinoma cells treated with lethal nsPEFs (10 pulses, 1 Hz, 300 ns 10–60 kV/cm) [375], electric field-dependent appearances of RCD indicators provided insight into death mechanisms. Caspase activation and PS externalization could be seen at lower electric fields than decreases in $\Delta\Psi_m$ and release of cytochrome c, suggesting an extrinsic mechanism for caspase activation without mitochondrial involvement. Consistent with this was the finding that decreases in BID, increases in t-BID, and release of cytochrome c were caspase dependent. Calcium-sensitive calpain(s), which can be stimulated in some apoptosis pathways [387], was also shown to be activated. BID cleavage was sensitive to a calpain inhibitor (and a caspase inhibitor) [390] and also sensitive to intracellular and extracellular calcium, implicating calpain(s) as part of apoptosis in E4 cell death.

As indicated above, data from several cell types suggested possible activation of multiple cell death pathways, but activation of caspases through non-mitochondrial pathways and involvement of extrinsic apoptosis regulators. In agreement with these observations, using a simple, bistable rate-equation-based model that is used to predict trends of cellular apoptosis following electric pulsing, it was predicted that an extrinsic mechanism would be more sensitive than the mitochondrial intrinsic pathway for electric pulse-induced cell apoptosis [388].

The importance of extrinsic apoptosis cell regulators was demonstrated as an explanation for inherent differences in sensitivities of Jurkat cells (more) and U937 cells (less) to nsPEFs to 100, 10 ns, 50 or 150 kV/cm pulses [389]. Measurement of basal expression levels indicated that U937 cells had a higher expression of c-FLIP, an extrinsic apoptosis inhibitor, while Jurkat cells had a higher expression of FasL, an extrinsic apoptosis agonist. When cells were exposed to lethal nsPEF (150 kV/cm) and siRNA was used to knockdown c-FLIP protein expression in U937 cells or knockdown FasL in Jurkat cells, U937 survival was reduced nearly 60 % while Jurkat survival improved by 40 %. This provided the first explanation for inherent cell type survival difference to nsPEFs as differences in expression levels of apoptosis-related proteins as opposed to physical cell characteristic, such as membrane or cytoskeletal structure. This is reminiscent of differences between type I and type II cells, where the later require amplification through mitochondria to activate caspases by overcome high levels of IAP, an intrinsic apoptosis antagonist [363]. This same study demonstrated extrinsic apoptosis in Jurkat cells showing caspase 8 activation without BID cleavage, cytochrome c release, or caspase 9 activation, which is typical of type I cells. This is noteworthy, since Jurkat cells are type II cells, which require amplification of apoptosis signals through the intrinsic mitochondrial pathway. Perhaps, nsPEFs play a role in formation of lipid rafts that allow efficient formation of DISC and activation of caspase 8.

The above results are in contrast to previous work in Jurkat cells demonstrating cytochrome c release and caspase 9- and caspase 3-dependent cell death, typical of intrinsic apoptosis [147, 390]. In this study using 10, 1 Hz, 60 ns, 10–60 kV/cm pulses with Jurkat cell clones that either did not express FADD, caspase 8, or APAF-1 [147], it was demonstrated that cytochrome *c* release was not caspase dependent, suggesting that its release was not due to caspase 8 activation through the DISC or FADD-mediated mechanisms. Furthermore, electric field-dependent cell death was identical for wild type, FADD-deficient, and caspase 8-deficient cells, indicating that nsPEF-induced cell death under these conditions did not require any FADD complex or caspase 8. It was shown in an APAF-deficient Jurkat clone, which did not respond to nsPEF with caspase 9 or caspase 3 activation, that cell death was caspase- and AFAF-1 dependent at lower electric fields, mimicking an intrinsic apoptosis mechanism, and caspase independent at higher electric fields demonstrating that nsPEF-induced cell death did not necessarily require active caspases under most intense conditions. It is highly likely that both regulated apoptosis (caspase dependent) and regulated necrosis (caspase independent) coexist during nsPEF-induced cell death in Jurkat cells. Based on findings from HCT-116 colon carcinoma, B16f10 melanoma, E4 squamous cell carcinoma, and Jurkat cells, as well as from other studies (see below), it also seems highly likely that nsPEF-induced RCD is dependent on nsPEF conditions as well as the cell type.

Using U937 cells, Pakhomova and colleagues [149] defined two modes of cell death after treatment with nsPEFs depending on the media for cell incubation. These studies either used trapezoidal waveforms with 600 pulses, 200 Hz, 300 ns, 7 kV/cm, or square waveforms at 50 pulses, 1 Hz, 60 ns 30–40 kV/cm. In the presence of cell culture media, cell death appeared to be due to plasma membrane permeabilization, water uptake, cell swelling, and membrane rupture by 40 min after pulsing, typical of signs of "necrosis." In contrast, when media contained sucrose, cell swelling was prevented and cell death was delayed. Compared to conditions without sucrose, the presence of sucrose facilitated caspase 3/7 activation, which was evident at 1 h and peaked at 3–6 h, and facilitated PARP cleavage, which increased by 2 h and peaked by 4 h. These findings were characteristic for both diverse types of pulse waveforms. These results indicated that under standard cell culture conditions, U937 cells died without significant caspase activation by osmotic stress-related cell death defined here as necrosis. This form of cell death has been called oncosis (Greek ónkos for swelling) [343]. However, it is not accidental cell death (ACD), because it was not immediate and was transiently rescued or modulated by sucrose, which led to apoptotic cell death. Thus, the predominant early "necrotic" cell death prevented or obscured the observation of apoptotic cell death. It is possible that both types of cell death mechanisms were operative, but one was faster than the other, and blunting the osmotic stress with sucrose allowed the slower apoptotic mechanism to fully develop.

In subsequent studies by this group, similar experiments were done with several cell types adhered to indium tin oxide (ITO)-coated glass slides that fit into cuvettes for nsPEF treatment [148]. Since ITO is biologically inert and exhibits high conductance with optical transparency, it provides a means to treat cells with nanosecond pulses and readily change media without potential artifacts due to cell damage by centrifugation. NsPEF treatment on ITO slides was highly efficient requiring about 20-fold fewer pulses. Treating cells with 20 pulses, 20 Hz, 300 ns 600–700 kV/cm, the author found that the mode of nsPEF-induced cell death (necrosis vs. apoptosis) was determined by the extracellular concentration of Ca^{2+}. There was a newly identified high Ca^{2+}-dependent necrotic cell death that was affected by a delayed, sudden, osmotic-independent pore expansion (or new pore formation); however, cell death was not caused by membrane rupture. This cell death mechanism is distinctly different than the osmotic-induced cell death characterized by swelling, but is not apoptotic cell death since caspase 3/7 activity was greatly reduced in the presence of high extracellular calcium. These studies demonstrate that there are multiple forms of cell death depending on the composition of the media and the levels of extracellular levels.

Other studies showed more directly that not all cell types died by apoptosis when exposed to nsPEFs. Unlike Jurkat cells, which readily exhibited apoptotic features in response to nsPEFs, nsPEF-treated HeLa S3 cells did not show caspase 3 activation or DNA ladder formation, both molecular signs of apoptotic cell death [146]. In contrast, these cells readily showed PARP-1-mediated PAR formation, a sign of regulated necrosis. When PARP-1 [poly(ADP ribose) (PAR) polymerase-1 becomes hyperactivated, it begins to oversynthesize branched, long poly (ADP ribose) (PAR). While PARP-1 is cleaved and inactivated by caspase 3 during apoptosis, hyperactivation and PAR formation lead to genotoxic and mitochondriotoxic stress causing loss of irreversible loss $\Delta\Psi_m$ and AIF release, overconsumption of NAD+ and ATP, and a dramatic energy crisis, leading to necrotic cell death. In these HeLa S3 cells, UV radiation-induced inactivation of PARP-1 and activation of caspase 3 indicate that apoptosis is functional program in these cells, but not activated by nsPEFs. In other studies these authors demonstrated that Ca^{2+} was required for nsPEF-induced PAR-mediated cell death, which was enhanced by the presence of the Ca^{2+} ionophore, ionomycin. They further showed that in the absence of Ca^{2+}, HeLa S3 cells were less susceptible to nsPEFs but could be forced to exhibit apoptotic signs of caspase 3 activation and PARP-1 cleavage with increasing pulse numbers [391]. Furthermore, nsPEF-induced regulated necrosis was not limited to HeLa cells but also occurred in K562 and HEK293 cells, with only limited signs of apoptosis markers (caspase 3 and PARP-1 cleavage). These studies demonstrate that nsPEF-induced cell death is cell type specific, but can be manipulated by altering conditions Ca^{2+} availability in HeLa S3 cells, demonstrating a characteristic of regulated cell death. PAR formation has been shown to be observed in parthanatos, a type of regulated necrosis.

A comparison of monopolar (MP) and bipolar (BP) pulses (polarity shift half way through pulse duration) with 600 ns pulse durations clearly demonstrated that BP pulses were less effective for membrane damage, permeability to ions, and cell lethality in three different cell types [392]. It took ten times more BPs to induce cell death than with MPs. This is in contract to conventional electroporation pulses (micro- and millisecond) where BPs are more effective for plasma membrane permeabilization [393] and DNA transfection while reducing cell death

[394]. Kotnik and colleagues also showed that BPs (500 + 500 μs and 1000 + 1000 μs) increased permeabilization of cells to bleomycin and Lucifer yellow without increased cell death [393]. It is reasoned that BPs reduce the holding time (passband) that would otherwise cause pore expansion, so BPs are less effective at causing pores to expand. These studies make clear that nsPEF-induced effects on the plasma membrane are critical for cell death induction.

4.10.7 Cellular Targets for nsPEF-Induced Regulated Cell Death

4.10.7.1 Plasma Membrane, Cell Ca^{2+}, and Mitochondria

The hypothesis that nsPEF have effects on intracellular structures prompted studies to investigate effects on cell organelles as targets for nsPEFs. It is generally agreed that plasma membrane permeabilization and influxes of Ca^{2+} and subsequent mitochondria Ca^{2+} overloading appear to be important determinants of cell death. Studies in N1-S1 hepatocellular carcinoma cells [233, 371] suggested that increases in intracellular Ca^{2+} were necessary, but not sufficient for cell death, unless there was a decrease in the $\Delta\Psi_m$, which was coincident with cell demise. This is similar to the "two-hit" hypothesis for cell death induced by ischemia reperfusion (Ca^{2+} and ROS) [395, 396], except the second hit, which is on mitochondria, is not likely due to generation of ROS [397]. Nevertheless, Ca^{2+} plays an important role in the nsPEF-triggered increase in ROS in BxPC-3 human pancreatic cells [145], and ROS appears to be required for formation of the apoptosome [398], the apoptotic assembly "station" for caspase 9 activation with cytochrome c, APAF-1, and dATP.

4.10.7.2 The Nucleus and DNA Have Been the Focus of a Number of Studies

Many studies have shown that nsPEFs induce DNA damage determined by TUNEL, comet assays, fragmentation by flow cytometry, and histone-2AX phosphorylation. The earliest and perhaps the most thorough study analyzing effects of nsPEF on DNA demonstrated that nsPEF-induced genotoxicity was cell type dependent, with nonadherent cells being more sensitive than adherent cells [399]. After a single 60 ns, 60 kV/cm pulse, unlike all but one of seven adherent cell types, nonadherent cells exhibited 10 % survival, induction of DNA damage, and a decrease in the number of cells reaching mitosis. Comet assays immediately after nsPEF treatment of Jurkat cells increased comet tail lengths 1.3- and 2.6-fold with one and ten pulses, respectively, with the same conditions indicated above, indicating that DNA damage was a direct effect of electric fields. HL60 cells also showed similar effects, albeit less striking. When DNA was isolated immediately after pulsing and analyzed by electrophoresis on agarose gels, a smear of DNA occurred in cells exposed to five pulses with conditions above while DNA from unexposed cells was present as a tight band, again indicating direct nsPEF-induced DNA damage. Examining lymphoblastoid cell line (LCL) from a homozygous patient with ataxia telangiectasia (AT), who are unable to repair DNA damage in response to radiation, chromatid aberrations were identified in 78 % of cells exposed to a single 600-ns, 60-kV/cm pulse, while only 22 % were observed in control cells. In all, 41 types of damage were observed in exposed cells including more chromatin gaps, breaks, and fragments than in control cells.

Another study demonstrated disruption of the actin cytoskeleton and nuclear membrane and damage to telomeres that appeared to decrease Jurkat cell survival exposed to single 60-ns 60-kV/cm pulses [251]. It is hypothesized that no nuclear damage was done to adherent cells under these conditions, presumably because their cytoskeletons absorb some of the impact of the electric field effects. In another study with Jurkat cells under "milder," nonlethal nsPEF conditions (single 60 ns, 10–25 kV/cm), there was a transient electric field-dependent decrease in the comet assay DNA migration. The authors suggested that their results were due to pulse-induced transient conformational change in the living cell nucleoprotein [400]. Others observed transiently disrupted cytoskeleton and transient increase in nuclear size in adherent HCT-116 cells exposed to three 60- or 300-ns, 60-kV/cm pulses [373].

4.10.8 In Vivo Studies Showing nsPEF-Induced Regulated Cell Death

Early studies with tumor tissues demonstrated that nsPEFs induced DNA fragmentation (TUNEL) and reduced fibrosarcoma tumor growth in a mouse model [233, 238, 371]. Using 10-ns or 300-ns pulses with electric field intensities of 280 and 30 kV/cm, respectively (near-equal energy densities), greater increases in DNA fragmentation were seen with the 300-ns condition, indicating that this effect, like effects in vitro, was independent of energy density. Fibrosarcoma tumors treated with three 10-ns pulses at 260 kV/cm exhibited twice as many TUNEL-positive tumor cells as sham-treated tumors. Tumors treated with 300-ns pulses and 75 kV/cm were 60 % smaller than sham-treated tumors, indicating that nsPEFs could likely be used for tumor treatment.

In a murine B16f10 melanoma model (hairless SKH-1 mice), Nuccitelli and coworkers [401] later showed that pulsed electric fields greater than 20 kV/cm, with risetimes of 30 ns and durations of 300 ns, caused tumor cell nuclei to rapidly shrink and tumor blood flow to stop in a temperature-independent manner (3 °C increase). Within 2 weeks, B16f10 tumors shrank by 90 % and a second treatment resulted in complete remission. In a later study [402], it was shown that after nsPEF ablation (300–600 pulses, 300 ns, 40–50 kV/cm 0.5 Hz), none of the melanomas recurred during a 4-month period. For complete remission, 24 % melanomas were eliminated after one treatment, 59 % required a second treatment, and 18 % required a third treatment. A microvessel density marker (CD31) was decreased significantly after treatment, agreeing with the decreased blood flow to treated tumors. The possible presence of apoptosis was suggested by a decrease in the antiapoptotic protein BCL-2.

It was later shown in this same model that after treatment with one hundred 300-ns pulses at 40 kV/cm, transient increases in histone 2AX phosphorylation (early DNA damage repair marker) coinciding with TUNEL-positive cells and pyknotic nuclei were observed. Caspase-positive cells were not observed until 6 h after pulsing, suggesting caspase-independent effects on DNA. Large DNA fragments, but not 180-bp fragmentation ladders, were observed, suggesting that apoptosis was incomplete. Anti-angiogenesis or anti-vascular effects were demonstrated by decreases in markers for vascular endothelial cells (CD31, CD35, and CD105) and growth factors (VEGF and PD-ECGF).

A study by X Chen et al. [403] showed that nsPEF effectively eliminated (75 %) and activated RCD in a mouse Hepa1-6 ectopic hepatocellular carcinoma (HCC) model with 100-ns, 65-kV/cm pulses (one treatment, 900 pulses, or three treatments 300 pulses each on alternate days), but not effectively with 100-ns, 33- or 50-kV/cm (37 %), or with 30-ns, $\leq$ 65-kV/cm pulses ($<$ 35 %). Under optimal conditions, nuclei rapidly shrunk (1–2 h) and became transiently TUNEL positive peaking at 3 h. Caspase 3/6/7 was transiently activated peaking at 3 h, but no more than 50 % of cells were caspase positive at that peak time. Phospho-BAD, which would be unavailable to promoting cytochrome c release, remained low. Anti-angiogenesis or anti-vascular effects were also demonstrated with CD34 and VEGF in this model.

Efficacy of nsPEF ablation was also shown in an orthotopic rat N1-S1 HCC model [404]. Data demonstrate 80–90 % when elimination with 1000, 100 ns, 50 kV/cm pulses were delivered at 1 Hz. Transient increases in active caspase 3 (6 h) and caspase 9 (2–6 h), but not in active caspase 8, indicating an intrinsic apoptosis mechanism(s) as well as caspase-independent mechanisms (cells negative for active caspase), which is in agreement with in vitro data. Interestingly, after ablation rats were resistant to challenge injections of the same cells, showing a vaccine-like effect that may be due to activation of an active immune response. Infiltration of immune cells and the presence of granzyme B expressing cells within days of treatment suggest an antitumor adaptive immune response. In other studies [405] nsPEF cleared UV-induced murine melanomas and was superior to tumor excision for accelerating secondary tumor rejection in immune-competent mice. An immune response was suggested by the presence of CD4+ T cells within treated tumors.

NsPEF treatment of basal cell carcinoma has shown promise in the first clinical trial of nsPEF ablation [406].

References

1. Cortese, B., Palamà, I.E., D'Amone, S., Gigli, G.: Influence of electrotaxis on cell behaviour. Integr. Biol. **6**, 817–830 (2014). doi:10.1039/C4IB00142G
2. Levin, M., Stevenson, C.G.: Regulation of cell behavior and tissue patterning by bioelectrical signals: challenges and opportunities for biomedical engineering. Annu. Rev. Biomed. Eng. **14**, 295–323 (2012). doi:10.1146/annurev-bioeng-071811-150114
3. Kotnik, T., Kramar, P., Pucihar, G., et al.: Cell membrane electroporation- part 1: the phenomenon. IEEE Electr. Insul. Mag. **28**, 14–23 (2012). doi:10.1109/MEI.2012.6268438
4. Klösgen, B., Reichle, C., Kohlsmann, S., Kramer, K.D.: Dielectric spectroscopy as a sensor of membrane headgroup mobility and hydration. Biophys. J. **71**, 3251–3260 (1996)
5. Tielrooij, K.J., Paparo, D., Piatkowski, L., et al.: Dielectric relaxation dynamics of water in model membranes probed by terahertz spectroscopy. Biophys. J. **97**, 2484–2492 (2009). doi:10.1016/j.bpj.2009.08.024
6. Feldman, Y., Ermolina, I., Hayashi, Y.: Time domain dielectric spectroscopy study of biological systems. IEEE Trans. Dielectr. Electr. Insul. **10**, 728–753 (2003). doi:10.1109/TDEI.2003.1237324
7. Vernier, P.T., Ziegler, M.J.: Nanosecond field alignment of head group and water dipoles in electroporating phospholipid bilayers. J. Phys. Chem. B **111**, 12993–12996 (2007). doi:10.1021/jp077148q
8. Kotnik, T., Bobanović, F., Miklavčič: Sensitivity of transmembrane voltage induced by applied electric fields—a theoretical analysis. Bioelectrochem. Bioenerg. **43**, 285–291 (1997). doi:10.1016/S0302-4598(97)00023-8
9. Kotnik, T., Miklavčič, D., Slivnik, T.: Time course of transmembrane voltage induced by time-varying electric fields—a method for theoretical analysis and its application. Bioelectrochem. Bioenerg. **45**, 3–16 (1998). doi:10.1016/S0302-4598(97)00093-7
10. Kotnik, T., Miklavčič, D.: Second-order model of membrane electric field induced by alternating external electric fields. IEEE Trans. Biomed. Eng. **47**, 1074–1081 (2000). doi:10.1109/10.855935
11. Hibino, M., Shigemori, M., Itoh, H., et al.: Membrane conductance of an electroporated cell analyzed by submicrosecond imaging of transmembrane potential. Biophys. J. **59**, 209–220 (1991)
12. Hibino, M., Itoh, H., Kinosita, K.: Time courses of cell electroporation as revealed by submicrosecond imaging of transmembrane potential. Biophys. J. **64**, 1789–1800 (1993)
13. Smith, K.C., Weaver, J.C.: Active mechanisms are needed to describe cell responses to submicrosecond, megavolt-per-meter pulses: cell models for ultrashort pulses. Biophys. J. **95**, 1547–1563 (2008). doi:10.1529/biophysj.107.121921
14. Kotnik, T., Miklavčič, D.: Theoretical evaluation of voltage inducement on internal membranes of biological cells exposed to electric fields. Biophys. J. **90**, 480–491 (2006). doi:10.1529/biophysj.105.070771
15. Kotnik, T., Pucihar, G.: Induced transmembrane voltage—theory, modeling, and experiments. In: Miklavčič, D., Markov, M.S., Pakhomov, A.G. (eds.) Advanced Electroporation Techniques in Biology and Medicine, pp. 51–70. CRC Press, Boca Raton (2010)
16. Asami, K., Takahashi, Y., Takashima, S.: Dielectric properties of mouse lymphocytes and erythrocytes. Biochim. Biophys. Acta. BBA – Mol. Cell Res. **1010**, 49–55 (1989). doi:10.1016/0167-4889(89)90183-3
17. Yao, C., Mo, D., Li, C., et al.: Study of transmembrane potentials of inner and outer membranes induced by pulsed-electric-field model and simulation. IEEE Trans. Plasma. Sci. **35**, 1541–1549 (2007). doi:10.1109/TPS.2007.905110
18. Retelj, L., Pucihar, G., Miklavčič, D.: Electroporation of intracellular liposomes using nanosecond electric pulses-a theoretical study. IEEE Trans. Biomed. Eng. **60**, 2624–2635 (2013). doi:10.1109/TBME.2013.2262177

19. Schoenbach, K.H., Beebe, S.J., Buescher, E.S.: Intracellular effect of ultrashort electrical pulses. Bioelectromagnetics **22**, 440–448 (2001). doi:10.1002/bem.71
20. Tekle, E., Oubrahim, H., Dzekunov, S.M., et al.: Selective field effects on intracellular vacuoles and vesicle membranes with nanosecond electric pulses. Biophys. J. **89**, 274–284 (2005). doi:10.1529/biophysj.104.054494
21. White, J.A., Blackmore, P.F., Schoenbach, K.H., Beebe, S.J.: Stimulation of capacitative calcium entry in HL-60 cells by nanosecond pulsed electric fields. J. Biol. Chem. **279**, 22964–22972 (2004). doi:10.1074/jbc.M311135200
22. Batista Napotnik, T., Reberšek, M., Kotnik, T., et al.: Electropermeabilization of endocytotic vesicles in B16 F1 mouse melanoma cells. Med. Biol. Eng. Comput. **48**, 407–413 (2010). doi:10.1007/s11517-010-0599-9
23. Krassowska, W., Filev, P.D.: Modeling electroporation in a single cell. Biophys. J. **92**, 404–417 (2007). doi:10.1529/biophysj.106.094235
24. Li, J., Tan, W., Yu, M., Lin, H.: The effect of extracellular conductivity on electroporation-mediated molecular delivery. Biochim. Biophys. Acta **1828**, 461–470 (2013). doi:10.1016/j.bbamem.2012.08.014
25. Pucihar, G., Kotnik, T., Valič, B., Miklavčič, D.: Numerical determination of transmembrane voltage induced on irregularly shaped cells. Ann. Biomed. Eng. **34**, 642–652 (2006). doi:10.1007/s10439-005-9076-2
26. Hu, Q., Viswanadham, S., Joshi, R.P., et al.: Simulations of transient membrane behavior in cells subjected to a high-intensity ultrashort electric pulse. Phys. Rev. E **71**, 031914 (2005). doi:10.1103/PhysRevE.71.031914
27. Gowrishankar, T.R., Smith, K.C., Weaver, J.C.: Transport-based biophysical system models of cells for quantitatively describing responses to electric fields. Proc. IEEE **101**, 505–517 (2013). doi:10.1109/JPROC.2012.2200289
28. Pucihar, G., Miklavčič, D., Kotnik, T.: A time-dependent numerical model of transmembrane voltage inducement and electroporation of irregularly shaped cells. IEEE Trans. Biomed. Eng. **56**, 1491–1501 (2009). doi:10.1109/TBME.2009.2014244
29. Susil, R., Semrov, D., Miklavčič, D.: Electric field induced transmembrane potential depends on cell density and organization. Electro. Magnetobiol. **17**, 391–399 (1998)
30. Pavlin, M., Pavselj, N., Miklavčič, D.: Dependence of induced transmembrane potential on cell density, arrangement, and cell position inside a cell system. IEEE Trans. Biomed. Eng. **49**, 605–612 (2002)
31. Pucihar, G., Kotnik, T., Teissié, J., Miklavčič, D.: Electropermeabilization of dense cell suspensions. Eur. Biophys. J. **36**, 173–185 (2007). doi:10.1007/s00249-006-0115-1
32. Ling, G., Gerard, R.W.: The normal membrane potential of frog sartorius fibers. J. Cell. Comp. Physiol. **34**, 383–396 (1949). doi:10.1002/jcp.1030340304
33. Neher, E., Sakmann, B.: Single-channel currents recorded from membrane of denervated frog muscle fibres. Nature **260**, 799–802 (1976). doi:10.1038/260799a0
34. Fluhler, E., Burnham, V.G., Loew, L.M.: Spectra, membrane binding, and potentiometric responses of new charge shift probes. Biochemistry (Mosc.) **24**, 5749–5755 (1985). doi:10.1021/bi00342a010
35. Gross, D., Loew, L.M., Webb, W.W.: Optical imaging of cell membrane potential changes induced by applied electric fields. Biophys. J. **50**, 339–348 (1986)
36. Pucihar, G., Kotnik, T., Miklavčič, D.: Measuring the induced membrane voltage with Di-8-ANEPPS. J. Vis. Exp. JoVE. (2009). doi:10.3791/1659
37. Frey, W., White, J.A., Price, R.O., et al.: Plasma membrane voltage changes during nanosecond pulsed electric field exposure. Biophys. J. **90**, 3608–3615 (2006). doi:10.1529/biophysj.105.072777
38. White, J.A., Pliquett, U., Blackmore, P.F., et al.: Plasma membrane charging of Jurkat cells by nanosecond pulsed electric fields. Eur. Biophys. J. **40**, 947–957 (2011). doi:10.1007/s00249-011-0710-7

39. Towhidi, L., Kotnik, T., Pucihar, G., et al.: Variability of the minimal transmembrane voltage resulting in detectable membrane electroporation. Electromagn. Biol. Med. **27**, 372–385 (2008). doi:10.1080/15368370802394644
40. Gabriel, B., Teissie, J.: Time courses of mammalian cell electropermeabilization observed by millisecond imaging of membrane property changes during the pulse. Biophys. J. **76**, 2158–2165 (1999)
41. Pucihar, G., Kotnik, T., Miklavčič, D., Teissié, J.: Kinetics of transmembrane transport of small molecules into electropermeabilized cells. Biophys. J. **95**, 2837–2848 (2008). doi:10.1529/biophysj.108.135541
42. He, H., Chang, D.C., Lee, Y.-K.: Using a micro electroporation chip to determine the optimal physical parameters in the uptake of biomolecules in HeLa cells. Bioelectrochem. Amst. Neth. **70**, 363–368 (2007). doi:10.1016/j.bioelechem.2006.05.008
43. Pakhomov, A.G., Gianulis, E., Vernier, P.T., et al.: Multiple nanosecond electric pulses increase the number but not the size of long-lived nanopores in the cell membrane. Biochim. Biophys. Acta. BBA – Biomembr. **1848**, 958–966 (2015). doi:10.1016/j.bbamem.2014.12.026
44. Vasilkoski, Z., Esser, A.T., Gowrishankar, T.R., Weaver, J.C.: Membrane electroporation: the absolute rate equation and nanosecond time scale pore creation. Phys. Rev. E **74**, 021904 (2006). doi:10.1103/PhysRevE.74.021904
45. Kotnik, T., Pucihar, G., Miklavčič, D.: Induced transmembrane voltage and its correlation with electroporation-mediated molecular transport. J. Membr. Biol. **236**, 3–13 (2010). doi:10.1007/s00232-010-9279-9
46. Smith, K.C.: A unified model of electroporation and molecular transport. Massachusetts Institute of Technology, http://dspace.mit.edu/bitstream/handle/1721.1/63085/725958797.pdf
47. Son, R.S., Smith, K.C., Gowrishankar, T.R., Vernier, P.T., Weaver, J.C.: J. Membr. Biol. **247**, 1209 (2014)
48. Pauly, H., Schwan, H.P.: Z Naturforsch **14B**, 125 (1959)
49. Gowrishankar, T.R., Weaver, J.C.: Proc. Natl. Acad. Sci. U. S. A. **100**, 3203 (2003)
50. Smith, K.C., Weaver, J.C.: IEEE Trans. Biomed. Eng. **59**, 1514 (2012)
51. Kinosita, K., Ashikawa, I., Saita, N., Yoshimura, H., Itoh, H., Nagayma, K., Ikegami, A.: Biophys. J. **53**, 1015 (1988)
52. Singer, S.J., Nicolson, G.L.: The fluid mosaic model of the structure of cell membranes. Science **175**, 720–731 (1972)
53. Abidor, I.G., Arakelyan, V.B., Chernomordik, L.V., Chizmadzhev, Y.A., Pastushenko, V.F., Tarasevich, M.R.: Electric breakdown of bilayer lipid membranes I. Main experimental facts and their qualitative discussion. Bioelectrochem. Bioenerg. **6**, 37–52 (1979)
54. Sugar, I.P.: The effects of external fields on the structure of lipid bilayers. J. Physiol. Paris **77**, 1035–1042 (1981)
55. Zimmermann, U., Scheurich, P., Pilwat, G., Benz, R.: Cells with manipulated functions: new perspectives for cell biology, medicine, and technology. Angew. Chem. Int. Ed. Engl. **20**, 325–344 (1981)
56. Neumann, E., Schaefer-Ridder, M., Wang, Y., Hofschneider, P.H.: Gene transfer into mouse lyoma cells by electroporation in high electric fields. EMBO J. **1**, 841–845 (1982)
57. Sugar, I.P., Neumann, E.: Stochastic model for electric field-induced membrane pores. Electroporation. Biophys. Chem. **19**, 211–225 (1984)
58. Bockmann, R.A., de Groot, B.L., Kakorin, S., Neumann, E., Grubmuller, H.: Kinetics, statistics, and energetics of lipid membrane electroporation studied by molecular dynamics simulations. Biophys. J. **95**, 1837–1850 (2008)
59. Levine, Z.A., Vernier, P.T.: Life cycle of an electropore: field-dependent and field-independent steps in pore creation and annihilation. J. Membr. Biol. **236**, 27–36 (2010)

60. Marracino, P., Amadei, A., Apollonio, F., D'Inzeo, G., Liberti, M., di Crescenzo, A., Fontana, A., Zappacosta, R., Aschi, M.: Modeling of chemical reactions in micelle: water-mediated keto-enol interconversion as a case study. J. Phys. Chem. B **115**, 8102–8111 (2011)
61. van der Ploeg, P., Berendsen, H.J.C.: Molecular dynamics simulation of a bilayer membrane. J. Chem. Phys. **76**, 3271–3276 (1982)
62. Egberts, E., Marrink, S.J., Berendsen, H.J.: Molecular dynamics simulation of a phospholipid membrane. Eur. Biophys. J. **22**, 423–436 (1994)
63. Tieleman, D.P., Leontiadou, H., Mark, A.E., Marrink, S.J.: Simulation of pore formation in lipid bilayers by mechanical stress and electric fields. J. Am. Chem. Soc. **125**, 6382–6383 (2003)
64. Gurtovenko, A.A., Vattulainen, I.: Pore formation coupled to ion transport through lipid membranes as induced by transmembrane ionic charge imbalance: atomistic molecular dynamics study. J. Am. Chem. Soc. **127**, 17570–17571 (2005)
65. Tarek, M.: Membrane electroporation: a molecular dynamics simulation. Biophys. J. **88**, 4045–4053 (2005)
66. Tieleman, D.P.: The molecular basis of electroporation. BMC Biochem. **5**, 10 (2004)
67. Ho, M.C., Levine, Z.A., Vernier, P.T.: Nanoscale, electric field-driven water bridges in vacuum gaps and lipid bilayers. J. Membr. Biol. **246**, 793–801 (2013)
68. Tokman, M., Lee, J.H., Levine, Z.A., Ho, M.C., Colvin, M.E., Vernier, P.T.: Electric field-driven water dipoles: nanoscale architecture of electroporation. PLoS One **8**, e61111 (2013)
69. Neumann, E., Rosenheck, K.: Permeability changes induced by electric impulses in vesicular membranes. J. Membr. Biol. **10**, 279–290 (1972)
70. Weaver, J.C.: Electroporation theory. Concepts and mechanisms. Methods Mol. Biol. **48**, 3–28 (1995)
71. Bernhardt, J., Pauly, H.: On the generation of potential differences across the membranes of ellipsoidal cells in an alternating electrical field. Biophysik **10**, 89–98 (1973)
72. Teissie, J., Golzio, M., Rols, M.P.: Mechanisms of cell membrane electropermeabilization: a minireview of our present (lack of ?) knowledge. Biochim. Biophys. Acta **1724**, 270–280 (2005)
73. Robello, M., Gliozzi, A.: Conductance transition induced by an electric field in lipid bilayers. Biochim. Biophys. Acta **982**, 173–176 (1989)
74. Teissie, J., Rols, M.P.: An experimental evaluation of the critical potential difference inducing cell-membrane electropermeabilization. Biophys. J. **65**, 409–413 (1993)
75. Rols, M.P., Teissie, J.: Electropermeabilization of mammalian cells. Quantitative analysis of the phenomenon. Biophys. J. **58**, 1089–1098 (1990)
76. Kotnik, T., Miklavcic, D.: Analytical description of transmembrane voltage induced by electric fields on spheroidal cells. Biophys. J. **79**, 670–679 (2000)
77. Kotnik, T., Pucihar, G., Rebersek, M., Miklavcic, D., Mir, L.M.: Role of pulse shape in cell membrane electropermeabilization. Biochim. Biophys. Acta **1614**, 193–200 (2003)
78. Gehl, J.: Electroporation: theory and methods, perspectives for drug delivery, gene therapy and research. Acta Physiol. Scand. **177**, 437–447 (2003)
79. Staal, L., Gilbert, R.: Generators and applicators: equipment for electroporation. In: Kee, S. T., Gehl, J., Lee, E.W. (eds.) Clinical Aspects of Electroporation. Springer, New York (2011)
80. Gothelf, A., Gehl, J.: Electroporation-based DNA delivery technology: methods for gene electrotransfer to skin. In: Rinaldi, M., Fioretti, D., Iurescia, S. (eds.) DNA Vaccines. Springer, New York (2014)
81. Tamzali, Y., Borde, L., Rols, M.P., Golzio, M., Lyazrhi, F., Teissie, J.: Successful treatment of equine sarcoids with cisplatin electrochemotherapy: a retrospective study of 48 cases. Equine Vet. J. **44**, 214–220 (2012)
82. Rols, M.P., Delteil, C., Serin, G., Teissie, J.: Temperature effects on electrotransfection of mammalian cells. Nucleic Acids Res. **22**, 540 (1994)
83. Rols, M.P., Golzio, M., Gabriel, B., Teissie, J.: Factors controlling electropermeabilisation of cell membranes. Technol. Cancer Res. Treat. **1**, 319–328 (2002)

84. Kinosita Jr., K., Tsong, T.Y.: Formation and resealing of pores of controlled sizes in human erythrocyte membrane. Nature **268**, 438–441 (1977)
85. Kinosita Jr., K., Tsong, T.Y.: Voltage-induced conductance in human erythrocyte membranes. Biochim. Biophys. Acta **554**, 479–497 (1979)
86. Rols, M.P., Teissie, J.: Electropermeabilization of mammalian cells to macromolecules: control by pulse duration. Biophys. J. **75**, 1415–1423 (1998)
87. Pucihar, G., Mir, L.M., Miklavcic, D.: The effect of pulse repetition frequency on the uptake into electropermeabilized cells in vitro with possible applications in electrochemotherapy. Bioelectrochemistry **57**, 167–172 (2002)
88. Rols, M.P., Teissie, J.: Flow cytometry quantification of electropermeabilization. Methods Mol. Biol. **91**, 141–147 (1998)
89. Escoffre, J.M., Portet, T., Favard, C., Teissie, J., Dean, D.S., Rols, M.P.: Electromediated formation of DNA complexes with cell membranes and its consequences for gene delivery. Biochim. Biophys. Acta **1808**, 1538–1543 (2011)
90. Orlowski, S., Belehradek Jr., J., Paoletti, C., Mir, L.M.: Transient electropermeabilization of cells in culture. Increase of the cytotoxicity of anticancer drugs. Biochem. Pharmacol. **37**, 4727–4733 (1988)
91. Portet, T., Camps I Febrer, F., Escoffre, J.M., Favard, C., Rols, M.P., Dean, D.S.: Visualization of membrane loss during the shrinkage of giant vesicles under electropulsation. Biophys. J. **96**, 4109–4121 (2009)
92. Chopinet, L., Roduit, C., Rols, M.P., Dague, E.: Destabilization induced by electropermeabilization analyzed by atomic force microscopy. Biochim. Biophys. Acta **1828**, 2223–2229 (2013)
93. Chernysh, A.M., Kozlova, E.K., Moroz, V.V., Borshagovskaya, P.Y., Bliznuk, U.A., Rysaeva, R.M.: Erythrocyte membrane surface after calibrated electroporation: visualization by atomic force microscopy. Bull. Exp. Biol. Med. **148**, 455–460 (2009)
94. Lopez, A., Rols, M.P., Teissie, J.: ^{31}P NMR analysis of membrane phospholipid organization in viable, reversibly electropermeabilized Chinese hamster ovary cells. Biochemistry **27**, 1222–1228 (1988)
95. Stulen, G.: Electric field effects on lipid membrane structure. Biochim. Biophys. Acta **640**, 621–627 (1981)
96. Mauroy, C., Portet, T., Winterhalder, M., Bellard, E., Blache, M.C., Teissie, J., Zumbusch, A., Rols, M.P.: Giant lipid vesicles under electric field pulses assessed by non invasive imaging. Bioelectrochemistry **87**, 253–259 (2012)
97. Silve, A., Leray, I., Mir, L.M.: Demonstration of cell membrane permeabilization to medium-sized molecules caused by a single 10 ns electric pulse. Bioelectrochemistry **87**, 260–264 (2012)
98. Golzio, M., Teissie, J., Rols, M.P.: Direct visualization at the single-cell level of electrically mediated gene delivery. Proc. Natl. Acad. Sci. U. S. A. **99**, 1292–1297 (2002)
99. Rols, M.P., Teissie, J.: Experimental evidence for the involvement of the cytoskeleton in mammalian cell electropermeabilization. Biochim. Biophys. Acta **1111**, 45–50 (1992)
100. Rols, M.P., Delteil, C., Golzio, M., Teissie, J.: Control by ATP and ADP of voltage-induced mammalian-cell-membrane permeabilization, gene transfer and resulting expression. Eur. J. Biochem. **254**, 382–388 (1998)
101. Escoffre, J.M., Bellard, E., Faurie, C., Sebai, S.C., Golzio, M., Teissie, J., Rols, M.P.: Membrane disorder and phospholipid scrambling in electropermeabilized and viable cells. Biochim. Biophys. Acta **1838**, 1701–1709 (2014)
102. Rosazza, C., Buntz, A., Riess, T., Woll, D., Zumbusch, A., Rols, M.P.: Intracellular tracking of single plasmid DNA-particles after delivery by electroporation. Mol. Ther. **21**, 2217–2226 (2013)
103. Mahrour, N., Pologea-Moraru, R., Moisescu, M.G., Orlowski, S., Leveque, P., Mir, L.M.: In vitro increase of the fluid-phase endocytosis induced by pulsed radiofrequency

electromagnetic fields: importance of the electric field component. Biochim. Biophys. Acta **1668**, 126–137 (2005)
104. Antov, Y., Barbul, A., Mantsur, H., Korenstein, R.: Electroendocytosis: exposure of cells to pulsed low electric fields enhances adsorption and uptake of macromolecules. Biophys. J. **88**, 2206–2223 (2005)
105. Teissie, J., Knutson, V.P., Tsong, T.Y., Lane, M.D.: Electric pulse-induced fusion of 3T3 cells in monolayer culture. Science **216**, 537–538 (1982)
106. Teissie, J., Rols, M.P.: Fusion of mammalian cells in culture is obtained by creating the contact between cells after their electropermeabilization. Biochem. Biophys. Res. Commun. **140**, 258–266 (1986)
107. Mekid, H., Mir, L.M.: In vivo cell electrofusion. Biochim. Biophys. Acta **1524**, 118–130 (2000)
108. Marrero, B., Heller, R.: The use of an in vitro 3D melanoma model to predict in vivo plasmid transfection using electroporation. Biomaterials **33**, 3036–3046 (2012)
109. Gibot, L., Wasungu, L., Teissie, J., Rols, M.P.: Antitumor drug delivery in multicellular spheroids by electropermeabilization. J. Control. Release **167**, 138–147 (2013)
110. Canatella, P.J., Black, M.M., Bonnichsen, D.M., McKenna, C., Prausnitz, M.R.: Tissue electroporation: quantification and analysis of heterogeneous transport in multicellular environments. Biophys. J. **86**, 3260–3268 (2004)
111. Chopinet, L., Wasungu, L., Rols, M.P.: First explanations for differences in electrotransfection efficiency in vitro and in vivo using spheroid model. Int. J. Pharm. **423**, 7–15 (2012)
112. Rols, M.P., Delteil, C., Golzio, M., Dumond, P., Cros, S., Teissie, J.: In vivo electrically mediated protein and gene transfer in murine melanoma. Nat. Biotechnol. **16**, 168–171 (1998)
113. Gibot, L., Rols, M.P.: Progress and prospects: the use of 3D spheroid model as a relevant way to study and optimize DNA electrotransfer. Curr. Gene Ther. **13**, 175–181 (2013)
114. Madi, M., Rols, M.P., Gibot, L.: Efficient in vitro electropermeabilization of reconstructed human dermal tissue. J. Membr. Biol. **248**, 903–908 (2015)
115. Neumann, E., Sowers, A.E., Jordan, C.A.: Electroporation and Electrofusion in Cell Biology. Plenum, New York (1989)
116. Saulis, G.: Kinetics of pore disappearance in a cell after electroporation. Biomed. Sci. Instrum. **35**, 409–414 (1999)
117. Saulis, G., Venslauskas, M.S., Naktinis, J.: Kinetics of pore resealing in cell membranes after electroporation. Bioelectrochem. Bioenerg. **26**, 1–13 (1991)
118. Smith, K.C., Gowrishankar, T.R., Esser, A.T., Stewart, D.A., Weaver, J.C.: Spatially distributed, dynamic transmembrane voltages of organelle and cell membranes due to 10 ns pulses: predictions of meshed and unmeshed transport network models. IEEE Trans. Plasma Sci. **34**, 1394–1404 (2006)
119. Gowrishankar, T.R., Weaver, J.C.: Electrical behavior and pore accumulation in a multicellular model for conventional and supra-electroporation. Biochem. Biophys. Res. Commun. **349**, 643–653 (2006)
120. Hu, Q., Joshi, R.P., Schoenbach, K.H.: Simulations of nanopore formation and phosphatidylserine externalization in lipid membranes subjected to a high-intensity, ultrashort electric pulse. Phys. Rev. E Stat. Nonlin. Soft Matter Phys. **72**, 031902 (2005)
121. Nesin, O.M., Pakhomova, O.N., Xiao, S., Pakhomov, A.G.: Manipulation of cell volume and membrane pore comparison following single cell permeabilization with 60- and 600-ns electric pulses. Biochim. Biophys. Acta **1808**, 792–801 (2011)
122. Pakhomov, A.G., Pakhomova, O.N.: Nanopores: a distinct transmembrane passageway in electroporated cells. In: Pakhomov, A.G., Miklavčič, D., Markov, M.S. (eds.) Advanced Electroporation Techniques in Biology and Medicine, pp. 178–194. CRC Press, Boca Raton (2010)

123. Bowman, A.M., Nesin, O.M., Pakhomova, O.N., Pakhomov, A.G.: Analysis of plasma membrane integrity by fluorescent detection of Tl(+) uptake. J. Membr. Biol. **236**, 15–26 (2010)
124. Glaser, R.W., Leikin, S.L., Chernomordik, L.V., Pastushenko, V.F., Sokirko, A.I.: Reversible electrical breakdown of lipid bilayers: formation and evolution of pores. Biochim. Biophys. Acta **940**, 275–287 (1988)
125. Gabai, V.L., Meriin, A.B., Mosser, D.D., Caron, A.W., Rits, S., Shifrin, V.I., Sherman, M.Y.: Hsp70 prevents activation of stress kinases. A novel pathway of cellular thermotolerance. J. Biol. Chem. **272**, 18033–18037 (1997)
126. Barros, L.F., Hermosilla, T., Castro, J.: Necrotic volume increase and the early physiology of necrosis. Comp. Biochem. Physiol. A Mol. Integr. Physiol. **130**, 401–409 (2001)
127. Barros, L.F., Stutzin, A., Calixto, A., Catalan, M., Castro, J., Hetz, C., Hermosilla, T.: Nonselective cation channels as effectors of free radical-induced rat liver cell necrosis. Hepatology **33**, 114–122 (2001)
128. Dyachok, O., Zhabyeyev, P., McDonald, T.F.: Electroporation-induced inward current in voltage-clamped guinea pig ventricular myocytes. J. Membr. Biol. **238**, 69–80 (2010)
129. Pakhomov, A.G., Bowman, A.M., Ibey, B.L., Andre, F.M., Pakhomova, O.N., Schoenbach, K.H.: Lipid nanopores can form a stable, ion channel-like conduction pathway in cell membrane. Biochem. Biophys. Res. Commun. **385**, 181–186 (2009)
130. Semenov, I., Xiao, S., Pakhomova, O.N., Pakhomov, A.G.: Recruitment of the intracellular Ca by ultrashort electric stimuli: The impact of pulse duration. Cell Calcium **54**, 145–150 (2013)
131. Semenov, I., Xiao, S., Pakhomov, A.G.: Primary pathways of intracellular Ca(2+) mobilization by nanosecond pulsed electric field. Biochim. Biophys. Acta **1828**, 981–989 (2013)
132. Okada, Y.: Ion channels and transporters involved in cell volume regulation and sensor mechanisms. Cell Biochem. Biophys. **41**, 233–258 (2004)
133. Okada, Y., Shimizu, T., Maeno, E., Tanabe, S., Wang, X., Takahashi, N.: Volume-sensitive chloride channels involved in apoptotic volume decrease and cell death. J. Membr. Biol. **V209**, 21–29 (2006)
134. Molleman, A.: Patch Clamping: An Introductory Guide to Patch Clamp Electrophysiology. Wiley, Padstow (2002)
135. Ibey, B.L., Xiao, S., Schoenbach, K.H., Murphy, M.R., Pakhomov, A.G.: Plasma membrane permeabilization by 60- and 600-ns electric pulses is determined by the absorbed dose. Bioelectromagnetics **30**, 92–99 (2009)
136. Varghese, A., Tenbroek, E.M., Coles Jr., J., Sigg, D.C.: Endogenous channels in HEK cells and potential roles in HCN ionic current measurements. Prog. Biophys. Mol. Biol. **90**, 26–37 (2006)
137. Ghamari-Langroudi, M., Bourque, C.W.: Ionic basis of the caesium-induced depolarisation in rat supraoptic nucleus neurones. J. Physiol. **536**, 797–808 (2001)
138. Pakhomov, A.G., Kolb, J.F., White, J.A., Joshi, R.P., Xiao, S., Schoenbach, K.H.: Long-lasting plasma membrane permeabilization in mammalian cells by nanosecond pulsed electric field (nsPEF). Bioelectromagnetics **28**, 655–663 (2007)
139. Andre, F.M., Rassokhin, M.A., Bowman, A.M., Pakhomov, A.G.: Gadolinium blocks membrane permeabilization induced by nanosecond electric pulses and reduces cell death. Bioelectrochemistry **79**, 95–100 (2010)
140. Pakhomov, A.G., Shevin, R., White, J.A., Kolb, J.F., Pakhomova, O.N., Joshi, R.P., Schoenbach, K.H.: Membrane permeabilization and cell damage by ultrashort electric field shocks. Arch. Biochem. Biophys. **465**, 109–118 (2007)
141. Berridge, M.J., Lipp, P., Bootman, M.D.: The versatility and universality of calcium signalling. Nat. Rev. Mol. Cell Biol. **1**, 11–21 (2000)
142. Tolstykh, G.P., Beier, H.T., Roth, C.C., Thompson, G.L., Ibey, B.L.: 600ns pulse electric field-induced phosphatidylinositol-bisphosphate depletion. Bioelectrochemistry **100**, 80–87 (2014)

143. Tolstykh, G.P., Beier, H.T., Roth, C.C., Thompson, G.L., Payne, J.A., Kuipers, M.A., Ibey, B. L.: Activation of intracellular phosphoinositide signaling after a single 600 nanosecond electric pulse. Bioelectrochemistry **94**, 23–29 (2013)
144. Zhivotovsky, B., Orrenius, S.: Calcium and cell death mechanisms: a perspective from the cell death community. Cell Calcium **50**, 211–221 (2011)
145. Nuccitelli, R., Lui, K., Kreis, M., Athos, B., Nuccitelli, P.: Nanosecond pulsed electric field stimulation of reactive oxygen species in human pancreatic cancer cells is Ca(2+)-dependent. Biochem. Biophys. Res. Commun. **435**, 580–585 (2013)
146. Morotomi-Yano, K., Akiyama, H., Yano, K.: Nanosecond pulsed electric fields induce poly (ADP-ribose) formation and non-apoptotic cell death in HeLa S3 cells. Biochem. Biophys. Res. Commun. **438**, 557–562 (2013)
147. Ren, W., Sain, N.M., Beebe, S.J.: Nanosecond pulsed electric fields (nsPEFs) activate intrinsic caspase-dependent and caspase-independent cell death in Jurkat cells. Biochem. Biophys. Res. Commun. **421**, 808–812 (2012)
148. Pakhomova, O.N., Gregory, B., Semenov, I., Pakhomov, A.G.: Calcium-mediated pore expansion and cell death following nanoelectroporation. Biochim. Biophys. Acta **1838**, 2547–2554 (2014)
149. Pakhomova, O.N., Gregory, B.W., Semenov, I., Pakhomov, A.G.: Two modes of cell death caused by exposure to nanosecond pulsed electric field. PLoS One **8**, e70278 (2013)
150. Ibey, B.L., Pakhomov, A.G., Gregory, B.W., Khorokhorina, V.A., Roth, C.C., Rassokhin, M. A., Bernhard, J.A., Wilmink, G.J., Pakhomova, O.N.: Selective cytotoxicity of intense nanosecond-duration electric pulses in mammalian cells. Biochim. Biophys. Acta **1800**, 1210–1219 (2010)
151. Pickard, B.G.: "second extrinsic organizational mechanism" for orienting cellulose: modelling a role for the plasmalemmal reticulum. Protoplasma **233**, 1–29 (2008)
152. Nick, P.: Microtubules and the tax payer. Protoplasma **249** (Special Issue Applied Plant Cell Biology), 81–94 (2012)
153. Frey, N., Klotz, J., Nick, P.: A kinesin with calponin-homology domain is involved in premitotic nuclear migration. J. Exp. Bot. **61**, 3423–3437 (2010)
154. Murata, T., Wada, M.: Effects of centrifugation on preprophase-band formation in Adiantum (1991)
155. Nick, P.: Signalling to the microtubular cytoskeleton in plants. Int. Rev. Cytol. **184**, 33–80 (1998)
156. Klotz, J., Nick, P.: A novel actin-microtubule cross-linking kinesin, NtKCH, functions in cell expansion and division. New Phytol. **193**, 576–589 (2012)
157. Parthasarathy, M.V.: F-actin architecture in coleoptile epidermal cells. Eur. J. Cell Biol. **39**, 1–12 (1985)
158. Grabski, S., Schindler, M.: Auxins and cytokinins as antipodal modulators of elasticity within the actin network of plant cells. Plant Physiol. **110**, 965–970 (1996)
159. Grabski, S., Arnoys, E., Busch, B., Schindler, M.: Regulation of actin tension in plant cells by kinases and phosphatases. Plant Physiol. **116**, 279–290 (1998)
160. Waller, F., Nick, P.: Response of actin microfilaments during phytochrome-controlled growth of maize seedlings. Protoplasma **200**, 154–162 (1997)
161. Sonobe, S., Shibaoka, H.: Cortical fine actin filaments in higher plant cells visualized by rhodamine-phalloidin after pretreatment with m-maleimidobenzoyl-N-hydroxysuccinimide ester. Protoplasma **48**, 80–86 (1989)
162. Thimann, K.V., Reese, K., Nachmikas, V.T.: Actin and the elongation of plant cells. Protoplasma **171**, 151–166 (1992)
163. Wang, Q.Y., Nick, P.: The auxin response of actin is altered in the rice mutant Yin-Yang. Protoplasma **204**, 22–33 (1998)
164. Sano, T., Higaki, T., Oda, Y., Hayashi, T., Hasezawa, S.: Appearance of actin microfilament 'twin peaks' in mitosis and their function in cell plate formation, as visualized in tobacco BY-2 cells expressing GFP–fimbrin. Plant J. **44**, 595–605 (2005)

165. Maisch, J., Nick, P.: Actin is involved in auxin-dependent patterning. Plant Physiol. **143**, 1695–1704 (2007)
166. Traas, J.A., Doonan, J.H., Rawlins, D.J., Shaw, P.J., Watts, J., Lloyd, C.W.: An actin network is present in the cytoplasm throughout the cell cycle of carrot cells and associates with the dividing (1987)
167. Durst, S., Hedde, P.N., Brochhausen, L., Nick, P., Nienhaus, G.U., Maisch, J.: Organization of perinuclear actin in live tobacco cells observed by PALM with optical sectioning. J. Plant Physiol. **141**, 97–108 (2014)
168. Fosket, D.E., Morejohn, L.C.: Structural and functional organization of tubulin. Annu. Rev. Plant. Physiol. Plant. Mol. Biol. **43**, 201–240 (1992)
169. Meagher, R.B., Mckinney, E.C., Vitale, A.V.: The evolution of new structures: clues from plant cytoskeletal genes. Trends Genet. **15**, 278–284 (1999)
170. Meagher, R.B.: Divergence and differential expression of actin gene families in higher plants. Int. Rev. Cytol. **125**, 139–163 (1991)
171. Silflow, C.D., Oppenheimer, D.G., Kopczak, S.D., Ploense, S.E., Ludwig, S.R., Haas, N., Snustad, D.P.: Plant tubulin genes: structure and differential expression during development. Dev. Genet. **8**, 435–460 (1987)
172. Vantard, M., Levilliers, N., Hill, A.M., Adoutte, A., Lambert, A.M.: Incorporation of Paramecium axonemal tubulin into higher plant cells reveals functional sites of microtubule assembly. Proc. Natl. Acad. Sci. U. S. A. **87**, 8825–8829 (1990)
173. Zhang, D., Waldsworth, P., Hepler, P.K.: Microtubule dynamics in living dividing plant cells: confocal imaging of microinjected fluorescent brain tubulin. Proc. Natl. Acad. Sci. U. S. A. **87**, 8820–8824 (1990)
174. Yuan, M., Shaw, P.J., Warn, R.M., Lloyd, C.W.: Dynamic reorientation of cortical microtubules from transverse to longitudinal, in living cells. Proc. Natl. Acad. Sci. U. S. A. **91**, 6050–6053 (1994)
175. Himmelspach, R., Wymer, C.L., Lloyd, C.W., Nick, P.: Gravity-induced reorientation of cortical microtubules observed in vivo. Plant J. **18**, 449–453 (1999)
176. Staiger, C.J., Poulter, N.S., Henty, J.L., Franklin-Tong, V.E., Blanchoin, L.: Regulation of actin dynamics by actin-binding proteins in pollen. J. Exp. Bot. **61**, 1969–1986 (2010)
177. Struk, S., Dhonukshe, P.: MAPs: cellular navigators for microtubule array orientations in Arabidopsis. Plant Cell Rep. **33**, 1–21 (2014)
178. Cai, G., Cresti, M.: Are kinesins required for organelle trafficking in plant cells? Front. Plant Sci. **3**, 170 (2012)
179. Sparkes, I.: Recent advances in understanding plant myosin function: life in the fast lane. Mol. Plant **4**, 805–812 (2011)
180. Geiger, B., Bershadsky, A.: Assembly and mechanosensory function of focal contacts. Curr. Opin. Cell Biol. **13**, 584–592 (2001)
181. Giancotti, G., Ruoslahti, E.: Integrin signalling. Science **285**, 1028–1032 (1999)
182. Canut, H., Carrasco, A., Galaud, J.-P., Cassan, C., Bouyssou, H., Vita, N., Ferrara, P., Pont-Lezica, R.: High affinity RGD-binding sites at the plasma membrane of Arabidopsis thaliana links the cell wall. Plant J. **16**, 63–71 (1998)
183. Wang, X., Zhua, L., Liu, B., Wang, J., Zhao, L., Yuan, M.: Arabidopsis microtubule associated protein 18 functions in directional cell growth by destabilizing cortical microtubules. Plant Cell **19**, 877–839 (2007)
184. Zaban, B., Maisch, J., Nick, P.: Dynamic actin controls polarity induction de novo in protoplasts. J. Int. Plant Biol. **55**, 142–159 (2013)
185. Baluška, F., Šamaj, J., Wojtaszek, P., Volkmann, D., Menzel, D.: Cytoskeleton-plasma membrane-cell wall continuum in plants. Emerging links revisited. Plant Physiol. **133**, 482–419 (2003)
186. Gittes, F., Mickey, B., Nettleton, J., Howard, J.: Flexural rigidity of microtubules and actin filaments measured from thermal fluctuations in shape. J. Cell Biol. **120**, 923–934 (1993)

187. Ingber, D.E.: Tensegrity l: cell structure and hierarchical systems biology. J. Cell Sci. **116**, 1157–1173 (2003)
188. Ingber, D.E.: Tensegrity II: how structural networks influence cellular information processing networks. J. Cell Sci. **116**, 1397–1403 (2003)
189. Green, P.B.: Organogenesis—a biophysical view. Annu. Rev. Plant. Physiol. Plant. Mol. Biol. **3**, 51–82 (1980)
190. Green, P.B.: Mechanism for plant cellular morphogenesis. Science **138**, 1401–1405 (1962)
191. Ledbetter, M.C., Porter, K.R.: A microtubule in plant cell fine structure. J. Cell Biol. **12**, 239–250 (1963)
192. Geitmann, A., Ortega, J.K.: Mechanics and modeling of plant cell growth. Trends Plant Sci. **14**, 467–478 (2009)
193. Nick, P.: Control of cell axis. In: Nick, P. (ed.) Plant Microtubules. Plant cell monogr, vol. 143, pp. 3–46. (2008)
194. Heath, I.B.: A unified hypothesis for the role of membrane bound enzyme complexes and microtubules in plant cell wall synthesis. J. Theor. Biol. **48**, 445–449 (1974)
195. Bichet, A., Desnos, T., Turner, S., Grandjean, O., Höfte, H.: BOTERO1 is required for normal orientation of cortical microtubules and anisotropic cell expansion in Arabidopsis. Plant J. **25**, 137–148 (2001)
196. Zhong, R., Burk, D.H., Morrison, W.H., Ye, Z.H.: A kinesin-like protein is essential for oriented deposition of cellulose microfibrils and cell wall strength. Plant Cell **14**, 3101–3117 (2002)
197. Paredez, A.R., Somerville, C.R., Ehrhardt, D.W.: Visualization of cellulose synthase demonstrates functional association with microtubules. Science **312**, 1491–1495 (2006)
198. Li, S., Lei, L., Somerville, C.R., Gua, Y.: Cellulose synthase interactive protein 1 (CSI1) links microtubules and cellulose synthase complexes. Proc. Natl. Acad. Sci. U. S. A. **109**, 185–190 (2012)
199. Los, D.A., Murata, N.: Membrane fluidity and its roles in the perception of environmental signals. Biochim. Biophys. Acta **1666**, 142–157 (2004)
200. Fischer, K., Schopfer, P.: Physical strain-mediated microtubule reorientation in the epidermis of gravitropically or phototropically stimulated maize coleoptiles. Plant J. **15**, 119–123 (1998)
201. Turing, A.M.: The chemical basis of morphogenesis. Philos. Trans. R. Soc. Lond. B Biol. **237**, 37–72 (1952)
202. Akhmanova, A., Steinmetz, M.O.: Tracking the ends: a dynamic protein network controls the fate of microtubule tips. Nat. Rev. Mol. Cell Biol. **9**, 309–322 (2008)
203. Savage, C., Hamelin, M., Culotti, J.G., Coulson, A., Albertson, D.G., Chalfie, M.: mec-7 is a β-tubulin gene required for the production of 15-protofilament microtubules in Caenorhabditis elegans. Genes Dev. **3**, 870–881 (1989)
204. Ding, J.P., Pickard, B.G.: Mechanosensory calcium-selective cation channels in epidermal cells. Plant J. **3**, 83–110 (1993)
205. Mazars, C., Thion, L., Thuleau, P., Graziana, A., Knight, M.R., Moreau, M., Ranjeva, R.: Organization of cytoskeleton controls the changes in cytosolic calcium of cold-shocked Nicotiana plumbaginifolia protoplasts. Cell Calcium **22**, 413–420 (1997)
206. Komis, G., Apostolakos, P., Galatis, B.: Hyperosmotic stress induces formation of tubulin macrotubules in root-tip cells of Triticum turgidum: their probable involvement in protoplast volume control. Plant Cell Physiol. **43**, 911–922 (2002)
207. Wang, S., Kurepa, J., Hashimoto, T., Smalle, J.A.: Salt stress induced disassembly of Arabidopsis cortical microtubule arrays involves 26S proteasome-dependent degradation of SPIRAL1. Plant Cell **23**, 3412–3427 (2011)
208. Guo, L., Devaiah, S.P., Narasimhan, R., Pan, X., Zhang, Y., Zhang, W., Wang, X.: Cytosolic glyceraldehyde-3-phosphate dehydrogenases interact with phospholipase Dδ to transduce hydrogen peroxide signals in the Arabidopsis response to stress. Plant Cell **24**, 2200–2212 (2012)

209. Nick, P.: Microtubules, and signalling in abiotic stress. Plant J. **75**, 309–323 (2013)
210. Mathur, J., Mathur, N., Hülskamp, M.: Simultaneous visualization of peroxisomes and cytoskeletal elements reveals actin and not microtubule-based peroxisome motility in plants. Plant Physiol. **128**, 1031–1045 (2002)
211. Kadota, A., Yamada, N., Suetsugu, N., Hirose, M., Saito, S., Shoda, K.: Short actin-based mechanism for light-directed chloroplast movement in Arabidopsis. Proc. Natl. Acad. Sci. U. S. A. **106**, 13106–13111 (2009)
212. Van Gestel, K., Kohler, R.H., Verbelen, J.P.: Plant mitochondria move on F-actin, but their positioning in the cortical cytoplasm depends on both F-actin and microtubules. J. Exp. Bot. **53**, 659–667 (2002)
213. Boevink, P., Oparka, K., Santa Cruz, S., Martin, B., Betteridge, A., Hawes, C.: Stacks on tracks: the plant Golgi apparatus traffics on an actin/ER network. Plant J. **15**, 441–447 (2002)
214. Rosazza, C., Escoffre, J.M., Zumbusch, A., Rols, M.P.: The actin cytoskeleton has an active role in the electrotransfer of plasmid DNA in mammalian cells. Mol. Ther. **19**, 913–921 (2011)
215. Sheahan, M.B., Rose, R.J., McCurdy, D.W.: Actin-filament-dependent remodeling of the vacuole in cultured mesophyll protoplasts. Protoplasma **230**, 141–152 (2007)
216. Waller, F., Riemann, M., Nick, P.: A role for actin-driven secretion in auxin-induced growth. Protoplasma **219**, 72–81 (2002)
217. Nick, P.: Probing the actin-auxin oscillator. Plant Signal. Behav. **5**, 4–9 (2010)
218. Nick, P., Han, M., An, G.: Auxin stimulates its own transport by actin reorganization. Plant Physiol. **151**, 155–167 (2009)
219. Jones, J.D., Dangl, J.L.: The plant immune system. Nature **444**, 323–329 (2006)
220. Qiao, F., Chang, X.L., Nick, P.: The cytoskeleton enhances gene expression in the response to the Harpin elicitor in grapevine. J. Exp. Bot. **61**, 4021–4031 (2010)
221. Chang, X., Heene, E., Qiao, F., Nick, P.: The phytoalexin resveratrol regulates the initiation of hypersensitive cell death in Vitis cells. PLoS One **6**, e26405 (2011)
222. Guan, X., Buchholz, G., Nick, P.: The cytoskeleton is disrupted by the bacterial effector HrpZ, but not by the bacterial PAMP flg22, in tobacco BY-2 cells. J. Exp. Bot. **64**, 1805–1816 (2013)
223. Gourlay, C.W., Ayscough, K.R.: The actin cytoskeleton: a key regulator of apoptosis and ageing? Nat. Rev. Mol. Cell Biol. **6**, 583–589 (2005)
224. Franklin-Tong, V.E., Gourlay, C.W.: A role for actin in regulating apoptosis/programmed cell death: evidence spanning yeast, plants and animals. Biochem. J. **413**, 389–404 (2008)
225. Smertenko, A., Franklin-Tong, V.E.: Organisation and regulation of the cytoskeleton in plant programmed cell death. Cell Death Diff. **18**, 1263–1270 (2011)
226. Smertenko, A., Bozhkov, P.: The life and death signalling underlying cell fate determination during somatic embryogenesis. Plant Cell Monogr. **22**, 131–178 (2014)
227. Mathur, J., Radhamony, R., Sinclair, A.M., Donoso, A., Dunn, N., Roach, E.: mEosFP-based green-to-red photoconvertible subcellular probes for plants. Plant Physiol. **154**, 1573–1587 (2010)
228. Opatrný, Z., Nick, P., Petrášek, J.: Plant Cell Strains in Fundamental Research and Applications Plant Cell Monographs, vol. 22, pp. 455–481. Springer, Heidelberg (2014)
229. Berghöfer, T., Eing, C., Flickinger, B., Hohenberger, P., Wegner, L., Frey, W., Nick, P.: Nanosecond electric pulses trigger actin responses in plant cells. Biochem. Biophys. Res. Commun. **387**, 590–595 (2009)
230. Hohenberger, P., Eing, C., Straessner, R., Durst, S., Frey, W., Nick, P.: Plant actin controls membrane permeability. BBA Membr. **1808**, 2304–2312 (2011)
231. Kühn, S., Liu, Q., Eing, C., Wüstner, R., Nick, P.: Nanosecond electric pulses target to a plant-specific kinesin at the plasma membrane. J. Membr. Biol. **246**, 927–938 (2013)
232. Campanoni, P., Blasius, B., Nick, P.: Auxin transport synchronizes the pattern of cell division in a tobacco cell line. Plant Physiol. **133**, 1251–1260 (2003)

233. Beebe, S.J., Fox, P.M., Rec, L.J., Somers, K., Stark, R.H., Schoenbach, K.H.: Nanosecond pulsed electric field (nsPEF) effects on cells and tissues: apoptosis induction and tumor growth inhibition. IEEE Trans. Plasma Sci. **30**, 286–292 (2002)
234. Buescher, E.S., Schoenbach, K.H.: Effects of submicrosecond, high intensity pulsed electric fields on living cells—intracellular electromanipulation. IEEE Trans. Dielectr. Electr. Insul. **10**, 5788–5794 (2003)
235. Chen, C., Smye, S.W., Robinson, M.P., Evans, J.A.: Membrane electroporation theories: a review. Med. Biol. Eng. Comput. **44**, 5–14 (2006)
236. Gowrishankar, T.R., Esser, A.T., Vasilkoski, Z.V., Smith, K.C., Weaver, J.C.: Microdosimetry for conventional and supra-electroporation in cells with organelles. Biochem. Biophys. Res. Commun. **310**, 1266–1276 (2006)
237. Schoenbach, K.H., Joshi, R.P., Chen, C., Kolb, J.F., Chen, N., Stacye, M., Blackmore, P., Buescher, E.S., Beebe, S.J.: Ultrashort electrical pulses open a new gateway into biological cells. Proc. IEEE **92**, 1122–1137 (2004)
238. Beebe, S.J., Fox, P.M., Rec, L.J., Willis, E.L., Schoenbach, K.H.: Nanosecond high-intensity pulsed electric fields induce apoptosis in human cells. FASEB J. **17**, 1493–1495 (2003)
239. Hu, Q., Joshi, P.R., Schoenbach, K.H.: Simulation of nanopore formation and phosphatidylserine externalization in lipid membranes subjected to a high-intensity, ultrashort electric pulses. Phys. Rev. E. **72**, 031902 (2005)
240. Flickinger, B., Berghöfer, T., Hohenberger, P., Eing, C., Frey, W.: Transmembrane potential measurements on plant cells using the voltage-sensitive dye ANNINE-6. Protoplasma **247**, 3–12 (2010)
241. Ibey, B.L., Roth, C.C., Pakhomov, A.G., Bernhard, J.A., Wilmink, G.J., Pakhomova, O.N.: Dose-dependent thresholds of 10-ns electric pulse induced plasma membrane disruption and cytotoxicity in multiple cell lines. PLoS One **6**, e15642 (2011)
242. Breton, M., Delemotte, L., Silve, A., Mir, L.M., Tarek, M.: Transport of siRNA through lipid membranes driven by nanosecond electric pulses: an experimental and computational study. J. Am. Chem. Soc. **134**, 13938–13941 (2012)
243. Preuss, M.L., Kovar, D.R., Lee, Y.R., Staiger, C.J., Delmer, D.P., Liu, B.: A plant-specific kinesin binds to actin microfilaments and interacts with cortical microtubules in cotton fibers. Plant Physiol. **136**, 945–3955 (2004)
244. Huang, B., Babcock, H., Zhuang, X.: Breaking the diffraction barrier: super-resolution imaging of cells. Cell **143**, 1047–1058 (2010)
245. Sparkes, I.A., Graumann, K., Martinière, A., Schoberer, J., Wang, P., Osterrieder, A.: Bleach it, switch it, bounce it, pull it: using lasers to reveal plant cell dynamics. J. Exp. Bot. **62**, 1–7 (2011)
246. Wolfe, J., Dowgert, M.F., Steponkus, P.L.: Dynamics of membrane exchange of the plasma membrane and the lysis of isolated protoplasts during rapid expansions in area. J. Membr. Biol. **86**, 127–138 (1985)
247. Liu, Q., Qiao, F., Ismail, A., Chang, X., Nick, P.: The plant cytoskeleton controls regulatory volume increase. BBA Membr. **1828**, 2111–2120 (2013)
248. Kusaka, N., Maisch, J., Nick, P., Hayashi, K.I., Nozaki, H.: Manipulation of intercellular auxin in a single cell by light with esterase-resistant caged auxins. Chembiochem **10**, 2195–2202 (2009)
249. Durst, S., Nick, P., Maisch, J.: Actin-depolymerizing factor 2 is involved in auxin dependent patterning. J. Plant Physiol. **170**, 1057–1066 (2013)
250. Thompson, G.L., Roth, C., Tolstykh, G., Kuipers, M., Ibey, B.L.: Disruption of the actin cortex contributes to susceptibility of mammalian cells to nanosecond pulsed electric fields. Bioelectromagnetics **35**, 262–272 (2014)
251. Stacey, M., Fox, P., Buescher, S., Kolb, J.: Nanosecond pulsed electric field induced cytoskeleton, nuclear membrane and telomere damage adversely impact cell survival. Bioelectrochemistry **82**, 131–134 (2011)

252. Kanthou, C., Kranjc, S., Sersa, G., Tozer, G., Zupanic, A., Cemazar, M.: The endothelial cytoskeleton as a target of electroporation-based therapies. Mol. Cancer Ther. **5**, 3145–3152 (2006)
253. Teissie, J., Rols, M.P.: Manipulation of cell cytoskeleton affects the lifetime of cell membrane electropermeabilization. Ann. N. Y. Acad. Sci. **720**, 98–110 (1994)
254. Chopinet, L., Batista-Napotnik, T., Montigny, A.: Nanosecond electric pulse effects on gene expression. J. Membr. Biol. **246**, 851–859 (2013)
255. Pakhomov, A.G., Xiao, S., Pakhomova, O.N., Semenov, I., Kuipers, M.A., Ibey, B.L.: Disassembly of actin structures by nanosecond pulsed electric field is a downstream effect of cell swelling. Bioelectrochemistry **100**, 88–95 (2014)
256. Gruenheid, S., Finlay, B.B.: Microbial pathogenesis and cytoskeletal function. Nature **422**, 775–781 (2003)
257. Marsh, M., Bron, R.: SFV infection in CHO cells: cell-type specific restrictions to productive virus entry at the cell surface. J. Cell Sci. **110**(Pt 1), 95–103 (1997)
258. Dauty, E., Verkman, A.S.: Actin cytoskeleton as the principal determinant of size-dependent DNA mobility in cytoplasm: a new barrier for non-viral gene delivery. J. Biol. Chem. **280**, 7823–7828 (2005)
259. Badding, M.A., Lapek, J.D., Friedman, A.E., Dean, D.A.: Proteomic and functional analyses of protein-DNA complexes during gene transfer. Mol. Ther. **21**, 775–785 (2013)
260. Chambers, R.: The micromanipulation of living cells. In: Moulton, F.R. (ed.) The Cell and Protoplasm, pp. 20–30. Science press, Washington, DC (1940)
261. Lukacs, G.L., Haggie, P., Seksek, O., Lechardeur, D., Freedman, N., Verkman, A.S.: Size-dependent DNA mobility in cytoplasm and nucleus. J. Biol. Chem. **275**, 1625–1629 (2000)
262. Seksek, O., Biwersi, J., Verkman, A.S.: Translational diffusion of macromolecule-sized solutes in cytoplasm and nucleus. J. Cell Biol. **138**, 131–142 (1997)
263. Badding, M.A., Dean, D.A.: Highly acetylated tubulin permits enhanced interactions with and trafficking of plasmids along microtubules. Gene Ther. **20**, 616–624 (2013)
264. Badding, M.A., Vaughan, E.E., Dean, D.A.: Transcription factor plasmid binding modulates microtubule interactions and intracellular trafficking during gene transfer. Gene Ther. **19**, 338–346 (2012)
265. Geiger, R.C., Taylor, W., Glucksberg, M.R., Dean, D.A.: Cyclic stretch-induced reorganization of the cytoskeleton and its role in enhanced gene transfer. Gene Ther. **13**, 725–731 (2006)
266. Vaughan, E.E., Dean, D.A.: Intracellular trafficking of plasmids during transfection is mediated by microtubules. Mol. Ther. **13**, 422–428 (2006)
267. Vaughan, E.E., Geiger, R.C., Miller, A.M., Loh-Marley, P.L., Suzuki, T., Miyata, N., Dean, D.A.: Microtubule acetylation through HDAC6 inhibition results in increased transfection efficiency. Mol. Ther. **16**, 1841–1847 (2008)
268. Mesika, A., Kiss, V., Brumfeld, V., Ghosh, G., Reich, Z.: Enhanced intracellular mobility and nuclear accumulation of DNA plasmids associated with a karyophilic protein. Hum. Gene Ther. **16**, 200–208 (2005)
269. Leopold, P.L., Kreitzer, G., Miyazawa, N., Rempel, S., Pfister, K.K., Rodriguez-Boulan, E., Crystal, R.G.: Dynein- and microtubule-mediated translocation of adenovirus serotype 5 occurs after endosomal lysis. Hum. Gene Ther. **11**, 151–165 (2000)
270. Miller, A.M., Munkonge, F.M., Alton, E.W., Dean, D.A.: Identification of protein cofactors necessary for sequence-specific plasmid DNA nuclear import. Mol. Ther. **17**, 1897–1903 (2009)
271. Munkonge, F.M., Amin, V., Hyde, S.C., Green, A.M., Pringle, I.A., Gill, D.R., Smith, J.W., Hooley, R.P., Xenariou, S., Ward, M.A., Leeds, N., Leung, K.Y., Chan, M., Hillery, E., Geddes, D.M., Griesenbach, U., Postel, E.H., Dean, D.A., Dunn, M.J., Alton, E.W.: Identification and functional characterization of cytoplasmic determinants of plasmid DNA nuclear import. J. Biol. Chem. **284**, 26978–26987 (2009)
272. Dean, D.A.: Import of plasmid DNA into the nucleus is sequence specific. Exp. Cell Res. **230**, 293–302 (1997)

273. Dean, D.A., Dean, B.S., Muller, S., Smith, L.C.: Sequence requirements for plasmid nuclear entry. Exp. Cell Res. **253**, 713–722 (1999)
274. Vacik, J., Dean, B.S., Zimmer, W.E., Dean, D.A.: Cell-specific nuclear import of plasmid DNA. Gene Ther. **6**, 1006–1014 (1999)
275. Langle-Rouault, F., Patzel, V., Benavente, A., Taillez, M., Silvestre, N., Bompard, A., Sczakiel, G., Jacobs, E., Rittner, K.: Up to 100-fold increase of apparent gene expression in the presence of Epstein-Barr virus oriP sequences and EBNA1: implications of the nuclear import of plasmids. J. Virol. **72**, 6181–6185 (1998)
276. Mesika, A., Grigoreva, I., Zohar, M., Reich, Z.: A regulated, NFkappaB-assisted import of plasmid DNA into mammalian cell nuclei. Mol. Ther. **3**, 653–657 (2001)
277. Degiulio, J.V., Kaufman, C.D., Dean, D.A.: The SP-C promoter facilitates alveolar type II epithelial cell-specific plasmid nuclear import and gene expression. Gene Ther. **17**, 541–549 (2010)
278. Sacramento, C.B., Moraes, J.Z., Denapolis, P.M., Han, S.W.: Gene expression promoted by the SV40 DNA targeting sequence and the hypoxia-responsive element under normoxia and hypoxia. Braz. J. Med. Biol. Res. **43**, 722–727 (2010)
279. Cramer, F., Christensen, C.L., Poulsen, T.T., Badding, M.A., Dean, D.A., Poulsen, H.S.: Insertion of a nuclear factor kappa B DNA nuclear-targeting sequence potentiates suicide gene therapy efficacy in lung cancer cell lines. Cancer Gene Ther. **19**, 675–683 (2012)
280. Wente, S.R., Rout, M.P.: The nuclear pore complex and nuclear transport. Cold Spring Harb. Perspect. Biol. **2**, a000562 (2010)
281. Wilson, G.L., Dean, B.S., Wang, G., Dean, D.A.: Nuclear import of plasmid DNA in digitonin-permeabilized cells requires both cytoplasmic factors and specific DNA sequences. J. Biol. Chem. **274**, 22025–22032 (1999)
282. Colin, M., Moritz, S., Fontanges, P., Kornprobst, M., Delouis, C., Keller, M., Miller, A.D., Capeau, J., Coutelle, C., Brahimi-Horn, M.C.: The nuclear pore complex is involved in nuclear transfer of plasmid DNA condensed with an oligolysine-RGD peptide containing nuclear localisation properties. Gene Ther. **8**, 1643–1653 (2001)
283. Sebestyén, M.G., Ludtke, J.L., Bassik, M.C., Zhang, G., Budker, V., Lukhtanov, E.A., Hagstrom, J.E., Wolff, J.A.: DNA vector chemistry: the covalent attachment of signal peptides to plasmid DNA. Nat. Biotechnol. **16**, 80–85 (1998)
284. Swindle, C.S., Zou, N., Van Tine, B.A., Shaw, G.M., Engler, J.A., Chow, L.T.: Human papillomavirus DNA replication compartments in a transient DNA replication system. J. Virol. **73**, 1001–1009 (1999)
285. Tang, Q., Bell, P., Tegtmeyer, P., Maul, G.G.: Replication but not transcription of simian virus 40 DNA is dependent on nuclear domain 10. J. Virol. **74**, 9694–9700 (2000)
286. Ondrej, V., Kozubek, S., Lukasova, E., Falk, M., Matula, P., Matula, P., Kozubek, M.: Directional motion of foreign plasmid DNA to nuclear HP1 foci. Chromosome Res. **14**, 505–514 (2006)
287. Gasiorowski, J.Z., Dean, D.A.: Intranuclear trafficking of episomal DNA is transcription-dependent. Mol. Ther. **15**, 2132–2139 (2007)
288. Kopp, K., Gasiorowski, J.Z., Chen, D., Gilmore, R., Norton, J.T., Wang, C., Leary, D.J., Chan, E.K., Dean, D.A., Huang, S.: Pol I transcription and pre-rRNA processing are coordinated in a transcription-dependent manner in mammalian cells. Mol. Biol. Cell **18**, 394–403 (2007)
289. Xu, M., Cook, P.R.: Similar active genes cluster in specialized transcription factories. J. Cell Biol. **181**, 615–623 (2008)
290. Larkin, J.D., Papantonis, A., Cook, P.R.: Promoter type influences transcriptional topography by targeting genes to distinct nucleoplasmic sites. J. Cell Sci. **126**, 2052–2059 (2013)
291. Alberts, B., Johnson, A., Lewis, J., Morgan, D., Raff, M., Roberts, K., Walter, P.: Molecular Biology of the Cell, 6th edn. Garland Publishing, New York (2014)
292. Kyriakis, J.M., Avruch, J.: Mammalian mitogen-activated protein kinase signal transduction pathways activated by stress and inflammation. Physiol. Rev. **81**, 807–869 (2001)

293. Morotomi-Yano, K., Akiyama, H., Yano, K.: Nanosecond pulsed electric fields activate MAPK pathways in human cells. Arch. Biochem. Biophys. **515**, 99–106 (2011)
294. Hardie, D.G.: AMP-activated protein kinase: an energy sensor that regulates all aspects of cell function. Genes Dev. **25**, 1895–1908 (2011)
295. Morotomi-Yano, K., Akiyama, H., Yano, K.: Nanosecond pulsed electric fields activate AMP-activated protein kinase: implications for calcium-mediated activation of cellular signaling. Biochem. Biophys. Res. Commun. **428**, 371–375 (2012)
296. Morotomi-Yano, K., Oyadomari, S., Akiyama, H., Yano, K.: Nanosecond pulsed electric fields act as a novel cellular stress that induces translational suppression accompanied by eIF2alpha phosphorylation and 4E-BP1 dephosphorylation. Exp. Cell Res. **318**, 1733–1744 (2012)
297. Wek, R.C., Jiang, H.Y., Anthony, T.G.: Coping with stress: eIF2 kinases and translational control. Biochem. Soc. Trans. **34**, 7–11 (2006)
298. Mir, L.M., Glass, L.F., Sersa, G., Teissie, J., Domenge, C., Miklavčič, D., Jaroszeski, M.J., Orlowski, S., Reintgen, D.S., Rudolf, Z., et al.: Effective treatment of cutaneous and subcutaneous malignant tumours by electrochemotherapy. Br. J. Cancer **77**(12), 2336–2342 (1998)
299. Mir, L.M., Bureau, M.F., Gehl, J., Rangara, R., Rouy, D., Caillaud, J.-M., Delaere, P., Branellec, D., Schwartz, B., Scherman, D.: High-efficiency gene transfer into skeletal muscle mediated by electric pulses. Proc. Natl. Acad. Sci. **96**(8), 4262–4267 (1999)
300. Larkin, J.O., Collins, C.G., Aarons, S., Tangney, M., Whelan, M., O'Reily, S., Breathnach, O., Soden, D.M., O'Sullivan, G.C.: Electrochemotherapy: aspects of preclinical development and early clinical experience. Ann. Surg. **245**(3), 469 (2007)
301. Davalos, R.V., Mir, L.M., Rubinsky, B.: Tissue ablation with irreversible electroporation. Ann. Biomed. Eng. **33**(2), 223–231 (2005)
302. Edd, J.F., Horowitz, L., Davalos, R.V., Mir, L.M., Rubinsky, B.: In vivo results of a new focal tissue ablation technique: irreversible electroporation. IEEE Trans. Biomed. Eng. **53**(7), 1409–1415 (2006)
303. Al-Sakere, B., Andre, F., Bernat, C., Connault, E., Opolon, P., Davalos, R.V., Rubinsky, B., Mir, L.M.: Tumor ablation with irreversible electroporation. PLoS One **2**(11), e1135 (2007)
304. Pucihar, G., Krmelj, J., Rebersek, M., Napotnik, T.B., Miklavčič, D.: Equivalent pulse parameters for electroporation. IEEE Trans. Biomed. Eng. **58**(11), 3279–3288 (2011). doi: 3210.1109/TBME.2011.2167232. Epub 2162011 Sep 2167236
305. Rubinsky, J., Onik, G., Mikus, P., Rubinsky, B.: Optimal parameters for the destruction of prostate cancer using irreversible electroporation. J. Urol. **180**(6), 2668–2674 (2008). Epub 2008 Oct 2631
306. Jiang, C., Shao, Q., Bischof, J.: Pulse timing during irreversible electroporation achieves enhanced destruction in a hindlimb model of cancer. Ann. Biomed. Eng. **1**, 1 (2014)
307. Appelbaum, L., Ben-David, E., Faroja, M., Nissenbaum, Y., Sosna, J., Goldberg, S.N.: Irreversible electroporation ablation: creation of large-volume ablation zones in in vivo porcine liver with four-electrode arrays. Radiology **270**(2), 416–424 (2014)
308. Diller, K.R.: Modeling of bioheat transfer processes at high and low temperatures. Adv. Heat Transf. **22**, 157–357 (1992)
309. Davalos, R.V., Rubinsky, B.: Temperature considerations during irreversible electroporation. Int. J. Heat Mass Transf. **51**(23), 5617–5622 (2008)
310. Thomsen, S., Pearce. J.A.: Thermal damage and rate processes in biologic tissues. In: Optical-Thermal Response of Laser-Irradiated Tissue, pp. 487–549. Springer, Dordrecht (2011)
311. Davalos, R.V., Rubinsky, B., Mir, L.M.: Theoretical analysis of the thermal effects during in vivo tissue electroporation. Bioelectrochemistry **61**(1–2), 99–107 (2003)
312. Neal II, R.E., Davalos, R.V.: The feasibility of irreversible electroporation for the treatment of breast cancer and other heterogeneous systems. Ann. Biomed. Eng. **37**(12), 2615–2625 (2009)

313. Županič, A., Miklavčič, D.: Tissue heating during tumor ablation with irreversible electroporation. Electrotechnol. Rev. **78**, 42–47 (2011)
314. Garcia, P.A., Rossmeisl, J.H., Neal, R.E., Ellis, T.L., Davalos, R.V.: A parametric study delineating irreversible electroporation from thermal damage based on a minimally invasive intracranial procedure. Biomed. Eng. Online **10**(1), 34 (2011)
315. Shafiee, H., Garcia, P.A., Davalos, R.V.: A preliminary study to delineate irreversible electroporation from thermal damage using the arrhenius equation. J. Biomech. Eng. **131** (7), 074509 (2009)
316. Garcia, P.A., Davalos, R.V., Miklavčič, D.: A numerical investigation of the electric and thermal cell kill distributions in electroporation-based therapies in tissue. PLoS One **9**(8), e103083 (2014)
317. Qin, Z., Jiang, J., Long, G., Lindgren, B., Bischof, J.C.: Irreversible electroporation: an in vivo study with dorsal skin fold chamber. Ann. Biomed. Eng. **41**(3), 619–629 (2013)
318. Garcia, P., Rossmeisl, J., Neal, R., Ellis, T., Olson, J., Henao-Guerrero, N., Robertson, J., Davalos, R.: Intracranial nonthermal irreversible electroporation: in vivo analysis. J. Membr. Biol. **236**(1), 127–136 (2010)
319. Long, G., Bakos, G., Shires, P.K., Gritter, L., Crissman, J.W., Harris, J.L., Clymer, J.W.: Histological and finite element analysis of cell death due to irreversible electroporation. Technol. Cancer Res. Treat. **13**(6), 561–569 (2014)
320. Neal 2nd, R.E., Rossmeisl Jr., J.H., Garcia, P.A., Lanz, O.I., Henao-Guerrero, N., Davalos, R. V.: Successful treatment of a large soft tissue sarcoma with irreversible electroporation. J. Clin. Oncol. Off. J. Am. Soc. Clin. Oncol. **29**(13), e372–e377 (2011)
321. Bower, M., Sherwood, L., Li, Y., Martin, R.: Irreversible electroporation of the pancreas: definitive local therapy without systemic effects. J. Surg. Oncol. **104**(1), 22–28 (2011)
322. Neal 2nd, R.E., Millar, J.L., Kavnoudias, H., Royce, P., Rosenfeldt, F., Pham, A., Smith, R., Davalos, R.V., Thomson, K.R.: In vivo characterization and numerical simulation of prostate properties for non-thermal irreversible electroporation ablation. Prostate **74**(5), 458–468 (2014)
323. Edd, J.F., Davalos, R.V.: Mathematical modeling of irreversible electroporation for treatment planning. Technol. Cancer Res. Treat. **6**(4), 275–286 (2007)
324. Ben-David, E., Ahmed, M., Faroja, M., Moussa, M., Wandel, A., Sosna, J., Appelbaum, L., Nissenbaum, I., Goldberg, S.N.: Irreversible electroporation: treatment effect is susceptible to local environment and tissue properties. Radiology **11**, 11 (2013)
325. Neal II, R.E., Singh, R., Hatcher, H.C., Kock, N.D., Torti, S.V., Davalos, R.V.: Treatment of breast cancer through the application of irreversible electroporation using a novel minimally invasive single needle electrode. Breast Cancer Res. Treat. **123**(1), 295–301 (2010)
326. Corovic, S., Zupanic, A., Kranjc, S., Al Sakere, B., Leroy-Willig, A., Mir, L.M., Miklavčič, D.: The influence of skeletal muscle anisotropy on electroporation: in vivo study and numerical modeling. Med. Biol. Eng. Comput. **48**(7), 637–648 (2010)
327. Neal II, R., Smith, R., Kavnoudias, H., Rosenfeldt, F., Ou, R., Mclean, C., Davalos, R., Thomson, K.: The effects of metallic implants on electroporation therapies: feasibility of irreversible electroporation for brachytherapy salvage. Cardiovasc. Interv. Radiol. **36**(6), 1638–1645 (2013)
328. Duck, F.A.: Physical Properties of Tissue: A Comprehensive Reference Book. Academic, New York (1990)
329. Neal II, R.E., Garcia, P.A., Robertson, J.L., Davalos, R.V.: Experimental characterization and numerical modeling of tissue electrical conductivity during pulsed electric fields for irreversible electroporation treatment planning. IEEE Trans. Biomed. Eng. **59**(4), 1076–1085 (2012). Epub 2012 Jan 1076
330. Osswald, K.: Messung der Leitfähigkeit und Dielektrizitätskonstante biologischer Gewebe und Flüssigkeiten bei kurzen Wellen. Akad. Verlagsges. (1937)
331. Mcrae, D.A., Esrick, M.A.: The dielectric parameters of excised EMT-6 tumours and their change during hyperthermia. Phys. Med. Biol. **37**(11), 2045 (1992)

332. Sel, D., Cukjati, D., Batiuskaite, D., Slivnik, T., Mir, L.M., Miklavčič, D.: Sequential finite element model of tissue electropermeabilization. IEEE Trans. Biomed. Eng. **52**(5), 816–827 (2005)
333. Ivorra, A., Rubinsky, B.: In vivo electrical impedance measurements during and after electroporation of rat liver. Bioelectrochemistry **70**(2), 287–295 (2007)
334. Pavšelj, N., Miklavčič, D.: Numerical modeling in electroporation-based biomedical applications. Radiol. Oncol. **42**(3), 159–168 (2008)
335. Ivorra, A., Mir, L.M., Rubinsky, B.: Electric field redistribution due to conductivity Changes during tissue electroporation: experiments with a simple vegetal model. In: IFMBE Proceedings, vol. 25/13, edn, pp. 59–62. Springer, Berlin/Heidelberg (2009)
336. Corovic, S., Lackovic, I., Sustaric, P., Sustar, T., Rodic, T., Miklavčič, D.: Modeling of electric field distribution in tissues during electroporation. Biomed. Eng. Online **12**(1), 16 (2013)
337. Neal, R.E., Garcia, P.A., Kavnoudias, H., Rosenfeldt, F., Mclean, C.A., Earl, V., Bergman, J., Davalos, R.V., Thomson, K.R.: In vivo irreversible electroporation kidney ablation: experimentally correlated numerical models. IEEE Trans. Biomed. Eng. **62**(2), 561–569 (2015). doi: 510.1109/TBME.2014.2360374. Epub 2362014 Sep 2360325
338. Gaylor, D.C., Prakah-Asante, K., Lee, R.C.: Significance of cell size and tissue structure in electrical trauma. J. Theor. Biol. **133**(2), 223–237 (1988)
339. Neal II, R.E., Rossmeisl Jr., J.H., Garcia, P.A., Lanz, O.I., Henao-Guerrero, N., Davalos, R. V.: Successful treatment of a large soft tissue sarcoma with irreversible electroporation. J. Clin. Oncol. **29**(13), e372–e377 (2011)
340. Wimmer, T., Srimathveeravalli, G., Gutta, N., Ezell, P.C., Monette, S., Maybody, M., Erinjery, J.P., Durack, J.C., Coleman, J.A., Solomon, S.B.: Planning irreversible electroporation in the porcine kidney: are numerical simulations reliable for predicting empiric ablation outcomes? Cardiovasc. Intervent. Radiol. **38**(1), 182–190 (2015)
341. Miklavčič, D., Semrov, D., Mekid, H., Mir, L.M.: A validated model of in vivo electric field distribution in tissues for electrochemotherapy and for DNA electrotransfer for gene therapy. Biochim. Biophys. Acta **1523**(1), 73–83 (2000)
342. Wimmer, T., Srimathveeravalli, G., Gutta, N., Ezell, P.C., Monette, S., Kingham, T.P., Maybody, M., Durack, J.C., Fong, Y., Solomon, S.B.: Comparison of simulation-based treatment planning with imaging and pathology outcomes for percutaneous CT-guided irreversible electroporation of the porcine pancreas: a pilot study. J. Vasc. Interv. Radiol. **24**(11), 1709–1718 (2013)
343. Majno, G., Joris, I.: Apoptosis, oncosis, and necrosis: an overview of cell death. Am. J. Pathol. **146**, 3–15 (1995)
344. Galluzzi, L., Bravo-San Pedro, J.M., Vitale, I., Aaronson, S.A., Abrams, J.M., Adam, D., et al. (2014a) Essential versus accessory aspects of cell death: recommendations of the NCCD 2015. Cell Death Differ. 2014 Sept 19. doi: 10.1038/cdd.2014.137
345. Galluzzi, L., Kepp, O., Krautwald, S., Kroemer, G., Linkermann, A.: Molecular mechanisms of regulated necrosis. Semin. Cell Dev. Biol. **35**, 24–32 (2014)
346. Kerr, J.F., Wyllie, A.H., Currie, A.R.: Apoptosis: a basic biological phenomenon with wide-ranging implications in tissue kinetics. Br. J. Cancer **26**, 239–257 (1972)
347. Degterev, A., Yuan, J.: Expansion and evolution of cell death programmes. Nat. Rev. Mol. Cell Biol. **9**, 378–390 (2008)
348. Arvanitis, M., Li, D.D., Lee, K., Mylonakis, E.: Apoptosis in C. elegans: lessons for cancer and immunity. Front Cell Infect. Microbiol. **3**, 67 (2013)
349. Galluzzi, L., Vitale, I., Abrams, J.M., Alnemri, E.S., Baehrecke, E.H., Blagosklonny, M.V., et al.: Molecular definitions of cell death subroutines: recommendations of the Nomenclature Committee on Cell Death 2012. Cell Death Differ. **19**, 107–120 (2012)
350. Yan, Q., Liu, J.P., Li, D.W.: Apoptosis in lens development and pathology. Differentiation **74**, 195–211 (2006)

351. Taylor, E.L., Rossi, A.G., Dransfield, I., Hart, S.P.: Analysis of neutrophil apoptosis. Methods Mol. Biol. **412**, 177–200 (2007)
352. Kroemer, G., El-Deiry, W.S., Golstein, P., Peter, M.E., Vaux, D., Vandenabeele, P., et al.: Classification of cell death: recommendations of the Nomenclature Committee on Cell Death. Cell Death Differ. **2**, 1463–1467 (2005)
353. Kroemer, G., Galluzzi, L., Vandenabeele, P., Abrams, J., Alnemri, E.S., Baehrecke, E.H., et al.: Classification of cell death: recommendations of the Nomenclature Committee on Cell Death 2009. Cell Death Differ. **16**, 3–11 (2009)
354. Galluzzi, L., Maiuri, M.C., Vitale, I., Zischka, H., Castedo, M., Zitvogel, L., Kroemer, G.: Cell death modalities: classification and pathophysiological implications. Cell Death Differ. **14**, 1237–1243 (2007)
355. Fulda, S., Gorman, A.M., Hori, O., Samali, A.: Cellular stress responses: cell survival and cell death. Int. J. Cell Biol. **2010**, 214074 (2010)
356. Clarke, P.G.: Developmental cell death: morphological diversity and multiple mechanisms. Anat. Embryol. (Berl.) **181**, 195–213 (1990)
357. Kroemer, G., Mariño, G., Levine, B.: Autophagy and the integrated stress response. Mol. Cell **40**, 280–293 (2010)
358. Eisenberg-Lerner, A., Bialik, S., Simon, H.U., Kimchi, A.: Life and death partners: apoptosis, autophagy and the cross-talk between them. Cell Death Differ. **16**, 966–975 (2009)
359. Ellis, H.M., Horvitz, H.R.: Genetic control of programmed cell death in the nematode C. elegans. Cell **44**, 817–829 (1986)
360. Scaffidi, C., Fulda, S., Srinivasan, A., Friesen, C., Li, F., Tomaselli, K.J., Debatin, K.M., Krammer, P.H., Peter, M.E.: Two CD95 (APO-1/Fas) signaling pathways. EMBO J. **17**, 1675–1687 (1998)
361. Muppidi, J.R., Siegel, R.M.: Ligand-independent redistribution of Fas (CD95) into lipid rafts mediates clonotypic T cell death. Nat. Immunol. **5**, 182–189 (2004)
362. Hengartner, M.O.: The biochemistry of apoptosis. Nature **407**, 770–776 (2000)
363. Jost, P.J., Grabow, S., Gray, D., McKenzie, M.D., Nachbur, U., Huang, D.C., Bouillet, P., Thomas, H.E., Borner, C., Silke, J., Strasser, A., Kaufmann, T.: XIAP discriminates between type I and type II FAS-induced apoptosis. Nature **460**, 1035–1039 (2009)
364. Willis, S.N., Adams, J.M.: Life in the balance: how BH3-only proteins induce apoptosis. Curr. Opin. Cell Biol. **17**, 617–625 (2005)
365. Shamas-Din, A., Brahmbhatt, H., Leber, B., Andrews, D.W.: BH3-only proteins: orchestrators of apoptosis. Biochim. Biophys. Acta **1813**, 508–520 (2011)
366. Tourneur, L., Chiocchia, G.: FADD: a regulator of life and death. Trends Immunol. **31**, 260–269 (2010)
367. Festjens, N., Vanden Berghe, T., Vandenabeele, P.: Necrosis, a well-orchestrated form of cell demise: signalling cascades, important mediators and concomitant immune response. Biochim. Biophys. Acta **1757**, 1371–1387 (2006)
368. Sosna, J., Voigt, S., Mathieu, S., Lange, A., Thon, L., Davarnia, P., Herdegen, T., Linkermann, A., Rittger, A., Chan, F.K., Kabelitz, D., Schütze, S., Adam, D.: TNF-induced necroptosis and PARP-1-mediated necrosis represent distinct routes to programmed necrotic cell death. Cell. Mol. Life Sci. **71**, 331–348 (2014)
369. Vandenabeele, P., Galluzzi, L., Vanden Berghe, T., Kroemer, G.: Molecular mechanisms of necroptosis: an ordered cellular explosion. Nat. Rev. Mol. Cell Biol. **11**, 700–714 (2010)
370. Schoenbach, K.H., Joshi, R.P., Stark, R.H., Dobbs, F., Beebe, S.J.: Bacterial decontamination of liquids with pulsed electric fields. IEEE Trans. Dielectr. Electr. Insul. **7**, 637–645 (2000)
371. Beebe, S.J., White, J.A., Blackmore, P.F., Deng, Y., Somers, K., Schoenbach, K.H.: Diverse effects of nanosecond pulsed electric fields on cells and tissues. DNA Cell Biol. **22**, 785–796 (2003)
372. Vernier, P.T., Aimin, L., Marcu, L., Craft, C.M., Gundersen, M.A.: Ultrashort pulsed electric fields induce membrane phospholipid translocation and caspase activation: differential

sensitivities of Jurkat T lymphoblasts and rat glioma C6 cells. IEEE Trans. Dielectr. Electr. Insul. **10**, 795–809 (2003)
373. Hall, E.H., Schoenbach, K.H., Beebe, S.J.: Nanosecond pulsed electric fields induce apoptosis in p53-wildtype and p53-null HCT116 colon carcinoma cells. Apoptosis **12**, 1721–1731 (2007)
374. Ford, W.E., Ren, W., Blackmore, P.F., Schoenbach, K.H., Beebe, S.J.: Nanosecond pulsed electric fields stimulate apoptosis without release of pro-apoptotic factors from mitochondria in B16f10 melanoma. Arch. Biochem. Biophys. **497**, 82–89 (2010)
375. Ren, W., Beebe, S.J.: An apoptosis targeted stimulus with nanosecond pulsed electric fields (nsPEFs) in E4 squamous cell carcinoma. Apoptosis **16**, 382–393 (2011)
376. Chen, X., Yin, S., Hu, C., Chen, X., Jiang, K., Ye, S., Feng, X., Fan, S., Xie, H., Zhou, L., Zheng, S.: Comparative study of nanosecond electric fields in vitro and in vivo on hepatocellular carcinoma indicate macrophage infiltration contribute to tumor ablation in vivo. PLoS One **9**, e86421 (2014)
377. Fadeel, B., Xue, D.: The ins and outs of phospholipid asymmetry in the plasma membrane: roles in health and disease. Crit. Rev. Biochem. Mol. Biol. **44**, 264–277 (2009)
378. Segawa, K., Kurata, S., Yanagihashi, Y., Brummelkamp, T.R., Matsuda, F., Nagata, S.: Caspase-mediated cleavage of phospholipid flippase for apoptotic phosphatidylserine exposure. Science **344**, 1164–1168 (2014)
379. Suzuki, J., Denning, D.P., Imanishi, E., Horvitz, H.R., Nagata, S.: Xk-related protein 8 and CED-8 promote phosphatidylserine exposure in apoptotic cells. Science **341**, 403–406 (2013)
380. Vernier, P.T., Ziegler, M.J., Sun, Y., Gundersen, M.A., Tieleman, D.P.: Nanopore-facilitated, voltage-driven phosphatidylserine translocation in lipid bilayers—in cells and in silico. Phys. Biol. **3**, 233–2147 (2006)
381. Tekle, E., Wolfe, M.D., Oubrahim, H., Chock, P.B.: Phagocytic clearance of electric field induced 'apoptosis-mimetic' cells. Biochem. Biophys. Res. Commun. **376**, 256–260 (2008)
382. Wang, Y., Dawson, V.L., Dawson, T.M.: Poly(ADP-ribose) signals to mitochondrial AIF: a key event in parthanatos. Exp. Neurol. **218**, 193–202 (2009)
383. Charriaut-Marlangue, C., Ben-Ari, Y.: A cautionary note on the use of the TUNEL stain to determine apoptosis. Neuroreport **7**, 61–64 (1995)
384. de Torres, C., Munell, F., Ferrer, I., Reventós, J., Macaya, A.: Identification of necrotic cell death by the TUNEL assay in the hypoxic-ischemic neonatal rat brain. Neurosci. Lett. **230**, 1–4 (1997)
385. Kanoh, M., Takemura, G., Misao, J., Hayakawa, Y., Aoyama, T., Nishigaki, K., Noda, T., Fujiwara, T., Fukuda, K., Minatoguchi, S., Fujiwara, H.: Significance of myocytes with positive DNA in situ nick end-labeling (TUNEL) in hearts with dilated cardiomyopathy: not apoptosis but DNA repair. Circulation **99**, 2757–2764 (1999)
386. Grasl-Kraupp, B., Ruttkay-Nedecky, B., Koudelka, H., Bukowska, K., Bursch, W., Schulte-Hermann, R.: In situ detection of fragmented DNA (TUNEL assay) fails to discriminate among apoptosis, necrosis, and autolytic cell death: a cautionary note. Hepatology **21**, 1465–1468 (1995)
387. Smith, M.A., Schnellmann, R.G.: Calpains, mitochondria, and apoptosis. Cardiovasc. Res. **96**, 32–37 (2012)
388. Song, J., Joshi, R.P., Beebe, S.J.: Cellular apoptosis by nanosecond, high-intensity electric pulses: model evaluation of the pulsing threshold and extrinsic pathway. Bioelectrochemistry **79**, 179–186 (2010)
389. Estlack, L.E., Roth, C.C., Thompson 3rd, G.L., Lambert 3rd, W.A., Ibey, B.L.: Nanosecond pulsed electric fields modulate the expression of Fas/CD95 death receptor pathway regulators in U937 and Jurkat Cells. Apoptosis **19**, 1755–1768 (2014)
390. Beebe, S.J., Sain, N.M., Ren, W.: Induction of cell death mechanisms and apoptosis by nanosecond pulsed electric fields (nsPEFs). Cells **2**, 136–162 (2013)

391. Morotomi-Yano, K., Akiyama, H., Yano, K.: Different involvement of extracellular calcium in two modes of cell death induced by nanosecond pulsed electric fields. Arch. Biochem. Biophys. **555–556**, 47–54 (2014)
392. Ibey, B.L., Ullery, J.C., Pakhomova, O.N., Roth, C.C., Semenov, I., Beier, H.T., Tarango, M., Xiao, S., Schoenbach, K.H., Pakhomov, A.G.: Bipolar nanosecond electric pulses are less efficient at electropermeabilization and killing cells than monopolar pulses. Biochem. Biophys. Res. Commun. **443**, 568–573 (2014)
393. Kotnik, T., Mir, L.M., Flisar, K., Puc, M., Miklavčič, D.: Cell membrane electropermeabilization by symmetrical bipolar rectangular pulses. Part I. Increased efficiency of permeabilization. Bioelectrochemistry **54**, 83–90 (2001)
394. Tekle, E., Astumian, R.D., Chock, P.B.: Electroporation by using bipolar oscillating electric field: an improved method for DNA transfection of NIH 3T3 cells. Proc. Natl. Acad. Sci. U. S. A. **88**, 4230–4234 (1991)
395. Jacobson, J., Duchen, M.R.: Mitochondrial oxidative stress and cell death in astrocytes—requirement for stored Ca^{2+} and sustained opening of the permeability transition pore. J. Cell Sci. **115**, 1175–1188 (2002)
396. Brookes, P.S., Yoon, Y., Robotham, J.L., Anders, M.W., Sheu, S.S.: Ca^{2+}, ATP, and ROS: a mitochondrial love-hate triangle. Am. J. Physiol. Cell Physiol. **287**, C817–C833 (2004)
397. Pakhomova, O.N., Khorokhorina, V.A., Bowman, A.M., Rodaitė-Riševičienė, R., Saulis, G., Xiao, S., Pakhomov, A.G.: Oxidative effects of nanosecond pulsed electric field exposure in cells and cell-free media. Arch. Biochem. Biophys. **527**, 55–64 (2012)
398. Sato, T., Machida, T., Takahashi, S., Iyama, S., Sato, Y., Kuribayashi, K., Takada, K., Oku, T., Kawano, Y., Okamoto, T., Takimoto, R., Matsunaga, T., Takayama, T., Takahashi, M., Kato, J., Niitsu, Y.: Fas-mediated apoptosome formation is dependent on reactive oxygen species derived from mitochondrial permeability transition in Jurkat cells. J. Immunol. **173**, 285–296 (2004)
399. Stacey, M., Stickley, J., Fox, P., Statler, V., Schoenbach, K., Beebe, S.J., Buescher, S.: Differential effects in cells exposed to ultra-short, high intensity electric fields: cell survival, DNA damage, and cell cycle analysis. Mutat. Res. **542**, 65–75 (2003)
400. Romeo, S., Zeni, L., Sarti, M., Sannino, A., Scarfı', M.R., et al.: DNA electrophoretic migration patterns change after exposure of jurkat cells to a single intense nanosecond electric pulse. PLoS One **6**, e28419 (2011)
401. Nuccitelli, R., Pliquett, U., Chen, X., Ford, W., James Swanson, R., Beebe, S.J., Kolb, J.F., Schoenbach, K.H.: Nanosecond pulsed electric fields cause melanomas to self-destruct. Biochem. Biophys. Res. Commun. **343**, 351–360 (2006)
402. Nuccitelli, R., Chen, X., Pakhomov, A.G., Baldwin, W.H., Sheikh, S., Pomicter, J.L., Ren, W., Osgood, C., Swanson, R.J., Kolb, J.F., Beebe, S.J., Schoenbach, K.H.: A new pulsed electric field therapy for melanoma disrupts the tumor's blood supply and causes complete remission without recurrence. Int. J. Cancer **125**, 438–445 (2009)
403. Chen, X., Zhuang, J., Kolb, J.F., Schoenbach, K.H., Beebe, S.J.: Long term survival of mice with hepatocellular carcinoma after pulse power ablation with nanosecond pulsed electric fields. Technol. Cancer Res. Treat. **11**, 83–93 (2012)
404. Chen, R., Sain, N.M., Harlow, K.T., Chen, Y.J., Shires, P.K., Heller, R., Beebe, S.J.: A protective effect after clearance of orthotopic rat hepatocellular carcinoma by nanosecond pulsed electric fields. Eur. J. Cancer **50**, 2705–2713 (2014)
405. Nuccitelli, R., Tran, K., Lui, K., Huynh, J., Athos, B., Kreis, M., Nuccitelli, P., De Fabo, E.C.: Pigment Cell Melanoma Res. **25**, 618–629 (2012)
406. Nuccitelli, R., Wood, R., Kreis, M., Athos, B., Huynh, J., Lui, K., Nuccitelli, P., Epstein Jr., E.H.: First-in-human trial of nanoelectroablation therapy for basal cell carcinoma: proof of method. Exp. Dermatol. **23**, 135–137 (2014)

Chapter 5
Medical Applications

Richard Heller, Justin Teissie, Marie-Pierre Rols, Julie Gehl, Gregor Sersa, Lluis M. Mir, Robert E. Neal II, Suyashree Bhonsle, Rafael Davalos, Stephen Beebe, Barbara Hargrave, Richard Nuccitelli, Chunqi Jiang, Maja Cemazar, Youssef Tamzali, and Natasa Tozon

Abstract Bioelectrics is a rapidly growing field at the intersection of the biological and physical sciences. Research has focused on understanding the basic interactions of pulse electric fields on biological systems, development of therapeutic and diagnostic approaches, and environmental applications. This chapter focuses on potential therapeutic and prophylactic applications of bioelectrics in human and veterinary medicine. Pulse electric fields can be ultrashort (picosecond to

R. Heller (✉) • S. Beebe • B. Hargrave • C. Jiang
Frank Reidy Research Center for Bioelectrics, Old Dominion University, Norfolk, VA, USA
e-mail: rheller@odu.edu

J. Teissie • M.-P. Rols
Institute of Pharmacology and Structural Biology, CNRS and University of Toulouse, Toulouse, France

J. Gehl
Center for Experimental Drug and Gene Electrotransfer, Copenhagen University Hospital, Herlev, Denmark

G. Sersa
Department of Experimental Oncology, Institute of Oncology Ljubljana, Ljubljana, Slovenia

L.M. Mir
Laboratory of Vectorology and Anticancer Therapeutics, CNRS and University of Paris-Sud, Villejuif, France

R.E. Neal II • S. Bhonsle • R. Davalos
Bioelectromechanical Systems Laboratory, Virginia Tech, Blacksburg, VA, USA

R. Nuccitelli
Pulse Biosciences, Inc., Burlingame, CA, USA

M. Cemazar
Department of Experimental Oncology, Institute of Oncology Ljubljana, Ljubljana, Slovenia

Faculty of Health Sciences, University of Primorska, Izola, Slovenia

Y. Tamzali
École Nationale Vétérinaire de Toulouse, Toulouse, France

N. Tozon
Veterinary Faculty, University of Ljubljana, Ljubljana, Slovenia

H. Akiyama, R. Heller (eds.), *Bioelectrics*, DOI 10.1007/978-4-431-56095-1_5

nanosecond) or longer (microsecond to millisecond), and dependent on the parameters of the applied pulse, the cellular effects on cells can be direct or indirect. Direct effect can include exciting cells or induction of apoptosis and/or necrosis. Indirect effects have been used to deliver molecules such as chemotherapeutics, nucleic acids, or protein into the cell interior. Therapeutic applications of bioelectrics have been developed for a large number of indications including many medical conditions including cancer, wound healing, ischemia, cardiovascular, and diabetes. The utilization of bioelectric-based medical techniques has recently surged, and the potential of such techniques has stemmed from a more wholistic understanding of the fundamental biophysical mechanisms driving underlying success.

Keywords Electrochemotherapy • Gene electrotransfer • Nanosecond pulse electric fields • Plasma medicine • Veterinary • Clinical trials • Wound healing • Cancer therapy

5.1 Overview

Richard Heller

5.1.1 Introduction

In the 1960s and 1970s, multiple studies evaluated the effects of electric fields on biological cells [1–9]. While this early work focused mainly on the effects on the membrane of isolated cells, it formed the basis for the more applied studies that followed. A more detailed description is provided in Sect. 1.2 of this textbook. Since the 1980s, significant amount of research has been focused on developing applications for pulsed electric fields and their effects on biological systems. This chapter will focus on medical applications. Medical applications related to bioelectrics can be broadly divided into two basic categories based on their mechanism of action: direct and indirect. Direct effects have typically included activating cells or inducing cell death. Indirect applications include delivering molecules such as plasmid DNA, chemotherapeutics, proteins, RNA, or other molecules into cells.

5.1.2 Direct Effects

Pulse electric fields can be utilized to manipulate cells. The three prototypical approaches utilize nanosecond pulse electric fields (nsPEF), microsecond pulse electric fields (μsPEF), or nonthermal plasma. Although some of these approaches have some overlap in their medical applications, there are differences in how each interacts with cells. Clinical testing has been initiated in some of these applications.

5.1.2.1 Nanosecond Pulse Electric Fields

Nanosecond pulse electric fields were originally evaluated for their potential to destroy cancer cells. Schoenbach, Beebe, and Buescher [10] reported in 2001 that nsPEF could influence intracellular function in mammalian cells. A year later, they reported that the application of nsPEF (75 kV/cm, 300 ns) could reduce the growth of murine fibrosarcoma allografts [11]. Since then the field has seen tremendous growth. Results from the clinical use of this approach were recently reported by Nuccitelli et al. in 2014. It was observed that exposing basal cell carcinomas to 100–1000 pulses of 100 ns in length and at an electric field intensity of 30 kV/cm resulted in complete ablation of seven of the ten lesions treated [12]. It is also interesting to note that focal ablation with nsPEF can be accomplished without resulting in a scar at the treatment site. Additional studies have been conducted in animal models to demonstrate that nsPEF could also be effectively employed to destroy tumors originating in internal organs. The approach has been successfully tested in models of hepatocellular carcinoma and pancreatic cancer [13–16].

There are many reports in the literature that suggest that nsPEF can be used to ablate or remove cells or tissues for a variety of applications including inactivating or destroying microorganisms [11, 17]. In addition, to cell destruction, nsPEF can be used to activate or electrically stimulate cells. Cardiomyocytes were excited following exposure to nsPEF through a mechanism that probably involved poration. The excitation was not seen in the absence of extracellular calcium [18]. Platelets can also be activated following exposure to nsPEF. Platelet-rich plasma or platelet gel can be used for a variety of applications including wound healing, bacterial decontamination, and as a potential treatment for myocardial infarcts. Platelet gel created following exposure to nsPEF was used to reduce the size of a myocardial infarct in an animal model and to improve left ventricular function [19, 20]. A study was performed that demonstrated platelet gel produced with nsPEF could reduce the number of microorganisms that are typically associated with wound infections [21]. In another study, platelet gel produced with nsPEF was able to increase blood flow to ischemic tissue [22]. Subnanosecond pulses could potentially be used for imaging malignancies [23] or for stimulating cells [24]. Nerve conduction could also be blocked utilizing high intensity nsPEF as a potential approach for controlling pain [25]. More detailed information on the use of submicrosecond pulse electric fields can be found in Sects. 5.6 and 5.7.

5.1.2.2 Microsecond Pulse Electric Fields

Cell ablation can also be accomplished using irreversible electroporation (IRE) which typically uses longer pulses in the range of microseconds. IRE was first reported in 2005 as a means to perform tissue ablation [26]. The mechanism of cell death induced by IRE is through disruption of the cell membrane. This can be accomplished without protein coagulation or causing significant alterations of collagen or other extracellular tissue components. Since there is minimal disruption to these critical stromal components, IRE can be performed to ablate tissue near sensitive structures. IRE can be administered in a nonthermal mode, which contributes significantly to the protection of vital structures within the ablation zone [27–29].

IRE is currently being utilized to ablate a number of soft-tissue targets, though predominately to treat solid tumors. Several clinical trials are being conducted for ablation of tumors in both human and veterinary applications. Clinical trials to evaluate IRE as a means to effectively ablate tumors have been conducted in the following organs: pancreas [30], kidney [31, 32], liver [33, 34], and prostate [35]. Gliomas were treated in veterinary patients [36]. Treatment of bone has been evaluated in preclinical studies. More detailed information on the use of IRE can be found in Sect. 5.4.

5.1.2.3 Plasma

Plasma technology has rapidly progressed and is being evaluated for potential biomedical applications. Plasma can be an ionized gas that consists of charged and neutral particles. The development of nonthermal atmospheric-pressure plasmas (NTAP) makes it possible to treat particularly heat-sensitive tissues. NTAPs are being evaluated for use in a variety of applications including decontamination of devices, wound care, tissue engineering, and dental treatments [37–39]. Work has included evaluating use related to periodontal treatment [40] and endodontic treatment including root canal [41, 42].

Wound care is an important target for the use of NTAPs. One major aspect is disinfection of the wound to prevent infection and to enhance the healing process. Several studies utilizing NTAPs have explored reducing the number of wound relevant microorganisms as well as biofilms. A clinical trial was conducted to evaluate the effectiveness on infected wounds [43]. Another important aspect of wound care is blood coagulation, and it was recently shown that this could be accomplished using a nonthermal approach [44]. The other aspect of wound care is to stimulate cell growth to facilitate healing. Both fibroblasts and endothelial cells have been evaluated [45–47]. Dental treatment has also been evaluated as a potential application of NTAPs where their use has been evaluated as a means to eliminate microbial biofilms. More detailed information on the use of plasma can be found in Sect. 5.8.

5.1.3 Indirect Effects

Pulse electric fields in the microsecond and millisecond range have been utilized to perform intracellular delivery of molecules. Pulses in this range will have an effect on the cell membrane which will facilitate transport of the molecule into the interior of the cell. The mechanisms of this transport are discussed in Sect. 5.2. Molecular delivery is a key component for multiple biomedical applications. The major use for this approach is small-molecule and nucleic acid delivery wherein delivery of both moieties has now been evaluated clinically.

5.1.3.1 Small-Molecule (Drug) Delivery

In the mid to late 1980s, it was demonstrated that a chemotherapeutic drug could be delivered to solid tumors in mouse models [48, 49]. This concept (electrochemotherapy, ECT) has proven to be very effective. Due to the physical nature of the delivery, it is possible to accomplish effective delivery to virtually any tumor type provided the tissue can be accessed for appropriate electrode placement [50]. Critical work has been accomplished to establish standard operating procedures [51]. An important component of ECT is the instrumentation and electrode arrays [52–55], which has led to vast improvements in the technology. One improvement has been the development of software that facilitates treatment planning [56]. Another interesting aspect of ECT is the effect the pulse fields have on the vasculature within the tumor [57, 58].

A major milestone for the use of pulse electric fields was the first clinical use of this technology. This occurred in the early 1990s when microsecond pulse electric fields were used to deliver bleomycin to treat squamous cell carcinoma [59]. Since then many clinical trials have been performed evaluating several cutaneous tumor targets including squamous cell carcinoma, melanoma, basal cell carcinoma, breast cancer, and Kaposi's sarcoma [60–63]. Response rates for individual lesions were in the 70–90 % range depending on tumor type, administration route, and size of the tumor. An analysis of the majority of trials conducted on cutaneous metastases has shown an overall efficiency of 75 % with a 47 % complete remission rate [64]. ECT is now approved in several European countries, and the National Institute for Health and Care Excellence in the United Kingdom has released guidance stating that ECT is efficient and without major safety concerns [65, 66]. Recently, internal targets including liver, pancreas, and bone have been evaluated [67–71]. ECT has also been evaluated in veterinary medicine [72]. Several chemotherapeutic agents have been evaluated in preclinical studies, but bleomycin and cisplatin have proven to be the most efficient to use with ECT. Recently, delivery of calcium directly to tumors using pulse electric fields was also shown to be effective in reducing or eliminating the tumor [73]. More detailed information on the use of ECT in human medicine can be found in Sect. 5.3 and in veterinary medicine in Sect. 5.9.

5.1.3.2 Nucleic Acid Delivery

A major aspect of successful gene therapy is effective delivery. As part of the delivery process, it is important to target the gene to the correct cell and to achieve the desired protein expression. The first demonstration of successful delivery of plasmid DNA using pulse electric fields (gene electrotransfer, GET) was in the skin, the liver, the muscle, and tumors [74–78]. Since then, GET has been shown to be an effective approach to deliver plasmid DNA to many tissue targets [79]. Protocol development utilizing GET not only includes proper selection of plasmid and transgene but also involves proper selection of pulse width, pulse number, applied potential (voltage), tissue target, and electrode configuration. While handling these many variables can be a daunting task, it allows better control over the eventual expression characteristics. The numerous successful applications tested in preclinical models demonstrate the advantages of GET. Studies have been conducted demonstrating potential therapeutic applicability including cancer, cardiovascular, wound healing, metabolic disorders, ischemia, and delivery of DNA vaccines [79].

GET has been successfully translated to clinical use [80]. The first clinical trial utilized a plasmid encoding interleukin-12 to treat melanoma was reported in 2008 [81]. Since that first trial, many studies have been initiated. A 2015 search of ClinicalTrials.GOV listed 69 trials being conducted that utilized GET to deliver plasmid DNA. The majority of the trials are designed for either treatment or prevention of infectious disease or the treatment of cancer. DNA vaccines and immunotherapy are the major approaches being tested [80]. Based on the number of successful preclinical studies, it is probable that the number of clinical trials will continue to increase. In addition to the delivery of plasmid DNA, other molecules such as siRNA, shRNA, mRNA, or RNAi have been delivered with electrotransfer [82–86]. The only clinical trials utilizing these molecules initiated thus far have been to deliver mRNA to either dendritic cells or cancer cells to express an antigen or a cytokine [87–92]. GET has also been used to treat cancer in veterinary medicine [93]. More detailed information on the use of GET in human medicine can be found in Sect. 5.5.

5.2 Electrotransfer Mechanism for Drug and Gene Delivery

Justin Teissie and Marie-Pierre Rols

This short section provides key information on the biophysical processes supporting the transport of exogenous molecules across the plasma membrane as a consequence of the delivery of calibrated electric field pulses applied to cells. This information can be used to optimize protocols for specific applications.

5.2.1 Introduction

The cell membrane can be locally and transiently permeabilized when a cell is submitted to well-calibrated electric pulses (see Sect. 4.4). This is a rather well-characterized process when the pulse duration is longer than a few microseconds ("classical" electropermeabilization) [94–97]. Different mechanisms may be present when nanosecond long pulses are delivered under very high voltage (>10 kV) but are described in other sections (see Sects. 5.6 and 4.5)

5.2.2 *Electrotransfer Associated Pharmaceutical Delivery*

Transport of charged molecules is obtained across the membrane allowing either loading of external molecules or extraction of intracellular compounds under "classical" electropermeabilization. A wide field of applications is already reported (5.3, 5.4, 5.5, 5.9, 6.4, 6.5, 6.9, 6.10).

Different mechanisms of transport are present depending on the molecules of interest and whether they are considered during or after the pulse application.

For technical reasons, among the different pulse configurations (capacitor discharge, sine wave, square pulses), we will report the state of the art for the square wave pulse, the technology currently in most frequent use (Fig. 5.1).

As described in Chap. 1, a region on the cell surface is brought to a permeabilized state during electric pulse delivery. When the cell is a sphere, the size of this region is directly related to the magnitude of the field strength *E*.

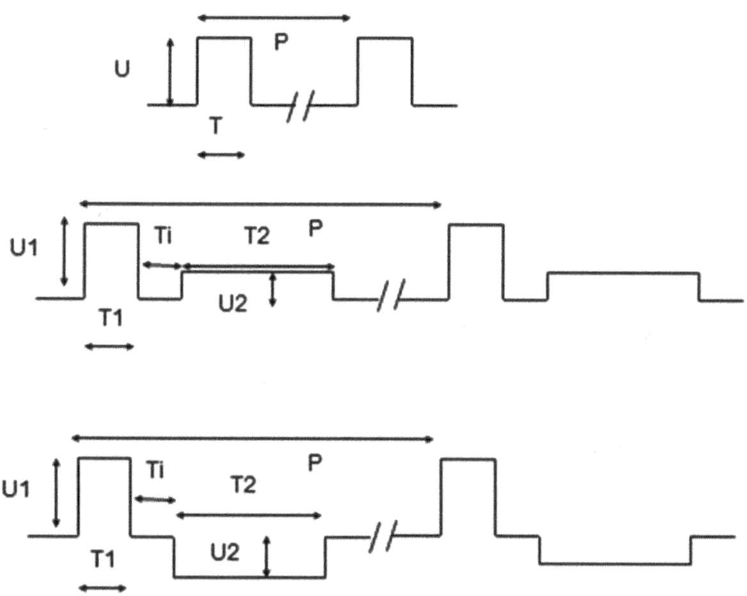

Fig. 5.1 Different profiles of electric pulses can be delivered to a cell suspension or on a tissue. *Top*: The most commonly used pulsing scheme is the repetition of similar pulses with a well-defined voltage U and duration T at a period P. *Middle*: To increase the electrophoretic drift, a train of double pulses is delivered with a period P. The double pulse is a high voltage U1 of short-duration T1 followed after a short interval Ti by a long pulse lasting T2 at a low voltage U2 keeping the same polarity as the first one. *Bottom*: To limit the electrochemical reactions at the interface between the electrodes and the solution, a more sophisticated train of double pulses is delivered with a period P. The double pulse is a high-voltage U1 of short-duration T1 followed after a short interval Ti by a long pulse lasting T2 at a low voltage U2 but with an inverted polarity as the first one

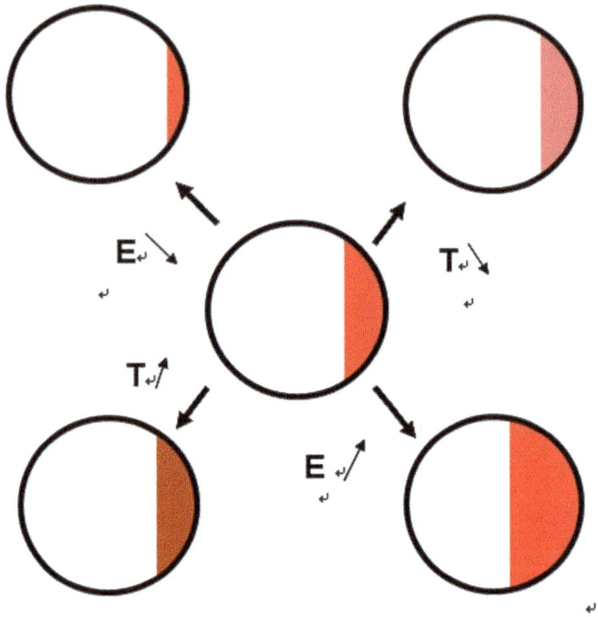

Fig. 5.2 The field strength controls the size of the permeabilized region (in *red*) on the cell surface, while the density of conducting defects (different levels in *red*) within the region is a function of the pulse duration

$$\mathrm{Ap} = \mathrm{A\ tot}(1 - E/\mathrm{Ep}) \tag{5.1}$$

where Ap is the area of the cap, A tot the surface of the cell (that depends on R the cell radius), and Ep the critical field intensity needed to trigger permeabilization (it depends on R).

The density of conducting defects in this cap is controlled by the single-pulse duration T, the number of pulses N in a train, and the delay P between each single pulse in a train of N pulses (Fig. 5.2).

Under reversible (or partly reversible) conditions, the permeabilization decays by three successive steps starting in the millisecond following the pulse, but only the slow one lasting several seconds or minutes is observed in routine practice.

$$P(t) = L + (P(0) - L)\exp(-t/\mathrm{tres}) \tag{5.2}$$

where $P(t)$ is the percentage of permeabilized cells at time t after the pulse train delivery, $P(0)$ is this percentage just after the pulse train delivery, L is the percentage of lyzed cells (irreversible permeabilization), and tres is the slow time constant for resealing.

Tres is dependent only on T, N, and P.

L and Tres are dependent on the temperature during the post pulse incubation as the post pulse repair mechanism is an active process.

5.2.3 *Electrotransfer Mechanism for Drug Delivery*

This part describes the events supporting the transfer of hydrophilic (charged) molecules with molecular weight less than 2–3 kD. The transport is of course modulated by the molecular characteristics of these compounds (size, shape, electrical charge).

Drugs (such as bleomycin and cisplatin), ions, and sugars are of concern.

5.2.3.1 During the Pulses

An electrophoretic drift is the mechanism driving transport during the application of an electric pulse. A direct transfer to the cytoplasm is present after a lag time that lasts a few microseconds after the onset of the pulse. The amount of molecules that are transferred is controlled by the electrophoretic mobility of the compound, the field strength, and the cumulated pulse duration, in addition to the molecule's permeability and extend of poration within the membrane. Electroosmosis may interfere with electrophoresis. From a practical point of view, this transport is increased by high field intensity (increase in the size of the permeabilized region), a high number of pulses, and long pulse durations (increase in the density of conductive defects), but a practical limit is the induced loss viability (L).

5.2.3.2 After the Pulses

Ions and small molecules are transported by diffusion following their concentration gradient as long as the membrane within the cap remains permeabilized. The resulting postpulse-induced permeabilization of cell membrane can be quantified in terms of the flow Fs of molecule S diffusing through the plasma membrane during the postpulse resealing. In the case of inflow, small molecules can then diffuse freely in the cytoplasm. Fick's law and experimental data indicate that the flow of molecules S at time t after the pulse is

$$\mathrm{Fs}(t) = \mathrm{Ps}\, x(N,T)A/2(1 - \mathrm{Ep}/E)\Delta S \exp(-t/\mathrm{tres}) \tag{5.3}$$

where

Ps is the permeation coefficient of the molecule S across the membrane, modulated by the molecular characteristics.

x depends on the number (N) and duration (T) of electric pulses. It represents the probability of permeabilization ($0 < x < 1$) (the density of induced defects).

A is the cell surface, E is the applied electric field intensity, and Ep is the field threshold for membrane permeabilization ($(1 - \mathrm{Ep}/E)$ reflects the cap where permeabilization is present).

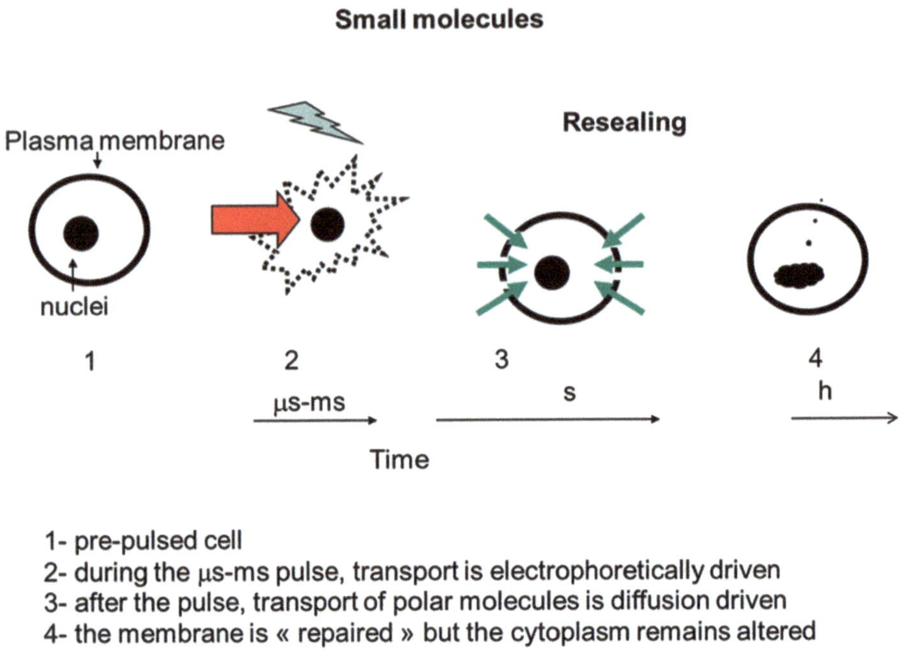

Fig. 5.3 The steps in the transport of drug induced by an electric pulse

ΔS is the concentration difference of S between the cell cytoplasm and external medium.

The final term reflects the membrane slow resealing where tres is the time constant of the membrane resealing process and t is the time after the electric pulse. The total accumulation is under the control of the pulse duration that acts on the density of defects (x) and the lifetime of resealing (tres). The transport is dependent on the nature of the target molecule (through Ps). Therefore, as Ps is dependent on its size, a larger transport is obtained for small molecules (Fig. 5.3).

5.2.3.3 Tips

The major part of the uploading is taking place during this postpulse process. The resealing is controlled by the cumulated pulse duration and the temperature. Long pulse with a low intensity (larger than Ep of course) should provide a larger uploading. The limit being that the viability must be preserved. A positive effect of the temperature improves the viability.

Transport can occur if the molecule is added after the pulse. But this can be controlled by the physiological reaction of the tissue (see the vascular lock effect in tumors).

The amount of drug that is uploaded is of course controlled by the concentration of drug present in the external solution close to the permeabilized cell. This is dependent on the concentration that was injected locally. But it is dependent on the external volume and appeared rather limited when cells are tightly packed. This is a very important parameter in tissues. The void volume is controlling the uptake.

5.2.4 Electrotransfer Mechanism for Small Oligonucleotide Delivery

The diffusion process described in 5.2.3 is a valid description for the transport of molecular weight of less than 2 kD. A different mechanism is present in the case of larger molecules such as oligonucleotides (siRNA) [98].

5.2.4.1 During the Pulses

An electrophoretic drift is the motor of transport. Oligonucleotides (siRNA) are directly transferred to the cytoplasm. The amount of molecules that are transferred is controlled by the electrophoretic mobility of the compound, the field strength, and the cumulated pulse duration (sum of all applied pulses).

Transport occurs only on the permeabilized cap that is facing the electrophoretic drift.

Changing the direction of the field (by using bipolar pulses or by moving the electrodes by 90°) increases the transport by inducing new caps (Fig. 5.4).

5.2.4.2 After the Pulses

No transmembrane transport of oligonucleotides is detected, but the molecules that are loaded due to the previous electric field transport can diffuse freely, allowing a homogeneous distribution in the cytoplasm (Fig. 5.5).

5.2.4.3 Tips

The oligonucleotide must be present during the pulse to be transported into the cytoplasm. Any hindrance to the electrophoretic movement of the oligonucleotides decreases the magnitude of the intracellular uptake. The magnitude of uploading is dependent on the external concentration of external oligonucleotides. Short pulses with high field strength can be used but are less effective than long pulses with lower field strength. The electrophoretic drift is indeed more limited. A similar mechanism is present for the transfer of proteins (antibodies).

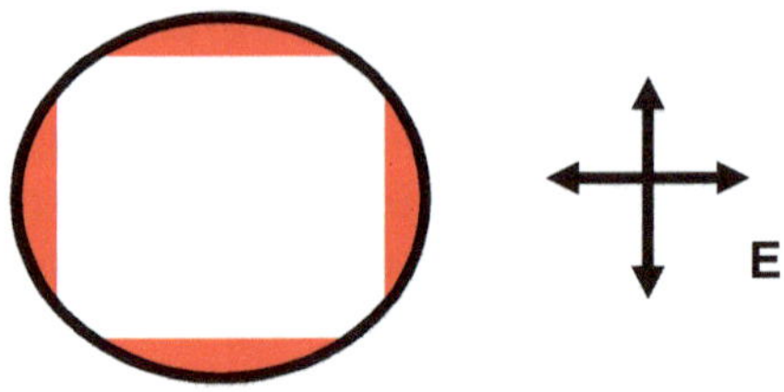

Fig. 5.4 The position of the electropermeabilized cap (in *red*) on the cell surface is dependent on the direction of the field. Changing this orientation will bring the formation of several caps

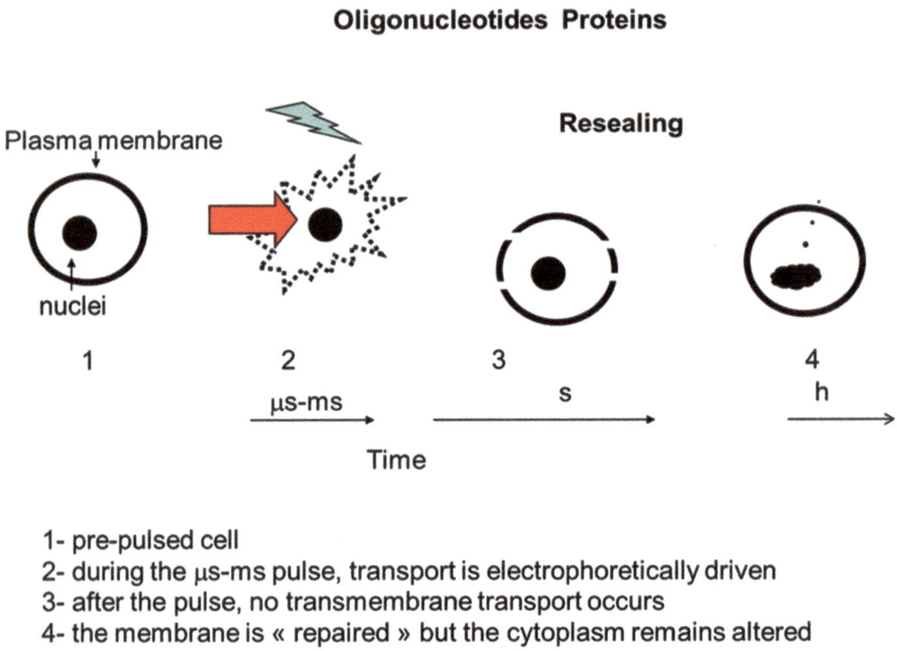

Fig. 5.5 Steps in the transport of oligonucleotides and proteins mediated by an electric field pulse

5.2.5 Electrotransfer Mechanism for Gene Delivery

Plasmids (pDNA) are huge macromolecules with a gyration radius of hundreds of nanometers. Their transport is supported by a different mechanism than for oligonucleotides [99–106].

5.2.5.1 During the Pulses

An electrophoretic force pushes the charged polyelectrolytes. The amount of molecules that are moved is controlled by the electrophoretic mobility of the compound, the field strength, and the cumulated pulse duration.

pDNAs are observed to form clusters on the cell surface on the side where they are electrophoretically pushed against the cell. No direct transfer of pDNA in the cytoplasm is observed.

Cluster formation occurs only on the permeabilized cap that is facing the electrophoretic drift.

Changing the direction of the field (by using bipolar pulses) or its orientation (by rotating the electrodes by a 90° angle) increases the number of clusters. But the amount of pDNA present in each single cluster is dependent on the delay P.

The dependence on the delay P was explained by a two-step process. The complex is formed during the pulse (sub ms) but in a metastable state that turns in an irreversible state slowly (on the tens of ms timescale).

The number of clusters remains unaffected by the number of pulses within a train of pulses. The number of pDNA within a given cluster is dependent on the number of pulses within the train.

5.2.5.2 After the Pulses

PDNAs are observed to be slowly translocated in the cytoplasm either as free molecules where they are transferred to the nucleus (see 4.7) or trapped within protein-coated vesicles. Electrotransferred DNA behaves as injected pDNA and is transported along the microtubules by a complex of proteins associated with molecular motors. Trapped pDNA exhibits the typical motion of endosomes or other cargo with intermittent phases of active transport and diffusion. Both are an active process that depends on the metabolic reserves of the cell. No transport is observed if the pDNA is added after the pulse.

The surface complex between the cell and the pDNA turns out to be protected from the action of externally added DNAses only after several seconds.

5.2.5.3 HV LV

Very short (0.1 ms) high field pulses are shown to trigger the formation of pDNA clusters but with a limited number of pDNA per aggregate. This results from the limited electrophoretic drift. More pDNA can be accumulated in the preformed clusters if the HV pulse is followed within a short delay (less than a few hundreds of ms) by a long (several tens of ms) low-field pulse (Fig. 5.6). This second pulse is delivered under a condition that cannot trigger a detectable permeabilization (no propidium iodide uptake). But it is associated with a large pDNA electrophoretic drift in the bulk.

This new protocol for pDNA transport (called HV LV) is shown to provide a significant level of expression even when using a low amount of pDNA.

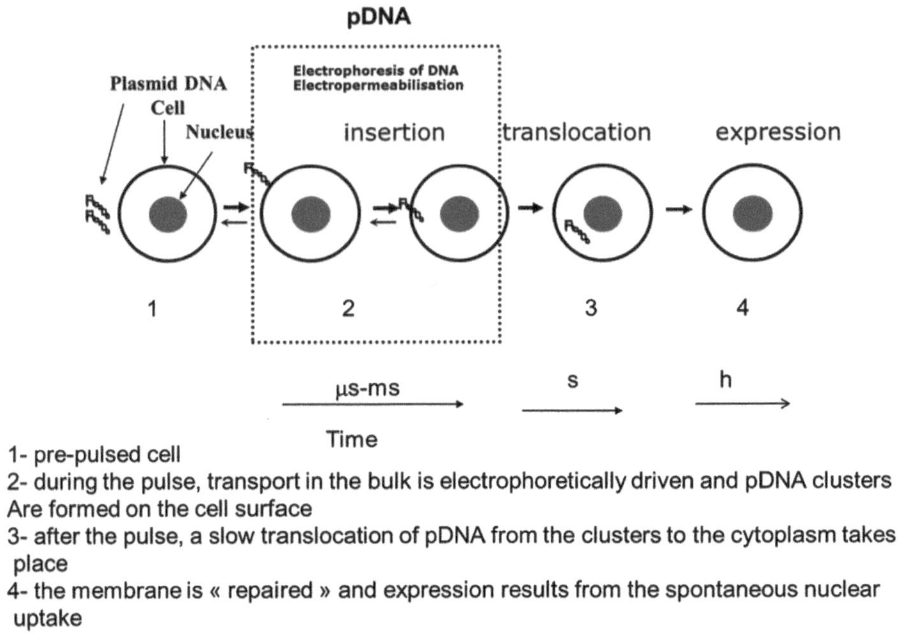

Fig. 5.6 Steps in the transport of pDNA mediated by an electric field pulse

5.2.5.4 Tips

The pDNA must be present during the pulse to be delivered into the cytoplasm.

The magnitude of uploading is dependent on the external concentration of external pDNA.

Short pulses with high field strength can be used but are less effective than long pulses with lower field strength.

The electrophoretic mobility of pDNA is not dependent on its number of base pairs.

Any hindrance to the electrophoretic movement of the pDNA decreases the magnitude of the intracellular uptake. This can be observed at high cell density or in a viscous buffer.

Expression of the pDNA is controlled by the viability of the pulsed population. A positive effect of an increase in temperature is observed. A postpulse addition of serum plays a protective effect. A high transport does not always result in a high level in expression.

5.2.6 *Endocytotic and Macropinocytotic Pathways*

The electric field pulse mediates a stress on the membrane. This results in different forms of endocytotic uptake. Therefore, some DNA aggregates undergo endosomal trafficking that allows them to remain in the cytoplasm while being relatively well protected from cytosolic DNAses. However, if DNA does not escape from endosomes before reaching the lysosomes, it will most probably be lost for gene expression. Endogenous molecules uptake results, but they remain trapped in the endosomes except if an escape mechanism is present.

The first indications of the induction of macropinocytosis in electropermeabilized cells came from the observation of electric field-induced ruffling and blebbing of the plasma membrane. Later studies of the transfer of proteins into cells following the application of electric field pulses also suggested a macropinocytosis-related mechanism.

5.2.7 *Transport in Tissues*

The transport is driven by the same physical mechanisms as on cells. The conclusion is therefore that the protocols must take into account the conclusions that were just described.

The field strength must be high enough at the level of the cell of interest (i.e., within the tissue). It is different from the external field applied to the tissue (to the organ). Detailed information is provided in Sect. 4.1 to obtain the corrections.

As the magnitude of transport (whatever the molecule of interest: drug, oligonucleotides, proteins, pDNA) is dependent on the local concentration, this is difficult to evaluate. It depends on the procedure of injection and on the tissue organization (void volume between cells). Concerning the large macromolecules (oligonucleotides, proteins, pDNA), they must be injected locally [107–116].

Transport is always controlled by the hindrance to diffusion or to mobility. The extracellular matrix therefore plays a critical role.

5.2.8 *Conclusion and Perspectives*

Transport supporting electrotransfer is clearly dependent on the nature of the molecules that are considered. The protocols to be used are therefore function of this conclusion. But another important parameter for medical application is to preserve the pulsed cell's viability. Drastic conditions can be associated to a highly effective transport but are destructive. The biological function is then lost.

5.3 Electrochemotherapy Basic Principles and Clinical Practice

Julie Gehl, Gregor Sersa, and Lluis M. Mir

5.3.1 Introduction

5.3.1.1 Basic Principles and Mechanisms of Action

All mammalian cells are surrounded by the cell membrane, and when an electric field is applied, water, which is a dipole, reorients, modifies its properties, and penetrates the cell membrane in such a way that a small bridge of water is formed in the lipid membrane. Very quickly a hydrophilic pore will form. This process is very brief, happening in a matter of a few nanoseconds [117].

Smaller hydrophilic molecules may enter through these pores by diffusion [118, 119]. As the cell has a negative resting potential, the greater transmembrane potential will be on the pole of the cell facing the positive electrode, but diffusion will also be present at the pole of the cell facing the negative electrode [97].

From an oncology point of view, the virtue of electrochemotherapy is that by creating a local electric field, it is possible to make cytotoxic drugs enter into cancer cells in a specific area delineated by the application of the electric field. Thus, for example, a drug like bleomycin may be administered systemically, the electric field applied to the tumor area(s), and a greatly enhanced uptake achieved in the exposed tumor area(s). Thus one single dose of the drug can become a highly effective but local treatment.

A multitude of anticancer agents are available, and many more are being approved or under development. For electrochemotherapy to work the differential magic of harming tumor cells within the electric field far more than normal cells outside the field, a central point is to use chemotherapeutic agents that would not normally pass the cell membrane. As discussed below, examples are bleomycin and cisplatin, but a number of drugs have been experimentally tested. It has been found that lipophilic drugs such as paclitaxel and etoposide [120, 121] cause no greater harm when cells are permeabilized because as lipophilic drugs they enter anyway. Amphiphilic drugs such as the anthracyclines (doxorubicin and others) have been shown to be either not or somewhat enhanced in efficacy when cell permeabilization is performed [120, 121], consistent with the fact that these amphiphilic drugs may differ in charge from neutral to charged molecules, altering their behavior in lipophilic environments. And finally hydrophilic drugs, unless a specific uptake mechanism is available, may be greatly enhanced in their activity through facilitated uptake over the permeabilized membrane [121–123].

The drugs currently used for electrochemotherapy are cisplatin and bleomycin.

Bleomycin is a large hydrophilic molecule discovered in 1966, from fungi [124]. Bleomycin is different from other chemotherapeutic agents in that it causes several assaults on the DNA; acting as an enzyme, it may cause 10–15 strand breaks per molecule and may also cause double-strand breaks, recognized as much more serious damage to the cell as repair of double-strand breaks is much more difficult [125]. Bleomycin has an almost negligible uptake without permeabilization, but is actually used in the treatment of testicular cancer and lymphoma in conjunction with other drugs; possibly the sensitivity of these cancers to chemotherapeutic agents may cause susceptibility and uptake of bleomycin. With electric pulses, the toxicity of bleomycin is enhanced 300- to 8000-fold, depending on method of investigation and incubation time [120–122]. This increase is exorbitantly high and allows electrochemotherapy to be used as a once-only treatment [51, 60, 61]. Due to the large therapeutic window, it is possible to administer bleomycin both intratumorally and intravenously. This is a great advantage because whereas some tumors are of limited size, other patients present with very large tumors where intravenous administration of the chemotherapeutic agent is necessary to be able to treat the patient in a single session.

Cisplatin is also a hydrophilic drug, but, in the absence of electric pulses, is taken up to a greater extent than bleomycin. Cisplatin is widely used both as a radiosensitizer in patients with cervical cancer and head and neck cancer and in a number of combination regimens, e.g., bladder cancer, testicular cancer, and head and neck cancer [126]. Cisplatin is an alkylating agent which crosslinks DNA, interfering with cell division. With electric pulses, the increase in cytotoxicity of cisplatin is lower, approximately eightfold [123], but for intratumoral injection, it has been found equally potent as bleomycin likely due to its potent cytotoxicity by itself alone [60].

Potentially also other drugs may be used, and a recent murine study has shown efficacy of mitomycin C in vivo [127]. Recent preclinical and ongoing clinical studies on the use of calcium are also showing promise [57, 73, 128], and future studies may indicate how these drugs will perform and indeed if there may also be other candidates.

Going from the in vitro to the in vivo setting, it is pertinent to discuss the pharmacokinetics of both intratumoral and intravenous injection. For intratumoral injection, it is important that pulses are applied immediately after drug injection, whereas some time must be allowed for diffusion when bleomycin is administered intravenously. An eight-minute interval after intravenous infusion is recommended [51, 60]. Local anesthesia generally contains epinephrine, causing vasoconstriction and reduced washout of drugs when given intratumorally; however, attention is needed; when the drug is injected intravenously, local anesthesia needs to be performed thereafter. The electroporation procedure itself also causes vasoconstriction, and so this will also reduce washout of drug just at the time of permeabilization [57, 58].

5.3.1.2 Importance of the Presence of the Electric Field

It was shown early on that the electric field alone does little damage (in reversible electroporation conditions) and that bleomycin and cisplatin alone have very limited effect [129]. From this also follows that it is very important for good antitumor effectiveness of electrochemotherapy; both an adequate electric field and a sufficient drug concentration are present at the same time in the treatment area.

When using plate electrodes, a homogeneous electric field can be obtained between the electrodes; however, limited penetration of the field into the deep tissue, and also the stratum corneum of the skin, weakens the electric field [130, 131]. The plate electrodes may be useful for small and superficial tumors [50, 62].

By far the most clinical treatments are being performed with needle electrodes. The most uniform field is achieved with linear array electrodes, where the parallell arranged needles in some way resemble the configuration of the plate electrodes. Linear arrays of electrodes are frequently used for smaller tumors and for tumors in the face because the linear electrodes leave smaller marks than the needle electrodes using higher voltages and also induce less pain [62].

Also other arrangements of electrodes are being used. A hexagonal array of seven electrodes with a diameter of approximately 1.5 cm activates two needles at a time, and the electric field consists of a multitude of paired electrodes being activated in a short time. This electrode is particularly appropriate for large lesions, such as chest wall recurrences of breast cancer [51, 62]. A needle electrode consisting of a ring of six electrodes where pairs of two are activated against each other and rotated in a circular fashion has been used in a number of studies on electrochemotherapy [81, 132] and is also currently used in gene therapy studies [53].

The voltage used in these electrode arrays must be adjusted to the electrode configuration [54, 130]. The plate electrodes which need to exert their effect through the stratum corneum are employed using a voltage to electrode distance ratio of 1.3 kV per centimeter [81]. For linear array and hexagonal electrodes, the voltage to electrode distance ratio has been 1 kV per centimeter, and for the circular array, electrode 1.3 kV per centimeter has been utilized [62]. When developing new electrodes, it is very important that electric field calculations are employed, in order to understand what the electric field is in the tissue [54, 130].

It has been suggested that nomenclature similar to that used in radiotherapy is utilized to express tissue volumes achieving relevant fields for reversible electroporation, as well as fields sufficient for irreversible electroporation. In this way, the efficiency of a particular electrode configuration relative to a defined tumor may be described [55].

5.3.2 Development of Appropriate Equipment and Standard Operating Procedures

A crucial development has been the translation from experimental prototype electrodes to clinically approved treatment systems. There are three major components in this process:

1. Production of a clinically approved generator. In order to obtain clinical approval, several standouts have to be met. To name just a few, there has to be a separation between the current in the mains and the current being administered via the patient treatment. The equipment must be able to deliver the correct pulses in a consistent and controlled way. Important user requests include being able to monitor delivery of the process during treatment. Production of clinically approved generators is a time-consuming and costly endeavor, and thus only a limited choice of approved generators exists [55].
2. Production of a range of clinically approved electrodes. Patients present with very different tumors to be treated; therefore, it is important to have more than one option in terms of electrodes in order to be able to help the individual patient. As an example, the patient presenting with a 40 cm wide chest wall recurrence may need a large electrode to facilitate treatment [62], whereas the patient presenting with a small facial lesion is best treated with a small electrode using low voltage so that it can be administered in local anesthesia, and the patient with an intramucosal lesion in the mouth is better treated with a flexible electrode that may be attached directly on a finger, allowing adequate positioning.
3. An important reason for the success of electrochemotherapy in Europe has been the development of standard operating procedures [62]. This allows physicians without prior knowledge of the field to start electrochemotherapy from a certain level of common understanding. In the 2006 standard operating procedures, a very detailed description is given of how to decide treatment strategy for patients and how to perform each of the different treatment options. Based on the experience obtained since 2006, revision of the standard operating procedures is in preparation.

5.3.3 Technology Use at Current Time

Currently electrochemotherapy is being used in more than 150 cancer centers across Europe. Patients treated have primarily had cutaneous metastasis of in particular malignant melanoma and breast cancer [61, 81, 133–136]. Also patients with primary skin cancer, where surgery and radiotherapy had either been exhausted or not possible, are being treated with electrochemotherapy [137–140]. Several European countries have now approved electrochemotherapy, and in 2013, the UK National Institute for Health and Care Excellence [65] stated in that guidance that electrochemotherapy is efficient and without major safety concerns. A recent meta-analysis for treatment of skin malignancies concluded that electrochemotherapy of cutaneous metastasis has an overall efficiency of 75 % with a 47 % complete remission rate [64].

Thus electrochemotherapy is increasingly being accepted as part of the treatment armamentarium. One of the major points about electrochemotherapy is that due to the high increase in cytotoxicity, this treatment is efficient in a variety of malignancies. The electrochemotherapy concept itself sustains this observation since cell electroporation, on which electrochemotherapy is based, is a very general phenomenon that affects all cell types. Thus treatment is applicable to a large number of patients.

The actual arrangements around the treatment of patients differ from center to center and country to country. For example, in Germany, where skin tumors are dealt with by oncologic dermatologists, it is usually these specialists that carry out electrochemotherapy. In other countries, it is the plastic surgeons that do the treatments with assistance from oncologists, and in yet other centers, the treatment resides with the oncologists in collaboration with surgeons, for example, plastic surgeons. In any case, it is widely accepted that patients for electrochemotherapy are reviewed in multidisciplinary teams. It is also obvious that acquiring an electric generator for a hospital makes sense when it is used for treatment of different tumor types, e.g., for colorectal cancer, skin cancer, and head and neck cancer.

5.3.4 Indications for Treatment

5.3.4.1 Treatment of Cutaneous Metastases

Treatment of cutaneous metastases has been the first and logical indication for electrochemotherapy, because electrode development for the skin is less of a challenge and because direct observation of treatment results is possible. The actual clinical presentation varies dramatically, from small malignant melanoma metastases to very large confluent chest wall recurrences. This has implications both for the treatment strategy, for the side effects, and for the expected outcome.

There is now considerable evidence on the treatment effectiveness of cutaneous metastases, from the first report [141], over numerous Phase I and II studies [60, 129, 133–137, 139, 142–144], to meta-analyses [64]. Responses are consistently high across these publications. High response rates have been reported for all investigated tumor histologies, which include malignant melanoma; planocellular carcinoma originating from, e.g., the skin, mucous membranes of the head, and neck; lung cancer; breast cancer; and other adenocarcinoma metastases from, e.g., colorectal cancer, bladder cancer, and renal cell carcinoma [61, 63, 133, 134, 139, 145, 146]. An initial study concluded that there was no difference between responses of different histologies [60]; however, it is likely that results from investigations of larger data sets may show differences. In investigating response across histologic diagnosis, it is very important that confounders such as tumor size are accounted for.

5.3.4.2 Treatment of Primary Skin Tumors

Primary skin tumors are predominantly basal cell carcinoma and planocellular carcinoma. In addition to dermatological treatments, surgery and radiotherapy are the cornerstones in treatment. In case of recurrence, or inoperability, electrochemotherapy has been successfully used [147–153], and an ongoing randomized study at the Cork Cancer Research Center in Ireland is investigating electrochemotherapy vs. surgery as the primary treatment for basal cell carcinoma.

5.3.4.2.1 Procedure and Protocol

The procedure of electrochemotherapy is really best described in the 2006 standard operating procedures [62]. However, procedures are briefly described below:

1. Discuss with the patient what the palliative need is, and address what the expected treatment outcome and side effects are likely to be. If the patient has a high pain score, address the need for adequate pain control. Cutaneous metastases may cause oozing, bleeding, odor, pain, disfigurement, and their growth needs to be controlled. Electrochemotherapy may alleviate one or more of these symptoms, and it is important to discuss with the patient, at the first visit, what the expectations of the patients are and to what extent the treating physician is confident that the expectations may be fulfilled.
2. For few and small tumors, plan local anesthesia and local injection of drug with use of linear array electrodes. For large or many tumors, plan general anesthesia with use of intravenous bleomycin, and choose either linear array (face, or relatively small treatment area) or hexagonal electrodes.
3. For i.v. injection, wait for 8 min to allow distribution of bleomycin into tumor tissue, for intratumoral injection pulse immediately after drug injection.

5.3.4.2.2 Retreatment

Treated lesions may recur, and new lesions may develop. Taking data from a recent meta-analysis, which of course give an average estimate for tumors that really present as a wide variety, approximately half of the treated tumors are not in complete remission after the first treatment [64]. Retreatment can be considered 1–2 months after electrochemotherapy when the maximal response is expected. Furthermore, new lesions may develop, heralding the need for additional treatment.

5.3.4.2.3 Side Effects

Application of local or general anesthesia during treatment is necessary to avoid treatment-related pain [60, 62]. It has been found that posttreatment pain is negligible for the majority of patients, but 13 % of patients had moderate, and another 13 % of patients had severe pain after electrochemotherapy [154]. Patients at risk for postprocedure pain may actually be identified before treatment by, e.g., pain level before treatment and previous irradiation in the treatment area [154], enabling institution of adequate pain medication plans.

Bleomycin is known to cause skin pigmentation, in particular in areas where (even minor) physical force is exerted. These types of minor trauma include patients itching themselves, removal of ECG electrodes, or adhesive surgical covers, as well as electrode use.

Bleomycin is also known to cause lung fibrosis, principally dependent on cumulative dose of bleomycin, but factors such as high oxygen concentration, fluid overload, and concomitant radiotherapy or thoracic surgery may contribute. One study investigated DLCO (lung diffusion capacity measured by CO diffusion) in patients with disseminated breast cancer did not find significant reduction, although it must be kept in mind that this was a small study [62].

5.3.4.2.4 Follow-Up

Patients are usually followed up twice in the first month after treatment and then at 2, 4, and 6 months. However, apart from the first two visits, the plan should be tailored to suit the needs of the individual patients.

Nursing follow-up is important, in particular in patients presenting with large metastatic lesions. Proper wound revision and dressings are a cornerstone in posttreatment care for the patient with large lesions, whereas patients presenting with small tumors may not need dressings, nursing assistance, or indeed any extra precautions.

Pain medication is only a need for approximately one of four patients—however, for these individuals, timely and thorough intervention is very important.

5.3.5 Ongoing Clinical Trials and Future Perspectives

The high response rates after once-only treatment for tumors of many histologies seen from the studies on treatment of metastases to the skin have led to a striving to develop electrodes that would be applicable for deep-seated tumors [56].

These different tissues require quite different approaches in terms of electrode development, for example, the bone is hard and the brain is very soft. In some organs, normal tissue damage is of great concern, e.g., in the brain, and in the colon, it is important to not have piercing electrodes due to risk of contaminated penetration.

For bone, rigid, penetrating electrodes have been developed, which may be inserted through the bone by drills by a technology already well established for bone surgery. Preclinical data, as well as clinical studies, have investigated this option, and electrochemotherapy may be a very interesting perspective for patients with bone metastases, as progression after radiotherapy can be a serious concern [71].

Liver metastases have both been the subject for a number of preclinical trials [155], and clinical studies using long individually placed needle electrodes have been completed [67, 156, 157], showing promising results.

Head and neck cancer includes tumors originating in the mucosa of the mouth and pharynx, and also for this indication, studies are completed [157–160], and new studies ongoing.

Pancreatic cancer, a disease with a high morbidity and mortality, may also be treated using electrochemotherapy, and this is being reported and further investigated in clinical trials [68].

Brain tumors, either secondary or primary, require a different setup due to the sensitive brain centers, the function of which must be respected in the placement of electrodes [69, 161].

Endoscopic electrodes for colorectal cancer have been developed, and these have the form of suction chambers—rather than needles—in order to obtain safe treatment of tumors by an endoscopic approach [70].

5.3.6 Randomized Trials

Most electrochemotherapy trials to date have been for patients without other treatment alternatives; for example, the NICE guidance states that patients be evaluated for electrochemotherapy when other options are not possible [65, 162]. In order to be considered for first-line treatment, new treatment must prove equivalent or superior safety and/or efficacy in randomized trials. The first of these randomized trials is emerging now, randomizing between surgery and electrochemotherapy for the treatment of basocellular carcinoma, a very common skin tumor. Further randomized trials are expected to commence, as the technology develops further.

5.3.7 Combination with Other Treatments

In clinical practice, electrochemotherapy is frequently used as single treatment modality. However, electrochemotherapy may be used in conjunction with, e.g., endocrine treatment, in cases where patients have stable disease in internal organs on current treatment, but experience disease progression in the skin. Also in this setting, electrochemotherapy has shown promising results and could be used on specific tumors in patients that have big nodules that are not expected to regress after systemic therapy or in drug resistant tumors, which recur after the systemic treatment [163].

Electrochemotherapy may be combined with immunotherapy, as electrochemotherapy may provide a danger signal that can be exploited. This has been demonstrated in vivo [164, 165], and clinical data are pointed in the same direction [166].

Furthermore, electrochemotherapy could be combined with radiation therapy, since bleomycin and cisplatin are good radiosensitizers. Extensive preclinical studies have been done [167, 168].

And last but not least, electrochemotherapy could be combined with gene therapy where electrochemotherapy would be "in situ" vaccination priming that could be boosted with IL-12 peritumoral therapy [166, 169, 170].

In conclusion, electrochemotherapy is at this time in routine clinical use for metastatic skin tumors, for patients where other treatments have failed or are not relevant for the patient. Further developments include use in internal organs, progression toward a first-line treatment usable in primary tumors, as well as combinations with other treatments.

5.4 Clinical Implications of Irreversible Electroporation: Clinical and Preclinical Findings

Robert E. Neal II, Suyashree Bhonsle, and Rafael Davalos

5.4.1 Introduction

Therapeutic application of irreversible electroporation (IRE) for killing diseased tissues was first demonstrated in 2007 [27], though it remains a field under extensive active development. IRE's clinical utility has been demonstrated in numerous organs for soft-tissue targets, particularly tumors. Where the cell death mechanisms for IRE relate to cell membrane disruption, the electric pulse protocols used to kill cells with IRE in a targeted volume of tissue are able to do so without inducing significant alterations of the extracellular components, notably collagen and other extracellular proteins. This enables IRE to be used in regions containing or near sensitive structures that make such targets contraindicated for other focal therapies or surgical resection. In this chapter, we discuss a number of IRE applications, particularly for cancer therapy, identifying preclinical and clinical data to determine the efficacy of IRE in these regions and also to consider findings regarding therapy morbidity.

5.4.2 Relevance and Considerations for Maintaining Nonthermal Ablation Modality

The active mechanism for inducing cell death from IRE relates to the disruption of cell membrane integrity and the cascade of effects that the induction of nanoscale defects has on the cell. The formation of these defects occurs due to cell exposure to electric fields capable of sufficiently altering the local membrane structure by way of a dramatically increased transmembrane potential and is typically applied using a series of 50 to several hundred brief (~100 μs) but intense square wave electric pulses delivered via electrodes inserted into or adjacent to the targeted tissue. The lethal electric field threshold relates to the strength of the electric field, as well as overall intensity of the collective pulse parameters used (pulse length, number of pulses, delivery rate).

A parallel occurrence from electric pulse delivery is the Joule heating of the tissue due to the conduction of electric current through the resistive tissue medium. In order to maintain IREs distinction from thermally based therapies, pulse protocols must be designed to ensure that the extent of Joule heating is sufficiently mitigated to prevent damage to the sensitive structures implicated in the morbidity profiles of other focal techniques, including thermal ablation, radiotherapy, or surgical resection. Thus, appropriately applied IRE ablation is able to kill the entire targeted region while controlling the extent of thermal effects to the sensitive structures.

A considerable number of publications exist in the literature that use numerical and experimental studies to address consideration of sufficient IRE without clinically limiting thermal damage. There are theoretical confirmations that IRE ablation occurred without thermal damage, which provide techniques to evaluate this [27, 29, 171, 172]. Furthermore, additional experimental studies have measured temperature during pulse delivery to examine the potential for thermal damage. This includes a clinical canine veterinary case, where maximum temperature rise was 2.4 °C immediately adjacent to the electrode with successful tumor remission [173], as well as an in vivo canine brain study, which showed ablation occurring with a maximum temperature change of 1.15 °C [28]. A more aggressive energy protocol in an in vivo porcine study using 3- and 4-needle protocols of 70 pulses, each 90 μs long, at a rate of 90 pulses/min determined a peak temperature rise to 57 and 79 °C for 3- and 4-needle protocols, respectively [174]. Consistent with the rapid decay in thermal effects away from the electrodes, max temperatures of 40 and 42 °C were measured 1 cm outside the electrode geometries, still within the IRE ablation lesion. Finally, an infrared camera monitored temperatures on an in vivo murine dorsal skin fold chamber to show a maximum temperature rise of

19 °C for 99 pulses, each 100 μs long delivered at 1 pulse/s [175]. These studies demonstrate that IRE ablation may occur at sublethal temperatures, but careful attention must be given to thermal effects to ensure they remain below those to cause damage to the sensitive structures. Several studies address this and suggest techniques to maintain modest temperature rise, including shorter pulse durations, alterations of pulse delivery into slower delivery rates or modulated pulse delivery, as well as proactive interventions, such as incorporating a phase change material (PCM) into the core of the electrodes that melts during treatment and absorbs heat out of the surrounding tissue as well as precooling the target tissue area prior to treatment [27, 176–179].

5.4.3 Preserved Patency and Function of Sensitive Structures and Their Implications

5.4.3.1 Structure-Specific Sparing

There are numerous relevant critical structures that may exist within or around targeted volumes for IRE therapy. The relative importance of each will vary with the targeted organ and cancer variety.

5.4.3.1.1 Blood Vessels

Extensive work has examined the effects of IRE on blood vessels. While IRE is shown to kill the endothelial cells and disrupt capillary-level vasculature, preservation of the collagen framework facilitates continued blood flow in large blood vessels [172, 180–182]. In an IRE study, Doppler ultrasound shows continued blood perfusion through a major vessel contained within a tissue region immediately after IRE treatment [183, 184]. Further, longitudinal studies have demonstrated the regeneration of endothelial cells within the affected regions of the blood vessels within 7 days [183] permitting full recovery and long-term function of the vessel.

5.4.3.1.2 Nerves

In an early in vivo study, it was shown that canine prostate tissue could be ablated while preserving the neurovascular bundles necessary for potency and continence [185]. The possibility of preserved neural function has since been explicitly investigated. Long-term in vivo studies on an IRE treated nerve showed that, despite an initial decrease in functionality, the nerve attained full recovery approximately 2 months later [186, 187].

5.4.3.1.3 Ductal Systems

In addition to vascularity and neurological implications that may be present for many organs targeted for IRE ablation, there are also many organ-specific ductal systems which also seem to indicate preserved function and patency while being contained within or adjacent to ablation zones. These include bile duct preservation in pancreatic and liver ablations [188, 189] collecting system and ureter in renal ablations [190] and urethra in prostate ablations [185]. The action behind this preservation likely relates to the structural organization of many ductal systems, which are composed of relatively low permeability connective tissue innervated with endothelial and epithelial cellularity. Although IRE will initially kill the cells within the ductal system architecture, preservation of the extracellular constituents permits continued function of these tissues and supports the recellularization of the systems over time [191].

5.4.3.2 Organ-Specific Sparing and Therapeutic Implications

In addition to the generalized sparing effects of IRE's nonthermal mechanism of action to induce cell death, different organ applications will exhibit unique implications in their relevance to successful therapy outcomes. This section addresses some of the more common IRE therapeutic applications and the preclinical and clinical data regarding the use of IRE for these targets.

5.4.3.2.1 Pancreas

One promising application for IRE ablation therapy is in the treatment of pancreas cancers. Locally advanced pancreatic cancers comprise approximately 40 % of the pancreatic tumor population [192]. Although these tumors have not fully metastasized, they have innervated and surrounded critical structures including the pancreatic duct and the superior mesenteric artery. Thus, tumors around this region cannot be safely removed without risking patient morbidity and mortality [193]. This leaves a large number of pancreatic cancers inoperable by the time of diagnosis. IRE's ability to spare the major vasculature, however, offers an approach to addressing the tumor in these regions, thus providing either a standalone treatment modality or one that can be used to augment the treatment margin of surgical resection.

Preclinical data have shown that IRE can be safely administered in the pancreas [194] with rapid reduction of inflammation and preservation of vascular structures [195]. Furthermore, IRE was shown to be effective in treating pancreatic ductal adenocarcinoma, an aggressive form of pancreatic cancer, in mouse models [196]. The data in this study indicated that IRE leads to an increase in survival (from 42 days in untreated mice to 88 days in the IRE-treated group) with 25 % of mice showing complete tumor regression.

Clinical data for IRE in the treatment of pancreatic cancer shows promising results for effectively treating the entire targeted tumor without damaging the critical superior mesenteric artery. An initial report on 27 patients who underwent IRE for palliation of their stage III pancreatic adenocarcinoma showed 100 % ablation success at the 90-day follow-up [30, 197]. A later clinical study was conducted by the same group on 139 locally advanced pancreatic cancer—stage III (LAP) patients to compare IRE to standard therapy of chemotherapy and chemoradiation therapy [198]. A comparison of IRE patients (IRE in conjunction with chemotherapy or chemoradiation) to standard therapy showed an improvement in local progression-free survival (14 vs. 6 months, $p = 0.01$), distant progression-free survival (15 vs. 9 months, $p = 0.02$), and overall survival (20 vs. 13 months, $p = 0.03$). In another study, fourteen patients with unresectable LAP were treated with percutaneous IRE [199]. The IRE treatment was well tolerated, and postprocedural scans showed preservation of vasculature.

5.4.3.2.2 Kidney

The increase of high-resolution diagnostic imaging is partially responsible for a rapid rise in the number of diagnoses of asymptomatic small renal masses, with over half of new cases detected incidentally. Frequent detection of low-risk renal tumors has encouraged care toward less-aggressive, nephron-sparing approaches [200]. Preservation of renal function in the affected kidney is vital for patients with solitary kidneys, comorbid conditions, multiple tumor sites, and genetic predisposition for recurrent bilateral renal tumors (von Hippel–Lindau disease). Partial nephrectomy still requires considerable invasiveness, while sensitivity of the ureter and high blood perfusion rate mitigate the efficacy of thermal therapies. These factors provide considerable value for IRE treatment of small renal masses.

Preclinical investigations into the ablation of kidneys show short- to medium-term preclinical studies investigating macro- and microscopic healthy renal tissue IRE effects show complete cell death of glomerular and tubule cortex structures while sparing major vasculature. Initial urothelial injury was noted, but the urinary system was completely regenerated 28 days post-IRE with the preservation of urothelial basement membranes and no urinoma evidence [201, 202]. Renal ablations showed a demarcation between necrotic and normal tissue, 24 and 36 h after IRE treatment with evidence of tubule degeneration. Two to 4 weeks later, there was evidence of regenerating tubules and pelvic epithelium with intact extracellular matrix and ablations turning to contracted scars. The urine-collecting system was essentially preserved with regenerated urothelial tissue [188, 189]. In addition, the vascular system was preserved in ex vivo perfused porcine kidneys as observed during angiographic contrast evaluation [190].

Early clinical investigations demonstrate the safety profile of IRE renal ablation performed immediately prior to resection and have reported promising early therapeutic potential for IRE as a standalone modality for renal tumor treatment [31, 32].

5.4.3.2.3 Liver

The size of the liver relative to typical tumors, coupled with its regenerative capacity, makes the liver a common target for focal ablation of tumors. IRE has been clinically applied in this environment, particularly for targeted regions near the inferior vena cava, gallbladder, and external sensitive organs such as bowel, pancreas, heart, and spleen. Further, the relative homogeneity and isotropy of hepatic organization make liver a suitable model for investigation of various aspects for IRE therapy.

The greatest collection of preclinical investigations regarding the effects from IRE therapy exists for liver ablations. Many of these studies are generalized to effects of IRE as a whole, such as characterization of the ultrasound and imaging effects [183, 203–205] as well as studies into different pulse parameters and electrode designs [178, 189, 206, 207]. In addition, preclinical studies have shown that IRE can be used to effectively treat liver tumors in rat [208] as well as rabbit models [209].

Clinical data regarding the use of IRE in patients has shown varied complete response rates from 50 to 98.1 % [32–34]. It has confirmed the ability to use IRE in the vicinity of sensitive adjacent organs, including bowel, heart, diaphragm, spleen, and inferior vena cava [210]. Early clinical investigations have shown outcomes with therapeutic efficacy ranging between 55 and 95 % [211], with outcome highly correlated to tumor size and thus complexity of ablation protocol [201, 202]. The robustness of liver has also shown the feasibility of using multiple IRE sessions to address multiple tumor nodules from less-aggressive cancers, such as a hepatic epithelioid hemangioendothelioma case study [190].

5.4.3.2.4 Bone

While IRE has primarily been evaluated for implications in soft-tissue ablation, several studies examine its effects on regions of bone. In [212], it was found that reversible and irreversible electroporation was possible in cancellous bone. Further, in a study performing IRE in porcine lumbar vertebrae, it was shown that IRE produces localized regions of well-demarcated necrosis with no detectable change in bone texture and limited neural toxicity [213], consistent with the transient Wallerian degeneration determined in [213].

5.4.3.2.5 Brain

IRE utility in the brain has been characterized extensively with preclinical and clinical canine models. The high vascularization and extreme sensitivity of adjacent neural tissue require precise treatment plans and blunt-tip electrodes to deliver the IRE pulses. Preclinical studies in canine patients have shown that IRE can be safely administered to the brain and lesion volume can be correlated to applied voltage

[29, 214, 215]. Studies also explored the therapeutic planning aspect and implementation of IRE for treating canines with spontaneous brain tumors [36].

A cohort of recent canine glioma veterinary patients has been treated with IRE. In an early case study, IRE attained a 75 % reduction in tumor volume within the first week post-IRE [36]. Following adjuvant fractionated radiotherapy, the tumor was determined to be in complete remission prior to the suggestive onset of early-delayed radiation encephalopathy, though recurrent glioma could not be excluded. In a recent examination of the cohort, it was found that of the patients that survived until discharge, Karnofsky performance scores were improved in 6 of 6 patients over pretreatment values, while seizure control improved in 5 of 6 [216].

5.4.3.2.6 Prostate

Optimal treatment strategies remain to be determined for low- to medium-risk organ-confined prostate cancer. While radical prostatectomy offers strong efficacy to tumor control of tumors that have not metastasized, it carries high rates of morbidity in regard to incontinence and impotence post-prostatectomy, including with robot-assisted procedures [217]. This morbidity results from damage to the urethra and sensitive neurovascular bundles at the perimeter of the prostate. Thermal and other focal therapies also risk such adverse effects, and efforts to mitigate these risks may jeopardize the efficacy of the treatment [218]. Where it remains to be well defined which prostate tumors require intervention, IRE may serve as an ideal focal or regional ablation approach to address identified tumor regions with significantly lower risk to potency and continence.

Preclinical studies, primarily in vivo canine prostate studies, have shown the ability to safely create ablation lesions while preserving integrity and regular system function. In an early study, IRE was shown to produce significant ablations in prostate while preserving the urethra and neurovascular bundle [185]. Further, the ability to produce IRE lesions in close proximity to the urethra and prostate capsule without notable effect to erectile and urinary function was demonstrated [219]. A further study investigated the implications of metallic seed implants within the prostate, identical to those used for brachytherapy [220]. This study found that the presence of the seeds did not significantly affect the ablation zone or thermal effects from the electric pulses, though it does not consider tissue changes in response to prior radiotherapy, such as fibrosis.

A stage I clinical safety study performed ablations in low- to medium-risk prostate cancer patients 3–4 weeks prior to their regularly scheduled prostatectomy [35]. This study showed the ability to attain lesions in the prostate and characterized the effective lethal electric field threshold for the pulse parameters used, which showed prostate tissue tends to be more resilient to IRE pulses than previously characterized tissues, such as kidney, liver, or brain. In addition, a study by Neal et al. [221] used IRE on low-, medium-, and high-risk prostate cancer patients, which showed encouraging toxicity profiles in patient outcomes, but did not assess oncologic outcome. Further expansion of this initial cohort to treatment of

34 patients with varying risk disease showed that of those with function prior to treatment, 24 of 24 patients maintaining continence and 19 of 20 patients preserving potency over a 6-month follow-up period, with 17 % showing suspicious residual disease. Further studies are underway to examine the oncologic efficacy and safety profile of IRE in the treatment of prostate cancer.

5.4.4 Imaging Findings and Implications in Planning and Determining IRE Outcomes

5.4.4.1 Secondary Effects and Additional Clinical Implications from IRE Clinical Therapy

Although IRE is able to destroy a targeted volume of tissue, with a rapid onset of initial cell death mechanisms, there remain several secondary effects and potential avenues of exploitation that may further expand the treatment zone both overall and selectively, as well as locally and systemically. Initial preclinical investigations into a selected number of these effects are presented below.

5.4.4.2 Immune Response and Activation

Where IRE is typically implemented by delivering electric pulses through electrodes inserted directly into the targeted tissue to induce the cell death, the affected tissue remains in situ following treatment. This permits the release of tumor-specific and tumor-associated antigens, as well as the various signals of cell distress into the treated and peripheral volumes. These molecules may have implications in the promotion of local and systemic immune responses, similar to those described from fields such as cryo-immunology [222]. Studies to exploit and encourage this response for electrochemotherapy have demonstrated remission of tumors contralateral to ones treated of the same cell line when immunomodulating molecules were added. In vivo IRE studies have shown immunocyte infiltration and lymphatic drainage following IRE ablation [213], and a veterinary case study showed a stronger tumor response than expected based on preclinical experiments using immunodeficient models. Further, in vivo murine studies showed changes in immune markers, cytokines, and immunocyte populations following tumor treatment with IRE. Finally, a comparative in vivo murine study showed improved local tumor response in immunocompetent versus athymic mice, including resistance to rechallenge with the same cancer cell line 18 days after IRE treatment [223]. Overall, the exploitation and encouragement of the immune response demonstrated from IRE treatment of tumors remains a promising yet relatively unexplored field for improving IRE therapy outcomes.

5.4.4.3 Adjuvant Augmentation of IRE Treatments

In addition to encouraging a more robust immune response to further encourage effective oncologic outcomes from IRE therapy, there are numerous other promising adjuvant approaches to increasing ablation zone and oncologic outcome from a given pulse protocol for IRE therapy. The most obvious of these is the inclusion of targeted chemotherapeutics already shown for ECT to selectively kill reversibly electroporated cells, where increases in effective treatment volume of 50 % have been predicted. In addition, the use of conventional therapies as adjuvants to IRE may improve patient outcomes, such as the inclusion of standard chemotherapy regimens to kill any distant micro-metastases while IRE kills the primary tumor, or the use of transarterial chemoembolization to further stress and promote death of tumor cells.

In addition to adding adjuvant therapies to improve patient outcome by superimposing or encouraging additional cell death locally and systemically, there are several studies examining potential IRE and reversible electroporation therapy augmentation by locally manipulating the properties of the cell membranes, to increase their susceptibility to electroporation and IRE. One approach that has been suggested, with implications primarily likely for in vitro applications, is the addition of DMSO in the cell membrane to alter the structural properties of the lipid bilayer [224]. An additional study has examined the sensitization of cells to electroporation within a given electric field by altering the electrochemical environment of cells in tissue in vitro with cationic and anionic substances. This study found markedly promising results with the inclusion of cationic anesthetics in the cellular environment, with particular benefits exhibited from procaine and lidocaine, where a 50 % reduction in the strength of the electric field is required to induce cell death with IRE. While it remains to be determined whether such a pronounced effect will translate to in vivo environments, it offers a valuable potential adjuvant to safely augment the treatment zone of IRE and other electroporation-based therapies.

5.4.5 Conclusion

The unique modality for inducing cell death from IRE enables targeted ablation of tissues without denaturing proteins or damaging vital structures implicated in the morbidity of surgical resection and thermal ablative technologies. This characteristic offers unique advantages for IRE treatment of diseases in an extensive array of applications and anatomical regions. This chapter focused on the preclinical and clinical findings regarding IRE ablation of tumors. In addition, secondary tissue-level effects from IRE when translated to diseased tissues in vivo offer several potential avenues for exploitation and enhancement of IRE ablation zones and therapeutic outcomes, either by encouraging an antitumor immune response, increasing the injury and cellular stresses that promote cell death, or manipulating the properties of the tissue environment containing the cells. Further work with ongoing studies will continue to determine and improve upon the oncologic efficacy of IRE therapy.

5.5 Gene Electrotransfer

Richard Heller

5.5.1 Introduction

The therapeutic potential of gene-based medicine is widely applicable. Gene transfer is being evaluated as a potential therapy for metabolic disorders, cardiovascular diseases, cancer, pulmonary diseases, and genetic disorders. In addition, there is the possibility to use this approach for vaccination for prophylactic applications. A critical element of successful gene therapy is effective delivery of the specific gene or cDNA. An important component of delivery is to target the gene to the right tissue/cells and in a manner that achieves appropriate expression that will result in the desired clinical response. There are essentially two main categories of gene delivery. One technique is to use a biological approach in which the gene of interest is delivered using a genetically engineered viral or bacterial vector. The alternative approach is to use plasmid DNA (nonviral) in which the delivery is done by simple injection or with chemical or physical assistance. When choosing which technique to use, efficiency and potential adverse effects are among the factors that should be considered.

For applications requiring high or long-term expression, such as in protein or gene replacement therapies, viral vectors are typically used [225–227]. For applications that require delivery of genes, encoding potentially toxic proteins or proteins to modulate the immune response and short-term and/or low levels of expression may be optimal; then a nonviral approach may be appropriate [228–231]. As improvements in biological and nonviral approaches have occurred, there is less of a clear distinction of which approach to use. However, if equivalent expression parameters can be achieved with either approach, then plasmid DNA-based gene transfer becomes more attractive because it removes the need for a biological vector. Nonviral delivery has distinct advantages over the use of a biological vector in particular an improved safety profile. Other advantages include reduce risk of integration into the genome, reduced immunogenicity, and decreased potential for environmental spread. The major obstacle for nonviral delivery has been the difficulty in obtaining reproducible delivery and achieving high expression levels. To overcome these issues, plasmids have been delivered with either chemical or physical approaches including lipid or polymer conjugation, particle-mediated delivery, hydrodynamic delivery, ultrasound, and electroporation [232–236].

Gene electrotransfer (GET) also known as electroporation, gene electroinjection, electrically mediated gene transfer, or electrogene transfer has emerged as a reliable physical method for delivering plasmid DNA [79]. In this section, the terms GET and electrotransfer will be used. Successful delivery using GET was initially demonstrated in the skin, the liver, the muscle, and tumors [74–78]. Because electrotransfer is a physical method, delivery is theoretically possible to any tissue or cell type. This aspect has been demonstrated as the effective delivery has been accomplished in many tissue types and animal species and with a diversity of plasmids [79, 236, 237]. GET has also been successfully translated to the clinic. The first trial utilized microsecond pulses to successfully treat metastatic melanoma using gene-based immunotherapy [81]. Clinical applications of GET are discussed below in Sect. 5.5.4. Delivery of plasmid DNA with GET has gained acceptance as a viable approach to achieve effective delivery.

5.5.2 *Protocol Development*

Development of a therapeutic application requires careful consideration of multiple factors. The first steps will be related to understanding the specific requirements of the treatment. In order to achieve the eventual clinical objective, what type of expression profile will be needed? This includes the levels, duration, and location of transgene expression. The required expression profile will dictate how to proceed with respect to selection of the appropriate target tissue and GET parameters [238]. Both safety and efficacy should be considered when developing the protocol.

Expression profile is a key component to developing a successful therapeutic protocol. As such, the utilization of GET is a distinct advantage. Due to advances in understanding the application of electric fields to deliver plasmid DNA, it is possible to control the delivery to obtain a desired expression pattern. In addition, GET is applicable to most tissues provided there is a means to access them in a manner that allows for proper placement of electrodes. Proper selection of the delivery site and the applied electrical parameters can impart control over the level and duration of expression of the transgene. While handling multiple variables to develop an effective protocol can appear to be a daunting task, it is this versatility that leads to the increased use of GET. An investigator that carefully selects delivery parameters including electrical conditions and tissue target can achieve the required transgene expression that could result in a successful therapeutic application.

5.5.2.1 Tissue Selection

An important criterion for GET delivery is access to the tissue. The majority of protocols utilizing GET have been targeted to muscle, skin, and cutaneous/subcutaneous tumors [79, 80, 112, 236, 239]. Other tissues such as liver, lung, heart, brain, kidney, and spleen have been used but require additional manipulation in order to access them [74, 240–249]. For internal tissues, typically surgery is utilized; however, more noninvasive approaches are being developed. Selection of the target tissue will depend on the specific therapeutic application. The organization of the tissue and the cells within the tissue should be considered when developing the delivery protocol as these factors could influence the electric field distribution [250, 251]. This assessment should include the amount of extracellular space between cells as how tightly packed the cells are can also influence the distribution of the plasmid DNA within the tissue [34, 35]. Other factors that will influence the expression characteristics are proteoglycan and collagen content and cell size and shape [252].

When designing a protocol, the investigator should determine what expression profile is needed. If the approach is to develop a protein replacement therapy that requires increased levels of the expressed protein in the circulation, then the muscle may be the appropriate choice. If the approach is to deliver a DNA vaccine, the target tissue is usually muscle or skin [253–256]. While both tissues have been used, skin may be more appropriate due to the number of antigen-presenting cells present in the skin [255, 257, 258]. If the protocol is to achieve a specific localized effect, then the delivery should be done to that specific site with parameters selected to achieve localized expression.

Delivery to a specific tissue will directly impact the expression profile not only with respect to levels and kinetics but if the expression will be localized or systemic. For example, muscle will typically yield high expression for a long time, although this can be modified by changing delivery parameters (Sect. 5.5.2.2). In addition, when delivering to the muscle, there will be increased levels in the circulation of the expressed protein. In contrast, delivery to the skin is typically localized with low levels of the expressed protein found in the circulation. Skin delivery is also typically short-lived.

5.5.2.2 Parameter Selection

After selection of the tissue to be targeted, the next critical aspect of protocol development is selection of GET parameters. The selection includes pulse length, number, duration, and frequency as well as the applied voltage and the waveform of the pulses. The levels and kinetics of expression will be determined by selecting combination of these parameters. The parameters that achieve a specific expression profile in one tissue may not equate to the same expression characteristics in a different tissue. As with selection of tissue, it is important to determine the requirements of the application prior to determining GET parameters. In addition, to the parameters mentioned above, another critical component of developing the protocol is the design and configuration of the electrodes. While pulse shape can be modified, typically rectangular-shaped pulses are used for in vivo GET protocols. There are several commercially available pulse generators that can be used. Most are relatively flexible with respect to setting, pulse width, amplitude, and number.

GET is considered to be a dual threshold phenomenon. The first threshold is the point at which the cell membrane is reversibly permeabilized and molecules can enter the cell. The second threshold is the point at which the process is irreversible and the cells die. Both thresholds are reached through a combination of pulse characteristics. The amplitude of a pulse is typically expressed as applied voltage across the distance between electrodes and is usually expressed as volts/cm. When designing and reporting on the protocol, it is important to include specifics regarding applied voltage and electrode gap. This enables other researchers to fully understand the specifics of the pulse amplitude used. The amplitude is typically referred to as high or low voltage with approximately 700 V/cm being the dividing line. Amplitudes above 700 V/cm are considered high voltage and below are low voltage. The amplitude needed for effective delivery is related to the tissue selected and the pulse width to be utilized. Pulse width is usually in the range of microseconds to milliseconds. For GET protocols, high-voltage pulses are typically associated with high voltage and millisecond pulse with low voltage. The final aspects are the number and repetition rate of the applied pulses. While a single pulse can be used, it is more typical for multiple pulses to be applied. The majority of protocols use a single-pulse type, although combination of micro- and millisecond pulses at different applied voltages and number of pulses has been used [250, 259]. More recently, a combination of nanosecond and millisecond has been shown to be effective for delivery [260].

For most tissues, it has been demonstrated that a protocol using microsecond pulse at high field strength will result in a relatively low expression [79, 236, 238]. In contrast, utilizing a pulse protocol with millisecond pulses at low-field strength will usually result in high gene expression [77, 261–263]. This is not always the case. As mentioned previously, the target tissue will influence the expression profile. It is important to match the pulse parameters to the tissue being delivered to. The selection of the appropriate pulsing protocol largely depends on the target tissue and the therapeutic application. For example, delivery to normal liver will require millisecond pulses at low voltage to achieve high expression [74]. However, when delivering to hepatocellular carcinoma in the liver, microsecond pulses at high voltage achieve higher expression [264].

When selecting pulse parameters, there is a balance between expression levels and damage. Typically, pulsing conditions that result in higher and/or longer expression are also ones that will equate to increased cellular damage to the exposed area. It is conceivable that the parameters that equate to higher expression are related to increased uptake of DNA by the cell and this has been shown to be associated with increased cell mortality. Depending on the specific application, some cellular damage could be tolerated. For example, if the protocol is for a cancer therapy, delivered to diseased tissue or for a life-threatening disease, then some cell or tissue damage would be acceptable provided it did not impact function. However, if delivery is to healthy tissue, the protocol should be developed to cause little to no tissue damage. Inflammation could also be induced following the administration of GET. In some cases, this could be beneficial as with immunotherapy or delivery of DNA vaccines.

5.5.2.3 Electrode Selection and Design

The design of the electrode can significantly influence the success of a GET protocol [79, 258, 262, 265–268]. Pulse parameters have to be determined based on the configuration of the electrode array with respect to the number and orientation of the electrodes [52, 80, 236, 238, 239, 268]. How the field is administered is directly related to the electrodes. The design of electrodes has matured significantly since the standard in vitro approach of placing cells in a chamber (cuvette) with two parallel electrodes and administering pulses [269, 270]. For in vivo delivery, electrode designs are either penetrating (needle) or noninvasive (surface). Initial electrode designs for in vivo use similar to the cuvette were noninvasive and were essentially two parallel plates on a handle. The first penetrating design was two parallel needles that were inserted into the tissue around the injection site.

Penetrating applicators are typically designed to efficiently administer the field to deeper parts of the tissue. The applicators typically contain two or more needles and are configured in a variety of patterns including circular, triangular, square, rectangular, or hexagonal. The firing pattern can be simple such as unidirectional from one or more electrodes to corresponding electrodes on opposite side of array. For this type of sequence, the array is usually configured as two electrodes or in a square or rectangular shape or two parallel rows of needles. There are also designs that address each electrode independently and can rotate or move the field to expose the tissue from different directions [52, 53, 100, 271–275]. Rotation of the fields has been shown to enhance cell permeabilization which could potentially increase distribution of expression within the tissue [53, 273]. Other iterations include insulating portions of the needles to focus the field within certain areas of the tissue [276, 277]. It is also possible to reduce the length of the needles or to use microneedles to target the fields to the skin [278–280].

Noninvasive applicators are designed for delivery to the skin or to surfaces of internal organs. Because these designs do not penetrate the tissue, there is a reduced potential for damage from the electrode itself. As with the needle arrays, this type of applicator can contain two or more electrodes. The basic design contains a pair of parallel plate electrodes on a handle or can be incorporated onto a tweezers or caliper. This type of design tends to yield a more homogenous field across the exposed tissue [281]. As with needles arrays, the nonpenetrating electrodes can be incorporated into multielectrode arrays. One advance is to use four plates instead of two which allows a 90° rotation of the fields [268]. Another design is to use a large number of electrodes, and these are typically surface electrodes that have blunt or rounded tips at the end of the electrode [266, 267, 282]. This design allows better distribution of the electric field. In addition, the design also allows for a smaller electrode gap which reduces the applied field reducing the potential for damage or discomfort from the pulses. Other designs can incorporate defibrillator pads as electrodes or one that could fit around a blood vessel [79, 283]. The arrays could also be fit into minimally invasive devices such as an endoscope [284, 285]. Recently, a catheter device that could be deployed via an intracranial route was tested for use in electrochemotherapy to deliver the chemotherapeutic bleomycin for treatment of brain metastases [69]. It is possible this design could also be used for plasmid DNA delivery.

For both the penetrating and noninvasive electrode designs, the simpler versions are available commercially. The more elaborate designs incorporating many electrodes are typically designed and produced in investigator's laboratories. The electrode array selected for GET will be dependent on the target tissue and the specific therapeutic application being developed.

5.5.2.4 Plasmid Design

In addition to electrotransfer parameters discussed above, expression can be manipulated and targeted based on the expression vector (plasmid construct). There are multiple approaches to enhance or control expression. These manipulations can be done to augment expression with respect to levels and longevity as well as to target specific cells. The plasmid contains specific sequences that serve as enhancer and promoter regions. These regions are important in obtaining efficient transcription of the transgene encoded on the expression vector. Other components that are typically part of the plasmid include origin of replication, gene encoding antibiotic resistance, transgene being delivered, poly-A signal, and multiple cloning sites for adding other sequences. Plasmids are typically delivered in a circular form and are usually supercoiled. There are instances when linearized DNA is used for delivery. There have been many studies looking at modifying various aspects of the plasmid structure to enhance expression or to better target where the expression will occur. Most plasmids used for gene transfer have a cytomegalovirus (CMV) promoter. The CMV promoter is used quite often because it is ubiquitous and can drive high unregulated expression of the transgene.

Studies have been performed to modify various aspects of the plasmid to either enhance, control, or target the expression [286, 287]. While expression can be limited to or targeted to a specific tissue by using a specific delivery protocol, it is also possible to limit expression within specific tissues by modifying the promoter. Studies have been conducted with tissue-specific promoters combined with electrotransfer. For example, skin can be targeted by using promoters specifically for keratinocytes or dendritic cells [79, 288]. This approach can restrict expression within the target tissue although there are some limitations. Expression with tissue-specific promoters is usually lower and in some cases may not be as restricted as desired. Other approaches include utilizing an inducible promoter. In this case, following delivery, expression is turned on by introduction of a specific ligand or stimulus [289, 290]. In this case, it is also possible that there will be some baseline expression even without the signal to turn on. Targeting can also be accomplished by using a DNA nuclear targeting sequence. This enhances nuclear uptake as well as restrict expression to specific cell types [102, 291, 292]. Modifications of plasmids have also included removing the antibiotic resistance gene [293–296]. Reduction of size of the plasmids to minicircles has also been shown to potentially enhance or extend expression [297–300]. Typically, delivery of plasmid DNA, although sometimes long lived, results in transient expression as integration into the genome is an extremely rare occurrence. However, the plasmid can be modified to include transposons to facilitate integration of the transgene. The sleeping beauty transposon system can be used without causing integration into host genes [301–305]. Another approach for long-term expression and safe integration is to use the integrase from phage phiC31. With the phiC31 integrase, site-specific integration of the transgene can be performed [306].

5.5.3 Preclinical Studies for Translation of GET

Selection of the appropriate transgene and plasmid construct is the first step. Once those aspects have been selected, it is important to test the construct in an appropriate animal model. Tissue selection will be dependent on the therapeutic application and if local or systemic expression is required. It is not unusual to test a protocol utilizing a plasmid encoding reporter genes. This will give some basic understanding of the expression characteristics of a specific pulsing protocol within the target tissue. However, expression of the therapeutic transgene may differ from the reporter genes. It is advisable to confirm the expression characteristics of the protocol with the actual therapeutic construct.

When initiating efficacy studies, a consideration of desired outcome should be performed. This will determine the level of success of a particular protocol. For a vaccine, level of antibody production, immune response, or level of protection would all be points to be considered. If a cancer therapy is being developed, the level of response would be a prime variable. This could be broken down to objective response or further to partial and complete responses. In these cases, one should also consider determining duration of responses and whether or not a protective response is achieved.

5.5.3.1 Delivery to Muscle

Muscle has been one of the most used targets for GET protocols. Muscle is highly vascularized and fibrous and is relatively accessible requiring minimally invasive procedures to accomplish delivery. Both penetrating and nonpenetrating electrodes have been utilized for delivery [307, 308]. Muscle cells are long lived which can potentially lead to long expression as well. Longevity and vascular feature make it a prime target for protocols requiring systemic release of the expressed protein. Muscle can be used as a protein factory for replacement therapies [253, 309]. The muscle has also been used as a target for delivery of DNA vaccines. In addition to being an effective means of delivering the plasmid, the electric fields may also induce some inflammation at the site of delivery which could act as an adjuvant. More detailed information can be found in review articles [111, 242, 254, 310].

5.5.3.2 Delivery to Skin

Skin is an excellent target for gene transfer protocols as it is easily accessible and is the largest organ within the body [311–316]. Delivery can be done in a noninvasive manner using nonpenetrating electrodes or minimally invasively using short penetrating electrodes. Skin is an excellent target for GET as the procedure at the site can be clearly observed for both the injection procedure and placement of the electrodes. Skin is an excellent target for delivery of DNA vaccines because of the presence of antigen-presenting cells such as dendritic cells. Protocols have been developed to test DNA vaccines for the prevention of infectious disease or for potential treatment of cancers [255, 257, 258, 317–319]. This site has also been used to enhance wound healing and for treating ischemic tissue [320–325].

5.5.3.3 Delivery to Tumors

Many studies have explored the use of gene transfer for the treatment of cancer. The approach has utilized both direct and indirect therapies. GET is an attractive approach because it allows for better control of the expressed transgene. This is important when delivering a transgene encoding a toxic protein or one that will stimulate the immune system. In both cases, the amount of expression should be regulated to achieve the desired response with minimal adverse effects. GET has been used to deliver several different types of transgenes. The strategies have included immune stimulation, agents that induce apoptosis, inhibition of angiogenesis, and toxic molecules that could reduce size of tumor [79, 239, 275]. Most protocols currently being tested have targeted cutaneous or subcutaneous tumors such as melanoma, squamous cell carcinoma, head and neck cancer, and breast cancer due to its accessibility. Direct delivery to internal tumors can be accomplished by combining with surgical procedures. As electrode design improves, it may be possible to access internal tumors using minimally invasive tools [69, 284, 285].

5.5.3.4 Delivery to Other Tissues

GET has been used to successfully deliver plasmids to many tissues. If the tissue is accessible to inject the plasmid and properly place the electrodes, then GET can be performed. Tissues that have been tested include liver, lung, heart, vasculature, eye, brain, testes, kidney, and spleen. Each of these tissues has their own unique challenges as to how to place the electrodes. In many cases, this will involve the use of either surgery or some form of invasive or minimally invasive approach. In this chapter, the possible applications for only a few of these tissues will be explored.

Liver has been used as a target tissue for both the potential treatment of metabolic disease and cancer. The plasmid can be introduced through direct injection into the tissue or via an intravascular delivery. As with other tissues, following injection of plasmid, the fields are applied through specially designed electrodes. Both penetrating needle electrodes and nonpenetrating electrodes (i.e., tweezertrodes) have been used [74, 326–328]. GET has also been used to deliver an agent to potentially prevent radiation-induced liver damage. In this study, the plasmid delivered encoded hepatocyte growth factor [329]. As more devices are developed to enhance access to the tissue, it is highly probable that more therapeutic protocols will be developed.

Lung is an important target for gene transfer protocols as it is a location for many genetic and acquired diseases. The difficulty for GET to the lung is the appropriate placement of the electrodes. However, effective GET protocols have been developed that provide efficient expression with minimal injury or inflammation of the airway [243, 244]. Initial studies utilized two flat plates that were placed across the chest of mice and rats to deliver electric pulses following inhalation of DNA solution [244]. This work has expanded to larger animal models including sheep and pigs. The electrode designs have also improved [245, 330, 331]. While the procedures were worked out utilizing reporter genes, more recent studies have tested the use of therapeutic genes. These later protocols were evaluating the ability to treat tumors, reduce inflammation, or improve lung function [332–334]. The application of pulse electric fields to the chest raises concerns related to potential disruption of cardiac function. However, in the studies conducted thus far, no signs of cardiac dysfunction have been observed [243, 244, 335].

As mentioned above, delivery of electric pulses to internal tissues raises the concern of cardiac damage or fibrillation. This would suggest that performing GET directly to the heart would not be possible. However, several studies have demonstrated that GET could be a useful tool for delivering plasmid DNA to myocardiocytes. GET has been tested in isolated cardiac cells and in ex vivo tissue preparations including Langendorff-perfused whole heart [240, 241, 336–338]. Evaluation of the effects of electric fields on the heart following defibrillation opened up the possibility of using gene transfer directly to the beating heart [338]. A key component to successfully perform GET to the beating heart is to synchronize the applied pulses with the electrocardiogram. The pulse needs to be applied during the QRS waves and be completed prior to the initiation of the T wave [240]. Additional critical components are electrode design, pulse parameters, and delivery site [241, 339]. GET has been used successfully to deliver plasmids encoding angiogenic factors directly to the beating heart [240, 241, 339]. The procedure was performed safely and resulted in significantly elevated expression of the delivered transgene. Electrotransfer of plasmid DNA directly to porcine heart was shown to be safe and effective and could provide an alternative approach for potential therapies for coronary diseases.

GET has been used to deliver plasmid DNA to several other tissues. An interesting target is blood vessels. Delivery could be attempted either through the use of catheters to target the inner surface of the vessel or externally utilizing surgical approach [243, 283, 340–342].

This method of gene transfer can also be used for the treatment of ocular diseases. Successful gene delivery to the cornea, retina, subretinal space, and ciliary muscle has been accomplished using GET. The protocols tested resulted in high levels of expression with minimal or no tissue damage or inflammation [343–347]. Success in this area could be a benefit for a variety of optical impairments.

5.5.4 Clinical Studies of GET

Gene therapy has the potential to be utilized as an effective treatment for many disorders. Clinical testing has been initiated for a variety of diseases including cancer, metabolic diseases, and cardiovascular diseases [79, 80, 236, 239]. In addition, gene-based approaches have been utilized for prophylactic or therapeutic vaccines for infectious diseases and cancer [80, 255, 348, 349]. The vast majority of gene therapy clinical trials are Phase I safety and tolerability studies. Many trials are not successful in Phase I due to either adverse reactions or low efficacy. Only about one third of these studies advance to Phase II studies and even fewer to Phase III.

The clinical use of GET has been steadily increasing. This is mainly due to reproducibility, high efficiency, potential effectiveness, and relatively low adverse effects. In preclinical studies, GET has been shown to be effective in delivering plasmid DNA to a large number of tissues; however, thus far in clinical studies the target tissues have been muscle, tumor, and skin. There are multiple steps in moving a potential therapy from the bench to the bedside. Efficacy of the approach needs to be evaluated in the appropriate preclinical model. In some cases, this may require two animal models. In addition to efficacy, a major aspect is to document safety. Prior to initiating a clinical trial, a toxicity evaluation should be performed. Evaluation should include hematological and histological data. The analysis should include relevant doses, number of treatments, and tissue target and include multiple time points [350]. It is also important to determine the distribution and persistence of the plasmid within the animal following therapy. If the preclinical studies demonstrate effectiveness and safety, the protocol can be submitted to the appropriate regulatory agencies for approval to initiate the clinical trial.

When developing a GET protocol, parameters should be selected to achieve the correct balance between expression of the transgene and tissue damage. This will be dependent on the tissue target and the expression profile required for the specific application. The advantage of using GET is that by manipulating the delivery parameters, a specific expression pattern can be achieved that can be used to obtain the desired clinical outcome [238]. The electrical parameters as well as electrode design were all discussed previously in this section [52, 119, 238, 274, 281, 351, 352].

It is possible to search and determine the status of clinical trials by using the searchable database Clinicaltrials.gov (www.clinicaltrials.gov). This is an outstanding resource that allows you to search based on one or more keywords. Conducting a search using the keywords "electroporation" and "electrotransfer" in May 2015 resulted in 122 trials being listed. Of these studies, 54 did not use gene-based therapy. Thirty nine were designed to evaluate irreversible electroporation to ablate tumors. Thirteen are designed to evaluate the use of electric fields to effectively deliver a chemotherapeutic drug such as bleomycin, and the other two are evaluating the delivery of calcium as an antitumor approach. These 54 studies are all focused on local ablation of solid tumors evaluating treatment of head and neck cancer, colorectal cancer, cutaneous malignancies, pancreatic cancer, brain metastases, keloids, and breast cancer. The other 69 clinical trials utilize a gene-based approach for therapy. Within this group of trials, three are follow-on studies, two are testing the safety of devices, and six are performed ex vivo on isolated cells prior to the modified cells being delivered to the patient. GET is performed directly to patients in the remaining 58 applications. Of the 64 studies evaluating the use of GET both with cell-based or in vivo delivery, 23 have been completed and 17 are actively recruiting. Forty one are in Phase I, 11 are listed as Phase I/II, and 12 are in Phase II. The major approach that is being tested is the delivery of DNA vaccines. The DNA vaccines are being tested for prevention of infectious diseases (30 trials) or treatment of cancer (16 trials). There are also 18 clinical trials evaluating antitumor therapies. Delivery to the muscle is the major target (49 trials) being utilized. Ten studies use intradermal delivery and seven use intratumor delivery. There are three studies comparing intramuscular and intradermal delivery.

5.5.4.1 Cancer

There are many trials evaluating potential therapies for cancer. A major approach is to use DNA vaccines to stimulate an immune response against existing disease [80, 112, 239]. Another immunotherapy approach is to deliver plasmids encoding cytokines or other immunostimulating molecules directly to the tumor or via an intramuscular route. Recently, three trials were initiated to evaluate GET as a means to deliver plasmid encoding an antiangiogenic factor [353]. The cell-based approaches are designed to stimulate an immune response by either modifying immune cells (dendritic or T cells) or modifying cancer cells [87–92]. There are nine trials testing therapies for melanoma, six trials for cervical cancer, four for squamous cell carcinoma, six for multiple types, two for prostate, and one each for Merkel cell carcinoma, chronic T-cell lymphoma, colorectal, B-cell malignancy, multiple myeloma, and leukemia cancers.

The first clinical trial conducted using GET was an immunotherapy approach utilizing a plasmid encoding interleukin-12 delivered directly to cutaneous melanoma tumors (NCT00323206). While this Phase I study was focused on safety and tolerability, efficacy was demonstrated as over 70 % of treated tumors went away and in three patients both treated and untreated tumors completely regressed [81]. Several studies utilizing cancer vaccines have been completed. One for melanoma utilized a plasmid encoding tyrosinase and observed immune stimulation with minimal toxicity [354]. Another study evaluated delivering a plasmid encoding a tetanus toxin domain fused to prostate-specific membrane antigen for patients with prostate cancer. Immune reactivity was also seen in this trial with limited toxicity [355, 356]. A second trial for prostate cancer patients is testing a plasmid encoding prostate-specific antigen (PSA). This one utilized an intradermal administration, while the previous one utilized an intramuscular delivery. While increases in PSA were noted, there were limited vaccine-specific responses [357]. Intramuscular administration was used to deliver a DNA vaccine targeting HPV16/18 in women with cervical intraepithelial neoplasia grade 2 or 3. Specific T-cell responses were observed with no dose-limiting toxicities [358].

5.5.4.2 Infectious Disease

A major area of growth for clinical use of GET is the delivery of DNA vaccines for infectious diseases. The use of DNA vaccines for this indication is appealing as it does not require pathogen cultivation as with conventional vaccines. It also reduces the risk of adverse reactivity to the vaccine and is a viable alternative to the use of live attenuated vaccines which have the potential for infection in immunosuppressed and immunocompromised individuals. Another major advantage is that if an outbreak occurs, DNA vaccines can be produced quickly. The major issue with the efficacy of DNA vaccines has been poor expression. GET has resolved this major issue [79, 80].

There are 29 trials listed on clinicaltrials.gov that are testing GET for delivery of DNA vaccines. There is one trial testing a vaccine against a parasitic agent and the other 28 are evaluating vaccines for viruses. Fourteen of the trials are testing HIV vaccines, four for influenza, three for hepatitis B, three for hepatitis C, one for Ebola, two for Hantaan and Puumala viruses, and one for Venezuelan equine encephalitis virus. The major administration route is intramuscular (23 trials) with four testing intradermal administration and two comparing the two routes of administration. Of the 29 trials listed, 13 have been completed and 6 are currently recruiting.

A major target for DNA vaccines with GET is HIV. Several studies have evaluated a variety of antigens. All have shown increased immunogenicity without any significant adverse effects [359–362]. The other aspect that has been evaluated has been tolerance to the administration of the electric pulses. The majority of patients thought the procedure was tolerable. Greater than 90 % thought that GET delivery of vaccines for life-threatening diseases was acceptable. This percentage dropped considerably if the vaccine would be for non-life-threatening diseases. Immune stimulation was also seen in trials testing vaccines for hepatitis viruses. In a Phase II trial for Hep B, the vaccine was tested either or alone or in combination with lamivudine therapy (100 mg daily, GlaxoSmithKline Pharmaceuticals, Suzhou, Jiangsu Province, China). The vaccine was well tolerated, and in the monotherapy, there was specific T-cell responses observed, and in the combination therapy, there was a suppression of Hep B DNA copy number in the serum [363]. In the hepatitis C vaccine studies, no serious adverse events were noted, and increased antibody and specific T-cell responses were observed [364].

5.5.5 Summary

Gene medicine has held great promise, but has been hampered by the lack of effective and safe delivery approaches. GET has emerged as a potential means to augment delivery in a safe and effective way. Because electrotransfer is a physical approach, it is possible to use it to deliver molecules to any accessible cell or tissue. In addition, based on tissue target and delivery parameters chosen, a specific expression pattern can be obtained. This flexibility when using GET to deliver plasmid DNA can increase the potential to achieve the desired clinical response by obtaining specific expression characteristics. This flexibility has helped fuel the growth of GET over the past decade. It has been recognized as an important nonbiological tool for performing effective gene transfer studies. Numerous studies have been performed that has documented the effectiveness of this approach [80, 112, 239, 255, 265, 349, 365, 366]. Interestingly, based on the delivery parameters used, it is possible to induce either a local or systemic response using GET. This approach is applicable to treating a variety of disease including metabolic diseases and cancers as well as to effectively deliver DNA vaccines for treatment or prevention of infectious diseases or cancer.

In this section, discussion has focused on the delivery of plasmid DNA. However, recent studies have shown that electrotransfer can also be used to effectively deliver RNAs for the treatment of disease. Studies have been performed to deliver siRNA, shRNA, mRNA, or RNAi. Plasmids encoding shRNA have been used to evaluate the potential effectiveness in treating cancer, increase in muscle mass, erectile dysfunction, and leukemogenesis [82–85]. Thus far, all of the studies evaluating siRNA and shRNA have been preclinical, and no clinical trials have been initiated. There have been three clinical trials initiated for evaluation using mRNA. In these studies, dendritic cell or cancer cells were modified with mRNA to express an antigen or a cytokine [87–92].

The use of GET continues to mature. Additional applications and targets are being reported on a regular basis. As clinical testing continues to increase and more results are reported, the true potential of this delivery approach will be determined. It should be noted that the use of electric fields to deliver DNA is relatively young. The first in vitro use was reported in 1982 [367], but the first in vivo studies were not reported to the early 1990s. It was not until the late 1990s that the first studies evaluating the therapeutic potential of in vivo GET were reported. It was almost 10 years later when the results from the first clinical trial were reported. Since then there has been a tremendous increase in publications reporting on the potential applications and additional tissue targets being tested. With the increase in clinical trials and positive reports, it is clear that GET can be used safely and has the potential to be an important tool in the creation of new gene medicines.

5.6 Nanosecond Pulse Application

Stephen Beebe and Barbara Hargrave

Applications of Bioelectrics for Cancer, Wound Healing, and Treatment of Heart Failure – A general overview of uses for bioelectrics includes applications that deliver DNA to cells and tissues, transiently or permanently alter cell membranes and other cellular elements, modulate cell signaling mechanisms, stimulate the release of cellular contents, disinfect, and decontaminate [368]. Therapeutic applications include treating cancer [11, 13–16, 369, 370], activating platelets to enhance wound healing [371], modulating cardiomyocyte action potentials [18], activating platelets in the preparation of platelet-rich plasma for the treatment of cardiac ischemic reperfusion injury [19, 20], inactivation of opportunistic wound pathogens on the skin [21], imaging malignancies [23], and controlling pain by blocking nerve conduction [25]. Other applications for nanosecond pulsed electric fields (nsPEFs) are discussed elsewhere that include biofouling [17, 372] and cold plasmas [37] for sterilizing liquids, foods, and surfaces; purifying air and water [373], and removing oxides of nitrogen and sulfur from diesel exhaust [374]. Another emerging field in bioelectrics that will be presented elsewhere includes plasma medicine, where room temperature atmospheric-pressure plasmas are used for sterilization, hygiene, and dental/medical applications [375].

In this section, applications of nanosecond pulses will be presented for treatment of cancer, wound healing, and cardiac failure.

5.6.1 Treating Skin Cancer

A wide range of stimuli with either plasma membrane, intracellular membrane, or other structural targets can be responsible for the induction of regulated cell death including apoptosis. This includes pharmacological agents, toxins, radiation, or simple removal of serum from the culture medium to trigger an apoptotic cascade. As discussed in one of the previous sections, nsPEFs joined the ranks of apoptosis inducing agents in 2002 [11, 376]. Thus, application of nsPEFs can be considered as a way to treat tumors without the additional requirement of chemotherapeutic agents, with no residual side effects following treatment.

It was shown that electric pulses (300 ns in duration and 40 kV/cm) caused total remission of B16f10 melanoma cancer in mice. When B16f10 tumors are treated in SKH-1 mice, they exhibited three rapid changes: (1) the average nuclei of tumor cells shrank by 50 % within an hour; (2) the vasculature to tumors was disrupted within a few minutes and was not restored for up to 2 weeks, depriving tumors of essential blood supply; and (3) tumors shrank by 90 % within 2 weeks. At this point, most of the roughly 200 treated tumors began to grow again, but if treated a second time, they completely disappeared. Long-term studies demonstrated that complete tumor regression occurred within an average of 47 days in the 17 animals treated. None of these melanomas recurred during the 4-month period after the initial melanoma had disappeared [369, 377]. Immunohistochemistry studies on B16f10 tumors clearly showed apoptotic effects of nsPEFs with time-dependent increases in executioner caspase 3/7 as well as increase of two markers for DNA damage, TUNNEL-positive cells, and histone 2AX phosphorylation [370], which is one of the earliest markers for DNA damage [378]. This was also addressed in a modeling paper, where apoptosis induction by nsPEFs was considered in terms of nanoporation of the plasma membrane [379].

In addition, decreases in microvessel density (MVD) markers were noted after nsPEF treatment [370]. After disruption of vasculature, tumor growth depends on the availability of nutrients, which requires formation of new capillaries from preexisting vessels in a process known as angiogenesis. After nsPEF treatment, decreases were observed in common MVD markers including CD31, a platelet endothelial cell adhesion molecule used as a panendothelial cell marker; CD34, an endothelial marker; and CD105, a proliferation-associated and hypoxia-inducible protein expressed in angiogenic endothelial cells. The decrease in MVD markers is consistent with the absence of a return of blood flow to nsPEF-treated B16f10 tumors [369].

Another study investigated murine BCCs developing in the skin as a result of weekly exposure of 5 Gy radiation (Cs_{137}) for 2–12 months [380]. These tumors in Ptch1(+/−)K14-Cre-ER p53 fl/fl mice are histologically similar to human BCCs as well as in their response to drug therapy. Tumors were treated with 300 pulses of 300 ns duration or 2700 pulses of 100 ns duration, all at 30 kV/cm and 5–7 pulses per second. For tumors that were fully encompassed by electrodes, 99.8 % were ablated after 4 weeks. Histological analysis suggested that a single treatment by nanoelectroablation did not completely ablate tumors in mice.

Yet another study treated carcinogen-induced cutaneous papilloma and squamous cell carcinoma with nsPEFs with durations of 7 ns and 14 ns and electric fields of 31 and 40 kV/cm using 50, 200, or 400 pulses [381]. One treatment with 200 pulses with durations of 14 ns, repetition rate of 50 Hz at electric fields of 31, or 40 kV/cm amplitude was sufficient to clear tumors (71.4 % and 87.5 % at respective electric fields) 1 week after treatment. For pulses with 7 ns durations, electric fields of 40 kV/cm as many as 400 pulses at 50 Hz were required to eliminate lesions (13.5 %) 1 week after a single treatment.

A study by Garon and colleagues [16] provided the first direct evidence that nsPEFs could effectively treat human skin cancer. A single case of a human basal cell carcinoma (BCC) exhibited complete remission and very little scarring after one treatment with nsPEF (200, 20 ns 43 kV/cm pulses). This single case indicated that this new therapy, which had proven very effective for treating mouse skin cancer, could be equally effective on humans. The first human trial with 10 BCCs on three subjects with 100–1000 pulses with a duration of 100 ns at 30 kV/cm in amplitude and a repetition rate of two pulses per second demonstrated significant success [12]. Nine tumors were ablated, seven completely ablated, and only one recurred by week 10. Importantly for visible skin lesions, no scars were visible at the healed sites of any of the successfully ablated BCC tumors.

5.6.2 Nanosecond Pulsed Electric Fields for Treating Other Cancers

While skin cancers are relatively easy targets for application of nsPEFs based on their locations, other cancers are also susceptible to ablation by nsPEFs. Ectopic mouse hepatocellular carcinoma (Hepa1-6 HCC) was also shown to be ablated by nsPEFs [15]. The pulses were delivered to tumors by a needle array or ring electrode. The needle electrode was a 4+1 needle array with four grounded perimeter needles forming a square and a high-voltage biased center needle. The ground needles are all about ~5 mm from the center needle. For the ring electrode, a coaxial ring applicator replaced the grounded needle electrodes. The inner ring diameter was ~8 mm, the same as the diagonal distance of the needle array. Most studies were carried out with the needle array. These studies considered whether 100 ns or 30 ns pulses were more effective when 900 pulses were delivered at an applied electric field of either 33, 50, or 65–68 kV/cm with needle electrodes in one treatment or with 300 pulses delivered three times with 48 h intervals. Results showed that the most effective treatments included 100 ns, 65–68 kV/cm pulses. There was little difference with the three treatment (66 %) vs. one treatment (75 %) protocol. Efficacy was low (33 % ablation) when three treatments were given with 100 ns at 33 or 50 kV/cm pulses. The needle electrode may have been better than the ring electrode (25 % vs. 15 % ablation, respectively) when 30 ns, 65–68 kV/cm pulses were delivered in three treatments, but again efficacy was low. Analyses of molecular mechanisms of action demonstrated that executioner caspase 3, 6, and 7 were transiently active in less than 50 % of cells (caspase 3 ~ 45 %, 7 ~ 30 %, 6 ~ 15 %) with all peak activities occurring 3 h after treatment. Thus, many, but not all, cells died while undergoing apoptosis; other cell death mechanisms were operative as ablation progressed. Evidence also indicated that BAD was not phosphorylated and thereby was possibly active to promote apoptosis. A delayed presence of TUNNEL-positive cells that coincided with caspase activation, both with transient peaks at 3 h after treatment, suggested that DNA damage was related to apoptosis as opposed to direct electric field damage. The appearances of nuclear condensation and pyknotic nuclei were consistent with TUNNEL findings. Like results in B16f10 melanoma cells, Hepal-6 HCC cells showed decreases in the endothelial cell marker CD34 and the angiogenic marker VEGF, both indicating diminishing vascularization in the treatment zone.

Another study evaluated ablation efficacy of nsPEFs on a highly metastatic human hepatocellular carcinoma cell line (HCCLM3) xenograft model on BALB/c nude mice, which is unable to produce T cells and is therefore immunodeficient [382]. The nsPEFs were delivered to subcutaneous tumors with a pair of electrodes with a needle anode placed within the tumor mass and a semiring cathode placed on the tumor periphery. Using 100 ns, 40 kV/cm pulses, decreases in cell viability were dependent on pulse number up to 40 pulses with no differences in maximum ablation up to 90 pulses (~75 %). In contrast to single vs. split treatments in Hepal-6 HCC [15], treating HCCLM3 HCC with 100 pulses 3 times at 48 h intervals (90 % ablation) was better than treating one time with 300 pulses (~45 % ablation) after about 8 weeks. The study suggests a reduction in metastasis, but this could be due to complete tumor clearance at the treatment site with no cells available for metastasis. No conclusion about immunity can be made in the immune-incompetent model.

Studies for treating pancreatic carcinoma in a murine xenograft model also demonstrated that nsPEFs or nanoelectroablation was effective to eliminate BxPC-3 and Capan-1 human pancreatic cells derived from pancreatic tumors [383]. The study determined that a minimum number of pulses for effective treatment were between 250 and 500 when 100 ns, 30 kV/cm pulses were applied. The electrode was a six-needle suction electrode composed of two parallel rows of three electrodes on each side. Using 500 or 1000 pulses, 17 of 19 tumors (89 %) completely regressed and 16 lived for 270–302 days when experiments were terminated.

While both pancreatic and HCC tumors have been successfully ablated with nsPEFs, neither tumor type was in its natural location. The first orthotopic, internal tumor studies were treated in a rat N1-S1 HCC model using 1000, 100 ns, 50 kV/cm pulses at 1 Hz [14]. Using 100, 300, or 500 pulses did not completely ablate these HCCs [13]. Eighty to ninty percent of N1-S1 HCC tumors were completely ablated for as long as 20 weeks when experiments were terminated. Both caspase 3 and caspase 9, but not caspase 8, were activated, indicating intrinsic apoptosis induction. Unlike caspase activities in Hepa1-6 HCC or B16f10 melanoma, caspases remained active for at least 6 h. As expected for intrinsic apoptosis, caspase 9 was activated before caspase 3. However, again, not all cells exhibited caspase activity, indicating that more than one cell death mechanism were activated. Rats with successfully ablated tumors failed to regrow tumors when challenged with a second injection of N1-S1 cells implanted in the same or different liver lobe that harbored the original tumor. Granzyme B-positive cells were present within the first day after treatment and continued to increase up to 9 days. This indicated the presence of activated innate and/or adaptive immune cells and an immune response. These data and other data are building a case that nsPEF ablation may induce an immune response that provides a vaccine-like effect. This suggests that efficacy of nsPEF ablation involves induction of regulated cell death programs as well as immunogenic mechanisms for tumor clearance and perhaps vaccination.

5.6.3 Treating Superficial Skin and Deeper Chronic Wounds with Platelet-Rich Plasma Supernatant Prepared Using nsPEFs

The requirement for wound healing following skin injury is essential for survival of all mammalian organisms. The skin is the largest organ in the body. It serves as a barrier to protect underlying tissues from the outside world, helps regulate temperature through sweating and changes in blood flow, and serves as an organ of sensation and as source of vitamin D synthesis. Following skin wounding, complex and concerted functions are required from resident skin cells, hematopoietic cells, and immune cells. Skin can be wounded in a number of ways. These include open wounds that separate the epidermis from the dermis as well as closed wounds such as blunt force contusions causing partial thickness skin injuries. Thermal wounds or burns are caused by heat or scalds, electricity, radiation, and sunburns and are graded as first, second, and third degree based on the wounding depth as superficial, superficial partial thickness/deep partial thickness, and full thickness, respectively [384]. Other types of skin wounds are chronic such as venous leg ulcers and diabetic foot ulcers caused by complications of diabetes mellitus. Still other wounds are iatrogenic such as sutured or stapled closed surgical wounds or wounds due to chemical skin peeling or plastic surgery. Regardless of the type of wound, repair involves four overlapping classic phases of healing [385–387]. These include hemostasis, which involves platelet aggregation in a mesh of cross-linked fibrin and release of growth factors and cytokines that "kick-starts" wound closure and the healing process; inflammation, which uses chemotactic signals that first recruit neutrophils and monocytes and later macrophages that remove dead and dying cells and prevent infection and septicemia and release angiogenic factors; proliferation, which involves new tissue formation by migration of keratinocytes, angiogenesis, and restoration of epithelial barrier; and remodeling, which involves restructuring tissue by matrix metalloproteinases on an acellular matrix background with collagen as a backbone.

Autologous platelet-rich plasma (PRP), which is prepared by concentrating platelets in vitro from the patient's blood, can serve as a therapeutic treatment for skin wounds. Before these procedures are presented, let's briefly review some information about platelets.

The major function of platelets is to contribute to hemostasis (to stop bleeding) as a first step for wound healing. Platelets are not really true cells, but fragments of megakaryocytes, which are large bone marrow cells. Hundreds of platelets are formed from a single megakaryocyte; platelets constitute the major "cell type" in blood. They have plate-like discoid shapes with smooth biconcave surfaces that express proteins on their surface that allow them to stick to breaks in blood vessels and aggregate. Platelets have invaginations and external openings in their plasma membranes to an open canalicular system for intake of stimulatory agonists and the release of effector substances and a closed-channel network of residual endoplasmic reticulum that form a dense tubular system. They contain alpha granules and dense granules that secrete factors that are important for clot (thrombus) formation and wound healing. They float in a quiescent (inactive) state through circulation along the endothelium of blood vessels until a wound or bleeding occurs. Then hemostasis occurs involving two major interactive processes of platelet activation and blood coagulation. Platelets undergo activation at a vessel wound site where they encounter subendothelial molecules such as collagen. Activation occurs by converting prothrombin to thrombin, which serves as an agonist for platelet aggregation in addition to thromboxane and ATP and ADP. These ligands bind to cognate receptors acting through G-protein signaling that lead to release of intracellular Ca^{2+}, depolarization, and influx of extracellular calcium and other ions. The platelets change shape, extending pseudopodia or long tentacles, and expose phosphatidylserine on the outer membrane surface providing negative charges and optimal surface for assembling coagulation complex. They increase their adhesiveness; release growth factors, chemokines, and proteases from granules; and activate receptors such as integrins on the platelet surface that bind fibrinogen. This leads to platelet aggregation (primary hemostasis) and participation in coagulation cascade that leads to fibrin formation (secondary hemostasis) and a resultant clot [394, 395].

The role of platelets goes beyond hemostasis and coagulation. Platelets are known to be involved in all stages of vessel repair including coagulation, immune cell recruitment, inflammation, wound healing, angiogenesis, and remodeling [396]. It is the contents of the α-granules and dense granules that are important for all of these processes. Secretion of more than 300 bioactive molecules from them is the most influential function of platelets not only at the thrombus site but beyond it. While it is beyond the scope of this chapter to discuss these events and their implication in detail, wound healing involves restoration of vascular integrity and angiogenesis, which requires coordinated proliferation and migration of smooth muscle cells, fibroblasts, and endothelial cells. Some of the most important mediators in these events are secreted from α-granules including platelet-derived growth factor (PDGF), vascular endothelial-derived growth factor (VEGF), epidermal growth factor (EGF), and transforming growth factor beta (TGFβ). PRP gels have been used in a number of clinical scenarios, from healing acute skin wounds and diabetic ulcers to regeneration of tendon, ligament, and nerve tissue [397]. These PRP gels are generally activated from the patient's blood using bovine thrombin as the agonist. However, another way to activate platelets that is safer and perhaps better is to stimulate them with nsPEFs as a non-ligand platelet activator.

To prepare nsPRP, PRP (platelets concentrated sevenfold from whole blood) is treated in electroporation cuvettes with five 300 ns pulses with electric field of 30 kV/cm by a pulse power device as previously described [371]. After platelet aggregates are visible, they are centrifuged and the supernatant is used as treatment.

In the hemostasis stage of wound healing, the aggregation of platelets plays a major role in preventing bleeding [388]. It has been shown that nsPEF has an effect on platelets similar to thrombin [371], an agonist that promotes aggregation. This process of aggregation is initiated by calcium leaking out of internal stores through nanopores and influx through nanopores in the plasma membrane. This is consistent with the well-known fact that aggregation of platelets by known agonists, such as thrombin, requires an increase in intracellular free calcium. However, mechanisms for calcium-induced aggregation of platelets by nsPEFs and thrombin are different [371]. Thrombin does this by increasing IP_3, which releases calcium from the ER and subsequently stimulates store-operated calcium channels (SOCC) in the plasma membrane. In contrast, nsPEFs cause smaller calcium increases due to internal release, yet produce greater influx through nanopores in the plasma membrane rather than through SOCC.

In addition to aggregation, an increase in platelet-derived growth factor (PDGF) release has been observed after washed platelets were pulsed with nsPEF [371]. It is likely that both PDGF and calcium are released from α-granules through nanopores by nsPEF-induced subcellular electromanipulation [10] along with other growth factors essential for wound healing. Wounds created in New Zealand White (NZW) rabbits and treated with nsPRP showed significantly greater blood flow in the wound area than wounds treated with vehicle (0.9 % NaCl) [19]. Two surgical wounds were created and analysis demonstrated the efficacy of PRP prepared using nsPEFs. A 3×8 cm skin flap wound was created on the dorsal surface of the animal. NsPRP or 0.9 % saline in a volume of 1 mL was placed on the wound and the flap sutured to its original position using 4-0 silk interrupted stitches. To investigate the effect of nsPEF-activated nsPRP on wound healing in the presence of ischemia, an incision was made on the medial thigh of the right hind limb. Ischemia was created by the removal of a 2 cm segment of the femoral artery. NsPRP supernatant was delivered continuously to the wound via an Alzet Osmotic pump measuring 3×0.7 cm. The pump was placed subcutaneously in the right thigh of the hind limb parallel to the ischemic region of the wound. The flow rate of the pump was 0.25 μL/h. The pump was filled with 250 μl of either nsPRP supernatant or saline and released over 28 days. Blood flow in the wound area was measured on day 0 before and after surgery, followed by postoperative days 3, 7, 14, 21, and 28 using laser Doppler imaging.

In addition to growth factor release, recent studies have also shown that nsPEF-induced activation of platelets has an antibacterial effect [21]. Human platelet-rich plasma supernatants were highly bactericidal against *A. baumannii* and moderately but significantly bactericidal against *S. aureus* in vitro and ex vivo in a skin model. *P. aeruginosa* was inactivated modestly but significantly ex vivo, but not in vitro.

5.6.4 Treating Myocardial Ischemia and Reperfusion Injury with Platelet-Rich Plasma Supernatant Prepared Using nsPEFs

When patients are admitted to a hospital with a myocardial infarction (heart attack), some of the myocardial tissue is damaged by ischemia (insufficient blood flow to the myocardium). The ischemia is generally caused by blockage of one or more of the coronary vessels and can result in dysfunction of one or more chambers of the heart, particularly the left ventricle (LV). The LV of the heart is the major pumping chamber and has the responsibility for distributing blood to other parts of the body. Although it is essential to restore coronary flow to the ischemic region (reperfusion), reperfusion, paradoxically, causes the production of reactive oxygen species (ROS), which can lead to substantial reperfusion injury (i.e., tissue death). This tissue loss in many cases increases the size of the infarct and starts a remodeling of the heart that after several years can lead to heart failure and sometimes death. So, a treatment for reperfusion injury could save thousands of lives every year. PRP has been used as a therapeutic treatment for ischemic reperfusion injury of the heart. The platelet concentrate is activated using nsPEFs producing nanosecond platelet-rich plasma (nsPRP). Currently, in most PRP preparations, platelets are activated to release proteins that are stored within their α-granules by exposing them to bovine or human thrombin. However, many principal actions of thrombin may be harmful to the heart. Thrombin is a multifunctional protease which is proapoptotic and proinflammatory [389] and may have adverse effects on cardiac myocytes, which are independent of its procoagulant activity [389–391]. It has been shown that in humans, thrombin generation during reperfusion after coronary artery bypass surgery is associated with postoperative myocardial damage [391–393] and that its thrombotic activity is only partially suppressed by heparin [390]. Also, it has been reported that PRP has the potential to increase the osteo-inductivity of demineralized bone matrix, but that its activation with bovine thrombin immediately prior to implantation significantly inhibits this activity [393], suggesting that thrombin may interfere with the efficacy of PRP. Of importance to the heart is the fact that thrombin has been shown to cause generation of ROS [390] and induction of apoptosis [392].

It was shown in a rabbit model that injection of 200 μl nsPRP into the ischemic area of the myocardium, in vivo, improved LV contractility and reduced infarct size. In addition, ROS production was reduced in H9c2 (rat heart) and HUVEC (human umbilical vein endothelial) cells in culture, pretreated with nsPRP for 24 h and then stimulated with hydrogen peroxide (H_2O_2). NsPRP significantly reduced mitochondrial depolarization in the H9c2 and HUVEC cells in response to H_2O_2. In a rabbit ischemia–reperfusion Langendorff model, work function (left ventricular pressure x heart rate) was significantly greater in hearts treated with nsPEF-activated PRP (nsPRP) than in hearts treated with thrombin-activated PRP or BSA-treated hearts [20]. Furthermore, the lysate from platelets activated using nsPEFs has been shown to support left ventricular contractility in rabbit hearts in vivo and in the Langendorff rabbit heart model following ischemia and reperfusion [19, 20]. In vitro, rabbit and swine nsPRP supernatant has been shown to reduce the production of reactive free radicals (ROS) and mitochondrial depolarization in both H9c2 and HUVEC cells in culture [19, 20]. Interestingly, there is preservation of endogenously produced antioxidant catalase (CAT) and superoxide dismutase (SOD) in nsPRP supernatant when platelets are activated with nsPEFs. This phenomenon is not observed when platelets are activated with bovine thrombin [20].

The mechanism(s) associated with the improved contractility of the left ventricle in the nsPRP-treated hearts remains to be more specifically determined and may involve interaction of several parameters. First, nsPRP reduces production of ROS in H9c2 and HUVEC cells stimulated with H_2O_2 in vitro and, therefore, may possibly reduce ROS in the ischemic heart during reperfusion. ROS are highly reactive chemical entities that can exert harmful effects on heart tissue when produced in concentrations that overwhelm the body's inherent antioxidant system. ROS have direct electrophysiological effects that contribute to arrhythmias [398] and are implicated in the pathogenesis of postischemic myocardial stunning (contractile dysfunction that is reversible). Myocardial cell death after ischemia and reperfusion results from necrosis and apoptosis which can be induced by ROS generated during ischemia and reperfusion, effectively increasing the size of the myocardial infarction. In addition to a reduction in ROS, significantly smaller infarct areas occurred in hearts treated with nsPRP. Another mechanism which nsPRP may use to protect the heart from further ischemic/reperfusion damage is suggested by data showing a reduction in mitochondrial depolarization in H9c2 and HUVEC cells treated with nsPRP. In the heart, mitochondria make up approximately 20–35 % of cardiomyocyte volume. Although mitochondria are a source of ROS, mitochondria also provide ROS defense mechanisms, including enzymatic antioxidants. When there is an imbalance between ROS generation and ROS scavenging, as is the case with ischemic reperfusion injury, accumulated ROS can alter the function of proteins, lipids, and DNA through structural modifications [399]. The nsPRP supernatant contained significantly greater amounts of CAT and SOD than did the PRP supernatant prepared with bovine thrombin [19].

NsPRP provides a novel approach to activate platelets in the preparation of PRP without red blood cells and immunoactive agents such as thrombin. This creates pores in the platelet membranes (electroporation) through which growth factors and cytokines can diffuse. The nsPRP supernatant is as effective, if not more so, as the PRP supernatant prepared using thrombin. It improves contractility of the left ventricle and reduces the size of the infarct in animals exposed to cardiac ischemia and reperfusion.

Applications of nsPEFs presently include treatment of cancer, skin wounds, and ischemia–reperfusion injury in the heart. As a therapy for local tumor ablation, nsPEFs have shown efficacy for several ectopic tumor models in skin and hepatocellular carcinoma tumors in liver and demonstrated safe and effective treatment of basal cell carcinoma in humans. It can be extended to treating internal tumors in organs such as the liver, lung, and pancreas. In treated tumors, nsPEF ablation or nanoelectroablation bypasses several cancer hallmarks by inducing regulated cell death mechanisms (evasion of apoptosis), promoting anti-vascular or antiangiogenic effects (inducing angiogenesis), and activating immune responses (evasion of immune surveillance). In this sense, nsPEFs can act as multiple therapeutic treatments targeting inherent cytotoxic mechanisms, vasculature and blood supply, and immune recognition. NsPEFs can also be used to activate platelets to treat wounds in skin after cosmetic surgery or mechanical injuries and have been used to treat cardiac tissue after ischemic–reperfusion injury, providing enhanced healing through platelet release of factors that promote tissue repair, growth, and vascularization. The presence of lower levels of ROS and elevated levels of antioxidants in nsPRP shifts the ratio in favor of antioxidants with greater ROS scavenging. NsPRP also avoids potential toxic side effects of thrombin-activated PRP, especially in the heart where it can act counter productively by generating ROS.

5.7 Nanoelectroablation for Tumor Therapy: Advantages and Challenges

Richard Nuccitelli

5.7.1 Introduction: How Do Nanosecond Pulses Differ from Microsecond Pulses?

Before diving into clinical applications and challenges of the application of nanosecond pulses, it is important to understand the fundamental differences between the microsecond and nanosecond pulse domains. Many of the previous sections have covered the use of microsecond pulsed electric fields for clinical applications. The membrane defects generated by these longer pulses are much larger than the 1 nm pores generated by nsPEF [400] and have been used to introduce large molecules and even DNA into cells [367, 401]. These longer pulses on the order of 1 kV/cm typically do not penetrate into the cell interior due to their slower risetime, and even if they did, the field strengths used are not large enough to electroporate the much smaller organelles. The central characteristics of nanosecond pulses are (1) they immediately penetrate into the cell interior because they are faster than the microsecond that it takes for intracellular ions to redistribute and block the field and (2) they are 1000-fold shorter than the microsecond pulses so they deliver much less energy and therefore less heat. This allows much larger field strengths to be applied that can generate water defects in organelle membranes as well as the plasma membrane. It takes about 500 mV to electroporate each membrane in an organelle so the field across a 1 μm wide organelle must be 1 V per 1 μm which is equivalent to 10 kV/cm. This field strength only delivers fractions of a Joule with nanosecond pulses but would heat the cell to hyperthermia temperatures if applied for microseconds. Therefore, the main characteristic that sets nanosecond pulsed fields apart from the microsecond domain is their ability to electroporate internal membranes. This unique feature first observed by Schoenbach's group [10] allows nanosecond pulses to be used to tease new responses out of cells such as the initiation of apoptosis and tumor ablation.

5.7.2 Preclinical Studies of Nanoelectroablation

The first indication that nanosecond pulsed electric fields (nsPEF) could be used to reduce tumor growth came from the pioneering work of Beebe and Schoenbach [11] in which they reported the slowing of growth of subdermal murine fibrosarcoma allografts following nsPEF treatment. They found that exposing the tumors to 5–7 pulses (75 kV/cm, 300 ns) on 2 days resulted in a 60 % reduction in size and weight. Since then more than 60 papers have been published describing various aspects of nanoelectroablation of tumors.

Much progress was made using the murine B16 melanoma model system to continue the tumor ablation studies. The murine SKH-1 strain was chosen as the host because as a hairless albino it has no pigment in its skin and light readily passed through the skin to allow the transillumination of subdermal tumors. When the skin is placed over the end of a vertical plastic cylinder with light shining through it on the stage of a stereoscope, the tumor and blood vessels are beautifully illuminated (Fig. 5.7). This is a very sensitive method for visualizing the nascent tumor and provides a much more precise measurement of the tumor area than the more common method of measurement with a caliper on the skin surface.

The tumor was treated with nsPEF by lifting a fold of skin containing the tumor away from the body and placing it between parallel plates of a clothespin-shaped electrode (ingeniously created by Uwe Pliquett) to apply a uniform electric field (see Fig. 5.8). Very little inhibition of tumor growth was observed when 10 kV/cm was applied using 10–100 pulses 300 ns long. However, increasing the field to 20 kV/cm began inhibiting growth when ten pulses were applied and 100–200 pulses resulted in shrinkage of the tumor [369]. Total remission could be observed when 100 pulses of 40 kV/cm were applied 3 days in a row. A reduction in blood

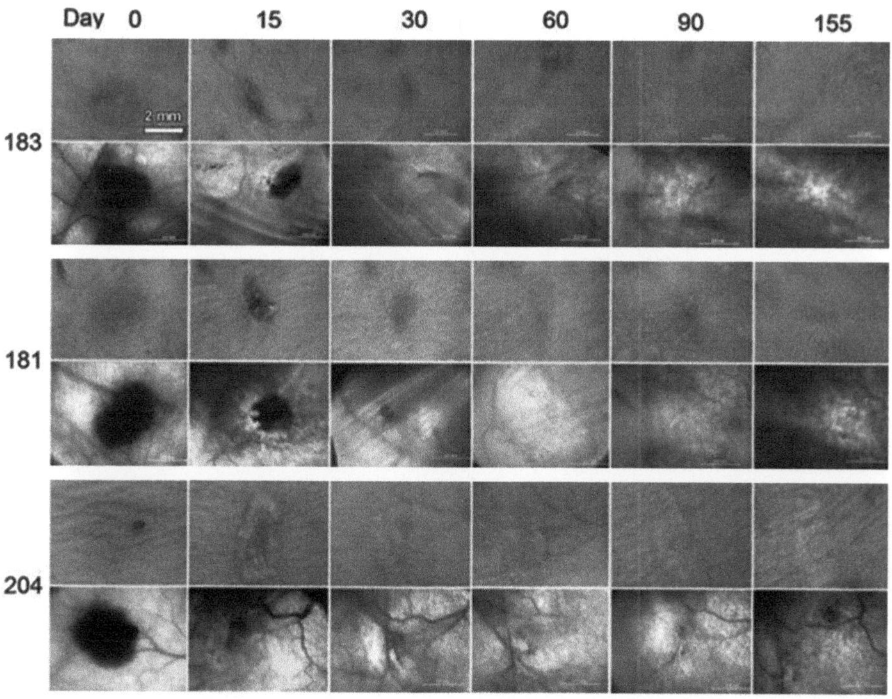

Fig. 5.7 Time course of melanoma nanoelectroablation. Each pair of rows shows a surface view (*top*) and transillumination view (*bottom*) of a melanoma at the day after treatment indicated at the *top* of each column. The animal number is shown to the *left* of each grouped pair. Mouse 183 and 181 were treated on day 0 with 600 pulses (45 kV/cm, 300 ns long) and again on day 18 with 300 pulses. Mouse 204 was only treated once on day 0 with 300 pulses (40 kV/cm, 90 A, 300 ns), and the tumor exhibited total remission following this single treatment. The scale bar in the upper left photo applies to all of these images (Reproduced with permission from Int. J. Cancer, 125 (2):438–445)

flow to the tumor was observed and pyknosis of nuclei within the tumor was evident by histology carried out by Xinhua Chen [402]. Measurement of the tumor temperature during pulsing by Uwe Pliquett indicated only a 2–3 °C increase.

During the past 9 years, studies on over 1000 rodents have established the safety and efficacy of nanoelectroablation. The main findings were:

1. Nanoelectroablated tumors do not recur in studies lasting up to nearly 1 year [15, 377, 383].
2. The nanoelectroablation mechanism involves a rapid increase in intracellular Ca^{2+} [403, 404], DNA fragmentation [377], the activation of reactive oxygen species generation [405, 406], and the initiation of apoptosis [376, 407].
3. Nanoelectroablation parameters were identified that could ablate allograft and xenograft murine tumors with a single, short treatment [408].

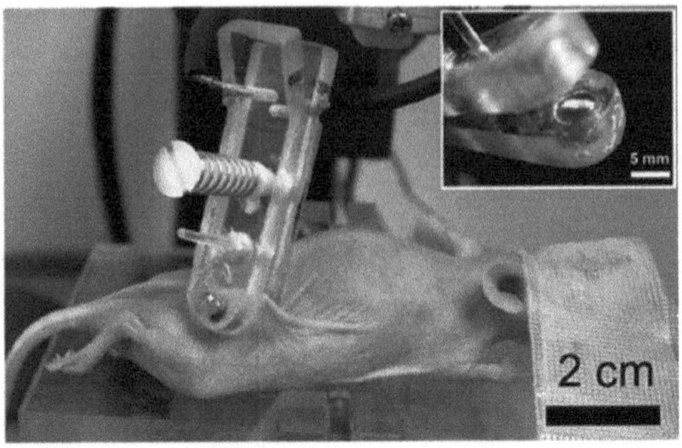

Fig. 5.8 Method for applying pulsed electric fields to subdermal allograft tumors. The electrode was composed of two polished metal plates that we slightly recessed in the ends of a spring-loaded clamp that sandwiched the skin containing the tumor between the two plates (Reproduced with permission from Biochem. Biophys. Res. Commun. 343: 351–360)

4. Endogenous murine tumors could also be ablated with similar pulse parameters using Ervin Epstein's murine BCC model [380] as well as Ed DeFabo's murine melanoma model [409].

5.7.3 Nanoelectroablation Stimulates a Systemic Immune Response

While treating allograft tumors, it was discovered that the growth rate of a second injected allograft tumor was much slower than that of the first tumor that had been nanoelectroablated in immunocompetent mice but not in immunodeficient mice [409]. Moreover, immunohistochemical studies indicated that $CD4^+$ T cells were present in untreated tumors removed from mice in which another tumor had been treated with nsPEF at either 19 or 32 days before removal. Since then more evidence for an immune response has come out [14, 382, 410]. In the strongest of these, Chen et al. used an orthotopic rat liver tumor and observed that in 23 rats the secondary tumor failed to grow after the first was nanoelectroablated. They also observed the presence of granzyme B-secreting cells which could implicate either T cells and the adaptive immune response or natural killer cells and the innate immune response.

We have used a similar orthotopic rat liver tumor model to study secondary liver tumor growth and also observe a much slower growth after nanoelectroablation of the first tumor. Rather than wait for 7 weeks before harvesting the liver lobe with the secondary tumor, we removed the lobe after 1 week. The secondary tumor was still present after that time but was much smaller than the primary was after the same growth period. However, when we depleted $CD8^+$ cytotoxic T cells prior to injecting the second tumor, it grew much faster suggesting that it was the cytotoxic

T cells that were inhibiting tumor growth [411]. This is the first evidence for the stimulation of an adaptive immune response to nanoelectroablated tumors.

5.7.4 *Clinical Studies*

There have only been two published reports of the nanoelectroablation of human skin lesions. The first was the ablation of a single basal cell carcinoma (BCC) by 200 pulses 20 ns in duration [16] and the second described the treatment of ten BCCs on three patients with between 100 and 1000 pulses (100 ns, 30 kV/cm, 2 pps) [12]. An electrode consisting of two parallel rows of six needles each (Fig. 5.9) with an ablation zone of 5 × 6 mm in size was used. A single treatment of 100 pulses was sufficient to cause most of these lesions to disappear during the ensuing several weeks without leaving a scar. Edema was present following treatment and a crust appeared over the treatment region within 2 days and began to fall off in 2 weeks. The skin was initially slightly pink in color and after a few weeks appeared normal such that some treated lesions could not be located several weeks after treatment without reference to a pretreatment map. Since the standard of care for BCC removal, surgery, or curettage usually leaves a scar, the absence of scarring following nanoelectroablation was a surprising result. About 14 weeks after treatment, we removed skin from the treated region and collected serial thin sections every 100 μm through the entire region. These H&E-stained sections were subjected to independent histological analysis by a dermatopathologist. Seven of the ten lesions were completely free of basaloid cells and two were partially ablated.

5.7.5 *Challenges for Clinical Applications*

The five main challenges to the successful application of nanoelectroablation technology to treat human disease are (1) the high electric fields required, (2) the electromagnetic interference (EMI) generated when applying these high field pulses, (3) the thickness of cables required to deliver the pulsed fields to specific tumor locations, (4) the large ablation zone required for human lesions, and (5) the need for a nanosecond pulse generator approved for human use.

The preclinical studies have indicated that the electric field strength required for ablation is at least 12 kV/cm in liver and 30 kV/cm in skin. The delivery of such large electric fields is a challenge because fields of this magnitude ionize the gas molecules in air, leading to corona and spark formation. That means that all metal surfaces that carry these high-voltage levels must be coated with a good insulator. Air must also be excluded at the interface between the delivery electrode and the tumor. This is usually accomplished by placing a nonconductive gel at the interface

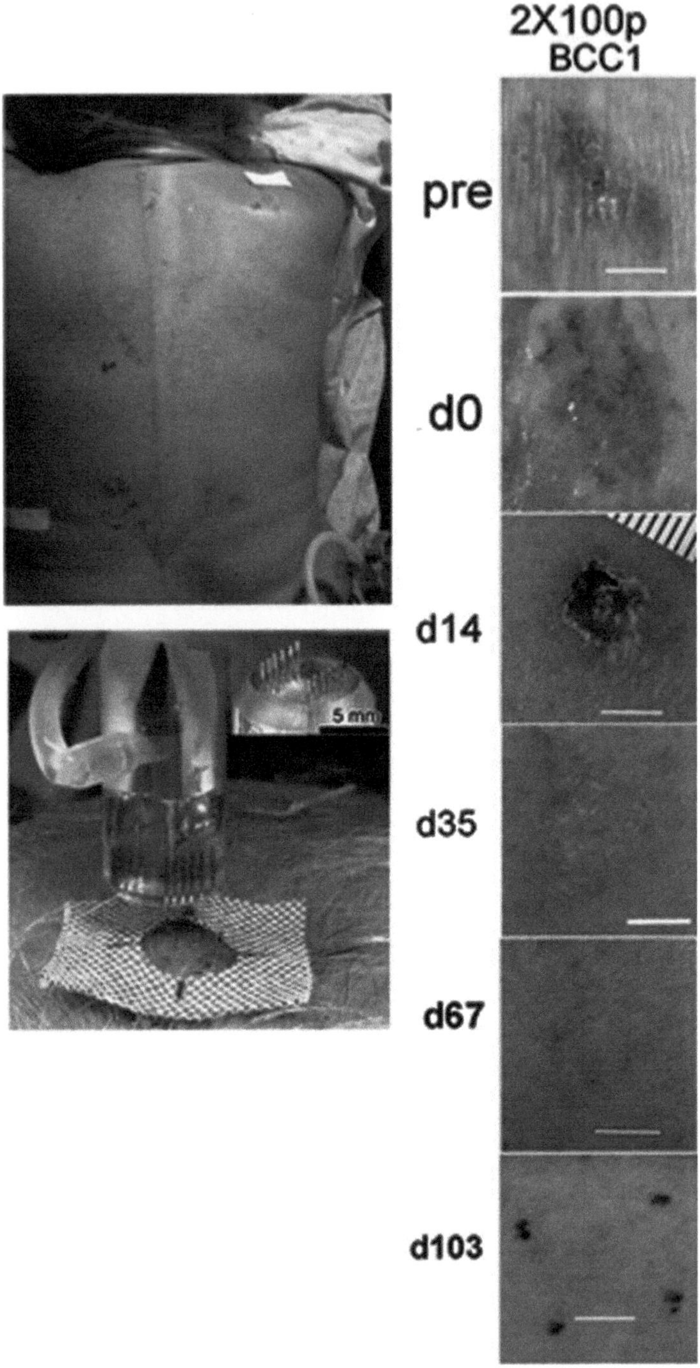

Fig. 5.9 Human basal cell carcinoma treated using the parallel needle array electrode pictured on the lower left. Since the lesion was larger than the electrode, we treated each half separately with 100 pulses each (30 kV/cm, 100 ns). Images of the lesion taken before and after treatment are

between the electrode and the skin or insuring that saline is present at the interface between the electrode and organ tumors.

The EMI generated by these ultrashort, high-voltage pulses can disrupt solid-state circuits so it must be attenuated. The pulse generator is typically housed in a grounded steel box to contain the EMI, but the cable connecting the pulse generator to the application electrode is an excellent antenna that must also be shielded usually with an outer conducting sheath. This is not as easy as it sounds because the pulse shape can be affected by the presence of this outer conductor.

This shielding also adds to the delivery cable thickness which is the third challenge for clinical applications. Since nanoelectroablation requires bipolar electrodes, the insulation between the two wires carrying the pulse to the electrodes must be able to withstand 30 kV. Most common insulators must be many millimeters thick to accomplish this. One of the best insulators is a polyimide film developed by DuPont called Kapton. However, even Kapton must be about 1 mm in thickness to avoid breaking down at these high fields. Therefore, the thickness of the delivery cable must be at least 2–3 mm.

As one example of the challenges we face for the delivery cable, let us consider one electrode delivery approach that we are implementing via an echo-endoscope. This instrument is used routinely to image tumors in the pancreas, liver, and kidney from inside the stomach by using an ultrasound transducer at its tip. A 4-ft-long aspiration needle fits in the working channel and can be used to penetrate through the stomach wall and into the tumor in order to aspirate a cell sample from tumors in organs located next to the stomach or digestive tract. By placing a delivery electrode into the working channel of the echo-endoscope, the pulses could be delivered to any organ accessible by the endoscope. This includes throat and stomach tumors as well as those in the pancreas or other organs imaged by ultrasound through the stomach wall. Once a tumor is detected in the ultrasound image, an electrode can be guided through the wall and into the imaged tumor. This approach has been successfully demonstrated with the ablation of 1 cc of pig pancreas without any side effects. It is also much less invasive than a laparotomy normally required for treating pancreatic carcinomas.

The fourth challenge to clinical application is the large ablation zone needed. Most human tumors that are currently treated using ablation modalities are between 1 and 3 cm^3 in size. All of the preclinical studies treated smaller tumors that were typically 0.5 cm wide. A fairly simple, two-needle electrode with a 5 mm spacing has an ablation zone that measures 1 cm by 0.5 cm. Therefore, in order to treat larger tumors, we would have to use several treatments while moving the electrode around or somehow increasing the size of the ablation zone. Since increasing the

Fig. 5.9 (continued) shown with the day when it is photographed indicated to the left of each photo. "d0" was taken right after treatment and shows some edema. By day 35 after treatment, the lesion was completely eliminated, and by day 103, the only way to know where it was originally located was by using a plastic overlay map made prior to treatment. The *white* scale bars are 5 mm long electric field magnitude is problematic, other approaches that might ablate regions exposed to lower field strengths are being explored.

The fifth challenge is developing a pulse generator and delivery system for human therapy. While there are many nanosecond pulse generators in use in laboratories around the world, none of them have been approved for human therapy. When electrodes are used to deliver these high-voltage pulses of 20–30 kV into the body to treat internal tumors, risk assessment and mitigation are required in the design and fabrication of the pulse generator and electrode delivery system. All components must meet the electrical safety standards for medical devices. These requirements add a great deal of expense in the development of the instrumentation which must then be approved for human applications by the FDA.

5.7.6 Targets for Nanoelectroablation

All cell lines treated with nsPEF to date have exhibited a similar ablation response providing sufficient energy was applied. Either apoptosis or necrosis can be obtained with a sufficient number of pulses. Generally, for a given field strength, apoptosis occurs at lower pulse numbers than necrosis. While some studies indicate that malignant cells might be more sensitive to nsPEF than nonmalignant cells [412, 413], all cell types can be ablated by nsPEF. This means that the possible targets for this therapy are limited only by one's ability to visualize them and position electrodes around them.

Most of the studies to date have treated skin tumors with very good success and have observed little or no scarring of the skin following nanoelectroablation. This suggests that nanoelectroablation might have advantages for cosmetic applications or lesion removal in exposed regions such as the face. The selection of the best tumor type for nanoelectroablation therapy will be influenced by several factors such as average tumor size, accessibility, tumor incidence, and competing therapies.

5.7.7 Advantages of Nanoelectroablation

Nanoelectroablation therapy has several advantages over the thermal ablation therapies of RFA, cryoablation, high intensity focused ultrasound (HIFU), and microwaves commonly used today. These therapies rely on heating the tissue to very high temperatures of 50–90 °C for 15 min or more to denature proteins and cause necrosis in the tissue. Due to nonuniform heat conduction in the treated tissue, the ablation zone does not have a well-defined or predictable boundary, and heat sinks such as vessels and ducts make it hard to ablate bordering tissue. Nanoelectroablation is nonthermal so it does not have these problems. Instead, it relies on large E fields to generate transient nanopores in both the plasma membrane and internal organelle membranes. These transient nanopores allow ion flow across cellular membranes that triggers a cascade of events leading to apoptosis or programmed cell death [407]. Apoptosis is the normal way that most of the cells in our body die when they are worn out and usually does not stimulate an immune response. However, nanoelectroablation triggers a different type of apoptosis called "immunogenic," and we are finding that this does result in the production of cytotoxic T cells that inhibit secondary tumor growth in mice and rats. If this immunogenic apoptosis is also found in humans, it will impart a very significant advantage to nanoelectroablation therapy.

An additional advantage of nanoelectroablation is that it immediately permeabilizes small blood vessels, inhibits blood flow to the tumor, and downregulates microvasculature endothelial components such as CD34 [15, 377]. The endothelial cell density in nanoelectroablated tumors is reduced by more than 90 % [370]. This inhibits angiogenesis and contributes to the death of the tumor.

5.7.8 Conclusion

Nanoelectroablation holds a great deal of promise for advancing tumor therapy. It has been shown to be effective at triggering apoptosis in every tissue and cell type to which it has been applied, and a single treatment lasting a few minutes is sufficient for complete ablation without recurrence. The first clinical trial treating basal cell carcinoma indicated a 78 % success rate without scarring. This lack of scarring is another important advantage over competing technologies, and if immunogenic apoptosis is triggered in human therapy, this modality will become very popular indeed.

5.8 Plasma Medicine

Chunqi Jiang

5.8.1 Introduction

Recent development of plasma technology for biomedical applications has given rise to the flourishing growth of a new field—plasma medicine, where the interactions between plasma and living cells, tissues, and organs, and the associated mechanisms are studied. Ionized gas plasmas are characterized as a quasineutral entity that consists of charged and neutral particles (e.g., ions, electrons, and excited species). In particular, nonthermal atmospheric-pressure plasmas (NTAPs) have made possible treatments of heat-sensitive biomaterials and living tissue. The energetic electrons in these plasmas collide with background atoms and molecules causing enhanced levels of dissociation, excitation, and ionization. Reactive neutral species and charged particles, generated by these collisions, interact with the materials under treatment, while the bulk gas remains near room temperature. Among the potential plasma agents for bacterial deactivation, reactive chemical species including reactive oxygen species (ROS) and reactive nitrogen species including NO and NO_2 (RNS) were considered to be playing crucial roles in the NTAP-induced antimicrobial process [37, 38]. The possible actions to biomaterials and the associated chemistry induced by ROS and RNS generated by NTAPs were assessed in detail by Graves [414]. Charged particles such as O^{2-} as well as electric fields may also importantly contribute to the antimicrobial process in some of the NTAPs [415]. In comparison, heat and UV radiation were not considered to play significant roles for bacterial inactivation by most of the NTAPs [383].

Extensive research and development of NTAP sources in the past decades have enabled their applications in rapid, contact-free sterilization of medical devices and biomaterials, as well as suggested their highly promising potential in dental treatment, wound care, tissue engineering, and cancer therapy. This section by no means covers all aspects of the field. Rather, it is a glimpse of the field with particular interest in NTAP applications for wound care and dental treatment. For further material on these and additional topics, the reader is referred to other reviews by Moreau et al. [416], Kong et al. [375], Weltmann et al. [417, 418], Lloyd et al. [419], Lu et al. [420], and Jiang [421].

5.8.2 Wound Care

A wound is a disruption of normal anatomic structure and function, resulting from pathologic processes beginning internally or externally to the involved organ (s) [422]. Wounds heal by various processes including coagulation, control of infection, resolution of inflammation, angiogenesis, fibroplasia, epithelialization, contraction, and remodeling [422, 423]. Wounds may be classified as acute wounds, which repair themselves or can be repaired in an orderly and timely process, and chronic wounds that do not [422]. Chronic wounds represent a worldwide problem, especially as the geriatric population increases. Bacterial infection is known to be one of the leading causes of the pathobiology leading to wound chronicity and delay of healing [423, 424]. NTAP with demonstrated in vitro antimicrobial effects against a broad spectrum of microorganisms would therefore be a potential disinfection tool assisting wound healing. NTAPs may have additional advantages in wound care with their unique properties of promoting blood coagulation and enhancing endothelial cell proliferation. Here we focus on three aspects of the recent applications of NTAPs in wound healing: wound disinfection, blood coagulation, and cell stimulation.

5.8.2.1 Wound Disinfection

Different research groups have observed the antimicrobial effects of NTAP against skin-relevant microorganisms and microbial biofilms in vitro [425–427]. One of the earliest clinical trials was conducted by Isbary et al. on 38 chronic infected wounds in 36 patients with 5 min daily cold atmospheric argon plasma treatment in addition to standard wound care [43]. The plasma was generated with a microwave-driven plasma device, called MicroPlaSter (originally developed by the Max Planck Institute, Germany), operated at 2.46 GHz, 86 W with an Ar flow of 2.2 SLPM. The distance of the sample surface to the plasma nozzle was kept at 2 cm during treatment. Prior to the plasma treatment, high pressure water jet or scalpel was used to clean all wounds of debris. Bacterial load of all control- and plasma-treated wounds was taken with swabs (for species identification) or nitrocellulose filters (for CFU counting) on different consecutive days. The plasma treatment was stopped if three consecutive daily negative bacterial swabs or nitrocellulose filters were obtained (wound was about to heal), or if the patient decided to stop the additional treatment. The authors showed diverse bacteria detected on wounds, among which *Enterobacteriaceae* and coagulase-negative streptococci (normal skin flora) were the most prevalent, each with 18 % of the total sampled bacterial load, followed by *Pseudomonas aeruginosa* (17 %) and *Staphylococcus aureus* (13 %). Methicillin-resistant *S. aureus* was also observed with a percentage of 3 %. A highly significant reduction in bacterial count of about 34 % in plasma treated compared with untreated areas. The authors noted that although most patients were also treated with systemic antibiotics, the reduction in bacterial load was, in most cases, less in the control wound than in the plasma-treated area. These results demonstrated the substantial potential of the cold atmospheric argon plasma treatment as a safe and painless new technique for wound disinfection and healing. Nevertheless, more clinical trials and long-term, follow-up studies are needed to assess the clinical efficacy and the treatment safety of the plasma method.

5.8.2.2 Blood Coagulation

The use of argon plasma coagulation (APC) to assist open surgery was first introduced in the late 1970s [428] and has become the most commonly used endoscopic coagulation technique since its adapted use in endoscopy in the 1990s [429–432]. Although the use of the argon plasma was commonly considered to induce a thermal effect (cauterization) to tissue in order to achieve devitalization of tissue and stanching of bleeding (hemostasis), the short duration (<1 μs) and low amplitude (<4 A) of the current pulses indicate the APC plasma is a nonthermal plasma [432]. A typical APC device is based on a radio frequency (RF)-driven

DBD configuration, where the discharges are generated between a stainless steel or tungsten electrode and the tissue to be treated [432]. The distance between electrode and tissue may be 2–20 mm. The active electrode, usually a thin (typically submillimeter in diameter) wire or spatula-shaped tip, is powered by 4 kV AC voltage at 350 kHz, running at a burst mode of typically 20 kHz. Argon flows at rates between 0.5 and 7 L/min through a dielectric tube that surrounds the electrode. Commercially available endoscopic APC systems include the System 7550™ with ABC® offered by Conmed corporation, Utica, NY, USA, and the APC 2 unit offered by ERBE Elektromedizin, Tübingen, Germany.

Recent studies by Fridman et al. suggested that a low-temperature air plasma achieved coagulation without inducing the thermal effect, rather by altering the protein or ion contents in blood, which promoted the coagulation process [44]. The experiments were conducted by exposing human blood drops to a DBD air plasma, generated by a RF-driven, electrically insulated "floating electrode" system. The authors proposed a model to indirectly derive the time dependence of the thrombin formation on the plasma treatment dosage. Thrombin is one of the proteins responsible for blood clot formation. They concluded that the thrombin formation time was reduced from ~30 s for normal blood to ~15 s for blood undergoing the DBD plasma treatment.

5.8.2.3 Cell Stimulation

When plasma comes into contact with mammalian tissues/cells, reactive plasma species may cause perturbations in the cellular structure, particularly the cell membrane, and induce degrees of structural and functional changes. Some changes may be encouraged for wound healing, while the other may be cautioned for potential damage of the host. The recent studies of NTAP interactions with different mammalian cells included fibroblasts and vascular cells and have all indicated that there is a dose dependence of the plasma methods and are important steps toward reliable and safe applications with NTAPs.

1. Fibroblasts
   Kieft and Stoffels et al. were among the earliest groups to conduct studies of the interactions of a RF "plasma needle" with living cells in culture [433, 434]. The so-called plasma needle in its later version for biomedical applications [45, 46] was a 0.3 mm diameter tungsten wire with a sharp tip, confined in a 4 mm inner diameter Perspex tube. The center needle served as the active electrode for the capacitively coupled RF discharges. The Perspex tube was filled with helium flow at 2 l/min 200 V peak-to-peak RF voltage at 13.56 MHz and was typically applied to the center needle to initiate the plasma (about 0.1–1 mm long) at the tip of the needle. It was advised that the optimal working distance of the plasma ranged from 1 to 3 mm above the surface to be treated, and the plasma power and treatment time not to exceed 150 mW and 60 s, respectively, to avoid hypothermia heating and to maintain the sample temperature under 43 °C. Chinese

hamster ovarian cells (CHO-K1) and human lung carcinoma cells (MR65) in suspension were subjected to plasma treatment applied at different dosages by varying the RF power level and the concentrations of molecular species. The authors showed that when applying the plasma under moderate conditions (e.g., 0.1 W and 3 % air in 2 SLPM He flow), the plasma induced cell detachment between the adjacent cells and the substrates without inducing cell death. The authors suggested that ROS and RNS played a role in the plasma-induced damage of cell adhesion molecules, cadherins, and integrins and caused the temporary situation of loss of cell contact and cell detachment from the substrate. A study of the long-term effect of the mammalian cell reaction after plasma was also conducted [40]. 3T3 mouse fibroblast cells were used for the study. The authors reported that 24 h after plasma treatment, more than 20 % of cells underwent apoptosis.

Lee et al. reported the suppression of hypertrophic scar generation in an animal model by plasma treatment [435]. A DBD plasma powered by 5 kV sinusoidal voltage at 4 kHz was ejected through two nozzles (20 × 1 mm) onto scars produced in a murine model. A mixture of He (3 SLPM) and air (50 SCCM) was used as the working gas for plasma generation. Eight-week-old C57/BL6 mice were subjected to a retractor to induce scars after the closure of the burn wound. The plasma treatment was performed 3 min each time and three times a week over 6 weeks. The authors observed a positive effect on the suppression of vascularization by the plasma stimulation and higher apoptotic cell death levels for mouse hypertrophic scar fibroblasts (HTSF) than normal fibroblasts. The authors further suggested that the plasma-induced scar suppression might be due to cellular apoptosis during the proliferative phase of wound healing and the early stage of scar generation, and cellular apoptosis was caused by oxidative stress of the ROS in plasma.

2. Endothelial (vascular) cells

   Using the same RF plasma needle source, Kieft et al. studied the interactions of plasma and mammalian endothelial and smooth muscle cells that are prevalent in arterial walls [47]. The specific cell types used in the study were bovine aortic endothelial cells (BAEC) and rat aortic smooth muscle cells. A cell viability assay using live/dead staining and fluorescence microscopy was used to examine the influence of the plasma treatment parameters on cells. The authors reported that the thickness of the liquid suspension layer covering the cells was the most important factor to achieve cell detachment and necrosis, compared to the plasma treatment time and the applied voltage. The results also suggested that decreasing the applied voltage could reduce the percentage of necrotic cells to below 10 % while still disrupting cell adhesion. However, this detachment was not observed in a later report in which the cells were treated indirectly by plasma through a permeable membrane [436]. Instead, a long-term behavior of vascular cells (endothelial and smooth muscle) after exposure to the RF plasma was observed. Six to ten hours after the indirect plasma treatment, apoptosis in low-dose plasma-treated smooth muscle cells was observed, but no immediate detachment occurred.

Kalghatgi et al. investigated the in vitro effect of a nonthermal DBD plasma on porcine aortic endothelial cells, to assess the potential application of the plasma for dose-dependent blood vessel growth/regression control [437]. The plasma source was based on a DBD configuration where the HV electrode was insulated with a thin layer of quartz dielectric that was separated from the sample by 2 mm. The sample consisted of the porcine aortic endothelial cells (PAEC) in the culture media on a glass cover slip that was placed on a ground base. Modulated AC voltages (20 kVp-p) at frequencies between 0.5 and 1.5 kHz were delivered to the HV electrode for the plasma generation. The authors showed that low dose of plasma treatment (e.g., up to 30 s or 4 J/cm^2) enhanced endothelial cell proliferation by a factor of 2 when compared to the untreated control cells, and higher dose of plasma treatment (60 s and higher or 8 J/cm^2) led to cell necrosis. The authors suggested that the enhancement of the cell proliferation was through fibroblast growth factor-2 (FGF2) release, which occurs only at cell injury or death, and the plasma induced the FGF2 release by the production of reactive oxygen species including OH radicals, H_2O_2, and HO_2.

5.8.3 *Dental Treatment*

The biggest challenge for conventional periodontal and endodontic treatments is to eliminate pathogenic microbial biofilms completely without causing damage or excessive loss of healthy tissues. Microbial biofilms, complex communities of microorganisms that are embedded in matrices of extracellular polymeric substance, are a common cause of numerous oral infections including dental caries, pulpitis, periodontitis, and periradicular lesions [438, 439]. Dental caries, for instance, are the localized destruction of tooth tissues by bacterial fermentation of dietary carbohydrates and are typically induced by cariogenic dental plaque, an example of microbial biofilm with a diverse microbial composition [440–442]. The acids produced by the fermentation of these dietary carbohydrates demineralize the enamel, which results in the formation of cavities. Treatment of carious lesions and cavity preparation were often achieved by removal of the infected and demineralized soft and hard tissues with mechanical, chemomechanical methods or laser ablation [443, 444]. To ensure the cavity free of bacteria, an excess of healthy tissue was often removed, which may have weakened the integrity of the remaining tooth [444]. Root canal treatment is described as the reduction or elimination of intracanal microbes and their byproducts from the root canal system, using endodontic instruments aided by antimicrobial agents, before filling [445, 446]. The formation of bacterial biofilms and the morphological complexity of the root canal system are the primary factors accounted for the complication or failure of the root canal treatment [447–449]. Conventional methods include mechanical instrumentation, antimicrobial irrigation, and removal of the smear layer in order to eliminate intracanal bacteria and debris and are not able to completely eliminate the postprocedure infection [450, 451]. Laser irradiation

[452] or laser-assisted photodynamic therapy [453, 454] was proposed in recent years as supplements to existing protocols for root canal disinfection, but still is not effective against all endodontic strains grown as biofilm phenotypes [455, 456], in addition to their high capital cost or causing possible tissue trauma in patients [457].

Streptococcus mutans, *Lactobacillus acidophilus*, and *Lactobacillus casei* are caries-associated bacteria and have been used often in periodontal studies [458–460]. *Enterococcus faecalis*, a gram-positive facultative bacterium, is a frequent and persistent isolate in teeth with failed root canal therapy and has been used extensively in endodontic studies [461]. *Candida albicans*, a yeast, is one of the predominant species isolated in different oral diseases including apical periodontitis and dental caries [462]. In this section, we will focus on the application of NTAPs for dental disinfection treatment of periodontal and endodontic diseases. The other cold plasma applications in dentistry such as biomaterial processing and tooth whitening [463] are not included here.

5.8.3.1 Periodontal Treatment

The RF plasma needle developed by Stoffels and Sladek et al. was among the first NTAP sources proposed for periodontal applications. The antimicrobial activity of the RF plasma against *Streptococcus mutans* biofilms was assessed and reported 90 % CFU (colony forming unit) reduction after 39–45 s helium plasma treatment [40]. Although this preliminary study showed that the plasma needle is capable of bacterial sterilization, quantitative assessment of the antimicrobial activity with respect to the plasma parameters, the biofilm initial conditions, and the substrate is needed. Goree et al. investigated a range of plasma parameters for the growth inhibition of S. mutans grown in nutrient agar and showed that the killing effect is sensitive to the applied voltages and the spacing distance between the substrate and the needle nozzle [464]. The presence of the radicals OH and O was verified using optical emission spectroscopy, implying these radicals may play important roles during the bactericidal process. Gonzalvo et al. used a molecular beam mass spectrometer (MBMS) system to conduct fractional number density measurements for the RF plasma "needle" operating at a range of discharge power up to 10 W [465]. The density of NO generated from the plasma increased substantially after the discharge power value exceeded 2 W. However, the power condition used for the diagnostics does not align very well with the plasma conditions applied in the previous biomedical studies, and the role of NO in the plasma-induced antimicrobial activity could not be directly derived.

More recently, NTAPs in the form of jets (or longer plasma plumes than the RF "plasma needle") powered by various power sources were developed for periodontal applications. Yang et al. used a DC-powered, atmospheric-pressure argon plasma brush [466] for oral bacterial deactivation [467]. Oral bacteria of *S. mutans* and *L. acidophilus* with an initial bacterial population density of 1.0–5.0 $\times$ 10^8 cfu/ml were seeded on various supporting media including P5 filter papers, glass slides, and PTFE films. The plasma exposure time for a 99.9999 % cell reduction was less than 15 s for *S. mutans* and within 5 min for *L. acidophilus*. It was found that the plasma deactivation efficiency was also dependent on the bacterial supporting media. Koban et al. tested the antifungal potential of different RF-driven plasma devices—a RF plasma jet (kINPen®09), hollow electrode DBD, and volume DBD—against *C. albicans* biofilms on titanium disks in vitro [468]. The plasma jet and the DBD plasmas were driven by high-frequency RF voltages (1.82 MHz for the jet, about 40 kHz for the DBDs) at power ranging from typical 3 W to 16 W. Electrode cooling was needed to prevent overheating of the treated subject. Two-day-old *C. albicans* biofilms cultivated on titanium disks were subject to plasma, CHX, and NaOCl treatment for total 10 min. The two chemical cleansers served as the positive controls. While the plasma jet only resulted in a one log reduction for initial cell concentrations between 10^6 CFU/ml and 10^7 CFU/ml, the DBD plasma achieved a log reduction factor between 3.1 (hollow electrode DBD) and 5.2 (volume DBD). The authors noted that the DBD plasmas exceeded the antifungal effects of CHX and NaOCl. However, the results of the above two experiments may not apply to the real oral environment where dentin or denture resin disks are better options as the substrate, which have cavities and porosities into which the biofilm could adhere.

Other NTAP jets were also investigated including microwave-powered and low-frequency plasma jets. Rupf et al. studied the antimicrobial effect of a 2.45 GHz microwave-powered NTAP jet for dental surface disinfection [469]. Dentin slices (8–10 mm^2) inoculated with 6 log10 CFU cm^{-2} *Lactobacillus casei*, *Streptococcus mutans*, *Candida albicans*, and *Escherichia coli* were irradiated with the $He/O_2/N_2$ (flow rates: 2.0/1.2/1.5 l/min) plasma jet for 0.3–0.9 s/mm^2 and resulted in 3–4 log10 intervals on the surface in comparison to the negative control. During the study, *S. mutans* revealed the strongest resistance to plasma jet irradiation on agar plates or dentin slices. Yamazaki et al. carried out experiments to evaluate the sterilization effects of a low-frequency atmospheric-pressure plasma jet on oral pathogenic microorganisms (*S. mutans*, *C. albicans*, and *E. faecalis*) [470]. The authors demonstrated that a 16 kHz plasma jet had sterilization effects mainly through reactive oxygen species (ROS) on oral pathogenic microorganisms present in both the solid and liquid phases.

5.8.3.2 Endodontic Treatment

Jiang et al. were among the earliest groups to develop a low-temperature pulsed "plasma dental probe" for root canal disinfection [41, 42]. A 2.5 cm long, 2 mm diameter pencil-like plasma jet was generated with a concentric tubular device that was typically powered with 100 ns, 4–8 kV voltage pulses at rates of 1–2 kHz. He/(1 %) O2 flow at 1–5 SLPM was passed through the tubular device to assist the initiation and sustaining of the plasma plume. The plasma-mediated antimicrobial effect was assessed against *Enterococcus faecalis* biofilms grown on hydroxyapatite or bovine dentin disks in vitro and saliva-derived multispecies biofilms inoculated in human root canals ex vivo [471, 472]. Treatment of dentin disks cultivated with *E. faecalis* monolayer biofilms with the plasma (average power 1$\leq$ W) for 5 min resulted in 92.4 % kill but not complete sterilization. Conspicuous biofilm disruption and cleared dentinal surfaces as well as partially biofilm removed surfaces were observed in the human root canals after the plasma treatment for 5 min. The authors suggested a plasma jet with smaller diameter, e.g., less than 1 mm, but with similar or longer length for better sterilization penetration in narrow and curving canals. They demonstrated a flexible plasma plume capable of filling up curving tubes, in simulation of root canals [473]. The feasibility of the plasma jet for root canal disinfection was tested using *E. faecalis* biofilm models in vitro, in which the biofilms were grown inside the simulated root canals or hydroxyapatite disks and applied to the plasma jet treatment [473, 474]. In addition, an ex vivo biofilm model using clinically obtained samples in human root canals was also used to assess the antimicrobial activity of the plasma jet [475]. Although the bactericidal effects are prominent, complete biofilm removal was not achieved, nor as effective as bleach (i.e., NaOCl) at relatively high concentrations (e.g., >6 %), which suggested further improvement of the plasma jet-based techniques was needed. The authors considered reactive plasma species such as ROS were playing an important role during the bactericidal process.

Lu et al. is another group actively developing NTAP jets for endodontic treatment. They used a submicrosecond voltage pulsed syringe needle for deactivation of *E. faecalis* bacteria [476]. The 3 cm long, 0.7 mm OD, and about 200 μm ID syringe needle-based configuration allowed insertion of the needle partially into the root canal and assumed the human gum as the ground to support strong reactive plasma species being generated inside the root canal. This needle jet was subsequently tested with an in vitro *E. faecalis* biofilm model grown in root canals [477]. Treatment for 5 min with the NTAP resulted in about one log reduction. Increasing the plasma treatment time up to 15 min resulted in nearly 3-log reduction, which was comparable with 2 % CHX for the same treatment time [477]. Later, Du and Lu et al. reported such plasma could be modified by flowing the He/O2 gas mixture through 2 % CHX to improve the antimicrobial activity [478]. This scheme was tested with an in vitro biofilm model; both *E. faecalis* biofilms and an endodontic multispecies biofilms grown on bovine dentin disks showed about 10 times higher killing rate by the modified plasma treatment compared with plasma or 2 % CHX treatment alone [478]. These studies further confirmed that NTAPs might prove to be a valuable alternative or complimentary method for endodontic treatment.

5.8.4 Concluding Remarks

The rapidly growing worldwide interest in the development and application of the NTAP sources and technology for life sciences has driven the new and fascinating field of plasma medicine, from its infancy toward maturity. As an interdisciplinary yet independent science of its own, plasma medicine specializes in the intersections of plasma science, engineering, biology, microbiology, chemistry, and biochemistry. While the demonstrated potential applications of plasma sources in biology, medicine, and dentistry suggest their indispensable value in technology innovation for life sciences, and some commercialized plasma products in specific medical fields encourage adaptation of similar plasma technology for more relevant applications, caution must be paid to assume the technology for any applications without a complete picture. Fortunately, the community has considered highly important to understand the short-term and long-term effects of a plasma for its intended application as well as the associated unwanted harmful side effects. Hence the challenge of developing the technology for specific applications with minimum negative effects would require fundamental studies of the chemistry and biochemistry at the gas and liquid phases and at the molecular and cellular levels, and these studies are yet to come.

5.9 Electrochemotherapy in Veterinary Oncology: A Routine Procedure for Treatment of Cutaneous Solid Tumors

Justin Teissie, Maja Cemazar, Youssef Tamzali, and Natasa Tozon

5.9.1 Introduction

The use of large animals, mainly dogs, cats, and horses, in clinical studies evaluating the effects of novel treatment approaches has many advantages over the use of small laboratory animals, i.e., rodents [479, 480]. Companion animals, specifically, share the same living environment with their owners and are thus subjected to similar environmental risk factors for the development of certain diseases. Additionally, they share certain anatomic and physiologic similarities with humans and can display similar clinical signs as affected humans: Furthermore, for certain types of cancer, it was demonstrated that the same (comparable) genetic mechanisms are responsible for the development of the disease. Dogs, cats, and horses have also longer lifespans than laboratory rodents and can therefore naturally reach the age commonly associated with the highest risk for cancer, which makes them especially valuable as natural models for research in oncology.

Veterinary clinical studies are performed on spontaneously occurring tumors in pets that are client owned. These tumors share many features with human tumors such as interindividual and intratumoral heterogeneity, the development of recurrent or resistant disease, and metastasis at relevant distant sites. Furthermore, due to pets' relatively long lifespans, long-term side effects and other possible limitations of novel treatment approaches can be detected more accurately than in rodent studies.

Experiments in large animal models therefore provide proof of principle and help discern the potential efficacy and safety of new treatment approaches, including electroporation-based treatment, which cannot be accurately determined in laboratory rodents [479–481].

5.9.1.1 Aim of the Section

The aim of this section is to define the operating procedures (OPs) in order to safely and conveniently treat cutaneous and subcutaneous tumors in pets and larger domestic animals (horses) by electrochemotherapy (ECT). To this end this section provides the reader with the basis for understanding the mechanisms of ECT as well as its possibilities as an antitumor treatment.

5.9.1.2 Chemotherapy of Malignant Tumors in Animals

Incidence of malignant tumors in companion animals is constantly increasing. In chemotherapy, the critical intracellular target for cytotoxicity of drugs such as cisplatin and bleomycin is DNA. Bleomycin causes breaks in DNA, whereas cisplatin forms DNA adducts. The cytotoxicity of drugs is dependent on their intracellular concentration which is controlled by their transmembrane transport. As in human cancer treatment, in veterinary medicine, standard treatment strategies do not always give satisfactory results.

In both cats and dogs, bleomycin is the chemotherapeutic agent indicated for treatment of some types of carcinomas, including SCC, mostly in combination with other drugs or modalities and at a recommended dosage of 2 mg/m^2 weekly or 125–200 mg of total cumulative dosage. It may cause acute toxicity with increased body temperature, anorexia, vomiting, and allergic reaction, rarely myelosuppression. Chronic toxicity may be presented with dermatological alterations (alopecia, rashes), stomatitis, pneumonitis, and pulmonary fibrosis [482, 483].

Cisplatin is primarily indicated for canine osteosarcoma treatment. There are many reports about sensitivity of different tumor types to cisplatin, but the nephrotoxicity, which could lead to acute or chronic renal failure, is a strong limitation of its systemic use. In addition, cisplatin, like other platinum-containing antitumor agents (e.g., carboplatin), causes gastrointestinal toxicity and myelosuppression. Cisplatin is contraindicated in cats, because of the extreme toxicity in these species in which even very low dose may lead in fatal pulmonary edema [482].

One of the perspective treatments in veterinary oncology is ECT, which was already tested and proven to be efficient in treatment of human tumors of different histologies (see Sect. 5.4).

5.9.1.3 Definition of ECT

ECT is a local drug delivery approach aimed at the treatment of cutaneous and subcutaneous tumor nodules of different histologies. ECT, via cell membrane permeabilizing electric pulses, potentiates the cytotoxicity of non-permeant or poorly permeant anticancer drugs with high intrinsic cytotoxicity such as bleomycin or cisplatin, at the site of electric pulse application.

5.9.1.4 Principle of ECT

As described in details in this book, electric field pulses can induce the transient permeabilization of cell plasma membrane (electropermeabilization, "electroporation").

The principal mechanism of ECT is to use electropermeabilization (electroporation) of cell membranes in tumors, to increase delivery of chemotherapeutic drugs with poor membrane permeability, by enabling more drugs to reach the intracellular targets. This increased drug uptake in tumors has been demonstrated for bleomycin and cisplatin: the accumulation of these two drugs in tumors was increased two- to fourfold, compared to tumors without electropulsation. Due to high cytotoxicity in situ of bleomycin or cisplatin, locally at the site of electric pulse application, low drug doses are needed, and consequently no systemic side effects of the drugs are observed. ECT can be effective as one-time treatment only, but in the case of partial tumor response, it can be repeated several times with improved effectiveness and no side effect.

Based on preclinical as well as clinical studies on ECT, this treatment was already used in the treatment in veterinary medicine in 1997 [484]. Mir and colleagues used ECT with bleomycin for treatment of cats with large soft-tissue sarcomas that suffered relapse after treatment with conventional therapies. Electric pulses were delivered after intravenous injection of bleomycin, using external surface electrodes, as well as needle-shaped electrodes that were designed to be inserted into the tumors for more effective electric field distribution in the tissue. The cat's lifespan increased significantly compared with control group of 11 untreated cats [484]. After this initial study, several other groups reported the effectiveness of electrochemotherapy in companion animals [72, 485–494, 501].

In the present chapter, we will focus our protocols on the use of approved generator of electric pulses for veterinary clinical applications, where the safety and the control of the delivery pulses are present as well as their digitized recording.

Furthermore, this section is describing drug delivery. Present investigations on gene therapy [495, 496] are out of the scope of this contribution.

This contribution will focus on trials with specific localization of tumors to the skin. The issues to be addressed include how to treat a large tumor volume and how to obtain good availability of cisplatin within the tumor. ECT, with or without concurrent tumor debulking, is an effective alternative for treatment of tumors in companion animals.

5.9.2 ECT Treatment

5.9.2.1 Generators of Electric Pulses

Most of the treatments were carried out with the electropulsator ELECTROvetS13 (BetaTech, Toulouse, France). Preset pulse parameters were used: 8 pulses, 100 μs each, voltage to electrode width ratio = 1300 V/cm, 500 Hz.

The S13 is composed of:

- A square wave voltage generator
- Integrated software with a touch-screen interface to select predesigned electric pulses
- A handheld electrode holder equipped with a pulse activating button, attached to the generator by a cord
- A foot switch that controls electric pulses delivery

Different designs of electrodes can be connected to the holder as described below.

ELECTROvet S13 meets European Union standards:

- 2006/95/CE from 12/12/2006 concerning electronic material used within specific voltage limits
- 2004/108/CE from 12/15/2004 concerning electromagnetic compatibility

It is compliant with FDA's regulation.

A detailed description of each treatment session can be saved in a .CSV file.

The other device that is approved for veterinary application is available from IGEA (Cliniporator) and is described elsewhere in the book. Cliniporator complies with electrical safety standards IEC 601-1, with the requirements of the European directives for medical devices 93/42/CEE and 2007/47/CEE, and it is marked CE0051 under the control of the Notification Body IMQ.

5.9.2.2 Electrodes

An important consideration is the appropriate choice of the optimal electrode configuration and pulse parameters for the particular target tissue. Electrode configuration affects electric field distribution in tissue. Empirical methods of design are developed frequently [497–499]. A safe approach is to compute in advance the electric field distribution in tissue by means of numerical modeling techniques [253]. Good contact between the electrodes and the skin must be assured by depilation and application of an ultrasound or similar gel to the treatment area.

5.9.2.2.1 Contact Wire Electrodes

Noninvasive electrodes were designed to face the specificity of the animal skin. Removable and reusable electrodes in L-form are contact wire electrodes that had a fixed width given by the distance between the two stainless steel rods. They were inserted at the end of an insulating holder which was held by the surgeon (Fig. 5.10).

Pulses were delivered after bringing the steel rods in contact with the skin above the tumor and the peritumoral tissues. At each position, pulses are delivered in two orthogonal directions per site just by rotating the holder (Fig. 5.11).

Multiple electro-treatments were applied by moving the electrodes on the tumor surface on adjacent positions. This allowed the treatment of the entire tumor surface even when it was several cm^2 large. The ergonomy of the electrodes allows fast movement along the tumor surface and facilitating change of orientation when in contact with the skin.

The field distribution was homogeneous laterally between the two electrodes but decayed in depth [500]. This led to a protocol where successive treatments (each other week) needed to be delivered. This sequence was chosen arbitrarily but appeared to be satisfactory.

5.9.2.2.2 Plate Electrodes

Parallel stainless steel electrodes: thickness, 1 mm; width, 7 mm; and length, 8 mm, with rounded tips and an inner distance between them of 7 mm. The electrodes (P-30-8B) are produced by IGEA [60] (Fig. 5.12).

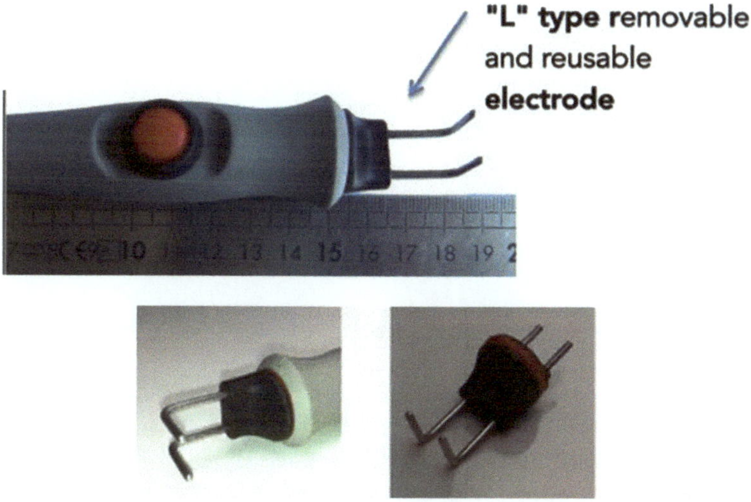

Fig. 5.10 Contact wire electrodes are noninvasive, removable, and reusable electrodes in L-shape that are used in with ELECTROvet S13

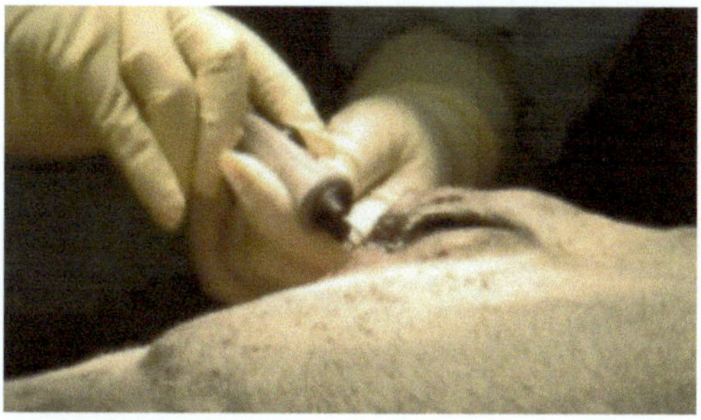

Fig. 5.11 Application of electric pulses onto the horse tumor using contact wire electrodes

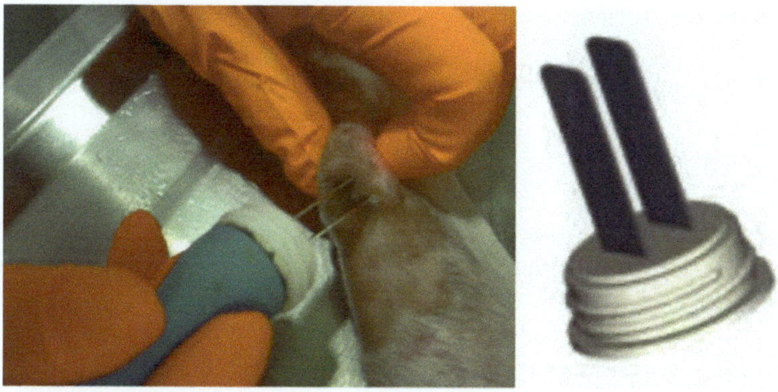

Fig. 5.12 Application of electric pulses onto the tumor using noninvasive, single-use, plate electrodes. The electrodes are mounted on a specific holder compatible with Cliniporator

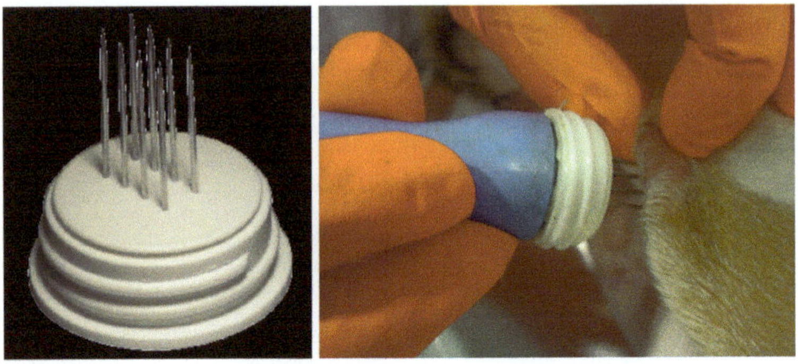

Fig. 5.13 Application of electric pulses onto the tumor using invasive, single-use, needle electrodes. The electrodes are mounted on a specific holder compatible with Cliniporator

Application of square wave electric pulses: pulse duration 100 μs, amplitude to electrode distance ratio 1300 V/cm, and frequency 1 Hz. Each run of electric pulses should be delivered in two trains of four pulses, with 1 s interval, in two perpendicular directions.

5.9.2.2.3 Needle Electrodes

Needle electrodes: four needles in a row, two rows, 4 mm apart. Application of square wave electric pulses: pulse duration 100 μs, amplitude to electrode distance ratio 1000 V/cm, and frequency 1 or 5000 Hz. Each run of electric pulses consisted of eight pulses. The electrodes N-20-4B are produced by IGEA [60] (Fig. 5.13).

5.9.3 Decision for Treatment

5.9.3.1 Owner Consent

When the problem is fully evaluated, full transparency is needed with the animal owner:

- Explain what ECT is and what could be expected from ECT in terms of clinical results.
- Explain what will be the treatment course: number of ECT sessions under general anesthesia, total duration of treatment, cost of treatment, and possibilities of recurrence.
- Obtain owner's written consent in the case of patient inclusion within a clinical trial of ECT, if patient is actually eligible.

5.9.3.2 Operating Environment

Room: any ambulatory room when ECT can be performed under deep sedation or any surgical room where general anesthesia is allowed

Personnel: trained ECT operator (DVM), anesthetist, and assistant

Required equipment:

For sedation and/or anesthesia: all required equipment according to the procedures for anesthesia

For the procedure:

- Electric pulse generator with plate/needle electrodes
- Cisplatin/bleomycin prepared in aseptic conditions
- Sterile syringes and needles
- Sterile gloves and special masks used with cytostatic drugs
- Protective glasses
- Sterile gauze pads
- Skin disinfectant pads
- Hair shaver
- Vernier caliper for measuring tumor diameter
- Digital camera to record pictures of the tumors
- Clock, pen, and paper
- Incineration container for waste

5.9.3.3 Operator and Surrounding Personnel Protection

Droplet exposure of the skin to antineoplastic drugs rarely causes acute problems. Cisplatin and bleomycin are not absorbed through intact skin. Cisplatin may cause only irritation.

Cisplatin and bleomycin aspersion could occur when injecting firm nodules or tumors.

The operator and surrounding assistants should protect themselves with protective glasses, masks, and gloves. In case of skin exposure to cisplatin or bleomycin solution, rinse immediately and abundantly with water.

In any case, biosafety measures should be set up according to a quality assurance program complying with local regulations about anticancer drug usage if any.

5.9.3.4 Pulsing Procedure

The interval between cisplatin or bleomycin administration and the application of electric pulses is considered 1–2 min.

5.9.3.5 Disposal of Contaminated Material

All contaminated materials (syringes, needles, vials, gauzes, gloves) should be disposed in a container intended for incineration.

5.9.3.6 Follow-Up

First two examinations are recommended every 14 days and after that monthly in order to evaluate the treatment effectiveness and possible local and systemic side effects.

At each visit, tumor nodules are measured with vernier caliper and photographed.

For evaluation of treatment response, the tumor size is calculated by the formula $V = ab^2\pi/6$ ("a" is the larger diameter of the tumor nodule and "b" the diameter of the tumor nodule perpendicular to "a"), in accordance with WHO response criteria. Response to the treatment is scored after 4 weeks and at the end of observation period, based on criteria of WHO Handbook for Reporting Results of Cancer Treatment:

- Complete response (CR) is determined when the tumor nodule is not palpable.
- Partial response (PR) is defined as a decrease of more than 50 % in the products of the largest perpendicular diameters of the measurable lesions.
- No change (NC) is defined as <50 % reduction and up to 25 % in the increase in the above measurements.
- Progressive disease (PD) is defined by an increase of more than 25 %.

Objective responses (OR) are determined by combining the number of CR and PR.

For all response determinations, a minimum of 4-week duration is required for determining each type of response. Observation time is calculated as the interval between the date of the first treatment and the date of the last examination of the patient.

5.9.4 *Cats and Dogs*

5.9.4.1 Patient Selection

This part covers the criteria that must be checked during the preinclusion visit for the treatment by ECT of patients with cutaneous and subcutaneous nodules. It will therefore allow to determine whether the patient is eligible or not and could benefit from this treatment and thus if patient can be enrolled by an operating trained team to deliver ECT.

5.9.4.1.1 History

A full history should be taken from the owner, and special attention should be paid to the following:

- General date about age, breed, sex, and presence of any concurrent diseases
- History of tumor (duration, any changes during course of disease)
- Evolution of the tumors with time
- Previous treatments and its results
- Recurrence and speed of recurrence
- Owner's expectations

5.9.4.1.2 Physical Examination

Tumor should be evaluated in detail in order to make possible clinical differential diagnosis.

Fine needle aspirate for cytology evaluation is recommended in order to determine the treatment area: in cases when potentially invasive tumor is presented, the wide margins should be treated (the decision about size of treated margins should be made according to recommendations for any other treatment for same type of tumor, e.g., mast cell tumor </=1 cm, 1 cm margin should be treated).

The number of nodules and their size should be evaluated in order to prepare treatment approach.

5.9.4.1.3 General Examination

A full general examination should be implemented in order to make staging of disease and to check:

- If there is no other concurrent disease which could preclude the treatment
- If the patient has history of any infection (e.g., Ehrlichial agent or other diseases according epidemiology in certain area) or immune-mediated disease which could alter hematological or coagulation changes, wherein the specific tests should be performed
- If the animal is able to withstand a series of deep sedation and/or general anesthesia
- If the patient has sensitivity to cisplatin or bleomycin

Diagnostic imaging techniques (radiography of thorax and ultrasound examination of abdomen) should be used in order to check any possible metastases according to the tumor type.

5.9.4.2 Laboratory Analysis

5.9.4.2.1 Routine Pre-deep Sedation and/or Pre-anesthesia Analysis

ECT is performed under deep sedation or general anesthesia.

Patient should be evaluated according to the American Society of Anesthesiologists' (ASA) physical status classification and according to tumor type because of possible paraneoplastic syndromes.

5.9.4.2.2 Hematology

Complete hematology analyses should be performed.

Additional specific analyses should be done:

- If the patient is taking any medicine which could potentially alter changes in coagulation ability, the coagulation profile should be checked.
- In the case using invasive treatment (needle electrodes), number of platelets should be >100 or in case <100, coagulation profile (PT, APTT, fibrinogen, D-dimer) should be normal.

5.9.4.2.3 Biochemistry

Blood urea nitrogen (BUN), creatinine, alanine aminotransferase (ALT), alkaline phosphatase (AF), and calcium should be made routinely; in specific cases, total protein, albumin, phosphor, sodium (Na), potassium (K), chloride (Cl), and lipase concentration tests should be performed.

5.9.4.2.4 Tumor(s) Histological Analysis

If the cytology diagnosis was made, the histology analysis is not necessary, except in case of tumor types where the histological grade is an important prognostic factor (e.g., MCT). The biopsy (incisional or "punch") can be made before ECT is performed. In any case, it is a good clinical practice to diagnose the tumor type.

5.9.4.3 Choice of the Treatment Modality

Cisplatin or bleomycin intratumorally is used for ECT in dogs and cats.

The treatment option is influenced by potential for better response to any of the drug choices for certain tumor types.

5.9.4.3.1 Size and Number of Nodules

Each tumor should be treated separately.

Assess location and size of each nodule and number of nodules to determine the treatment protocol.

5.9.4.3.2 Tumor Volume

Calculate the volume of tumor nodules by the formula $V = ab^2\pi/6$ ("*a*" is the larger diameter of the tumor nodule and "*b*" the diameter of the tumor nodule perpendicular to "*a*").

5.9.4.4 Operating Modality

5.9.4.4.1 Choice of the Treatment Modality (Table 5.1)

Table 5.1 Selection of operating modality and type of electrodes is based on the number of tumor nodules and their size

	Operating modality (OP)		Type of electrodes	
Number, volume and location of tumor nodules	A	B	**plate**	**needle**
1 – 3 nodules < 1 cm³ located anywhere, except head				
> 3 nodules, any volume located anywhere, except head				
any number, any volume located on head				

Deep sedation = OP A; general anesthesia = OP B

5.9.4.4.2 Deep Sedation

Medetomidine (0.08 mg/kg) or double combination medetomidine (0.025 mg/kg) with Torbugesic (0.1 mg/kg) can be used in dogs or follow standard procedures for deep sedation in the specific hospital.

5.9.4.4.3 Anesthesia

Premedication: combination of acepromazine (0.02 mg/kg) and methadone (2 mg/kg).

Thirty minutes later: general anesthesia induced using thiopental (5 mg/kg) or propofol (1 mg/kg), endotracheal tube placement, and maintenance under anesthesia with isoflurane in oxygen delivered via a T-piece anesthetic circuit.

Fluid support treatment during the anesthesia is recommended using Hartmann's solution at rate 10 ml/kg/h and analgesia with single dose of nonsteroidal analgesic drug (e.g., carprofen 2–4 mg/kg). After treatment, patients are kept in the clinic for about 2–4 h. Standard procedures for general anesthesia in specific hospital can be followed.

5.9.4.4.4 Injection of Cisplatin/Bleomycin

ECT consists of intratumoral administration of cisplatin or bleomycin and exposure of tumor nodules to electric pulses.

Cisplatin—cis-diamminedichloroplatinum II—is dissolved in distillate water at a concentration 2 mg/mL and is given intratumorally.

Bleomycin is dissolved in physiological saline at a concentration 3 mg/mL and is given intratumorally.

The interval between cisplatin or bleomycin administration and the application of electric pulses is 1–2 min.

Intratumoral cisplatin concentration used is 2 mg/ml and doses are in Table 5.2.

Intratumoral bleomycin concentration is 3000 IU/ml and doses are in Table 5.3.

Injection procedure is critical. The presence of the product in the target tissues determines ECT efficiency. When injecting, the sign of good tissue retention is whitening of the tissue.

Some leakage of solution could occur through the holes produced by the needle in the tumor or in necrotizing tissues during further ECT sessions. Reaching 2 mg/cm^3 (cisplatin) or 3 mg/cm^3 (bleomycin) concentrations could be then purely theoretical in those cases, and the goal is to try to retain as much solution as possible in the target tissues. In addition, electric pulses should be applied as quickly as possible in order to obtain the vasoconstricting effect which contributes to solution retention.

Table 5.2 The appropriate injection volume of cisplatin according to the tumor size

Tumor volume ($V = ab^2\pi/6$)	1 cm^3	>1 cm^3
CDDP dose, 2 mg/ml	1 ml (2 mg)/cm^3 of tumor	0.5–1 ml (1–2 mg)/cm^3 of tumor

Table 5.3 The appropriate injection volume of bleomycin according to the tumor size

Tumor volume ($V = ab^2\pi/6$)	1 cm^3	>1 cm^3
BLM dose, 3000 IU/ml	1 ml (3000 IU)/cm^3 of tumor	0.5–1 ml (1500–3000 IU)/cm^3 of tumor

5.9.4.4.5 Pulsing Procedure

5.9.4.4.5.1 Choice of Electrodes, Pulse Parameters, and Frequency

Different electroporation protocols can be used and are described in the previous parts of Sect. 5.9.2.

5.9.4.5 Animal Responses

Our group started to apply ECT in veterinary medicine in 1999. The objective was to introduce ECT with cisplatin into veterinary medicine, where there is a need for inexpensive and effective treatment of cutaneous and subcutaneous tumors of various histological types. In our first clinical study, the response to ECT was assessed on tumor nodules in three cats with mammary adenocarcinoma and fibrosarcoma and in seven dogs with mammary adenocarcinoma, cutaneous mast cell tumor, hemangioma, hemangiosarcoma, adenocarcinoma glandulae paranalis, and neurofibroma. All together 24 tumor nodules of different sizes were treated: five with cisplatin injected intratumorally and 19 with ECT with intratumoral administration of cisplatin. ECT with cisplatin had a good antitumor effect on all tumors treated. The average size 4 weeks after treatment was greatly reduced (0.01 cm^3) compared to those treated by intratumoral cisplatin injection alone (3.0 cm^3). Collectively, ECT-treated tumors responded with 84 % objective responses, whereas only one tumor partially responded to cisplatin treatment alone. Evaluated by contingency table, the response to treatment with ECT was significantly better than that of the cisplatin-treated group ($p = 0.014$). Furthermore, there was a significant prolongation of the duration of response in ECT-treated tumors ($p = 0.046$) [485].

In the clinical study performed on a large group of dogs with perianal tumors (28 dogs with 76 nodules), we elaborated further on ECT with intratumoral injection of cisplatin and also tested ECT with intratumorally injected bleomycin. ECT with bleomycin was used in tumors unresponsive to treatment with cisplatin. In order to compare antitumor effectiveness of ECT with regard to different tumor volumes, tumors were divided into two groups: group with tumors smaller than 1 cm^3 and group with tumors bigger than 1 cm^3. ECT of tumors smaller than 1 cm^3 resulted in 86 % objective responses 4 weeks after the treatment with 57 % of complete responses and 29 % of partial responses. At the end of observation period for each tumor ranging from 4 to 34 months, the percentage of complete responses increased to 88 %, and in the rest of the tumors, partial response was obtained (12 %). In the second group of tumors (tumor volume more than 1 cm^3), ECT was less effective. Four weeks after the treatment, objective response was obtained in 75 % of tumors (complete response = 12.5 %, partial response = 62.5 %). The same percentage of objective responses was obtained also at the end of observation period with 22 % of complete responses and 55.5 % of partial responses [501].

It's worth mentioning that the treatment with cisplatin or bleomycin given intratumorally alone or combined with electric pulses did not result in any local or systemic toxicity. We noticed partial necrosis of the tumors after a week and ulceration with the formation of a superficial scab, which fell off within 5 weeks. After treatment, none of the animals suffered from a local or systemic infection. In addition, the treatment had no effect on blood tests or biochemistry of the treated animals [501].

A clinical study was also performed in cats with squamous cell carcinoma (SCC). Eleven cats with 17 superficial SCC nodules in different clinical stages (ranging from Tis to T4), located on nasal planum (6 of 11), pinnae (3 of 11), and both locations (2 of 11), were included in the prospective nonrandomized study. Sixteen of seventeen SCC nodules were treated with ECT (15 of 16 with single session and in one case with two sessions); one nodule was surgically removed. Altogether 81.8 % (9 of 11) of CR for cats and 87.5 % (14/16) of CR for nodules were achieved, lasting from 2 months up to more than 3 years. Only two of nine cats in which CR was initially observed had a recurrence 2 and 8 months after ECT. In the remaining two cats with highly infiltrative spread into adjacent tissues, progression of the disease was observed, and both were euthanized 4 and 5 months after the procedure. Also in cats, electrochemotherapy was well tolerated, and no evident local or systemic side effects were observed [494].

5.9.5 Horse Sarcoids

5.9.5.1 Preparation of the Patients

5.9.5.1.1 Inclusion Criteria

The animals should be able to withstand general anesthesia with normal hematological and biochemical profiles. The maximum size of sarcoids should not exceed 2 cm diameter.

5.9.5.1.2 Surgical Preparation of the Patients

The animals are treated under general anesthesia of short duration either with injectable agents or with volatile inhalants. As an example of injectable protocol, ECT treatment was conducted under short-duration general anesthesia with i.v. agents: romifidine (0.08 mg/kg bwt) and diazepam (0.04 mg/kg bwt) as premedication prior to induction with ketamine (2.2 mg/kg bwt). Anesthesia was maintained using a continuous infusion of romifidine, ketamine, and guaifenesin.

For the treatment of periorbital lesions, a sensory block of the eyelids was performed with 2 % lidocaine in order to limit the head movements during the electric pulse application.

Depending on the number of tumors to be treated, anesthesia ranges from 15 min for one tumor to 40 min for several tumors. Several successive treatments are performed with a 2-week interval.

5.9.5.2 Treatment

Firstly, the antimitotic drug cisplatin is injected intratumorally, and secondly, 5 min after, the electric pulses are applied by bringing contact electrodes in contact with the skin.

5.9.5.2.1 Antimitotic Drug Injection

Sterile commercial cisplatin aqueous solution at 1 mg/ml concentration is used. It is, after aseptic preparation of the tumor site, intratumorally injected in a standardized manner (0.2–0.3 ml every 0.6 cm) by using "luer-lock" needles [502]. This dose is lower than the 3.3 mg/ml dosage used in cisplatin oil emulsions. Total volume injected is ½ of tumor volume. The skin is infiltrated on the tumor sides. The theoretical calculated total dose of cisplatin is 0.3 mg of cisplatin per cm^3 of target tissue. In our experimental trials, a NaCl solution was used. A multisite intratumoral injection procedure was selected to attempt to obtain a homogeneous tumor irrigation by the drug. This limits the amount of drug which is needed due to the weight of horses.

5.9.5.2.2 Application of Electric Pulses

Less than 5 min after the drug injection, the electrical treatment is applied by bringing contact electrodes in contact with the skin. The procedure with contact wire electrodes is described in 5.9.2.2.1. Needle electrodes (as described in 5.9.2.2.3) cannot be used due to the resistance of the horse skin to punching.

5.9.5.2.3 Number of Successive Treatments

The number of ECT sessions depends on the surface and volume of tumors to treat: the number of treatments required for effective therapy is significantly related to tumor size (the larger the tumors, the higher the number of treatments). In the same manner, when ECT is used following surgery, the number of sessions is significantly related to the excision quality (increasing number in case of incomplete surgical excision) and with the healing mode (increasing number in case of secondary wound closure).

5.9.5.3 Animal Responses

5.9.5.3.1 Immediate Responses

Muscular contractions are observed during each pulse application. But as a 1 ms delay is preset, only one contraction is observed. The amplitude of these movements is more pronounced when the application is close to a limb root and at the ear level. These effects are expected when electrical stimulation is used.

The color of the tissue in the electro-treated region was observed to change. This appears to be related to the previous observation that electric pulses induced a temporary shutdown of the blood flow [503].

5.9.5.3.2 Long-Term Responses

In the days following an ECT treatment, a slightly edematous reaction is detected for lesions located on thin skin regions on some horses. No other adverse reaction was observed in a retrospective study.

Animals should be evaluated every 2 weeks during the course of treatment.

Tumor response is assessed at follow-up examinations undertaken at 1, 3, and 6 months after the last ECT treatment.

5.9.5.3.3 Response Rates at 4-Year Follow-Up

In a retrospective study, the results on a per tumor basis were 99.5 % (193 of 194) (95 % CI: 0.971–0.999) of relapse-free tumors after a mean of 2.81 ECT treatments. Results on a per animal basis were 97.9 % (47 of 48) (95 % CI: 0.889–0.999) of relapse-free animals after a mean of 3.62 ECT sessions [493].

References

1. Coster, H.G.: A quantitative analysis of the voltage-current relationships of fixed charge membranes and the associated property of "punch-through". Biophys. J. **5**(5), 669–686 (1965)
2. Sale, A.J.H., Hamilton, W.A.: Effects of high electric fields on microorganisms: I. Killing of bacteria and yeasts. Biochim. Biophys. Acta Gen. Subj. **148**(3), 781–788 (1967)
3. Sale, A.J.H., Hamilton, W.A.: Effects of high electric fields on micro-organisms: III. Lysis of erythrocytes and protoplasts. Biochim. Biophys. Acta Biomembr. **163**(1), 37–43 (1968)

4. Pohl, H.A., Crane, J.S.: Dielectrophoresis of cells. Biophys. J. **11**(9), 711–727 (1971)
5. Crowley, J.M.: Electrical breakdown of bimolecular lipid membranes as an electromechanical instability. Biophys. J. **13**(7), 711–724 (1973)
6. Zimmermann, U., Pilwat, G., Riemann, F.: Dielectric breakdown of cell membranes. Biophys. J. **14**(11), 881–899 (1974)
7. Hamilton, W.A., Sale, A.J.H.: Effects of high electric fields on microorganisms II. Mechanism of action of the lethal effect. Biochim. Biophys. Acta **148**, 789–800 (1967)
8. Neumann, E., Rosenheck, K.: Permeability changes induced by electric impulses in vesicular membranes. J. Membr. Biol. **10**, 279–290 (1972)
9. Kinosita Jr., K., Tsong, T.Y.: Hemolysis of human erythrocytes by transient electric field. Proc. Natl. Acad. Sci. U. S. A. **74**, 1923–1927 (1977)
10. Schoenbach, K.H., Beebe, S.J., Buescher, E.S.: Intracellular effect of ultrashort electrical pulses. Bioelectromagnetics **22**(6), 440–448 (2001)
11. Beebe, S.J., Fox, P.M., Rec, L.H., Buescher, E.S., Somers, K., Schoenbach, K.H.: Nanosecond pulsed electric field (nsPEF) effects on cells and tissues: apoptosis induction and tumor growth inhibition. IEEE. T. Plasma Sci. **30**, 286–292 (2002)
12. Nuccitelli, R., Wood, R., Kreis, M., Athos, B., Huynh, J., Lui, K., Nuccitelli, P., Epstein Jr., E.H.: First-in-human trial of nanoelectroablation therapy for basal cell carcinoma: proof of method. Exp. Dermatol. **23**, 135–137 (2014)
13. Chen, R., Chen, X., Beebe, S.J.: Nanosecond Pulsed Electric Field (nsPEF) ablation as an alternative or adjunct to surgery for treatment of cancer. Surg. Curr. Res. **S12**, 005 (2013)
14. Chen, R., Sain, N.M., Harlow, K.T., Chen, Y.J., Shires, P.K., Heller, R., Beebe, S.J.: A protective effect after clearance of orthotopic rat hepatocellular carcinoma by nanosecond pulsed electric fields. Eur. J. Cancer **50**(15), 2705–2713 (2014)
15. Chen, X., Zhuang, J., Kolb, J.F., Schoenbach, K.H., Beebe, S.J.: Long term survival of mice with hepatocellular carcinoma after pulse power ablation with nanosecond pulsed electric fields. Technol. Cancer Res. Treat. **11**, 83–93 (2012)
16. Garon, E.B., Sawcer, D., Vernier, P.T., Tang, T., Sun, Y., Marcu, L., Gundersen, M.A., Koeffler, H.P.: In vitro and in vivo evaluation and a case report of intense nanosecond pulsed electric field as a local therapy for human malignancies. Int. J. Cancer **121**, 675–682 (2007)
17. Chalise, P.R., Perni, S., Shama, G., Novac, B.M., Smith, R., et al.: Lethality mechanisms in Escherichia coli induced by intense sub-microsecond electrical pulses. Appl. Phys. Lett. **89**, 153902 (2006)
18. Wang, S., Chen, J., Chen, M.T., Vernier, P.T., Gundersen, M.A., Valderrábano, M.: Cardiac myocyte excitation by ultrashort high-field pulses. Biophys. J. **96**, 1640–1648 (2009)
19. Hargrave, B., Li, F.: Platelet Rich Plasma (PRP) Reduces Reactive Oxygen Species (ROS) production and mitochondrial depolarization in H9c2 cells in culture, reduces infarct size and improves left ventricular mechanical function in the rabbit langendorff heart and in the rabbit heart in vivo. J. Extra Corp. Technol. **44**(4), 198–204 (2012)
20. Hargrave, B.Y.: Autologous Platelet rich plasma (Platelet Gel): an appropriate intervention for salvaging cardiac myocytes under oxidative stress after myocardial infarction. Anat. Physiol. **4**, 1 (2014)
21. Edelblute, C.M., Amy, L., Donate, A.L., Hargrave, B.Y., Heller, L.C.: Human Platelet Gel Supernatant Inactivates Opportunistic Wound Pathogens on Skin Platelets, Online: Informa UK Ltd, 1–41, London, England (2014)
22. Hargrave, B., Li, F.: Nanosecond pulse electric field activated-platelet rich plasma enhances the return of blood flow to large and ischemic wounds in a rabbit model. Physiol. Rep. **3**(7), 1–9 (2015)
23. Kolb, J.F., Xiao, S., Camp, J.T., Migliaccio, M., Bajracharya, C., Schoenbach, K.H.: Sub-nanosecond electrical pulses for medical therapies and imaging. Proceedings of the Fourth European Conference Antennas and Propagation, pp. 1–5 (2010)
24. Semenov, I., Xiao, S., Kang, D., Schoenbach, K.H., Pakhomov, A.G.: Cell stimulation and calcium mobilization by picosecond electric pulses. Bioelectrochemistry **105**, 65–71 (2015)

25. Joshi, R.P., Mishra, A., Song, J., Pakhomov, A., Schoenbach, K.H.: Simulation studies of ultrashort, high-intensity electric pulse induced action potential block in whole-animal nerves. IEEE. T. Plasma Sci. **55**, 1391–1398 (2008)
26. Davalos, R.V., Mir, L., Rubinsky, B.: Tissue ablation with irreversible electroporation. Ann. Biomed. Eng. **33**(2), 223–231 (2005)
27. Al-Sakere, B., Andre, F., Bernat, C., Connault, E., Opolon, P., Davalos, R.V., Rubinsky, B., Mir, L.M.: Tumor ablation with irreversible electroporation. PLoS ONE **2**(11), e1135 (2007)
28. Garcia, P.A., Rossmeisl Jr., J.H., Neal 2nd, R.E., Ellis, T.L., Davalos, R.V.: A parametric study delineating irreversible electroporation from thermal damage based on a minimally invasive intracranial procedure. Biomed. Eng. Online **10**(1), 34 (2011)
29. Garcia, P.A., Rossmeisl Jr., J.H., Neal II, R.E., Ellis, T.L., Olson, J.D., Henao-Guerrero, N., Robertson, J., Davalos, R.V.: Intracranial nonthermal irreversible electroporation: in vivo analysis. J. Membr. Biol. **236**(1), 127–136 (2010)
30. Martin 2nd, R.C., McFarland, K., Ellis, S., Velanovich, V.: Irreversible electroporation therapy in the management of locally advanced pancreatic adenocarcinoma. J. Am. Coll. Surg. **215**(3), 361–369 (2012)
31. Pech, M., Janitzky, A., Wendler, J.J., Strang, C., Blaschke, S., Dudeck, O., Ricke, J., Liehr, U. B.: Irreversible electroporation of renal cell carcinoma: a first-in-man phase I clinical study. Cardiovasc. Intervent. Radiol. **34**(1), 132–138 (2011)
32. Thomson, K.R., Cheung, W., Ellis, S.J., Federman, D., Kavnoudias, H., Loader-Oliver, D., Roberts, S., Evans, P., Ball, C., Haydon, A.: Investigation of the safety of irreversible electroporation in humans. J. Vasc. Interv. Radiol.: JVIR **22**(5), 611–621 (2011)
33. Kingham, T.P., Karkar, A.M., D'Angelica, M.I., Allen, P.J., DeMatteo, R.P., Getrajdman, G. I., Sofocleous, C.T., Solomon, S.B., Jarnagin, W.R., Fong, Y.: Ablation of perivascular hepatic malignant tumors with irreversible electroporation. J. Am. Coll. Surg. **215**(3), 379–387 (2012)
34. Golberg, A., Yarmush, M.L.: Nonthermal irreversible electroporation: fundamentals, applications, and challenges. Biomed. Eng. IEEE. Trans. **60**(3), 707–714 (2013)
35. Cheung, W., Kavnoudias, H., Roberts, S., Szkandera, B., Kemp, W., Thomson, K.R.: Irreversible electroporation for unresectable hepatocellular carcinoma: initial experience and review of safety and outcomes. Technol. Cancer Res. Treat. **12**, 233–241 (2013)
36. Garcia, P.A., Pancotto, T., Rossmeisl Jr., J.H., Henao-Guerrero, N., Gustafson, N.R., Daniel, G.B., Robertson, J.L., Ellis, T.L., Davalos, R.V.: Non-thermal irreversible electroporation (N-TIRE) and adjuvant fractionated radiotherapeutic multimodal therapy for intracranial malignant glioma in a canine patient. Technol. Cancer Res. Treat. **10**(1), 73–83 (2011)
37. Laroussi, M.: Low temperature plasma-based sterilization: overview and state-of-the-art. Plasma Process. Polym. **2**, 391–400 (2005)
38. Lu, X., Ye, T., Cao, Y.G., Sun, Z.Y., Xiong, Q., Tang, Z.Y., Xiong, Z.L., Hu, J., Jiang, Z.H., Pan, Y.: The roles of the various plasma agents in the inactivation of bacteria. J. Appl. Phys. **104**, 053309-1-5 (2008)
39. Graves, D.B.: The emerging role of reactive oxygen and nitrogen species in redox biology and some implications for plasma applications to medicine and biology. J. Phys. D. Appl. Phys. **45**, 263001-1-42 (2012)
40. Kieft, I.E., Kurdi, M., Stoffels, E.: Reattachment and apoptosis after plasma-needle treatment of cultured cells. IEEE. T. Plasma Sci. **34**, 1331–1336 (2006)
41. Jiang, C., Vernier, P.T., Chen, M.T., Wu, Y.H., Wang, L.L., Gundersen, M.A.: Low Energy Nanosecond Pulsed Plasma Sterilization for Endodontic Applications, Proceedings of the 2008 I.E. International Power Modulators and High Voltage Conference, pp. 77–79 (2008)
42. Jiang, C.Q., Chen, M.T., Gorur, A., Schaudinn, C., Jaramillo, D.E., Costerton, J.W., Sedghizadeh, P.P., Vernier, P.T., Gundersen, M.A.: Nanosecond pulsed plasma dental probe. Plasma Process. Polym. **6**, 479–483 (2009)
43. Isbary, G., Morfill, G., Schmidt, H.U., Georgi, M., Ramrath, K., Heinlin, J., Karrer, S., Landthaler, M., Shimizu, T., Steffes, B., Bunk, W., Monetti, R., Zimmermann, J.L., Pompl,

R., Stolz, W.: A first prospective randomized controlled trial to decrease bacterial load using cold atmospheric argon plasma on chronic wounds in patients. Br. J. Dermatol. **163**, 78–82 (2010)
44. Fridman, G., Peddinghaus, M., Ayan, H., Fridman, A., Balasubramanian, M., Gutsol, A., Brooks, A., Friedman, G.: Blood coagulation and living tissue sterilization by floating-electrode dielectric barrier discharge in air. Plasma Chem. Plasma P. **26**, 425–442 (2006)
45. Sladek, R.E.J., Stoffels, E., Walraven, R., Tielbeek, P.J.A., Koolhoven, R.A.: Plasma treatment of dental cavities: a feasibility study. IEEE. T. Plasma Sci. **32**, 1540–1543 (2004)
46. Sladek, R.E.J., Filoche, S.K., Sissons, C.H., Stoffels, E.: Treatment of Streptococcus mutans biofilms with a nonthermal atmospheric plasma. Lett. Appl. Microbiol. **45**, 318–323 (2007)
47. Kieft, I.E., Darios, D., Roks, A.J.M., Stoffels, E.: Plasma treatment of mammalian vascular cells: a quantitative description. IEEE. T. Plasma Sci. **33**, 771–775 (2005)
48. Okino, M., Mohri, H.: Effects of a high-voltage electrical impulse and an anticancer drug on in vivo growing tumors. Jpn. J. Cancer Res. **78**, 1319–1321 (1987)
49. Mir, L.M., Banoun, H., Paoletti, C.: Introduction of definite amounts of nonpermeant molecules into living cells after electropermeabilization: direct access to the cytosol. Exp. Cell Res. **175**, 15–25 (1988)
50. Gothelf, A., Mir, L.M., Gehl, J.: Electrochemotherapy: results of cancer treatment using enhanced delivery of bleomycin by electroporation. Cancer Treat. Rev. **29**(5), 371–387 (2003)
51. Mir, L.M., Gehl, J., Sersa, G., Collins, C.G., Garbay, J.R., Billard, V., et al.: Standard operating procedures of the electrochemotherapy: instructions for the use of bleomycin or cisplatin administered either systemically or locally and electric pulses delivered by the Cliniporator (TM) by means of invasive or non-invasive electrodes. EJC Suppl. **4**(11), 14–25 (2006)
52. Gilbert, R., Jaroszeski, M.J., Heller, R.: Novel electrode designs for electrochemotherapy. Biochem. Biophys. Acta **1334**, 9–14 (1997)
53. Miklavcic, D., Beravs, K., Semrov, D., Cemazar, M., Demsar, F., Sersa, G.: The importance of electric field distribution for effective in vivo electroporation of tissues. Biophys. J. **74**(5), 2152–2158 (1998)
54. Mahmood, F., Gehl, J.: Optimizing clinical performance and geometrical robustness of a new electrode device for intracranial tumor electroporation. Bioelectrochemistry **81**, 10–16 (2011)
55. Staal, L.G., Gilbert, R.: In: Kee, S., Gehl, J., Lee, E. (eds.) Generators and Applicators: Equipment for Electroporation, pp. 45–65. Springer, New York (2011)
56. Miklavcic, D., Sersa, G., Brecelj, E., Gehl, J., Soden, D., Bianchi, G., et al.: Electrochemotherapy: technological advancements for efficient electroporation-based treatment of internal tumors. Med. Biol. Eng. Comput. **50**(12), 1213–1225 (2012)
57. Sersa, G., Krzic, M., Sentjurc, M., Ivanusa, T., Beravs, K., Kotnik, V., et al.: Reduced blood flow and oxygenation in SA-1 tumours after electrochemotherapy with cisplatin. Br. J. Cancer **87**(9), 1047–1054 (2002)
58. Gehl, J., Skovsgaard, T., Mir, L.M.: Vascular reactions to in vivo electroporation: characterization and consequences for drug and gene delivery. Biochim. Biophys. Acta **1569**(1–3), 51–58 (2002)
59. Mir, L.M., Belehradek, M., Domenge, C., Orlowski, S., Poddevin, B., Belehradek Jr., J., Schwaab, G., Luboinski, B., Paoletti, C.: Electrochemotherapy, a new antitumor treatment: first clinical trial. C. R. Acad. Sci. III **313**, 613–618 (1991)
60. Marty, M., Sersa, G., Garbay, J.R., Gehl, J., Collins, C.G., Snoj, M., et al.: Electrochemotherapy – an easy, highly effective and safe treatment of cutaneous and subcutaneous metastases: Results of ESOPE (European Standard Operating Procedures of Electrochemotherapy) study. EJC Suppl. **4**(11), 3–13 (2006)

61. Matthiessen, L.W., Chalmers, R.L., Sainsbury, D.C., Veeramani, S., Kessell, G., Humphreys, A.C., et al.: Management of cutaneous metastases using electrochemotherapy. Acta Oncol. **50**(5), 621–629 (2011)
62. Matthiessen, L.W., Johannesen, H.H., Hendel, H.W., Moss, T., Kamby, C., Gehl, J.: Electrochemotherapy for large cutaneous recurrence of breast cancer: a phase II clinical trial. Acta Oncol. **51**(6), 713–721 (2012)
63. Heller, R., Jaroszeski, M.J., Reintgen, D.S., Puleo, C.A., DeConti, R.C., Gilbert, R.A., et al.: Treatment of cutaneous and subcutaneous tumors with electrochemotherapy using intralesional bleomycin. Cancer **83**(1), 148–157 (1998)
64. Spratt, D.E., Spratt, E.A.G., Wu, S.: Efficacy of skin-directed therapy for cutaneous metastases from advanced cancer: a meta-analysis. J. Clin. Oncol. **32**, 3144–3155 (2014)
65. National Institute for H, Care E.: Electrochemotherapy for metastases in the skin from tumours of non-skin origin and melanoma. http://publicationsnice.org.uk/electrochemotherapy-for-metastases-in-the-skin-from-tumours-of-non-skin-origin-and-mela noma-ipg446 (2013)
66. Cadossi, R., Ronchetti, M., Cadossi, M.: Locally enhanced chemotherapy by electroporation: clinical experiences and perspective of use of electrochemotherapy. Future Oncol. **10**, 877–890 (2014)
67. Edhemovic, I., Brecelj, E., Gasljevic, G., Marolt Music, M., Gorjup, V., Mali, B., et al.: Intraoperative electrochemotherapy of colorectal liver metastases. J. Surg. Oncol. **110**(3), 320–327 (2014)
68. Granata, V., Fusco, R., Piccirillo, M., Palaia, R., Petrillo, A., Izzo, F.: Electrochemotherapy in locally advanced pancreatic cancer: preliminary results. Int. J. Surg. **18**, 230–236 (2015)
69. Agerholm-Larsen, B., Iversen, H.K., Ibsen, P., Moller, J.M., Mahmood, F., Jensen, K.S., et al.: Preclinical validation of electrochemotherapy as an effective treatment for brain tumors. Cancer Res. **71**, 3753–3762 (2011)
70. Bourke, M.G., Salwa, S., Sadadcharam, M., Forde, P., O'Sullivan, G.C., Soden, D.: Development of an endoscopically delivered ablative treatment for gastrointestinal tumours. Ir. J. Med. Sci. **179**, S361–S61 (2010)
71. Fini, M., Salamanna, F., Parrilli, A., Martini, L., Cadossi, M., Maglio, M., et al.: Electrochemotherapy is effective in the treatment of rat bone metastases. Clin. Exp. Metastasis **30**(8), 1033–1045 (2013)
72. Cemazar, M., et al.: Electrochemotherapy in veterinary oncology. J. Vet. Intern. Med. **22**(4), 826–831 (2008)
73. Frandsen, S.K., Gissel, H., Hojman, P., Tramm, T., Eriksen, J., Gehl, J.: Direct therapeutic applications of calcium electroporation to effectively induce tumor necrosis. Cancer Res. **72** (6), 1336–1341 (2012)
74. Heller, R., Jaroszeski, M., Atkin, A., Moradpour, D., Gilbert, R., Wands, J., Nicolau, C.: In vivo gene electroinjection and expression in rat liver. FEBS Lett. **389**(3), 225–228 (1996)
75. Aihara, H., Miyazaki, J.: Gene transfer into muscle by electroporation in vivo. Nat. Biotechnol. **16**(9), 867–870 (1998). doi:10.1038/nbt0998-867
76. Titomirov, A.V., Sukharev, S., Kistanova, E.: In vivo electroporation and stable transformation of skin cells of newborn mice by plasmid DNA. Biochim. Biophys. Acta **1088**(1), 131–134 (1991)
77. Rols, M.P., Delteil, C., Golzio, M., Dumond, P., Cros, S., Teissie, J.: In vivo electrically mediated protein and gene transfer in murine melanoma. Nat. Biotechnol. **16**(2), 168–171 (1998). doi:10.1038/nbt0298-168
78. Niu, G., Heller, R., Catlett-Falcone, R., Coppola, D., Jaroszeski, M., Dalton, W., Jove, R., Yu, H.: Gene therapy with dominant-negative Stat3 suppresses growth of the murine melanoma B16 tumor in vivo. Cancer Res. **59**(20), 5059–5063 (1999)
79. Young, J.L., Dean, D.A.: Electroporation-mediated gene delivery. Adv. Genet. **89**, 49–88 (2015)
80. Heller, R., Heller, L.C.: Gene electrotransfer clinical trials. Adv. Genet. **89**, 235–262 (2015)

81. Daud, A.I., DeConti, R.C., Andrews, S., Urbas, P., Riker, A.I., Sondak, V.K., Munster, P.N., Sullivan, D.M., Ugen, K.E., Messina, J.L., Heller, R.: Phase I trial of interleukin-12 plasmid electroporation in patients with metastatic melanoma. J. Clin. Oncol. **26**, 5896–5903 (2008)
82. Yang, L.J., Yu, W.D., Du, J.B., Chao, S., Chen, M.X., Zhao, H.H., Guo, J.Z.: Overexpression or knock-down of runt-related transcription factor 1 affects BCR-ABL-induced proliferation and migration in vitro and leukemogenesis in vivo in mice. Chin. Med. J. (Engl.) **122**(3), 331–337 (2009)
83. Magee, T.R., Artaza, J.N., Ferrini, M.G., Vernet, D., Zuniga, F.I., Cantini, L., Reisz-Porszasz, S., Rajfer, J., Gonzalez-Cadavid, N.F.: Myostatin short interfering hairpin RNA gene transfer increases skeletal muscle mass. J. Gene Med. **8**(9), 1171–1181 (2006)
84. Magee, T.R., Kovanecz, I., Davila, H.H., Ferrini, M.G., Cantini, L., Vernet, D., Zuniga, F.I., Rajfer, J., Gonzalez-Cadavid, N.F.: Antisense and short hairpin RNA (shRNA) constructs targeting PIN (Protein Inhibitor of NOS) ameliorate aging-related erectile dysfunction in the rat. J. Sex. Med. **4**(3), 633–643 (2007)
85. Takahashi, Y., Nishikawa, M., Takakura, Y.: Suppression of tumor growth by intratumoral injection of short hairpin RNA-expressing plasmid DNA targeting beta-catenin or hypoxia-inducible factor 1alpha. J. Control. Release **116**(1), 90–95 (2006)
86. Dolinsek, T., Markelc, B., Sersa, G., Coer, A., Stimac, M., Lavrencak, J., Brozic, A., Kranjc, S., Cemazar, M.: Multiple delivery of siRNA against endoglin into murine mammary adenocarcinoma prevents angiogenesis and delays tumor growth. PLoS ONE **8**(3), e58723 (2013)
87. Benteyn, D., Van Nuffel, A.M., Wilgenhof, S., Corthals, J., Heirman, C., Neyns, B., Thielemans, K., Bonehill, A.: Characterization of CD8+ T-cell responses in the peripheral blood and skin injection sites of melanoma patients treated with mRNA electroporated autologous dendritic cells (TriMixDC-MEL). Biomed. Res. Int. **2013**, 976383 (2013)
88. Van Nuffel, A.M., Benteyn, D., Wilgenhof, S., Corthals, J., Heirman, C., Neyns, B., Thielemans, K., Bonehill, A.: Intravenous and intradermal TriMix-dendritic cell therapy results in a broad T-cell response and durable tumor response in a chemorefractory stage IV-M1c melanoma patient. Cancer Immunol. Immunother. **61**(7), 1033–1043 (2012)
89. Van Nuffel, A.M., Benteyn, D., Wilgenhof, S., Pierret, L., Corthals, J., Heirman, C., van der Bruggen, P., Coulie, P.G., Neyns, B., Thielemans, K., Bonehill, A.: Dendritic cells loaded with mRNA encoding full-length tumor antigens prime CD4+ and CD8+ T cells in melanoma patients. Mol. Ther. **20**(5), 1063–1074 (2012)
90. Wilgenhof, S., Corthals, J., Van Nuffel, A.M., Benteyn, D., Heirman, C., Bonehill, A., Thielemans, K., Neyns, B.: Long-term clinical outcome of melanoma patients treated with messenger RNA-electroporated dendritic cell therapy following complete resection of metastases. Cancer Immunol. Immunother. **64**(3), 381–388 (2015)
91. Wilgenhof, S., Van Nuffel, A.M., Corthals, J., Heirman, C., Tuyaerts, S., Benteyn, D., De Coninck, A., Van Riet, I., Verfaillie, G., Vandeloo, J., Bonehill, A., Thielemans, K., Neyns, B.: Therapeutic vaccination with an autologous mRNA electroporated dendritic cell vaccine in patients with advanced melanoma. J. Immunother. **34**(5), 448–456 (2011)
92. Beatty, G.L., Haas, A.R., Maus, M.V., Torigian, D.A., Soulen, M.C., Plesa, G., Chew, A., Zhao, Y., Levine, B.L., Albelda, S.M., Kalos, M., June, C.H.: Mesothelin-specific chimeric antigen receptor mRNA-engineered T cells induce anti-tumor activity in solid malignancies. Cancer Immunol. Res. **2**(2), 112–120 (2014)
93. Pavlin, D., Cemazar, M., Sersa, G., Tozon, N.: IL-12 based gene therapy in veterinary medicine. J. Transl. Med. **10**, 234-1-10 (2012)
94. Neumann, E., Sowers, A.E., Jordan, C.A.: Electroporation and Electrofusion in Cell Biology. Plenum press, York (1989)
95. Weaver, J.C., Chizmadzhev, Y.A.: Theory of electroporation: a review. Bioelectrochem. Bioenerg. **41**, 135–160 (1996)

96. Teissié, J., Golzio, M., Rols, M.P.: Mechanisms of cell membrane electropermeabilization: a mini-review of our present (lack of?) knowledge. Biochim. Biophys. Acta **2005**, 270–280 (1724)
97. Tekle, E., Astumian, R.D., Chock, P.B.: Selective and asymmetric molecular transport across electroporated cell membranes. Proc. Natl. Acad. Sci. U. S. A. **91**, 11512–11516 (1994)
98. siRNA, Paganin-Gioanni, A., et al.: Direct visualization at the single-cell level of siRNA electrotransfer into cancer cells. Proc. Natl. Acad. Sci. U. S. A. **108**(26), 10443–10447 (2011)
99. pDNA, Neumann, E., Shaefer-Ridder, M., Wang, Y., Hofschneider, P.H.: Gene transfer into mouse lyoma cells by electroporation in high electric fields. EMBO J. **1**, 841–845 (1982)
100. Tekle, E., Astumian, R.D., Chock, P.B.: Electroporation by using bipolar oscillating electric field: an improved method for DNA transfection of NIH 3T3 cells. Proc. Natl. Acad. Sci. U. S. A. **88**, 4230–4234 (1991)
101. Golzio, M., Teissié, J., Rols, M.P.: Direct visualization at the single-cell level of electrically mediated gene delivery. Proc. Natl. Acad. Sci. U. S. A. **99**, 1292–1297 (2002)
102. Escoffre, J.M., Portet, T., Favard, C., Teissié, J., Dean, D.S., Rols, M.P.: Electromediated formation of DNA complexes with cell membranes and its consequences for gene delivery. Biochim. Biophys. Acta. **1808**(6), 1538–1543 (2011)
103. Rosazza, C., Buntz, A., Rieß, T., Wöll, D., Zumbusch, A., Rols, M.P.: Intracellular tracking of single-plasmid DNA particles after delivery by electroporation. Mol. Ther. **21**(12), 2217–2226 (2013)
104. Golzio, M., Mora, M.P., Raynaud, C., Delteil, C., Teissie, J., Rols, M.P.: Control by osmotic pressure of voltage-induced permeabilization and gene transfer in mammalian cells. Biophys. J. **74**, 3015–3022 (1998)
105. Faurie, C., Phez, E., Golzio, M., Vossen, C., Lesbordes, J.C., Delteil, C., Teissié, J., Rols, M. P.: Effect of electric field vectoriality on electrically mediated gene delivery in mammalian cells. Biochim. Biophys. Acta **1665**(1–2), 92–100 (2004)
106. Kanduser, M., Miklavcic, D., Pavlin, M.: Mechanisms involved in gene electrotransfer using high- and low-voltage pulse—an in vitro study. Bioelectrochemistry **74**, 265–271 (2009)
107. Cemazar, M., Sersa, G.: Electrotransfer of therapeutic molecules into tissues. Curr. Opin. Mol. Ther. **9**(6), 554–562 (2007)
108. siRNA, Golzio, M., et al.: Inhibition of gene expression in mice muscle by in vivo electrically mediated siRNA delivery. Gene Ther. **12**(3), 246–251 (2005)
109. Golzio, M., et al.: In vivo gene silencing in solid tumors by targeted electrically mediated siRNA delivery. Gene Ther. **14**(9), 752–759 (2007)
110. pDNA, Cemazar, M., Golzio, M., Sersa, G., Rols, M.P., Teissié, J.: Electrically-assisted nucleic acids delivery to tissues in vivo: where do we stand? Curr. Pharm. Des. **12**(29), 3817–3825 (2006)
111. Hojman, P.: Basic principles and clinical advancements of muscle electrotransfer. Curr. Gene Ther. **10**(2), 128–138 (2010)
112. Bodles-Brakhop, A.M., Heller, R., Draghia-Akli, R.: Electroporation for the delivery of DNA-based vaccines and immunotherapeutics. Curr. Clin. Dev. Mol. Ther. **17**(4), 585–592 (2009)
113. Satkauskas, S., Bureau, M.F., Puc, M., Mahfoudi, A., Scherman, D., Miklavcic, D., Mir, L. M.: Mechanisms of in vivo DNA electrotransfer: respective contributions of cell electropermeabilization and DNA electrophoresis. Mol. Ther. **5**, 133–140 (2002)
114. André, F.M., Gehl, J., Sersa, G., Préat, V., Hojman, P., Eriksen, J., Golzio, M., Cemazar, M., Pavselj, N., Rols, M.P., Miklavcic, D., Neumann, E., Teissié, J., Mir, L.M.: Efficiency of high- and low-voltage pulse combinations for gene electrotransfer in muscle, liver, tumor, and skin. Hum. Gene Ther. **19**(11), 1261–1271 (2008)
115. Cemazar, M., Golzio, M., Sersa, G., Escoffre, J.M., Coer, A., Vidic, S., Teissie, J.: Hyaluronidase and collagenase increase the transfection efficiency of gene electrotransfer in various murine tumors. Hum. Gene Ther. **23**(1), 128–137 (2012)

116. McMahon, J.M., Signori, E., Wells, K.E., Fazio, V.M., Wells, D.J.: Optimisation of electrotransfer of plasmid into skeletal muscle by pretreatment with hyaluronidase – increased expression with reduced muscle damage. Gene Ther. **8**(16), 1264–1270 (2001)
117. Levine, Z.A., Vernier, P.T.: Life cycle of an electropore: field-dependent and field-independent steps in pore creation and annihilation. J. Membr. Biol. **236**(1), 27–36 (2010)
118. Gabriel, B., Teissie, J.: Direct observation in the millisecond time range of fluorescent molecule asymmetrical interaction with the electropermeabilized cell membrane. Biophys. J. **73**(5), 2630–2637 (1997)
119. Canatella, P.J., Karr, J.F., Petros, J.A., Prausnitz, M.R.: Quantitative study of electroporation-mediated molecular uptake and cell viability. Biophys. J. **80**, 755–764 (2001)
120. Gehl, J., Skovsgaard, T., Mir, L.M.: Enhancement of cytotoxicity by electropermeabilization: an improved method for screening drugs. Anti-Cancer Drugs **9**(4), 319–325 (1998)
121. Jaroszeski, M.J., Dang, V., Pottinger, C., Hickey, J., Gilbert, R., Heller, R.: Toxicity of anticancer agents mediated by electroporation in vitro. Anti-Cancer Drugs **11**, 201–208 (2000)
122. Orlowski, S., Belehradek Jr., J., Paoletti, C., Mir, L.M.: Transient electropermeabilization of cells in culture. Increase of the cytotoxicity of anticancer drugs. Biochem. Pharmacol. **37**(24), 4727–4733 (1988)
123. Sersa, G., Cemazar, M., Miklavcic, D.: Antitumor effectiveness of electrochemotherapy with cis-diamminedichloroplatinum(II) in mice. Cancer Res. **55**(15), 3450–3455 (1995)
124. Umezawa, H., Maeda, K., Takeuchi, T., Okami, Y.: New antibiotics, bleomycin A and B. J. Antibiot. (Tokyo) **19**(5), 200–209 (1966)
125. Tounekti, O., Kenani, A., Foray, N., Orlowski, S., Mir, L.M.: The ratio of single- to double-strand DNA breaks and their absolute values determine cell death pathway. Br. J. Cancer **84** (9), 1272–1279 (2001)
126. Vermorken, J.B., Mesia, R., Rivera, F., Remenar, E., Kawecki, A., Rottey, S., et al.: Platinum-based chemotherapy plus cetuximab in head and neck cancer. N. Engl. J. Med. **359**, 1116–1127 (2008)
127. Vasquez, J.L., Ibsen, P., Lindberg, H., Gehl, J.: In vitro and in vivo experiments on electrochemotherapy for bladder cancer. J. Urol. **193**(3), 1009–1015 (2015)
128. Frandsen, S.K., Gissel, H., Hojman, P., Eriksen, J., Gehl, J.: Calcium electroporation in three cell lines: a comparison of bleomycin and calcium, calcium compounds, and pulsing conditions. Biochim. Biophys. Acta **2014**, 1204–1208 (1840)
129. Glass, L.F., Pepine, M.L., Fenske, N.A., Jaroszeski, M., Reintgen, D.S., Heller, R.: Bleomycin-mediated electrochemotherapy of metastatic melanoma. Arch. Dermatol. **132** (11), 1353–1357 (1996)
130. Gehl, J., Sorensen, T.H., Nielsen, K., Raskmark, P., Nielsen, S.L., Skovsgaard, T., et al.: In vivo electroporation of skeletal muscle: threshold, efficacy and relation to electric field distribution. Biochim. Biophys. Acta **1428**(2–3), 233–240 (1999)
131. Gothelf, A., Mahmood, F., Dagnaes-Hansen, F., Gehl, J.: Efficacy of transgene expression in porcine skin as a function of electrode choice. Bioelectrochemistry **82**, 95–102 (2011)
132. Heller, R., Jaroszeski, M.J., Glass, L.F., Messina, J.L., Rapaport, D.P., DeConti, R.C., et al.: Phase I/II trial for the treatment of cutaneous and subcutaneous tumors using electrochemotherapy. Cancer **77**(5), 964–971 (1996)
133. Sersa, G., Stabuc, B., Cemazar, M., Miklavcic, D., Rudolf, Z.: Electrochemotherapy with cisplatin: clinical experience in malignant melanoma patients. Clin. Cancer Res. **6**(3), 863–867 (2000)
134. Campana, L.G., Valpione, S., Mocellin, S., Sundararajan, R., Granziera, E., Sartore, L., et al.: Electrochemotherapy for disseminated superficial metastases from malignant melanoma. Br. J. Surg. **99**, 821–830 (2012)
135. Quaglino, P., Mortera, C., Osella-Abate, S., Barberis, M., Illengo, M., Rissone, M., et al.: Electrochemotherapy with intravenous bleomycin in the local treatment of skin melanoma metastases. Ann. Surg. Oncol. **15**, 2215–2222 (2008)

136. Sersa, G., Stabuc, B., Cemazar, M., Miklavcic, D., Rudolf, Z.: Electrochemotherapy with cisplatin: the systemic antitumour effectiveness of cisplatin can be potentiated locally by the application of electric pulses in the treatment of malignant melanoma skin metastases. Melanoma Res. **10**(4), 381–385 (2000)
137. Glass, L.F., Jaroszeski, M., Gilbert, R., Reintgen, D.S., Heller, R.: Intralesional bleomycin-mediated electrochemotherapy in 20 patients with basal cell carcinoma. J. Am. Acad. Dermatol. **37**(0190-9622 SB - IM), 596–599 (1997)
138. Curatolo, P., Rotunno, R., Miraglia, E., Mancini, M., Calvieri, S., Giustini, S.: Complete remission of Merkel cell carcinoma treated with electrochemotherapy and etoposide. G. Ital. Dermatol. Venereol. **148**(3), 310–311 (2013)
139. Curatolo, P., Quaglino, P., Marenco, F., Mancini, M., Nardo, T., Mortera, C., et al.: Electrochemotherapy in the treatment of Kaposi sarcoma cutaneous lesions: a two-center prospective phase II trial. Ann. Surg. Oncol. **19**(1), 192–198 (2012)
140. Curatolo, P., Miraglia, E., Rotunno, R., Calvieri, S., Giustini, S.: Electrochemotherapy: a valid treatment for Gorlin-Goltz syndrome. Acta Dermatovenerol. Croat. **21**(2), 132–133 (2013)
141. Belehradek, M., Domenge, C., Luboinski, B., Orlowski, S., Belehradek Jr., J., Mir, L.M.: Electrochemotherapy, a new antitumor treatment. First clinical phase I-II trial. Cancer **72** (12), 3694–3700 (1993)
142. Campana, L.G., Valpione, S., Falci, C., Mocellin, S., Basso, M., Corti, L., et al.: The activity and safety of electrochemotherapy in persistent chest wall recurrence from breast cancer after mastectomy: a phase-II study. Breast Cancer Res. Treat. **134**, 1169–1178 (2012)
143. Campana, L.G., Mali, B., Sersa, G., Valpione, S., Giorgi, C.A., Strojan, P., et al.: Electrochemotherapy in non-melanoma head and neck cancers: a retrospective analysis of the treated cases. Br. J. Oral Maxillofac. Surg. **52**(10), 957–964 (2014)
144. Campana, L.G., Galuppo, S., Valpione, S., Brunello, A., Ghiotto, C., Ongaro, A., et al.: Bleomycin electrochemotherapy in elderly metastatic breast cancer patients: clinical outcome and management considerations. J. Cancer Res. Clin. Oncol. **140**(9), 1557–1565 (2014)
145. Kubota, Y., Mir, L.M., Nakada, T., Sasagawa, I., Suzuki, H., Aoyama, N.: Successful treatment of metastatic skin lesions with electrochemotherapy. J. Urol. **160**(4), 1426–26 (1998)
146. Wasungu, L., Marty, A.L., Bureau, M.F., Kichler, A., Bessodes, M., Teissie, J., et al.: Pre-treatment of cells with pluronic L64 increases DNA transfection mediated by electrotransfer. J. Control. Release **149**(2), 117–125 (2011)
147. Sersa, G., Stabuc, B., Cemazar, M., Jancar, B., Miklavcic, D., Rudolf, Z.: Electrochemotherapy with cisplatin: potentiation of local cisplatin antitumour effectiveness by application of electric pulses in cancer patients. Eur. J. Cancer **34**(8), 1213–1218 (1998)
148. Glass, L.F., Fenske, N.A., Jaroszeski, M., Perrott, R., Harvey, D.T., Reintgen, D.S., et al.: Bleomycin-mediated electrochemotherapy of basal cell carcinoma. J. Am. Acad. Dermatol. **34**(1), 82–86 (1996)
149. Salwa, S.P., Bourke, M.G., Forde, P.F., O'Shaughnessy, M., O'Sullivan, S.T., Kelly, E.J., et al.: Electrochemotherapy for the treatment of ocular basal cell carcinoma; a novel adjunct in the disease management. J. Plast. Reconstr. Aesthet. Surg. **67**(3), 403–406 (2014)
150. Rodriguez-Cuevas, S., Barroso-Bravo, S., Almanza-Estrada, J., Cristobal-Martinez, L., Gonzalez-Rodriguez, E.: Electrochemotherapy in primary and metastatic skin tumors: phase II trial using intralesional bleomycin. Arch. Med. Res. **32**(4), 273–276 (2001)
151. Gargiulo, M., Papa, A., Capasso, P., Moio, M., Cubicciotti, E., Parascandolo, S.: Electrochemotherapy for non-melanoma head and neck cancers: clinical outcomes in 25 patients. Ann. Surg. **255**(6), 1158–1164 (2012)
152. Richetta, A.G., Curatolo, P., D'Epiro, S., Mancini, M., Mattozzi, C., Giancristoforo, S., et al.: Efficacy of electrochemotherapy in ulcerated basal cell carcinoma. Clin. Ter. **162**(5), 443–445 (2011)

153. Fantini, F., Gualdi, G., Cimitan, A., Giannetti, A.: Metastatic basal cell carcinoma with squamous differentiation. Arch. Dermatol. **144**(9), 1186–1188 (2008)
154. Quaglino, P., Matthiessen, L.W., Curatolo, P., Muir, T., Bertino, G., Kunte, C., et al.: Predicting patients at risk for pain associated with electrochemotherapy. Acta Oncol. **54**(3), 298–306 (2015)
155. Ramirez, L.H., Orlowski, S., An, D., Bindoula, G., Dzodic, R., Ardouin, P., et al.: Electrochemotherapy on liver tumours in rabbits. Br. J. Cancer **77**(12), 2104–2111 (1998)
156. Edhemovic, I., Gadzijev, E.M., Brecelj, E., Miklavcic, D., Kos, B., Zupanic, A., et al.: Electrochemotherapy: a new technological approach in treatment of metastases in the liver. Technol. Cancer Res. Treat. **10**(5), 475–485 (2011)
157. Panje, W.R., Sadeghi, N.: Endoscopic and electroporation therapy of paranasal sinus tumors. Am. J. Rhinol. **14**(3), 187–191 (2000)
158. Panje, W.R., Hier, M.P., Garman, G.R., Harrell, E., Goldman, A., Bloch, I.: Electroporation therapy of head and neck cancer. Ann. Otol. Rhinol. Laryngol. **107**(9), 779–785 (1998)
159. Goldfarb, P., Biel, M., Hanna, E., Houck, J., Klotch, C., Nathan, C., et al.: A Phase II Study Using Electroporation (EPT) and Intratumoral Bleomycin in Patients with Recurrent Head and Neck Cancer: A Safe and Active Treatment Approach. ASCO Proceedings (1999)
160. Allegretti, J.P., Panje, W.R.: Electroporation therapy for head and neck cancer including carotid artery involvement. Laryngoscope **111**(1), 52–56 (2001)
161. Linnert, M., Agerholm-Larsen, B., Mahmood, F., Iversen, H.K., Gehl, J.: Electrochemotherapy for primary and secondary brain tumors. In: Hayat MA, editor. Tumors of the Central Nervous System, pp. 195–206 (2011)
162. (NICE) NIfHaCE: Electrochemotherapy for primary basal cell carcinoma and primary squamous cell carcinoma. http://www.nice.org.uk (2014)
163. Valpione, S., Campana, L.G., Pigozzo, J., Chiarion-Sileni, V.: Consolidation electrochemotherapy with bleomycin in metastatic melanoma during treatment with dabrafenib. Radiol. Oncol. **49**(1), 71–74 (2015)
164. Mir, L.M., Orlowski, S., Poddevin, B., Belehradek Jr., J.: Electrochemotherapy tumor treatment is improved by interleukin-2 stimulation of the host's defenses. Eur. Cytokine Netw. **3**(3), 331–334 (1992)
165. Calvet, C.Y., Famin, D., Andre, F.M., Mir, L.M.: Electrochemotherapy with bleomycin induces hallmarks of immunogenic cell death in murine colon cancer cells. Oncoimmunology **3**, e28131 (2014)
166. Andersen, M.H., Gehl, J., Reker, S., Pedersen, L.O., Becker, J.C., Geertsen, P., et al.: Dynamic changes of specific T cell responses to melanoma correlate with IL-2 administration. Semin. Cancer Biol. **13**(1044-579X), 449–459 (2003)
167. Maxim, P.G., Carson, J.J., Ning, S., Knox, S.J., Boyer, A.L., Hsu, C.P., et al.: Enhanced effectiveness of radiochemotherapy with tirapazamine by local application of electric pulses to tumors. Radiat. Res. **162**, 185–193 (2004)
168. Kranjc, S., Cemazar, M., Grosel, A., Sentjurc, M., Sersa, G.: Radiosensitising effect of electrochemotherapy with bleomycin in LPB sarcoma cells and tumors in mice. BMC Cancer **5**, 115 (2005)
169. Gehl, J., Andersen, M.H., Straten, P.T., Geertsen, P.: Tumor autovaccination by electrochemotherapy followed by low-dose Il-2 in advanced malignant melanoma: efficient with low toxicity. J Clin Oncol; 20 (Proceedings, Am Soc Clin Oncol) (2001)
170. Sersa, G., Teissie, J., Cemazar, M., Signori, E., Kamensek, U., Marshall, G.D., et al.: Electrochemotherapy of tumors as in situ vaccination boosted by immunogene electrotransfer. Cancer Immunol. Immunother. **64**, 1315–1327 (2015)
171. Edd, J.F., Horowitz, L., Davalos, R.V., Mir, L.M., Rubinsky, B.: In vivo results of a new focal tissue ablation technique: irreversible electroporation. IEEE. Trans. Biomed. Eng. **53**(7), 1409–1415 (2006)
172. Phillips, M.A., Narayan, R., Padath, T., Rubinsky, B.: Irreversible electroporation on the small intestine. Br. J. Cancer **106**(3), 490–495 (2012)

173. Neal II, R.E., Rossmeisl Jr., J.H., Garcia, P.A., Lanz, O.I., Henao-Guerrero, N., Davalos, R. V.: Successful treatment of a large soft tissue sarcoma with irreversible electroporation. J. Clin. Oncol. **29**(13), e372–e377 (2011)
174. Wagstaff, P.G.K., de Bruin, D.M., van den Bos, W., Ingels, A., van Gemert, M.J.C., Zondervan, P.J., Verdaasdonk, R.M., van Lienden, K.P., van Leeuwen, T.G., de la Rosette, J.J.: Irreversible electroporation of the porcine kidney: temperature development and distribution. Urol. Oncol. Semin. Orig. Invest. **33**(4), 168e1–168.e7 (2015)
175. Qin, Z., Jiang, J., Long, G., Lindgren, B., Bischof, J.C.: Irreversible electroporation: an in vivo study with dorsal skin fold chamber. Ann. Biomed. Eng. **41**(3), 619–629 (2013)
176. Becker, S.M., Kuznetsov, A.V.: Thermal damage reduction associated with in vivo skin electroporation: a numerical investigation justifying aggressive pre-cooling. Int. J. Heat Mass Transf. **50**(1), 105–116 (2007)
177. Arena, C.B., Mahajan, R.L., Rylander, M.N., Davalos, R.V.: An experimental and numerical investigation of phase change electrodes for therapeutic irreversible electroporation. J. Biomech. Eng. **135**(11), 111009 (2013)
178. Appelbaum, L., Ben-David, E., Faroja, M., Nissenbaum, Y., Sosna, J., Goldberg, S.N.: Irreversible electroporation ablation: creation of large-volume ablation zones in in vivo porcine liver with four-electrode arrays. Radiology **270**(2), 416–424 (2014)
179. Jiang, C., Shao, Q., Bischof, J.: Pulse timing during irreversible electroporation achieves enhanced destruction in a hindlimb model of cancer. Ann. Biomed. Eng. **43**(4), 887–895 (2015)
180. Maor, E., Ivorra, A., Leor, J., Rubinsky, B.: The effect of irreversible electroporation on blood vessels. Technol. Cancer Res. Treat. **6**(4), 307–312 (2007)
181. Lee, E.W., Chen, C., Prieto, V.E., Dry, S.M., Loh, C.T., Kee, S.T.: Advanced hepatic ablation technique for creating complete cell death: irreversible electroporation 1. Radiology **255**(2), 426–433 (2010)
182. Rossmeisl Jr., J.H., Garcia, P.A., Roberston, J.L., Ellis, T.L., Davalos, R.V.: Pathology of non-thermal irreversible electroporation (N-TIRE)-induced ablation of the canine brain. J. Vet. Sci. **14**(4), 433–440 (2013)
183. Appelbaum, L., Ben-David, E., Sosna, J., Nissenbaum, Y., Goldberg, S.N.: US findings after irreversible electroporation ablation: radiologic-pathologic correlation. Radiology **262**(1), 117–125 (2012)
184. Ortiz, M.V., Davalos, R.V.: Cell electroporation mechanisms and preclinical foundation for focal therapy. In: Polascik, J., (ed.) Imaging and Focal Therapy of Early Prostate Cancer. Springer, pp. 309–329 (2013)
185. Onik, G., Mikus, P., Rubinsky, B.: Irreversible electroporation: implications for prostate ablation. Technol. Cancer Res. Treat. **6**(4), 295–300 (2007)
186. Li, W., Fan, Q., Ji, Z., Qiu, X., Li, Z.: The effects of irreversible electroporation (IRE) on nerves. PLoS ONE **6**(4), e18831 (2011)
187. Schoellnast, H., Monette, S., Ezell, P.C., Maybody, M., Erinjeri, J.P., Stubblefield, M.D., Single, G., Solomon, S.B.: The delayed effects of irreversible electroporation ablation on nerves. Eur. Radiol. **23**(2), 375–380 (2013)
188. Bower, M., Sherwood, L., Li, Y., Martin, R.: Irreversible electroporation of the pancreas: definitive local therapy without systemic effects. J. Surg. Oncol. **104**(1), 22–28 (2011)
189. Charpentier, K.P., Wolf, F., Noble, L., Winn, B., Resnick, M., Dupuy, D.E.: Irreversible electroporation of the liver and liver hilum in swine. HPB **13**(3), 168–173 (2011)
190. Deodhar, A., Monette, S., Single Jr., G.W., Hamilton Jr., W.C., Thornton, R., Maybody, M., Coleman, J.A., Solomon, S.B.: Renal tissue ablation with irreversible electroporation: preliminary results in a porcine model. Urology **77**(3), 754–760 (2011)
191. Wendler, J.J., Pech, M., Porsch, M., Janitzky, A., Fischbach, F., Buhtz, P., Vogler, K., Huhne, S., Borucki, K., Strang, C., Mahnkopf, D., Ricke, J., Liehr, U.B.: Urinary tract effects after multifocal nonthermal irreversible electroporation of the kidney: acute and chronic

monitoring by magnetic resonance imaging, intravenous urography and urinary cytology. Cardiovasc. Intervent. Radiol. **35**(4), 921–926 (2012)
192. Willett, C.G., Czito, B.G., Bendell, J.C., Ryan, D.P.: Locally advanced pancreatic cancer. J. Clin. Oncol. **23**(20), 4538–4544 (2005)
193. Von Hoff, D.D., Evans, D.B., Hruban, R.H.: Pancreatic Cancer. Jones & Bartlett Learning, Burlington (2005)
194. Charpentier, K.P., Wolf, F., Noble, L., Winn, B., Resnick, M., Dupuy, D.E.: Irreversible electroporation of the pancreas in swine: a pilot study. HPB **12**(5), 348–351 (2010)
195. Pantuck, A.J., Zisman, A., Belldegrun, A.S.: The changing natural history of renal cell carcinoma. J. Urol. **166**(5), 1611–1623 (2001)
196. José, A., Sobrevals, L., Ivorra, A., Fillat, C.: Irreversible electroporation shows efficacy against pancreatic carcinoma without systemic toxicity in mouse models. Cancer Lett. **317** (1), 16–23 (2012)
197. Martin, R.C.G., Kwon, D., Chalikonda, S., Sellars, M., Kortz, E., Scoggins, C.R., Watkins K. T., McMasters, K.: Treatment of 200 Locally Advanced (Stage III) Pancreatic Adenocarcinoma Patients with Irreversible Electroporation: Safety and Efficacy. American Surgical Association 136th Annual Meeting (2015).
198. Martin 2nd, R.C., McFarland, K., Ellis, S., Velanovich, V.: Irreversible electroporation in locally advanced pancreatic cancer: potential improved overall survival. Ann. Surg. Oncol. **20**(Suppl 3), S443–S449 (2013)
199. Narayanan, G., Hosein, P.J., Arora, G., Barbery, K.J., Froud, T., Livingstone, A.S., Franceschi, D., Rocha Lima, C.M., Yrizarry, J.: Percutaneous irreversible electroporation for downstaging and control of unresectable pancreatic adenocarcinoma. J. Vasc. Interv. Radiol. **23**(12), 1613–1621 (2012)
200. Phillips, M., Maor, E., Rubinsky, B.: Principles of tissue engineering with nonthermal irreversible electroporation. J. Heat Transf. **133**(1), 011004 (2011)
201. Tracy, C.R., Kabbani, W., Cadeddu, J.A.: Irreversible electroporation (IRE): a novel method for renal tissue ablation. BJU Int. **107**(12), 1982–1987 (2011)
202. Wendler, J.J., Porsch, M., Hühne, S., Baumunk, D., Buhtz, P., Fischbach, F., Pech, M., Mahnkopf, D., Kropf, S., Roessner, A., Ricke, J., Schostak, M., Liehr, U.B.: Short- and mid-term effects of irreversible electroporation on normal renal tissue: an animal model. Cardiovasc. Intervent. Radiol. **36**(2), 512–520 (2013)
203. Lee, E.W., Loh, C.T., Kee, S.T.: Imaging guided percutaneous irreversible electroporation: ultrasound and immunohistological correlation. Technol. Cancer Res. Treat. **6**(4), 287–293 (2007)
204. Schmidt, C.R., Shires, P., Mootoo, M.: Real-time ultrasound imaging of irreversible electroporation in a porcine liver model adequately characterizes the zone of cellular necrosis. HPB **14**(2), 98–102 (2012)
205. Liu, Y., Xiong, Z., Zhou, W., Hua, Y., Li, C., Yao, C.: Percutaneous ultrasound-guided irreversible electroporation: a goat liver study. Oncol. Lett. **4**(3), 450–454 (2012)
206. Ben-David, E., Appelbaum, L., Sosna, J., Nissenbaum, I., Goldberg, S.N.: Characterization of irreversible electroporation ablation in in vivo porcine liver. Am. J. Roentgenol. **198**(1), W62–W68 (2012)
207. Rubinsky, B., Onik, G., Mikus, P.: Irreversible electroporation: a new ablation modality – clinical implications. Technol. Cancer Res. Treat. **6**(1), 37–48 (2007)
208. Guo, Y., Zhang, Y., Klein, R., Nijm, G.M., Sahakian, A.V., Omary, R.A., Yang, G.Y., Larson, A.C.: Irreversible electroporation therapy in the liver: longitudinal efficacy studies in a rat model of hepatocellular carcinoma. Cancer Res. **70**(4), 1555–1563 (2010)
209. Lee, E.W., Wong, D., Tafti, B.A., Prieto, V., Totonchy, M., Hilton, J., Dry, S., Cho, S., Loh, C.T., Kee, S.T.: Irreversible electroporation in eradication of rabbit VX2 liver tumor. J. Vasc. Interv. Radiol. **23**(6), 833–840 (2012)
210. Ginzburg, S., Uzzo, R., Kutikov, A.: The role of minimally invasive surgery in multifocal renal cell carcinoma. Curr. Urol. Rep. **13**(3), 202–210 (2012)

211. Venkatesan, A.M., Wood, B.J., Gervais, D.A.: Percutaneous ablation in the kidney. Radiology **261**(2), 375–391 (2011)
212. Fini, M., Tschon, M., Alberghini, M., Bianchi, G., Mercuri, M., Campanacci, L., Cavani, F., Ronchetti, M., de Terlizzi, F., Cadossi, R.: Cell electroporation in bone tissue. In: Kee, S.T., Gehl, J., Lee, E.W. (eds.) Clinical Aspects of Electroporation, pp. 115–127. Springer, New York (2011)
213. Tam, A.L., Abdelsalam, M.E., Gagea, M., Ensor, J.E., Moussa, M., Ahmed, M., Goldberg, S. N., Dixon, K., McWatters, A., Miller, J.J., Srimathveeravalli, G., Solomon, S.B., Avritscher, R., Wallace, M.J., Gupta, S.: Irreversible electroporation of the lumbar vertebrae in a porcine model: is there clinical-pathologic evidence of neural toxicity? Radiology **272**(3), 709–719 (2014)
214. Garcia, P.A., Rossmeisl, Jr J.H., Ellis, T.L., Davalos, R.V.: Nonthermal irreversible electroporation as a focal ablation treatment for brain cancer. In: Hayat, M.A. (ed.) Tumors of the Central Nervous System, vol. 12, pp. 171–182. Springer, Dordrecht (2014)
215. Ellis, T.L., Garcia, P.A., Rossmeisl Jr., J.H., Henao-Guerrero, N., Robertson, J., Davalos, R. V.: Nonthermal irreversible electroporation for intracranial surgical applications: laboratory investigation. J. Neurosurg. **114**(3), 681–688 (2011)
216. Rossmeisl Jr., J.H., Garcia, P., Pancotto, T.E., Robertson, J.L., Henao-Geurrero, N., Neal 2nd, R.E., Ellis, T.L., Davalos, R.V.: Safety and feasibility of the NanoKnife system for irreversible electroporation ablative treatment of canine spontaneous intracranial gliomas. J. Neurosurg. **123**(4), 1008–1018 (2015)
217. Ari Hakimi, A., Feder, M., Ghavamian, R.: Minimally invasive approaches to prostate cancer: a review of the current literature. Urol. J. **4**(3), 130–137 (2009)
218. Nguyen, C.T., Jones, J.S.: Focal therapy in the management of localized prostate cancer. BJU Int. **107**(9), 1362–1368 (2011)
219. Neal II, R.E., Cheung, W., Kavnoudias, H., Thomson, K.R.: Spectrum of imaging and characteristics for liver tumors treated with irreversible electroporation. J. Biomed. Sci. Eng. **5**(12A), 813–818 (2012)
220. Scheffer, S.R., Nave, H., Korangy, F., Schlote, K., Pabst, R., Jaffee, E.M., Manns, M.P., Greten, T.F.: Apoptotic, but not necrotic, tumor cell vaccines induce a potent immune response in vivo. Int. J. Cancer **103**(2), 205–211 (2003)
221. Neal II, R.E., Kavnoudias, H., Cheung, W., Golebiowski, B., McLean, C.A., Thomson, K.R.: Hepatic epithelioid hemangioendothelioma treated with irreversible electroporation and antibiotics. J. Clin. Oncol. **31**(27), e422–e426 (2013)
222. Sabel, M.S.: Cryo-immunology: a review of the literature and proposed mechanisms for stimulatory versus suppressive immune responses. Cryobiology **58**(1), 1–11 (2009)
223. Neal II, R.E., Rossmeisl Jr., J.H., Robertson, J.L., Arena, C.B., Davis, E.M., Singh, R.N., Stallings, J., Davalos, R.V.: Improved local and systemic anti-tumor efficacy for irreversible electroporation in immunocompetent versus immunodeficient mice. PLoS ONE **8**(5), e64559 (2013)
224. Jiang, C., Qin, Z., Bischof, J.: Membrane-targeting approaches for enhanced cancer cell destruction with irreversible electroporation. Ann. Biomed. Eng. **42**(1), 193–204 (2014)
225. Vannucci, L., Lai, M., Chiuppesi, F., Ceccherini-Nelli, L., Pistello, M.: Viral vectors: a look back and ahead on gene transfer technology. New Microbiol. **36**(1), 1–22 (2013)
226. Wirth, T., Parker, N., Yla-Herttuala, S.: History of gene therapy. Gene **525**(2), 162–169 (2013)
227. Kaufmann, K.B., Buning, H., Galy, A., Schambach, A., Grez, M.: Gene therapy on the move. EMBO. Mol. Med. **5**(11), 1642–1661 (2013)
228. Ramamoorth, M., Narvekar, A.: Non viral vectors in gene therapy – an overview. J. Clin. Diagn. Res. **9**(1), GE01–GE06 (2015)
229. Yin, H., Kanasty, R.L., Eltoukhy, A.A., Vegas, A.J., Dorkin, J.R., Anderson, D.G.: Non-viral vectors for gene-based therapy. Nat. Rev. Genet. **15**(8), 541–555 (2014)
230. Alsaggar, M., Liu, D.: Physical methods for gene transfer. Adv. Genet. **89**, 1–24 (2015)

231. Li, S.D., Huang, L.: Gene therapy progress and prospects: non-viral gene therapy by systemic delivery. Gene Ther. **13**(18), 1313–1319 (2006)
232. Song, S., Shen, Z., Chen, L., Brayman, A.A., Miao, C.H.: Explorations of high-intensity therapeutic ultrasound and microbubble-mediated gene delivery in mouse liver. Gene Ther. **18**(10), 1006–1014 (2011)
233. Noble, M.L., Kuhr, C.S., Graves, S.S., Loeb, K.R., Sun, S.S., Keilman, G.W., Morrison, K.P., Paun, M., Storb, R.F., Miao, C.H.: Ultrasound-targeted microbubble destruction-mediated gene delivery into canine livers. Mol. Ther. **21**(9), 1687–1694 (2013)
234. Bonamassa, B., Hai, L., Liu, D.: Hydrodynamic gene delivery and its applications in pharmaceutical research. Pharm. Res. **28**(4), 694–701 (2011)
235. Suda, T., Liu, D.: Hydrodynamic gene delivery: its principles and applications. Mol. Ther. **15** (12), 2063–2069 (2007)
236. Heller, L.C., Heller, R.: In vivo electroporation for gene therapy. Hum. Gene Ther. **17**(9), 890–897 (2006)
237. Gothelf, A., Gehl, J.: Gene electrotransfer to skin; review of existing literature and clinical perspectives. Curr. Gene Ther. **10**(4), 287–299 (2010)
238. Shirley, S.A., Lundberg, C.G., Li, F., Burcus, N., Heller, R.: Controlled gene delivery can enhance therapeutic outcome for cancer immune therapy for melanoma. Curr. Gene Ther. **15** (1), 32–43 (2015)
239. Heller, L.C., Heller, R.: Electroporation gene therapy preclinical and clinical trials for melanoma. Curr. Gene Ther. **10**(4), 312–317 (2010)
240. Marshall Jr., W.G., Boone, B.A., Burgos, J.D., Gografe, S.I., Baldwin, M.K., Danielson, M. L., Larson, M.J., Caretto, D.R., Cruz, Y., Ferraro, B., Heller, L.C., Ugen, K.E., Jaroszeski, M. J., Heller, R.: Electroporation-mediated delivery of a naked DNA plasmid expressing VEGF to the porcine heart enhances protein expression. Gene Ther. **17**(3), 419–423 (2010)
241. Hargrave, B., Downey, H., Strange Jr., R., Murray, L., Cinnamond, C., Lundberg, C., Israel, A., Chen, Y.J., Marshall Jr., W., Heller, R.: Electroporation-mediated gene transfer directly to the swine heart. Gene Ther. **20**(2), 151–157 (2013)
242. Dean, D.A.: Nonviral gene transfer to skeletal, smooth, and cardiac muscle in living animals. Am. J. Physiol. Cell Physiol. **289**(2), C233–C245 (2005)
243. Dean, D.A.: Electroporation of the vasculature and the lung. DNA Cell Biol. **22**(12), 797–806 (2003)
244. Dean, D.A., Machado-Aranda, D., Blair-Parks, K., Yeldandi, A.V., Young, J.L.: Electroporation as a method for high-level nonviral gene transfer to the lung. Gene Ther. **10**(18), 1608–1615 (2003)
245. Zhou, R., Norton, J.E., Zhang, N., Dean, D.A.: Electroporation-mediated transfer of plasmids to the lung results in reduced TLR9 signaling and inflammation. Gene Ther. **14**(9), 775–780 (2007)
246. Ding, X.F., Ma, D.L., Zhang, Q., Peng, W., Fan, M., Suo, W.Z.: Progress of in vivo electroporation in the rodent brain. Curr. Gene Ther. **14**(3), 211–217 (2014)
247. Murakami, T., Sunada, Y.: Plasmid DNA gene therapy by electroporation: principles and recent advances. Curr. Gene Ther **11**(6), 447–456 (2011)
248. Isaka, Y.: Gene therapy targeting kidney diseases: routes and vehicles. Clin. Exp. Nephrol. **10** (4), 229–235 (2006)
249. Scherman, D., Bigey, P., Bureau, M.F.: Applications of plasmid electrotransfer. Technol. Cancer Res. Treat. **1**(5), 351–354 (2002)
250. Andre, F.M., Gehl, J., Sersa, G., Preat, V., Hojman, P., Eriksen, J., Golzio, M., Cemazar, M., Pavselj, N., Rols, M.P., Miklavcic, D., Neumann, E., Teissie, J., Mir, L.M.: Efficiency of high- and low-voltage pulse combinations for gene electrotransfer in muscle, liver, tumor, and skin. Hum. Gene Ther. **19**(11), 1261–1271 (2008)
251. Corovic, S., Lackovic, I., Sustaric, P., Sustar, T., Rodic, T., Miklavcic, D.: Modeling of electric field distribution in tissues during electroporation. Biomed. Eng. Online **12**, 16 (2013)

252. Mesojednik, S., Pavlin, D., Sersa, G., Coer, A., Kranjc, S., Grosel, A., Tevz, G., Cemazar, M.: The effect of the histological properties of tumors on transfection efficiency of electrically assisted gene delivery to solid tumors in mice. Gene Ther. **14**(17), 1261–1269 (2007)
253. Trollet, C., Bloquel, C., Scherman, D., Bigey, P.: Electrotransfer into skeletal muscle for protein expression. Curr. Gene Ther. **6**(5), 561–578 (2006)
254. Trollet, C., Scherman, D., Bigey, P.: Delivery of DNA into muscle for treating systemic diseases: advantages and challenges. Methods Mol. Biol. **423**, 199–214 (2008)
255. Gothelf, A., Gehl, J.: What you always needed to know about electroporation based DNA vaccines. Hum. Vaccin Immunother. **8**(11), 1694–1702 (2012)
256. Hirao, L.A., Wu, L., Khan, A.S., Satishchandran, A., Draghia-Akli, R., Weiner, D.B.: Intradermal/subcutaneous immunization by electroporation improves plasmid vaccine delivery and potency in pigs and rhesus macaques. Vaccine **26**(3), 440–448 (2008)
257. Donate, A., Heller, R.: Assessment of delivery parameters with the multi-electrode array for development of a DNA vaccine against Bacillus anthracis. Bioelectrochemistry **94C**, 1–6 (2013)
258. Donate, A., Coppola, D., Cruz, Y., Heller, R.: Evaluation of a novel non-penetrating electrode for use in DNA vaccination. PLoS ONE **6**(4), e19181 (2011)
259. Hojman, P., Gissel, H., Andre, F.M., Cournil-Henrionnet, C., Eriksen, J., Gehl, J., Mir, L.M.: Physiological effects of high- and low-voltage pulse combinations for gene electrotransfer in muscle. Hum. Gene Ther. **19**(11), 1249–1260 (2008)
260. Guo, S., Jackson, D.L., Burcus, N.I., Chen, Y.J., Xiao, S., Heller, R.: Gene electrotransfer enhanced by nanosecond pulsed electric fields. Mol. Ther. Methods Clin. Dev. **1**, 14043 (2014)
261. Bigey, P., Bureau, M.F., Scherman, D.: In vivo plasmid DNA electrotransfer. Curr. Opin. Biotechnol. **13**(5), 443–447 (2002)
262. Gehl, J.: Electroporation: theory and methods, perspectives for drug delivery, gene therapy and research. Acta Physiol. Scand. **177**(4), 437–447 (2003)
263. Hojman, P., Zibert, J.R., Gissel, H., Eriksen, J., Gehl, J.: Gene expression profiles in skeletal muscle after gene electrotransfer. BMC Mol. Biol. **8**, 56 (2007)
264. Heller, L., Jaroszeski, M.J., Coppola, D., Pottinger, C., Gilbert, R., Heller, R.: Electrically mediated plasmid DNA delivery to hepatocellular carcinomas in vivo. Gene Ther. **7**(10), 826–829 (2000)
265. Andre, F.M., Mir, L.M.: Nucleic acids electrotransfer in vivo: mechanisms and practical aspects. Curr. Gene Ther. **10**(4), 267–280 (2010)
266. Guo, S., Israel, A.L., Basu, G., Donate, A., Heller, R.: Topical gene electrotransfer to the epidermis of hairless guinea pig by non-invasive multielectrode array. PLoS ONE **8**(8), e73423 (2013)
267. Heller, R., Cruz, Y., Heller, L.C., Gilbert, R.A., Jaroszeski, M.J.: Electrically mediated delivery of plasmid DNA to the skin, using a multielectrode array. Hum. Gene Ther. **21**(3), 357–362 (2010)
268. Heller, L.C., Jaroszeski, M.J., Coppola, D., McCray, A.N., Hickey, J., Heller, R.: Optimization of cutaneous electrically mediated plasmid DNA delivery using novel electrode. Gene Ther. **14**(3), 275–280 (2007)
269. Potter, H.: Electroporation in biology: methods, applications, and instrumentation. Anal. Biochem. **174**(2), 361–373 (1988)
270. Potter, H., Weir, L., Leder, P.: Enhancer-dependent expression of human kappa immunoglobulin genes introduced into mouse pre-B lymphocytes by electroporation. Proc. Natl. Acad. Sci. U. S. A. **81**(22), 7161–7165 (1984)
271. Faurie, C., Rebersek, M., Golzio, M., Kanduser, M., Escoffre, J.M., Pavlin, M., Teissie, J., Miklavcic, D., Rols, M.P.: Electro-mediated gene transfer and expression are controlled by the life-time of DNA/membrane complex formation. J Gene Med **12**(1), 117–125 (2010)

272. Kotnik, T., Mir, L.M., Flisar, K., Puc, M., Miklavcic, D.: Cell membrane electropermeabilization by symmetrical bipolar rectangular pulses. Part I. Increased efficiency of permeabilization. Bioelectrochemistry **54**(1), 83–90 (2010)
273. Rols, M.P., Teissie, J.: Electropermeabilization of mammalian cells to macromolecules: control by pulse duration. Biophys. J. **75**(3), 1415–1423 (1998)
274. Wolf, H., Rols, M.P., Boldt, E., Neumann, E., Teissie, J.: Control by pulse parameters of electric field-mediated gene transfer in mammalian cells. Biophys. J. **66**(2 Pt 1), 524–531 (1994)
275. Cemazar, M., Golzio, M., Sersa, G., Hojman, P., Kranjc, S., Mesojednik, S., Rols, M.P., Teissie, J.: Control by pulse parameters of DNA electrotransfer into solid tumors in mice. Gene Ther. **16**(5), 635–644 (2009)
276. Kotnik, T., Pucihar, G., Miklavcic, D.: Induced transmembrane voltage and its correlation with electroporation-mediated molecular transport. J. Membr. Biol. **236**(1), 3–13 (2010)
277. Sel, D., Mazeres, S., Teissie, J., Miklavcic, D.: Finite-element modeling of needle electrodes in tissue from the perspective of frequent model computation. IEEE Trans. Biomed. Eng. **50** (11), 1221–1232 (2003)
278. Singh, N., Kalluri, H., Herwadkar, A., Badkar, A., Banga, A.K.: Transcending the skin barrier to deliver peptides and proteins using active technologies. Crit. Rev. Ther. Drug Carrier Syst. **29**(4), 265–298 (2012)
279. Prausnitz, M.R., Mikszta, J.A., Cormier, M., Andrianov, A.K.: Microneedle-based vaccines. Curr. Top. Microbiol. Immunol. **333**, 369–393 (2009)
280. Daugimont, L., Baron, N., Vandermeulen, G., Pavselj, N., Miklavcic, D., Jullien, M.C., Cabodevila, G., Mir, L.M., Preat, V.: Hollow microneedle arrays for intradermal drug delivery and DNA electroporation. J. Membr. Biol. **236**(1), 117–125 (2010)
281. Gehl, J., Mir, L.M.: Determination of optimal parameters for in vivo gene transfer by electroporation, using a rapid in vivo test for cell permeabilization. Biochem. Biophys. Res. Commun. **261**(2), 377–380 (1999)
282. Ferraro, B., Heller, L.C., Cruz, Y.L., Guo, S., Donate, A., Heller, R.: Evaluation of delivery conditions for cutaneous plasmid electrotransfer using a multielectrode array. Gene Ther. **18** (5), 496–500 (2011)
283. Martin, J.B., Young, J.L., Benoit, J.N., Dean, D.A.: Gene transfer to intact mesenteric arteries by electroporation. J. Vasc. Res. **37**(5), 372–380 (2000)
284. Soden, D., Larkin, J., Collins, C., Piggott, J., Morrissey, A., Norman, A., Dunne, C., O'Sullivan, G.C.: The development of novel flexible electrode arrays for the electrochemotherapy of solid tumour tissue. (Potential for endoscopic treatment of inaccessible cancers). Conf. Proc. IEEE. Eng. Med. Biol. Soc. **5**, 3547–3550 (2004)
285. Soden, D.M., Larkin, J.O., Collins, C.G., Tangney, M., Aarons, S., Piggott, J., Morrissey, A., Dunne, C., O'Sullivan, G.C.: Successful application of targeted electrochemotherapy using novel flexible electrodes and low dose bleomycin to solid tumours. Cancer Lett. **232**(2), 300–310 (2006)
286. Tolmachov, O.: Designing plasmid vectors. Methods Mol. Biol. **542**, 117–129 (2009)
287. Gill, D.R., Pringle, I.A., Hyde, S.C.: Progress and prospects: the design and production of plasmid vectors. Gene Ther. **16**(2), 165–171 (2009)
288. Vandermeulen, G., Richiardi, H., Escriou, V., Ni, J., Fournier, P., Schirrmacher, V., Scherman, D., Preat, V.: Skin-specific promoters for genetic immunisation by DNA electroporation. Vaccine **27**(32), 4272–4277 (2009)
289. Hojman, P., Eriksen, J., Gehl, J.: Tet-On induction with doxycycline after gene transfer in mice: sweetening of drinking water is not a good idea. Anim. Biotechnol. **18**(3), 183–188 (2007)
290. Hojman, P., Gissel, H., Gehl, J.: Sensitive and precise regulation of haemoglobin after gene transfer of erythropoietin to muscle tissue using electroporation. Gene Ther. **14**(12), 950–959 (2007)

291. Dean, D.A.: Cell-specific targeting strategies for electroporation-mediated gene delivery in cells and animals. J. Membr. Biol. **246**(10), 737–744 (2013)
292. Vaughan, E.E., DeGiulio, J.V., Dean, D.A.: Intracellular trafficking of plasmids for gene therapy: mechanisms of cytoplasmic movement and nuclear import. Curr. Gene Ther. **6**(6), 671–681 (2006)
293. Marie, C., Vandermeulen, G., Quiviger, M., Richard, M., Preat, V., Scherman, D.: pFARs, plasmids free of antibiotic resistance markers, display high-level transgene expression in muscle, skin and tumour cells. J. Gene Med. **12**(4), 323–332 (2010)
294. Mignon, C., Sodoyer, R., Werle, B.: Antibiotic-free selection in biotherapeutics: now and forever. Pathogens **4**(2), 157–181 (2015)
295. Mairhofer, J., Cserjan-Puschmann, M., Striedner, G., Nobauer, K., Razzazi-Fazeli, E., Grabherr, R.: Marker-free plasmids for gene therapeutic applications – lack of antibiotic resistance gene substantially improves the manufacturing process. J. Biotechnol. **146**(3), 130–137 (2010)
296. Mairhofer, J., Pfaffenzeller, I., Merz, D., Grabherr, R.: A novel antibiotic free plasmid selection system: advances in safe and efficient DNA therapy. Biotechnol. J. **3**(1), 83–89 (2008)
297. Mayrhofer, P., Blaesen, M., Schleef, M., Jechlinger, W.: Minicircle-DNA production by site specific recombination and protein-DNA interaction chromatography. J. Gene Med. **10**(11), 1253–1269 (2008). doi:10.1002/jgm.1243
298. Bigger, B.W., Tolmachov, O., Collombet, J.M., Fragkos, M., Palaszewski, I., Coutelle, C.: An araC-controlled bacterial cre expression system to produce DNA minicircle vectors for nuclear and mitochondrial gene therapy. J. Biol. Chem. **276**(25), 23018–23027 (2001)
299. Broll, S., Oumard, A., Hahn, K., Schambach, A., Bode, J.: Minicircle performance depending on S/MAR-nuclear matrix interactions. J. Mol. Biol. **395**(5), 950–965 (2010)
300. Chen, Z.Y., He, C.Y., Ehrhardt, A., Kay, M.A.: Minicircle DNA vectors devoid of bacterial DNA result in persistent and high-level transgene expression in vivo. Mol. Ther. **8**(3), 495–500 (2003)
301. Izsvak, Z., Ivics, Z.: Sleeping beauty transposition: biology and applications for molecular therapy. Mol. Ther. **9**(2), 147–156 (2004)
302. Izsvak, Z., Ivics, Z., Plasterk, R.H.: Sleeping Beauty, a wide host-range transposon vector for genetic transformation in vertebrates. J. Mol. Biol. **302**(1), 93–102 (2000)
303. Swierczek, M., Izsvak, Z., Ivics, Z.: The Sleeping Beauty transposon system for clinical applications. Expert. Opin. Biol. Ther. **12**(2), 139–153 (2012)
304. Aronovich, E.L., McIvor, R.S., Hackett, P.B.: The Sleeping Beauty transposon system: a non-viral vector for gene therapy. Hum. Mol. Genet. **20**(R1), R14–R20 (2011)
305. Hackett Jr., P.B., Aronovich, E.L., Hunter, D., Urness, M., Bell, J.B., Kass, S.J., Cooper, L.J., McIvor, S.: Efficacy and safety of Sleeping Beauty transposon-mediated gene transfer in preclinical animal studies. Curr. Gene Ther. **11**(5), 341–349 (2011)
306. Olivares, E.C., Hollis, R.P., Chalberg, T.W., Meuse, L., Kay, M.A., Calos, M.P.: Site-specific genomic integration produces therapeutic Factor IX levels in mice. Nat. Biotechnol. **20**(11), 1124–1128 (2002)
307. Lucas, M.L., Heller, R.: Immunomodulation by electrically enhanced delivery of plasmid DNA encoding IL-12 to murine skeletal muscle. Mol. Ther. **3**(1), 47–53 (2001)
308. Mir, L.M., Bureau, M.F., Rangara, R., Schwartz, B., Scherman, D.: Long-term, high level in vivo gene expression after electric pulse-mediated gene transfer into skeletal muscle. C. R. Acad. Sci. III **321**(11), 893–899 (1998)
309. Bloquel, C., Fabre, E., Bureau, M.F., Scherman, D.: Plasmid DNA electrotransfer for intracellular and secreted proteins expression: new methodological developments and applications. J. Gene Med. **6**(Suppl 1), S11–S23 (2004)
310. McMahon, J.M., Wells, D.J.: Electroporation for gene transfer to skeletal muscles: current status. BioDrugs **18**(3), 155–165 (2004)
311. Khavari, P.A., Krueger, G.G.: Cutaneous gene therapy. Dermatol. Clin. **15**(1), 27–35 (1997)

312. Khavari, P.A., Rollman, O., Vahlquist, A.: Cutaneous gene transfer for skin and systemic diseases. J. Intern. Med. **252**(1), 1–10 (2002)
313. Ariza, M.E., Williams, M.V., Wong, H.K.: Targeting IL-17 in psoriasis: from cutaneous immunobiology to clinical application. Clin. Immunol. **146**(2), 131–139 (2013)
314. Vicentini, F.T., Borgheti-Cardoso, L.N., Depieri, L.V., de Macedo, M.D., Abelha, T.F., Petrilli, R., Bentley, M.V.: Delivery systems and local administration routes for therapeutic siRNA. Pharm. Res. **30**(4), 915–931 (2013)
315. Geusens, B., Strobbe, T., Bracke, S., Dynoodt, P., Sanders, N., Van Gele, M., Lambert, J.: Lipid-mediated gene delivery to the skin. Eur. J. Pharm. Sci. **43**(4), 199–211 (2011)
316. Kim, Y.C., Jarrahian, C., Zehrung, D., Mitragotri, S., Prausnitz, M.R.: Delivery systems for intradermal vaccination. Curr. Top. Microbiol. Immunol. **351**, 77–112 (2012)
317. Medi, B.M., Singh, J.: Skin targeted DNA vaccine delivery using electroporation in rabbits II. Safety. Int. J. Pharm. **308**(1–2), 61–68 (2006)
318. Hooper, J.W., Golden, J.W., Ferro, A.M., King, A.D.: Smallpox DNA vaccine delivered by novel skin electroporation device protects mice against intranasal poxvirus challenge. Vaccine **25**(10), 1814–1823 (2007)
319. Roos, A.K., Moreno, S., Leder, C., Pavlenko, M., King, A., Pisa, P.: Enhancement of cellular immune response to a prostate cancer DNA vaccine by intradermal electroporation. Mol. Ther. **13**(2), 320–327 (2006)
320. Ferguson, M., Byrnes, C., Sun, L., Marti, G., Bonde, P., Duncan, M., Harmon, J.W.: Wound healing enhancement: electroporation to address a classic problem of military medicine. World J. Surg. **29**(Suppl 1), S55–S59 (2005)
321. Marti, G., Ferguson, M., Wang, J., Byrnes, C., Dieb, R., Qaiser, R., Bonde, P., Duncan, M.D., Harmon, J.W.: Electroporative transfection with KGF-1 DNA improves wound healing in a diabetic mouse model. Gene Ther. **11**(24), 1780–1785 (2004)
322. Steinstraesser, L., Lam, M.C., Jacobsen, F., Porporato, P.E., Chereddy, K.K., Becerikli, M., Stricker, I., Hancock, R.E., Lehnhardt, M., Sonveaux, P., Preat, V., Vandermeulen, G.: Skin electroporation of a plasmid encoding hCAP-18/LL-37 host defense peptide promotes wound healing. Mol. Ther. **22**(4), 734–742 (2014)
323. Gothelf, A., Hojman, P., Gehl, J.: Therapeutic levels of erythropoietin (EPO) achieved after gene electrotransfer to skin in mice. Gene Ther. **17**(9), 1077–1084 (2010)
324. Ferraro, B., Cruz, Y.L., Baldwin, M., Coppola, D., Heller, R.: Increased perfusion and angiogenesis in a hindlimb ischemia model with plasmid FGF-2 delivered by noninvasive electroporation. Gene Ther. **17**(6), 763–769 (2010). doi:10.1038/gt.2010.43
325. Ferraro, B., Cruz, Y.L., Coppola, D., Heller, R.: Intradermal delivery of plasmid VEGF(165) by electroporation promotes wound healing. Mol. Ther. **17**(4), 651–657 (2009)
326. Suzuki, T., Shin, B.C., Fujikura, K., Matsuzaki, T., Takata, K.: Direct gene transfer into rat liver cells by in vivo electroporation. FEBS Lett. **425**(3), 436–440 (1998)
327. Jaichandran, S., Yap, S.T., Khoo, A.B., Ho, L.P., Tien, S.L., Kon, O.L.: In vivo liver electroporation: optimization and demonstration of therapeutic efficacy. Hum. Gene Ther. **17**(3), 362–375 (2006)
328. Sakai, M., Nishikawa, M., Thanaketpaisarn, O., Yamashita, F., Hashida, M.: Hepatocyte-targeted gene transfer by combination of vascularly delivered plasmid DNA and in vivo electroporation. Gene Ther. **12**(7), 607–616 (2005)
329. Chi, C.H., Liu, I.L., Lo, W.Y., Liaw, B.S., Wang, Y.S., Chi, K.H.: Hepatocyte growth factor gene therapy prevents radiation-induced liver damage. World J. Gastroenterol.: WJG **11**(10), 1496–1502 (2005)
330. Pringle, I.A., McLachlan, G., Collie, D.D., Sumner-Jones, S.G., Lawton, A.E., Tennant, P., Baker, A., Gordon, C., Blundell, R., Varathalingam, A., Davies, L.A., Schmid, R.A., Cheng, S.H., Porteous, D.J., Gill, D.R., Hyde, S.C.: Electroporation enhances reporter gene expression following delivery of naked plasmid DNA to the lung. J. Gene Med. **9**(5), 369–380 (2007)

331. Gazdhar, A., Bilici, M., Pierog, J., Ayuni, E.L., Gugger, M., Wetterwald, A., Cecchini, M., Schmid, R.A.: In vivo electroporation and ubiquitin promoter – a protocol for sustained gene expression in the lung. J. Gene Med. **8**(7), 910–918 (2006)
332. Machado-Aranda, D., Adir, Y., Young, J.L., Briva, A., Budinger, G.R., Yeldandi, A.V., Sznajder, J.I., Dean, D.A.: Gene transfer of the Na+, K+-ATPase beta1 subunit using electroporation increases lung liquid clearance. Am. J. Respir. Crit. Care Med. **171**(3), 204–211 (2005)
333. Mutlu, G.M., Machado-Aranda, D., Norton, J.E., Bellmeyer, A., Urich, D., Zhou, R., Dean, D.A.: Electroporation-mediated gene transfer of the Na+, K+ -ATPase rescues endotoxin-induced lung injury. Am. J. Respir. Crit. Care Med. **176**(6), 582–590 (2007)
334. Gazdhar, A., Fachinger, P., van Leer, C., Pierog, J., Gugger, M., Friis, R., Schmid, R.A., Geiser, T.: Gene transfer of hepatocyte growth factor by electroporation reduces bleomycin-induced lung fibrosis. Am. J. Physiol. Lung Cell. Mol. Physiol. **292**(2), L529–L536 (2007)
335. Davies, J.C., Alton, E.W.: Airway gene therapy. Adv. Genet. **54**, 291–314 (2005)
336. Harrison, R.L., Byrne, B.J., Tung, L.: Electroporation-mediated gene transfer in cardiac tissue. FEBS Lett. **435**(1), 1–5 (1998)
337. Wang, Y., Bai, Y., Price, C., Boros, P., Qin, L., Bielinska, A.U., Kukowska-Latallo, J.F., Baker Jr., J.R., Bromberg, J.S.: Combination of electroporation and DNA/dendrimer complexes enhances gene transfer into murine cardiac transplants. Am. J. Transplant. **1**(4), 334–338 (2001)
338. Nikolski, V.P., Efimov, I.R.: Electroporation of the heart. Eur.: Eur. Pacing Arrhythmias Card Electrophysiol: J. Work. Groups Card. Pacing Arrhythmias Card. Cell. Electrophysiol. Eur. Soc. Cardiol. **7**(Suppl 2), 146–154 (2005)
339. Hargrave, B., Strange Jr., R., Navare, S., Stratton, M., Burcus, N., Murray, L., Lundberg, C., Bulysheva, A., Li, F., Heller, R.: Gene electro transfer of plasmid encoding vascular endothelial growth factor for enhanced expression and perfusion in the ischemic swine heart. PLoS ONE **9**(12), e115235 (2014)
340. Seidler, R.W., Allgauer, S., Ailinger, S., Sterner, A., Dev, N., Rabussay, D., Doods, H., Lenter, M.C.: In vivo human MCP-1 transfection in porcine arteries by intravascular electroporation. Pharm. Res. **22**(10), 1685–1691 (2005)
341. Matsumoto, T., Komori, K., Shoji, T., Kuma, S., Kume, M., Yamaoka, T., Mori, E., Furuyama, T., Yonemitsu, Y., Sugimachi, K.: Successful and optimized in vivo gene transfer to rabbit carotid artery mediated by electronic pulse. Gene Ther. **8**(15), 1174–1179 (2001)
342. Miyahara, T., Koyama, H., Miyata, T., Shigematsu, H., Inoue, J.-I., Takato, T., Nagawa, H.: Inflammatory responses involving tumor necrosis factor receptor-associated factor 6 contribute to in-stent lesion formation in a stent implantation model of rabbit carotid artery. J. Vasc. Surg. **43**(3), 592–600 (2006)
343. Touchard, E., Berdugo, M., Bigey, P., El Sanharawi, M., Savoldelli, M., Naud, M.C., Jeanny, J.C., Behar-Cohen, F.: Suprachoroidal electrotransfer: a nonviral gene delivery method to transfect the choroid and the retina without detaching the retina. Mol. Ther. **20**(8), 1559–1570 (2012). doi:10.1038/mt.2011.304
344. Touchard, E., Bloquel, C., Bigey, P., Kowalczuk, L., Jonet, L., Thillaye-Goldenberg, B., Naud, M.C., Scherman, D., de Kozak, Y., Benezra, D., Behar-Cohen, F.: Effects of ciliary muscle plasmid electrotransfer of TNF-alpha soluble receptor variants in experimental uveitis. Gene Ther. **16**(7), 862–873 (2009)
345. Touchard, E., Kowalczuk, L., Bloquel, C., Naud, M.C., Bigey, P., Behar-Cohen, F.: The ciliary smooth muscle electrotransfer: basic principles and potential for sustained intraocular production of therapeutic proteins. J. Gene Med. **12**(11), 904–919 (2010)
346. Bejjani, R.A., Andrieu, C., Bloquel, C., Berdugo, M., BenEzra, D., Behar-Cohen, F.: Electrically assisted ocular gene therapy. Surv. Ophthalmol. **52**(2), 196–208 (2007)
347. Bloquel, C., Bejjani, R., Bigey, P., Bedioui, F., Doat, M., BenEzra, D., Scherman, D., Behar-Cohen, F.: Plasmid electrotransfer of eye ciliary muscle: principles and therapeutic efficacy using hTNF-alpha soluble receptor in uveitis. FASEB J. **20**(2), 389–391 (2006)

348. Kutzler, M.A., Weiner, D.B.: DNA vaccines: ready for prime time? Nat. Rev. Genet. **9**(10), 776–788 (2008)
349. Sardesai, N.Y., Weiner, D.B.: Electroporation delivery of DNA vaccines: prospects for success. Curr. Opin. Immunol. **23**(3), 421–429 (2011)
350. Heller, L., Merkler, K., Westover, J., Cruz, Y., Coppola, D., Benson, K., Daud, A., Heller, R.: Evaluation of toxicity following electrically mediated interleukin-12 gene delivery in a B16 mouse melanoma model. Clin. Cancer Res. **12**(10), 3177–3183 (2006)
351. Cemazar, M., Sersa, G., Wilson, J., Tozer, G.M., Hart, S.L., Grosel, A., Dachs, G.U.: Effective gene transfer to solid tumors using different nonviral gene delivery techniques: electroporation, liposomes, and integrin-targeted vector. Cancer Gene Ther. **9**(4), 399–406 (2002)
352. Canatella, P.J., Prausnitz, M.R.: Prediction and optimization of gene transfection and drug delivery by electroporation. Gene Ther. **8**(19), 1464–1469 (2001)
353. Spanggaard, I., Snoj, M., Cavalcanti, A., Bouquet, C., Sersa, G., Robert, C., Cemazar, M., Dam, E., Vasseur, B., Attali, P., Mir, L.M., Gehl, J.: Gene electrotransfer of plasmid antiangiogenic metargidin peptide (AMEP) in disseminated melanoma: safety and efficacy results of a phase I first-in-man study. Hum. Gene Ther. Clin. Dev. **24**(3), 99–107 (2013)
354. Yuan, J., Ku, G.Y., Adamow, M., Mu, Z., Tandon, S., Hannaman, D., Chapman, P., Schwartz, G., Carvajal, R., Panageas, K.S., Houghton, A.N., Wolchok, J.D.: Immunologic responses to xenogeneic tyrosinase DNA vaccine administered by electroporation in patients with malignant melanoma. J. Immunother. Cancer **1**, 20 (2013)
355. Chudley, L., McCann, K., Mander, A., Tjelle, T., Campos-Perez, J., Godeseth, R., Creak, A., Dobbyn, J., Johnson, B., Bass, P., Heath, C., Kerr, P., Mathiesen, I., Dearnaley, D., Stevenson, F., Ottensmeier, C.: DNA fusion-gene vaccination in patients with prostate cancer induces high-frequency CD8(+) T-cell responses and increases PSA doubling time. Cancer Immunol. Immunother. **61**(11), 2161–2170 (2012)
356. Low, L., Mander, A., McCann, K.J., Dearnaley, D., Tjelle, T.E., Mathiesen, I., Stevenson, F. K., Ottensmeier, C.H.: DNA vaccination with electroporation induces increased antibody responses in patients with prostate cancer. Hum. Gene Ther. **20**(11), 1269–1278 (2009)
357. Eriksson, F., Totterman, T., Maltais, A.K., Pisa, P., Yachnin, J.: DNA vaccine coding for the rhesus prostate specific antigen delivered by intradermal electroporation in patients with relapsed prostate cancer. Vaccine **31**(37), 3843–3848 (2013)
358. Bagarazzi, M.L., Yan, J., Morrow, M.P., Shen, X., Parker, R.L., Lee, J.C., Giffear, M., Pankhong, P., Khan, A.S., Broderick, K.E., Knott, C., Lin, F., Boyer, J.D., Draghia-Akli, R., White, C.J., Kim, J.J., Weiner, D.B., Sardesai, N.Y.: Immunotherapy against HPV16/18 generates potent TH1 and cytotoxic cellular immune responses. Sci. Transl. Med. **4**(155), 155ra138 (2012)
359. Kopycinski, J., Cheeseman, H., Ashraf, A., Gill, D., Hayes, P., Hannaman, D., Gilmour, J., Cox, J.H., Vasan, S.: A DNA-based candidate HIV vaccine delivered via in vivo electroporation induces CD4 responses toward the alpha4beta7-binding V2 loop of HIV gp120 in healthy volunteers. Clin. Vaccine Immunol. **19**(9), 1557–1559 (2012)
360. Vasan, S., Hurley, A., Schlesinger, S.J., Hannaman, D., Gardiner, D.F., Dugin, D.P., Boente-Carrera, M., Vittorino, R., Caskey, M., Andersen, J., Huang, Y., Cox, J.H., Tarragona-Fiol, T., Gill, D.K., Cheeseman, H., Clark, L., Dally, L., Smith, C., Schmidt, C., Park, H.H., Kopycinski, J.T., Gilmour, J., Fast, P., Bernard, R., Ho, D.D.: In vivo electroporation enhances the immunogenicity of an HIV-1 DNA vaccine candidate in healthy volunteers. PLoS ONE **6**(5), e19252 (2011)
361. Dolter, K.E., Evans, C.F., Ellefsen, B., Song, J., Boente-Carrera, M., Vittorino, R., Rosenberg, T.J., Hannaman, D., Vasan, S.: Immunogenicity, safety, biodistribution and persistence of ADVAX, a prophylactic DNA vaccine for HIV-1, delivered by in vivo electroporation. Vaccine **29**(4), 795–803 (2011)
362. Kalams, S.A., Parker, S.D., Elizaga, M., Metch, B., Edupuganti, S., Hural, J., De Rosa, S., Carter, D.K., Rybczyk, K., Frank, I., Fuchs, J., Koblin, B., Kim, D.H., Joseph, P., Keefer, M.

C., Baden, L.R., Eldridge, J., Boyer, J., Sherwat, A., Cardinali, M., Allen, M., Pensiero, M., Butler, C., Khan, A.S., Yan, J., Sardesai, N.Y., Kublin, J.G., Weiner, D.B.: Safety and comparative immunogenicity of an HIV-1 DNA vaccine in combination with plasmid interleukin 12 and impact of intramuscular electroporation for delivery. J. Infect. Dis. **208** (5), 818–829 (2013)
363. Yang, F.Q., Yu, Y.Y., Wang, G.Q., Chen, J., Li, J.H., Li, Y.Q., Rao, G.R., Mo, G.Y., Luo, X. R., Chen, G.M.: A pilot randomized controlled trial of dual-plasmid HBV DNA vaccine mediated by in vivo electroporation in chronic hepatitis B patients under lamivudine chemotherapy. J. Viral Hepat. **19**(8), 581–593 (2012)
364. Weiland, O., Ahlen, G., Diepolder, H., Jung, M.C., Levander, S., Fons, M., Mathiesen, I., Sardesai, N.Y., Vahlne, A., Frelin, L., Sallberg, M.: Therapeutic DNA vaccination using in vivo electroporation followed by standard of care therapy in patients with genotype 1 chronic hepatitis C. Mol. Ther. **21**(9), 1796–1805 (2013)
365. Aurisicchio, L., Mancini, R., Ciliberto, G.: Cancer vaccination by electro-gene-transfer. Expert Rev. Vaccines **12**(10), 1127–1137 (2013). doi:10.1586/14760584.2013.836903
366. Fioretti, D., Iurescia, S., Fazio, V.M., Rinaldi, M.: In vivo DNA electrotransfer for immunotherapy of cancer and neurodegenerative diseases. Curr. Drug Metab. **14**(3), 279–290 (2013)
367. Neumann, E., Schaefer-Ridder, M., Wang, Y., Hofschneider, P.H.: Gene transfer into mouse lyoma cells by electroporation in high electric fields. EMBO J. **1**(7), 841–845 (1982)
368. Beebe, S.J.: Bioelectrics in basic science and medicine: impact of electric fields on cellular structures and functions. J. Nanomed. Nanotechnol. **4**, 163 (2013)
369. Nuccitelli, R., Pliquett, U., Chen, X., Ford, W., James Swanson, R., Beebe, S.J., Kolb, J.F., Schoenbach, K.H.: Nanosecond pulsed electric fields cause melanomas to self-destruct. Biochem. Biophys. Res. Commun. **343**, 351–360 (2006)
370. Chen, X., Kolb, J.F., Swanson, R.J., Schoenbach, K.H., Beebe, S.J.: Apoptosis initiation and angiogenesis inhibition: melanoma targets for nanosecond pulsed electric fields. Pigment Cell Melanoma Res. **23**, 554–563 (2010)
371. Zhang, J., Blackmore, P.F., Hargrave, B.Y., Xiao, S., Beebe, S.J., Schoenbach, K.H.: Nanosecond pulse electric field (nanopulse): a novel non-ligand agonist for platelet activation. Arch. Biochem. Biophys. **471**, 240–248 (2008)
372. Schoenbach, K.H., Katsuki, S., Stark, R.H., Buescher, E.S., Beebe, S.J.: Bioelectrics-new applications for pulsed power technology. IEEE. Trans. Plasma Sci. **30**, 293–300 (2002)
373. Malik, M.A., Xiao, S., Schoenbach, K.H.: Scaling of surface-plasma reactors with a significantly increased energy density for NO conversion. J. Hazard. Mater. **209**, 293–298 (2012)
374. Malik, M.A.: Water purification by plasmas: which reactors are most energy efficient? Plasma Chem. Plasma Process. **30**, 21–31 (2010)
375. Kong, M.G., Kroesen, G., Morfill, G., Nosenko, T., Shimizu, T., et al.: Plasma medicine: an introductory review. New J. Phys. **11**, 115012 (2009)
376. Beebe, S.J., Fox, P.M., Rec, L.J., Willis, E.L., Schoenbach, K.H.: Nanosecond, high-intensity pulsed electric fields induce apoptosis in human cells. FASEB J. **17**, 1493–1495 (2003)
377. Nuccitelli, R., Chen, X., Pakhomov, A.G., Baldwin, W.H., Sheikh, S., et al.: A new pulsed electric field therapy for melanoma disrupts the tumor's blood supply and causes complete remission without recurrence. Int. J. Cancer **125**, 438–445 (2009)
378. Rogakou, E.P., Pilch, D.R., Orr, A.H., Ivanova, V.S., Bonner, W.M.: DNA double-stranded breaks induce histone H2AX phosphorylation on serine 139. J. Biol. Chem. **273**, 5858–5868 (1998)
379. Esser, A.T., Smith, K.C., Gowrishankar, T.R., Weaver, J.C.: Towards solid tumor treatment by nanosecond pulsed electric fields. Technol. Cancer Res. Treat. **8**(4), 289–306 (2009)
380. Nuccitelli, R., Tran, K., Athos, B., Kreis, M., Nuccitelli, P., Chang, K.S., Epstein Jr., E.H., Tang, J.Y.: Nanoelectroablation therapy for murine basal cell carcinoma. Biochem. Biophys. Res. Commun. **424**, 446–450 (2012)
381. Yin, D., Yang, W.G., Weissberg, J., Goff, C.B., Chen, W., Kuwayama, Y., Leiter, A., Xing, H., Meixel, A., Gaut, D., Kirkbir, F., Sawcer, D., Vernier, P.T., Said, J.W., Gundersen, M.A.,

Koeffler, H.P.: Cutaneous papilloma and squamous cell carcinoma therapy utilizing nanosecond pulsed electric fields (nsPEF). PLoS ONE **7**(8), e43891 (2012)
382. Yin, S., Chen, X., Hu, C., Zhang, X., Hu, Z., Yu, J., Feng, X., Jiang, K., Ye, S., Shen, K., Xie, H., Zhou, L., James Swanson, R., Zheng, S.: Nanosecond pulsed electric field (nsPEF) treatment for hepatocellular carcinoma: a novel locoregional ablation decreasing lung metastasis. Cancer Lett. **346**, 85–291 (2014)
383. Nuccitelli, R., Huynh, J., Lui, K., Wood, R., Kreis, M., Athos, B., Nuccitelli, P.: Nanoelectroablation of human pancreatic carcinoma in a murine xenograft model without recurrence. Int. J. Cancer **132**, 1933–1939 (2013)
384. Vaughn, L., Beckel, N.: Severe burn injury, burn shock, and smoke inhalation injury in small animals. Part 1: burn classification and pathophysiology. J. Vet. Emerg. Crit. Care (San Antonio) **22**(2), 179–186 (2012)
385. Martin, P.: Wound healing – aiming for perfect skin regeneration. Science **276**, 75–81 (1997)
386. Martin, P., Leibovich, S.J.: Inflammatory cells during wound repair: the good, the bad and the ugly. Trends Cell Biol. **15**, 599–607 (2005)
387. Gurtner, G.C., Werner, S., Barrandon, Y., Longaker, M.T.: Wound repair and regeneration. Nature **453**, 314–321 (2008)
388. Yeaman, M.R.: Platelets: at the nexus of antimicrobial defense. Nat. Rev. Microbiol. **12**, 426–437 (2014)
389. Raivio, P., Lassila, R., Petaja, J.: Thrombin in myocardial ischemia-reperfusion during cardiac surgery. Ann. Thorac. Surg. **88**, 318–325 (2009)
390. Raivio, P., Kuitunen, A., Suojaranta-Ylinen, R., Lassila, R., Petaja, J.: Thrombin generation during reperfusion after coronary artery bypass surgery associates with postoperative myocardial damage. J. Throm. Homeostat. **4**, 1523–1529 (2006)
391. Edmunds, L.H., Colman, R.W.: Thrombin during cardiopulmonary bypass. Ann. Thorac. Surg. **82**, 2315–2322 (2006)
392. Lopez, J.J., Salido, G.M., Gómez-Arteta, E., et al.: Thrombin induces apoptotic events through the generation of reactive oxygen species in human platelets. J. Thromb. Haemost. **5**(6), 1283–1291 (2007)
393. Han, B., Woodell-May, J., Ponticiello, M., et al.: The effect of thrombin activation of platelet-rich plasma on demineralized bone matrix osteoinductivity. J. Bone. Joint Surg. **91**, 1459–1470 (2009)
394. Thon, J.N., Italiano, J.E.: Platelets: production, morphology and ultrastructure. Handb. Exp. Pharmacol. **210**, 3–22 (2012)
395. Cimmino, G., Golino, P.: Platelet biology and receptor pathways. J. Cardiovasc. Transl. Res. **6**, 299–309 (2013)
396. Golebiewska, E.M., Poole, A.W.: Platelet secretion: from haemostasis to wound healing and beyond. Blood Rev. **29**(3), 153–162 (2015)
397. Nurden, A.T., Nurden, P., Sanchez, M., Andia, I., Anitua, E.: Platelets and wound healing. Front. Biosci. **13**, 3532–3548 (2008)
398. Kevin, L.G., Enis Novalija, F., David, F., Stowe, D.F.: Reactive oxygen species as mediators of cardiac injury and protection: the relevance to anesthesia practice. Anesth. Analg. **101**, 1275–1287 (2005)
399. Elahi, M.M., Kong, Y.X., Matata, B.M.: Oxidative stress as a mediator of cardiovascular disease. Oxid. Med. Cell Longev. **2**(5), 259–269 (2009)
400. Vernier, P.T., Sun, Y., Gundersen, M.A.: Nanoelectropulse-driven membrane perturbation and small molecule permeabilization. BMC Cell Biol. **7**, 37 (2006)
401. Wong, T.K., Neumann, E.: Electric field mediated gene transfer. Biochem. Biophys. Res. Commun. **107**, 584–587 (1982)
402. Chen, X., Swanson, R.J., Kolb, J.F., Nuccitelli, R., Schoenbach, K.H.: Histopathology of normal skin and melanomas after nanosecond pulsed electric field treatment. Melanoma Res. **19**, 361–371 (2009)

403. Vernier, P.T., Sun, Y., Marcu, L., Salemi, S., Craft, C.M., Gundersen, M.A.: Calcium bursts induced by nanosecond electric pulses. Biochem. Biophys. Res. Commun. **310**, 286–295 (2003)
404. White, J.A., Blackmore, P.F., Schoenbach, K.H., Beebe, S.J.: Stimulation of capacitative calcium entry in HL-60 cells by nanosecond pulsed electric fields. J. Biol. Chem. **279**, 22964–22972 (2004)
405. Pakhomova, O.N., Khorokhorina, V.A., Bowman, A.M., Rodaite-Riseviciene, R., Saulis, G., Xiao, S., Pakhomov, A.G.: Oxidative effects of nanosecond pulsed electric field exposure in cells and cell-free media. Arch. Biochem. Biophys. **527**, 55–64 (2012)
406. Nuccitelli, R., Lui, K., Kreis, M., Athos, B., Nuccitelli, P.: Nanosecond pulsed electric field stimulation of reactive oxygen species in human pancreatic cancer cells is Ca^{2+}-dependent. Biochem. Biophys. Res. Commun. **435**, 580–585 (2013)
407. Beebe, S.J., Sain, N.M., Ren, W.: Induction of cell death mechanisms and apoptosis by nanosecond pulsed electric fields (nsPEFs). Cells **2**, 136–162 (2013)
408. Nuccitelli, R., Tran, K., Sheikh, S., Athos, B., Kreis, M., Nuccitelli, P.: Optimized nanosecond pulsed electric field therapy can cause murine malignant melanomas to self-destruct with a single treatment. Int. J. Cancer **127**, 1727–1736 (2010)
409. Nuccitelli, R., Tran, K., Lui, K., Huynh, J., Athos, B., Kreis, M., Nuccitelli, P., De Fabo, E.C.: Non-thermal nanoelectroablation of UV-induced murine melanomas stimulates an immune response. Pigment Cell Melanoma Res. **25**, 618–629 (2012)
410. Chen, X., Yin, S., Hu, C., Chen, X., Jiang, K., Ye, S., Feng, X., Fan, S., Xie, H., Zhou, L., Zheng, S.: Comparative study of nanosecond electric fields in vitro and in vivo on hepatocellular carcinoma indicate macrophage infiltration contribute to tumor ablation in vivo. PLoS ONE **9**, e86421 (2014)
411. Nuccitelli, R., Berridge, J.C., Mallon, Z., Kreis, M., Athos, B., Nuccitelli, P.: Nanoelectroablation of rat orthotopic hepatocellular carcinoma triggers a CD8-dependent adaptive immune response. PLoS ONE **10**(7), e0134364 (2015)
412. Ibey, B.L., Roth, C.C., Pakhomov, A.G., Bernhard, J.A., Wilmink, G.J., Pakhomova, O.N.: Dose-dependent thresholds of 10-ns electric pulse induced plasma membrane disruption and cytotoxicity in multiple cell lines. PLoS ONE **6**, e15642 (2011)
413. Yang, W., Wu, Y.H., Yin, D., Koeffler, H.P., Sawcer, D.E., Vernier, P.T., Gundersen, M.A.: Differential sensitivities of malignant and normal skin cells to nanosecond pulsed electric fields. Technol. Cancer Res. Treat. **10**, 281–286 (2011)
414. Graves, D.B.: The emerging role of reactive oxygen and nitrogen species in redox biology and some implications for plasma applications to medicine and biology. J. Phys. D. Appl. Phys. **45**, 263001 (2012)
415. Babaeva, N.Y., Ning, N., Graves, D.B., Kushner, M.J.: Ion activation energy delivered to wounds by atmospheric pressure dielectric-barrier discharges: sputtering of lipid-like surfaces. J. Phys. D. Appl. Phys. **45**, 115203 (2012)
416. Moreau, M., Orange, N., Feuilloley, M.G.J.: Non-thermal plasma technologies: new tools for bio-decontamination. Biotechnol. Adv. **26**, 610–617 (2008)
417. Weltmann, K.D., Kindel, E., von Woedtke, T., Hahnel, M., Stieber, M., Brandenburg, R.: Atmospheric-pressure plasma sources: prospective tools for plasma medicine. Pure Appl. Chem. **82**, 1223–1237 (2010)
418. Weltmann, K.D., Polak, M., Masur, K., von Woedtke, T., Winter, J., Reuter, S.: Plasma processes and plasma sources in medicine. Contrib. Plasm. Phys. **52**, 644–654 (2012)
419. Lloyd, G., Friedman, G., Jafri, S., Schultz, G., Fridman, A., Harding, K.: Gas plasma: medical uses and developments in wound care. Plasma Process. Polym. **7**, 194–211 (2010)
420. Lu, X., Naidis, G.V., Laroussi, M., Ostrikov, K.: Guided ionization waves: theory and experiments. Phys. Rep. **540**, 123–166 (2014)
421. Jiang, C. Emerging applications of plasmas in medicine: fashion vs. efficacy. In: Chu, P.K., Lu, X. (eds.) Low Temperature Plasma Technology: Methods and Applications. Taylor & Francis Group, CRC Press (2013)

422. Lazarus, G.S., Cooper, D.M., Knighton, D.R., Margolis, D.J., Pecoraro, R.E., Rodeheaver, G., Robson, M.C.: Definitions and guidelines for assessment of wounds and evaluation of healing. Arch. Dermatol. **130**, 489–493 (1994)
423. Robson, M.C.: Wound infection. A failure of wound healing caused by an imbalance of bacteria. Surg. Clin. North Am. **77**, 637–650 (1997)
424. Tarnuzzer, R.W., Schultz, G.S.: Biochemical analysis of acute and chronic wound environments. Wound Repair Regen. **4**, 321–325 (1996)
425. Park, B.J., Lee, D.H., Park, J.C., Lee, I.S., Lee, K.Y., Hyun, S.O., Chun, M.S., Chung, K.H.: Sterilization using a microwave-induced argon plasma system at atmospheric pressure. Phys. Plasmas **10**, 4539–4544 (2003)
426. Montie, T.C., Kelly-Wintenberg, K., Roth, J.R.: An overview of research using the one atmosphere uniform glow discharge plasma (OAUGDP) for sterilization of surfaces and materials. IEEE. T. Plasma Sci. **28**, 41–50 (2000)
427. Lee, M.H., Park, B.J., Jin, S.C., Kim, D., Han, I., Kim, J., Hyun, S.O., Chung, K.H., Park, J. C.: Removal and sterilization of biofilms and planktonic bacteria by microwave-induced argon plasma at atmospheric pressure. New J. Phys. **11**, 115022 (2009)
428. Morrison, JCF: Electrosurgical method and apparatus for initiating an electrical discharge in an inert gas flow. In: U.S. Patent (ed.) U.S. Patent, U.S.A. (1977)
429. Farin, G., Grund, K.E.: Technology of argon plasma coagulation with particular regard to endoscopic applications. Endosc. Surg. Allied Technol. **2**, 71–77 (1994)
430. Grund, K.E., Storek, D., Farin, G.: Endoscopic argon plasma coagulation (APC) first clinical experiences in flexible endoscopy. Endosc. Surg. Allied Technol. **2**, 42–46 (1994)
431. Vargo, J.J.: Clinical applications of the argon plasma coagulator. Gastrointest. Endosc. **59**, 81–88 (2004)
432. Raiser, J., Zenker, M.: Argon plasma coagulation for open surgical and endoscopic applications: state of the art. J. Phys. D. Appl. Phys. **39**, 3520–3523 (2006)
433. Stoffels, E., Kieft, I.E., Sladek, R.E.J.: Superficial treatment of mammalian cells using plasma needle. J. Phys. D. Appl. Phys. **36**, 2908–2913 (2003)
434. Kieft, I.E., Broers, J.L.V., Caubet-Hilloutou, V., Slaaf, D.W., Ramaekers, F.C.S., Stoffels, E.: Electric discharge plasmas influence attachment of cultured CHO k1 cells. Bioelectromagnetics **25**, 362–368 (2004)
435. Lee, D.H., Lee, J.O., Jeon, W., Choi, I.G., Kim, J.S., Jeong, J.H., Kang, T.C., Seo, C.H.: Suppression of scar formation in a murine burn wound model by the application of non-thermal plasma. Appl. Phys. Lett. **99**, 203701-1-3 (2011)
436. Stoffels, E., Roks, A.J.M., Deelmm, L.E.: Delayed effects of cold atmospheric plasma on vascular cells. Plasma Process. Polym. **5**, 599–605 (2008)
437. Kalghatgi, S.U., Fridman, G., Cooper, M., Nagaraj, G., Peddinghaus, M., Balasubramanian, M., Vasilets, V.N., Gutsol, A.F., Fridman, A., Friedman, G.: Mechanism of blood coagulation by nonthermal atmospheric pressure dielectric barrier discharge plasma. IEEE. T. Plasma Sci. **35**, 1559–1566 (2007)
438. Costerton, J.W., Stewart, P.S., Greenberg, E.P.: Bacterial biofilms: a common cause of persistent infections. Science **284**, 1318–1322 (1999)
439. Estrela, C., Sydney, G.B., Figueiredo, J.A., Estrela, C.R.: Antibacterial efficacy of intracanal medicaments on bacterial biofilm: a critical review. J. Appl. Oral Sci. **17**, 1–7 (2009)
440. Moore, W.E., Moore, L.V.: The bacteria of periodontal diseases. Periodontol. **5**, 66–77 (1994)
441. Paster, B.J., Boches, S.K., Galvin, J.L., Ericson, R.E., Lau, C.N., Levanos, V.A., Sahasrabudhe, A., Dewhirst, F.E.: Bacterial diversity in human subgingival plaque. J. Bacteriol. **183**, 3770–3783 (2001)
442. Marsh, P.D.: Microbiologic aspects of dental plaque and dental caries. Dent. Clin. N. Am. **43**, 599–614, v–vi (1999)
443. Yip, H.K., Samaranayake, L.P.: Caries removal techniques and instrumentation: a review. Clin. Oral Invest. **2**, 148–154 (1998)

444. Banerjee, A., Watson, T.F., Kidd, E.A.: Dentine caries excavation: a review of current clinical techniques. Br. Dent. J. **188**, 476–482 (2000)
445. Buchanan, L.S.: Cleaning and shaping the root canal system: negotiating canals to the termini. Dent. Today **13**(76), 78–81 (1994)
446. Schilder, H.: Cleaning and shaping the root canal. Dent. Clin. N. Am. **18**, 269–296 (1974)
447. Chavez de Paz, L.E.: Redefining the persistent infection in root canals: possible role of biofilm communities. J. Endod. **33**, 652–662 (2007)
448. Chavez De Paz, L.E., Dahlen, G., Molander, A., Moller, A., Bergenholtz, G.: Bacteria recovered from teeth with apical periodontitis after antimicrobial endodontic treatment. Int. Endod. J. **36**, 500–508 (2003)
449. Nair, P.N., Henry, S., Cano, V., Vera, J.: Microbial status of apical root canal system of human mandibular first molars with primary apical periodontitis after "one-visit" endodontic treatment. Oral Surg. Oral Med. Oral Pathol. Oral Radiol. Endod. **99**, 231–252 (2005)
450. Nair, P.N., Sjogren, U., Krey, G., Kahnberg, K.E., Sundqvist, G.: Intraradicular bacteria and fungi in root-filled, asymptomatic human teeth with therapy-resistant periapical lesions: a long-term light and electron microscopic follow-up study. J. Endod. **16**, 580–588 (1990)
451. Sjogren, U., Figdor, D., Persson, S., Sundqvist, G.: Influence of infection at the time of root filling on the outcome of endodontic treatment of teeth with apical periodontitis. Int. Endod. J. **30**, 297–306 (1997)
452. Moritz, A., Schoop, U., Goharkhay, K., Jakolitsch, S., Kluger, W., Wernisch, J., Sperr, W.: The bactericidal effect of Nd:YAG, Ho:YAG, and Er:YAG laser irradiation in the root canal: an in vitro comparison. J. Clin. Laser Med. Surg. **17**, 161–164 (1999)
453. Bergmans, L., Moisiadis, P., Huybrechts, B., Van Meerbeek, B., Quirynen, M., Lambrechts, P.: Effect of photo-activated disinfection on endodontic pathogens ex vivo. Int. Endod. J. **41**, 227–239 (2008)
454. Soukos, N.S., Chen, P.S., Morris, J.T., Ruggiero, K., Abernethy, A.D., Som, S., Foschi, F., Doucette, S., Bammann, L.L., Fontana, C.R., Doukas, A.G., Stashenko, P.P.: Photodynamic therapy for endodontic disinfection. J. Endod. **32**, 979–984 (2006)
455. Bergmans, L., Moisiadis, P., Teughels, W., Van Meerbeek, B., Quirynen, M., Lambrechts, P.: Bactericidal effect of Nd:YAG laser irradiation on some endodontic pathogens ex vivo. Int. Endod. J. **39**, 547–557 (2006)
456. Noiri, Y., Katsumoto, T., Azakami, H., Ebisu, S.: Effects of Er:YAG laser irradiation on biofilm-forming bacteria associated with endodontic pathogens in vitro. J. Endod. **34**, 826–829 (2008)
457. Dederich, D.N., Bushick, R.D.: Lasers in dentistry: separating science from hype. J. Am. Dent. Assoc. **135**, 204–212; quiz 229 (2004)
458. Ahmady, K., Marsh, P.D., Newman, H.N., Bulman, J.S.: Distribution of Streptococcus mutans and Streptococcus sobrinus at subsites in human approximal dental plaque. Caries Res. **27**, 135–139 (1993)
459. Babaahmady, K.G., Challacombe, S.J., Marsh, P.D., Newman, H.N.: Ecological study of Streptococcus mutans, Streptococcus sobrinus and Lactobacillus spp. at sub-sites from approximal dental plaque from children. Caries Res. **32**, 51–58 (1998)
460. Badet, C., Thebaud, N.B.: Ecology of lactobacilli in the oral cavity: a review of literature. Open Microbiol. J. **2**, 38–48 (2008)
461. Sedgley, C.M., Lennan, S.L., Clewell, D.B.: Prevalence, phenotype and genotype of oral enterococci. Oral Microbiol. Immunol. **19**, 95–101 (2004)
462. Waltimo, T.M., Sen, B.H., Meurman, J.H., Orstavik, D., Haapasalo, M.P.: Yeasts in apical periodontitis. Crit. Rev. Oral Biol. Med. **14**, 128–137 (2003)
463. Lee, H.W., Nam, S.H., Mohamed, A.A.H., Kim, G.C., Lee, J.K.: Atmospheric pressure plasma jet composed of three electrodes: application to tooth bleaching. Plasma Process. Polym. **7**, 274–280 (2010)
464. Goree, J., Liu, B., Drake, D., Stoffels, E.: Killing of S-mutans bacteria using a plasma needle at atmospheric pressure. IEEE. T. Plasma Sci. **34**, 1317–1324 (2006)

465. Gonzalvo, Y.A., Whitmore, T.D., Rees, J.A., Seymour, D.L., Stoffels, E.: Atmospheric pressure plasma analysis by modulated molecular beam mass spectrometry. J. Vac. Sci. Technol. A **24**, 550–553 (2006)
466. Duan, Y.X., Huang, C., Yu, Q.S.: Cold plasma brush generated at atmospheric pressure. Rev. Sci. Instrum. **78**, 015104-1-5 (2007)
467. Yang, B., Chen, J.R., Yu, Q.S., Li, H., Lin, M.S., Mustapha, A., Hong, L.A., Wang, Y.: Oral bacterial deactivation using a low-temperature atmospheric argon plasma brush. J. Dent. **39**, 48–56 (2011)
468. Koban, I., Matthes, R., Hubner, N.O., Welk, A., Meisel, P., Holtfreter, B., Sietmann, R., Kindel, E., Weltmann, K.D., Kramer, A., Kocher, T.: Treatment of Candida albicans biofilms with low-temperature plasma induced by dielectric barrier discharge and atmospheric pressure plasma jet. New J. Phys. **12**, 073039-1-12 (2010)
469. Rupf, S., Lehmann, A., Hannig, M., Schafer, B., Schubert, A., Feldmann, U., Schindler, A.: Killing of adherent oral microbes by a non-thermal atmospheric plasma jet. J. Med. Microbiol. **59**, 206–212 (2010)
470. Yamazaki, H., Ohshima, T., Tsubota, Y., Yamaguchi, H., Jayawardena, J.A., Nishimura, Y.: Microbicidal activities of low frequency atmospheric pressure plasma jets on oral pathogens. Dent. Mater. J. **30**, 384–391 (2011)
471. Jiang, C., Schaudinn, C., Jaramillo, D.E., Gundersen, M.A., Costerton, J.W.: A sub-microsecond pulsed plasma jet for endodontic biofilm disinfection. In: Machala, Z., Hensel, K., Akishev, Y. (eds.) Plasma for Bio-Decontamination, Medicine and Food Security. Springer, Heidelberg (2012)
472. Jiang, C.Q., Chen, M.T., Schaudinn, C., Gorur, A., Vernier, P.T., Costerton, J.W., Jaramillo, D.E., Sedghizadeh, P.P., Gundersen, M.A.: Pulsed atmospheric-pressure cold plasma for endodontic disinfection. IEEE. T. Plasma Sci. **37**, 1190–1195 (2009)
473. Jiang, C., Schaudinn, C.: A curving bactericidal plasma needle. IEEE. T. Plasma Sci. **39**, 2966–2967 (2011)
474. Jiang, C., Schaudinn, C., Jaramillo, D.E., Webster, P., Costerton, J.W.: In vitro antimicrobial effect of a cold plasma jet against enterococcus faecalis biofilms. ISRN Dent. **2012**, 295736 (2012)
475. Schaudinn, C., Jaramillo, D., Freire, M.O., Sedghizadeh, P.P., Nguyen, A., Webster, P., Costerton, J.W., Jiang, C.: Evaluation of a nonthermal plasma needle to eliminate ex vivo biofilms in root canals of extracted human teeth. Int. Endod. J. **46**, 930–937 (2013)
476. Lu, X.P., Cao, Y.G., Yang, P., Xiong, Q., Xiong, Z.L., Xian, Y.B., Pan, Y.: An RC plasma device for sterilization of root canal of teeth. IEEE. T. Plasma Sci. **37**, 668–673 (2009)
477. Du, T., Ma, J., Yang, P., Xiong, Z., Lu, X., Cao, Y.: Evaluation of antibacterial effects by atmospheric pressure nonequilibrium plasmas against Enterococcus faecalis biofilms in vitro. J. Endod. **38**, 545–549 (2012)
478. Du, T., Shi, Q., Shen, Y., Cao, Y., Ma, J., Lu, X., Xiong, Z., Haapasalo, M.: Effect of modified nonequilibrium plasma with chlorhexidine digluconate against endodontic biofilms in vitro. J. Endod. **39**, 1438–1443 (2013)
479. Casal, M., Haskins, M.: Large animal models and gene therapy. Eur. J. Hum. Genet. **14**(3), 266–272 (2006)
480. Khanna, C., Lindblad-Toh, K., Vail, D., London, C., Bergman, P., Barber, L., Breen, M., Kitchell, B., McNeil, E., Modiano, J.F., Niemi, S., Comstock, K.E., Ostrander, E., Westmoreland, S., Withrow, S.: The dog as a cancer model. Nat. Biotechnol. **24**(9), 1065–1066 (2006)
481. Ranieri, G., Panteleo, M., Piccinno, M., Roncetti, M., Mutinati, M., Patruno, R., Rizzo, A., Sciorsci, R.L.: Tyrosine kinase inhibitors (TKIs) in human and pet tumours with special reference to breast cancer: a comparative review. Crit. Rev. Oncol. Hematol. **88**(2), 293–308 (2013)
482. Marconato, L.: Chemioterapici utilizzati in medicina veterinaria. In: Marconato, L., e Del Piero, F. (eds.) Oncologia medica dei piccoli animali, 1st edn, pp. 108–144. Poletto editore, Millano (2005)
483. Plumb, D.C.: Veterinary Drug Handbook, 7th edn, p. 1208. PharmaVet, Stocholm (2011)

484. Mir, L.M., Devauchelle, P., Quintin-Colonna, F., Delisle, F., Doliger, S., Fradelizi, D., Belehradek Jr., J., Orlowski, S.: First clinical trial of cat soft-tissue sarcomas treatment by electrochemotherapy. Br. J. Cancer **76**(12), 1617–1622 (1997)
485. Tozon, N., Sersa, G., Cemazar, M.: Electrochemotherapy: potentiation of local antitumour effectiveness of cisplatin in dogs and cats. Anticancer Res. **21**(4A), 2483–2488 (2001)
486. Rols, M.P., Tamzali, Y., Teissié, J.: Electrochemotherapy of horses. A preliminary clinical report. Bioelectrochemistry **55**(1–2), 101–105 (2002)
487. Spugnini, E.P., Porrello, A.: Potentiation of chemotherapy in companion animals with spontaneous large neoplasms by application of biphasic electric pulses. J. Exp. Clin. Cancer Res. **22**(4), 571–580 (2003)
488. Spugnini, E.P., Filliponni, M., Romani, L., Dotinsky, I., Mudrov, N., Baroni, A., Ruocco, E., Laieta, M.T., Montesarchio, V., Cassandro, R., Citro, G., Baldi, A.: Local control and distant metastasis after electrochemotherapy of a canine anal melanoma. In Vivo **21**(5), 897–899 (2007)
489. Kodre, V., Cemazar, M., Pecar, J., Sersa, G., Cor, A., Tozon, N.: Electrochemotherapy compared to surgery for treatment of canine mast cell tumours. In Vivo **23**(1), 55–62 (2009)
490. Spugnini, E.P., Citro, G., Baldi, A.: Electrochemotherapy in veterinary oncology part I: solid tumors. In: Spugnini, E.P., Baldi, A. (eds.) Electroporation in Laboratory and Clinical Investigations, pp. 245–255. Nova Science Publishers, New York (2011). cap. 12
491. Spugnini, E.P., Baldi, A., Citro, G.: Electrochemotherapy in veterinary oncology part II: round cell tumors. In: Spugnini, E.P., Baldi, A. (eds.) Electroporation in Laboratory and Clinical Investigations, pp. 257–264. Nova Science Publishers, New York (2011). cap. 13
492. Spugnini, E.P., Vincenzi, B., Citro, G., Dotsinsky, I., Mudrov, T., Baldi, A.: Evaluation of Cisplatin as an electrochemotherapy agent for the treatment of incompletely excised mast cell tumors in dogs. J. Vet. Intern. Med. **25**(2), 407–411 (2011)
493. Tamzali, Y., Borde, L., Rols, M.P., Golzio, M., Lyazrhi, F., Teissie, J.: Successful treatment of equine sarcoids with cisplatin electrochemotherapy: a retrospective study of 48 cases. Equine Vet. J. **44**(2), 214–220 (2012)
494. Tozon, N., Pavlin, D., Sersa, G., Dolinsek, T., Cemazar, M.: Electrochemotherapy with intravenous bleomycin injection: an observational study in superficial squamous cell carcinoma in cats. J. Feline Med. Surg. **16**(4), 291–299 (2014)
495. Reed, S.D., Fulmer, A., Buckholtz, J., Zhang, B., Cutera, J., Shiomitsu, K., Li, S.: Bleomycin/interleukin-12 electrochemogene therapy for treating naturally occurring spontaneous neoplasms in dogs. Cancer Gene Ther. **17**(7), 457–464 (2010)
496. Pavlin, D., Cemazar, M., Cor, A., Sersa, G., Poqacnik, A., Tozon, N.: Electrogene therapy with interleukin-12 in canine mast cell tumors. Radiol. Oncol. **45**(1), 31–39 (2010)
497. Liu, F., Huang, L.A.: Syringe electrode device for simultaneous injection of DNA and electrotransfer. Mol. Ther. **5**(3), 323–328 (2002)
498. Spugnini, E.P., Citro, G., Porrello, A.: Rational design of new electrodes for electrochemotherapy. J. Exp. Clin. Cancer Res. **24**(2), 245–254 (2005)
499. Tjelle, T.E., Salte, R., Mathiesen, I., Kjeken, R.: A novel electroporation device for gene delivery in large animals and humans. Vaccine **24**(21), 4667–4670 (2006)
500. Mazeres, S., Sel, D., Golzio, M., Pucihar, G., Tamzali, Y., Miklavcic, D., Teissie, J.: Non invasive contact electrodes for in vivo localized cutaneous electropulsation and associated drug and nucleic acid delivery. J. Control Release **134**(2), 125–131 (2009), Mar 4 2009. ISSN 1873-4995 (Electronic) 0168-3659 (Linking). Disponível em: http://www.ncbi.nlm.nih.gov/pubmed/19084039
501. Tozon, N., Kodre, V., Sersa, G., Cemazar, M.: Effective treatment of perianal tumors in dogs with electrochemotherapy. Anticancer Res. **25**(2A), 839–845 (2005)
502. Theon, A.P., Pascoe, J.R., Carlson, G.P., Krag, D.N.: Intratumoral chemotherapy with cisplatin in oily emulsion in horses. J. Am. Vet. Med. Assoc. **202**(2), 261–267 (1993)
503. Sersa, G., Cemazar, M., Miklavcic, D., Chaplin, D.J.: Tumor blood flow modifying effect of electrochemotherapy with bleomycin. Anticancer Res. **19**(5B), 4017–4022 (1999)

Chapter 6
Environmental Applications, Food and Biomass Processing by Pulsed Electric Fields

Wolfgang Frey, Christian Gusbeth, Takashi Sakugawa, Martin Sack, Georg Mueller, Juergen Sigler, Eugene Vorobiev, Nikolai Lebovka, Ignacio Álvarez, Javier Raso, Loree C. Heller, Muhammad A. Malik, Christian Eing, and Justin Teissie

Abstract Pulsed electric field (PEF) treatment is a physical method which exhibits specific advantages over conventional processing in various applications and was proven for feasibility on pilot and industrial scale. For bacterial inactivation in wastewater and liquid food and for eradication of *Cyanobacteria* in surface waters, PEF-based techniques are demonstrated to be energy saving and persistent in efficacy without adding harmful chemicals and in particular do not cause adverse effects to food matrices or to the aquatic environment. For component extraction, the specific advantages of PEF treatment, i.e., low heat influx, low-energy demand, and selectivity of compound release, promote PEF processing in winemaking,

W. Frey (✉) • C. Gusbeth • M. Sack • G. Mueller • C. Eing
Karlsruhe Institute of Technology, Institute for Pulsed Power and Microwave Technology, Eggenstein-Leopoldshafen, Germany
e-mail: wolfgang.frey@kit.edu

T. Sakugawa
Institute of Pulsed Power Science, Kumamoto University, Kumamoto, Japan

J. Sigler
State Institute for Viticulture and Oenology, Freiburg, Germany

E. Vorobiev
Sorbonne Universités, Université de Technologie de Compiègne, Compiègne, France

N. Lebovka
F. D. Ovcharenko Institute of Biocolloidal Chemistry, Kiev, Ukraine

I. Álvarez • J. Raso
Food Science Department, University of Zaragoza, Zaragoza, Spain

L.C. Heller • M.A. Malik
Frank Reidy Research Center for Bioelectrics, Old Dominion University, Norfolk, VA, USA

J. Teissie
Institute of Pharmacology and Structural Biology, CNRS and Université de Toulouse, UMR 5089, Toulouse, France

H. Akiyama, R. Heller (eds.), *Bioelectrics*, DOI 10.1007/978-4-431-56095-1_6

extraction of sugar from sugar beets and valuable components from fruits and vegetables, and PEF-downstream processing of microalgae. These promising applications of PEF processing will be introduced in more detail in the following sections.

Keywords Bacterial inactivation • Food preservation • Cell component extraction • Microalgae downstream processing

6.1 Overview

Pulsed electric field (PEF) treatment is a physical method which exhibits specific advantages over conventional processing in various applications and was proven for feasibility on pilot and industrial scale for large mass-flow applications in energy technology, environmental technology, and biotechnology.

For bacterial inactivation of wastewater, PEF treatment was shown to exhibit high inactivation efficiency without adding harmful chemicals and proven not to cause any adverse effects to the aquatic environment. Studies on sustainability revealed that the inactivation rate is not impaired by repetitive treatments of bacterial populations already multiply exposed to PEFs. Moreover, by combining mild thermal pretreatment and PEFs, the electrical energy demand for bacterial inactivation is competitively low.

The technique was also demonstrated to be highly efficient for eradication of unwanted *Cyanobacteria* in surface waters. For this application, it was shown at demo-scale that an autarkic treatment facility can efficiently control *Cyanobacteria* populations in lakes.

Another specific advantage of PEF treatment, the comparably low heat transfer to the processed medium, is exerted in liquid food preservation, maintaining flavor and nutritional value of foods.

For component extraction, the specific advantages of PEF treatment are low heat influx, low-energy demand, and selectivity of compound release. Existing PEF processing facilities in winemaking, extraction of sugar from sugar beets, and valuable component extraction from fruits and vegetables nowadays operate on a mass-flow scale of 1–10 tons/h and larger.

Recent activities in PEF-downstream processing of microalgae have proven a high-energy efficiency of PEF treatment in comparison to conventional techniques and furthermore revealed the fractionating ability of PEF processing, enabling separation and simultaneous recovery of lipids and water-soluble compounds at low-energy demand.

These promising applications of PEF processing are introduced in more detail in the following sections.

6.2 Disinfection of Hospital Wastewater by PEF Treatment

Christian Gusbeth and Wolfgang Frey

6.2.1 Introduction

The latest report by the World Health Organization 2014 reveals that the dissemination of antibiotic-resistant bacteria is a global issue and that "this serious threat is no longer a prediction for the future, it is happening right now in every region of the world and has the potential to affect anyone, of any age, in any country."

Hospital wastewaters are known to be one important source of the dissemination of infectious agents, especially antibiotic multiresistant pathogens, into the downstream aquatic systems [1–5, 44]. In contrast to chemical contaminants that were metabolized by a number of biological and physic-chemical reactions, harmful microorganisms are able to proliferate in an appropriate milieu to high densities. Moreover, when they reach the aquatic environment system, these resistant bacteria contribute to the variety of resistance genes and enable the genetic exchange between many different bacterial genera in nature [6, 7]. Beyond that, resistance determinants most probably were acquired also by pathogenic bacteria from this pool of resistance genes in other microbial genera, including antibiotic-producing organisms.

Therefore, it is essential to retard the development of antibiotic resistance and to prolong the effective lifetime of valuable antimicrobial agents until bacteria adapted to them. This can be done, on the one hand, by enhancing the overall medical care and hygiene and, on the other hand, by preventing the dissemination of resistant bacteria. Disinfection is considered to be the primary mechanism for the inactivation/destruction of pathogenic and resistant organisms to prevent the spread of waterborne diseases to downstream users and the environment [8]. Whenever antibiotic-resistant bacteria will be released in the environmental aquatic system, the chance for gene transfer and implicit antibiotic resistance transfer will be increased. During an outbreak of infections in a hospital, the hospital wastewater effluent becomes a major source for pathogens and opportunistic bacteria. As a consequence, a direct contact of humans with pathogenic agents cannot be excluded, and the risk of contamination of water resources increases with the diversity of harmful bacteria [43]. Once the pathogens have reached surface water, the bacterial dissemination is irreversible. The dissemination of antibiotic-resistant bacteria can be stopped by a variety of disinfection methods like ultraviolet (UV) radiation, thermal treatment (pasteurization), and chlorination. It is known that the disinfection techniques applied have demonstrated disadvantages like generation of toxic disinfection by-products (DBP) during chemical disinfection or reduced efficiency in liquids with high turbidity in case of UV radiation. An alternative disinfection technique for clinical wastewater is the pulsed electric field treatment (PEF) that is based on the process of electroporation in cells [9–12]. Previous work on applications of PEF demonstrated the feasibility of PEF treatment to improve the shelf life of juice, milk, soup, and other liquid foods [13–15]. Owing to the nonchemical and nonthermal character of PEF treatment, no toxic DBP can be expected. The chemical stability of flavors and vitamins of PEF-treated food was demonstrated for different kinds of fruit juices [16]. It was also shown that a large variety of undesirable microorganisms, among them Gram-negative, Gram-positive bacteria and fungi, were effectively reduced by PEF treatment [17].

6.2.2 Conventional Disinfection Methods Versus PEF Treatment

The majority of the conventional water disinfection methods are known from water purification plants. These methods have been intensively used and continuously developed for decades. Therefore, most of them are economic and effective in bacterial reduction. Owing to the different consistency of freshwater and hospital wastewater, which has an increased organic and particle load, some or most of these methods cannot be used for hospital wastewater effluents.

Water chlorination is the most widely used disinfection method for municipal water purification plants and swimming pools. It consists in adding chlorine as gas (Cl_2), sodium hypochlorite solution, or other chlorine compounds to water. As a strong oxidizing agent, chlorine is highly efficient in killing pathogenic agents causing many diseases such as cholera, typhoid fever, or dysentery [18]. It was shown that the required concentrations of disinfectants were higher in the case of heavily contaminated wastewater than in uncontaminated, if similar disinfection levels have to be achieved [19]. It has, however, some drawbacks including minimum contact time and the need to handle, store, and dispense hazardous chemicals accurately in case of fast-changing flow rates; chlorine residuals can prolong and affect the environmental surface water. One significant drawback of chlorination is that chlorine reacts with organic compounds in water, causing unwanted DBP such as halogenated organics [20], which are harmful and/or carcinogenic (trihalomethanes and haloacetic acids). This excludes its use for hospital wastewater disinfection.

Ozone disinfection is an effective method to inactivate harmful microorganisms in drinking water. Ozone (O_3) is an unstable inorganic molecule, decomposing in di-oxygen and one atom of oxygen providing a powerful oxidizing agent. Similar to chlorination, it consists in adding ozone directly to water. Due to the instability of ozone, it must be created on site to ensure a high level of bacterial inactivation. Owing to the short half-life of ozone and the limited production of harmful DBP in comparison to chlorination, ozone disinfection is considered a feasible disinfection technology [21, 22]. It was found that the half-life of O_3 in water ranges from 14 h (without presence of phosphate and carbonate) to 8 min and strongly depends on water impurities [23, 24]. In a medium such as wastewater, O_3 reacts with the complex organic compounds (e.g., saturated hydrocarbons, amines, and aromatic compounds), destroys them, and forms by-products such as acids, aldehydes, bromates, ketones, and peroxides [25–27]. In addition, it was shown [28] that the use of ozone in marine-based aquaculture systems has been limited because of the risk to form the carcinogenic bromate, which is formed during the oxidation of naturally occurring bromide by ozone. However, the presence of water impurities was found to be a major limiting factor of ozonation for wastewater applications [29].

UV disinfection is nowadays a widely used method for water disinfection and became more and more a common method in municipal drinking water treatment in the last decades. It is based on irradiation of water with UV light at a wavelength of 254 nm (generated by low- and medium-pressure mercury arc lamps), which is near the absorption peak of nucleotide bases of RNA and DNA. The UV radiation damages the bacterial DNA of microorganisms and retards their ability to reproduce. In continuous operation, an effective bacterial reduction of Gram-negative and Gram-positive bacteria such as enteric bacteria and *Enterococci* could be demonstrated [30]. However, some authors found a recovery of the bacterial population 24–48 h after UV treatment [31]. In addition, RNA repair mechanisms that were demonstrated in *E. coli* might lead to a long-term UV adaptation of bacteria. In case of hospital wastewater, the high turbidity of the wastewater impedes application of this disinfection method.

Pulsed electric field treatment is a nonthermal physical method for liquid food preservation and water disinfection. In recent years, there has been increasing interest for a potential use of PEF treatment for inactivation of pathogenic microorganisms carried by hospital wastewater effluents. Although this method was intensively investigated for food preservation over decades, there is less experience for wastewater disinfection. The benefit of PEF processing of food lies in the extended shelf life without the loss of flavor and nutrition value (fresh-like liquid food). PEF can be used for processing liquid and semiliquid (mash-like) food products with low electrical conductivity ($<4\ \mathrm{mS \cdot cm^{-1}}$), but also for drinking water and wastewater. The fluid is disinfected on-site in the PEF-treatment zone, and there is no prolonged effect such as in case of chlorination. Rather, no systemic DBP generation is expected after PEF disinfection of hospital wastewater. Commercial scale continuous PEF systems for processing volumes ranging from 100 to $5000\ \mathrm{l\ h^{-1}}$ are available [32–34]. This equipment is scalable when solid-state pulse power generator modules are used, which can be stacked in series and parallel to provide the required/desired generator power. According to the state of the art, PEF processing would add only \$0.10 $\mathrm{l^{-1}}$ to the final treatment costs [35]. For wastewater purification plants with a large throughput of 1 million liters of water per day, the operating and maintenance costs increase with the throughput flow, negatively affecting the economic efficiency of this method. On the other hand, due to the high efficiency in bacterial inactivation and the small equipment size, PEF processing is a safe method for local hotspot disinfection of hospital wastewater effluents (Table 6.1), especially for combating temporary outbreaks of infections.

Table 6.1 Advantages and disadvantages of different technologies for wastewater disinfection. The total equipment costs and operating and maintenance costs were calculated on the basis of a treatment volume flow of 4000 $m^3 \cdot h^{-1}$ and a disinfection efficiency of at least 99.9 % (3 log). Cost estimates for PEF-treatment facilities are only available on pilot scale and are based on real prototype costs. Please note that alternative technology costs are adopted from industrial municipal water processing technology which may not be transferable to or applicable for hospital wastewater treatment

	Chlorination	Ozonation	UV irradiation	PEF treatment
Advantages	Well-established technology	Very strong disinfectant	Effective in inactivating viruses, bacteria and spores	Physical process, no chemical impact on the liquid being treated
	Reliable and effective against a wide spectrum of pathogenic organisms	Low half-life (no toxic residuals)	Effective in low turbidity water	No harmful DBP
	Chlorine residual can prolong disinfection	Effective for a large range of microbes	Low-energy consumption	Effective also in high turbidity water
		No harmful DBP (except bromides)	Physical process, no chemical disinfectant	No residual disinfectant in the water or any other effect that can be harmful
		No residual disinfectant in the water	No residual disinfectant or effect that can be harmful	
Disadvantages	Highly corrosive and toxic	Unstable, must be manufactured on-site	Limited by turbidity	Ineffective for spores and viruses
	Toxic chlorine residuals	Does not provide a long-term-effective inactivation	Regeneration and regrowth of bacteria	Limited experience in wastewater treatment technology
	Hazardous DBP from oxidation of the organic matter in wastewater	Limited experience in wastewater treatment	Degradation of molecules	Energy efficiency is limited by conductivity to <4 $mS \cdot cm^{-1}$
			Fouling of tubes	
			Replacement of lamps after 12,000 working hours	
Efficacy	More cost-effective than either UV or ozone disinfection	Must be coupled with filtration to minimize potential regrowth	Efficiency depends on the characteristics of the wastewater	Requires specific energy 120 $J \cdot g^{-1}$ at 30 °C
	Higher chlorine doses required in the presence of high chlorine-binding materials	50–200 $g \cdot m^{-3}$ Ozone concentration required	Cost-competitive with chlorination when a dechlorination step is included	Less energy consumption, 40 $J \cdot g^{-1}$, when coupled with thermal treatment (60 °C)

(continued)

Table 6.1 (continued)

	Chlorination	Ozonation	UV irradiation	PEF treatment
	5–20 $mg \cdot l^{-1}$ chlorine concentration		900 $J \cdot m^{-2}$ specific energy consumption	
Equipment	Flexible dosing control	High operation and maintenance costs	Requires less space than other methods	High investment costs
	Chlorination/dechlorination systems are more complex		User-friendly for operators	High operating and maintenance costs
			Has a shorter contact time (20–30 s)	Short contact time
Total equipment costs (4000 $m^3 \cdot h^{-1}$)	1,270,000 ($)	400,000 ($)	250,000 ($)	200,000 ($) (1 $m^3 \cdot h^{-1}$)
Operating and maintenance costs	80,000 ($)	100,000 ($)	20,000 ($)	100,000 ($)* (1 $m^3 \cdot h^{-1}$)

6.2.3 Reduction Efficiencies of Bacteria During PEF Treatment of Clinical Wastewater

Since PEF treatment is a novel method in the context of disinfection of hospital wastewater, investigation of the efficiency of the PEF processing depending on the specific bacterial population in wastewater is necessary. Considering the fact that there is no "standard" wastewater, this can be done by processing wastewater, sampled from different municipal wastewater purification plant and hospital effluents. By regular sampling during the year, the variety of bacterial population and the influence of weather and season are covered. One problem is that the cell density and cell population strongly vary, complicating comparability and conclusiveness of experimental results.

Bacterial Load of Hospital Wastewater Enterococci/Streptococci and Enterobacteriaceae The bacterial load of hospital wastewater *Enterococci/Streptococci* and *Enterobacteriaceae* can be found in all biofilms from hospital wastewater effluents. *Enterococci* and *Enterobacteriaceae* are naturally occurring microorganisms from human and animal intestines. The *Enterococci/Streptococci* loads and the presence of resistant bacteria are higher in hospital biofilms than in effluxes of municipal sewage. The largest number of cefazolin- and cefotaxime-resistant *Enterobacteriaceae* is measured in biofilms from hospital wastewater, 54 % and 17 %, respectively. Typical bacterial densities of *E. faecium* and *P. aeruginosa* being below 10^5 $cells \cdot ml^{-1}$ were found by Volkman et al. [5, 43] and Schwartz et al. [4] in wastewater, sampled at the outlet of the hospital wastewater system, close to the public sewer system.

Hospital Wastewater Parameters The conductivity of wastewater has a great influence on the configuration of a PEF system that has to be employed. Wastewater conductivity varies in the range between 0.5 and 2.0 mS·cm^{-1}, depending on time and site of sampling. In the case of hospital wastewater from a sewer of the university hospital of the city of Mainz, the measured conductivity was between 1.0 and 1.7 mS·cm^{-1} at a temperature ranging from 7 to 14 °C [1]. For lab experiments, depending on the original wastewater parameters and the employed pulse generator, an adjustment of the conductivity of the model wastewater by dilution or addition of buffer is possible. Usually the wastewater pH is in the range of six to eight, whereas the turbidity at 600 nm is 0.7.

Performance of PEF Processing For the evaluation of the PEF performance, the parameters used for food processing serve as a reference for performing inactivation studies on hospital wastewater. The main parameters under investigation are pulsed electric field amplitude $\boldsymbol{E}$ (5–100 kV·cm^{-1}), pulse duration $\boldsymbol{t}$ (0.2–100 μs), and specific energy, which can be adjusted by the number of pulses $\boldsymbol{N}$. The pulse shape (rectangular, exponentially decaying, unipolar, or bipolar) and the frequency play a minor role in improving the inactivation performance. Bipolar pulse protocols are proven to be advantageous for pulsed electric field processing in general since it decreases electrochemical erosion of treatment electrodes [36, 37]. Based on these studies, the following *standard treatment* conditions have been defined: specific treatment energy 120 J·ml^{-1}, electric field amplitude 80 kV·cm^{-1}, and pulse duration 1–2 μs.

In general, the density of bacteria in hospital wastewater is not high enough to demonstrate the PEF efficiency at inactivation rates of several orders of magnitude. Therefore, for the evaluation of the PEF technique, with respect to its efficiency in reducing *Enterococci* and *Pseudomonads* as target organisms, an increased cell density is considered necessary. For this purpose, model wastewater spiked with specific bacteria strains with certain cell densities can be used. Prior to spiking with target organisms, hospital wastewater must be disinfected at 121 °C for 15 min in order to eliminate the original population that could interfere with the subsequent analysis. *P. aeruginosa* as a representative of Gram-negative bacteria, well-known as an opportunistic human pathogen with high incidences in hospitals, and the enteric Gram-positive bacterium *E. faecium*, are identified as target organisms. A main observation is that cell inactivation is not directly proportional to a specific energy input. The efficiency of PEF processing, investigated on *P. putida*—a widespread wastewater microorganism and nonpathogenic and safe pseudomonas species—was similar to the results obtained from food processing (Fig. 6.1). Also in this case, the inactivation kinetics deviates from the first-order kinetics, and it seems that the bacterial inactivation saturates at around 6 log reduction when pulse numbers range between 120 and 200 pulses. Furthermore, within a pulse duration range of 100–10 μs and at invariable suspension conductivity, inactivation does not depend on pulse duration in case the external field amplitude is high enough to provide fast membrane charging at the entire cell surface [38]. The dissipated energy and bacterial reduction follow a dose–response relationship, and the inactivation rate scales with the specific energy (~E·t^2) [38]. It has to be

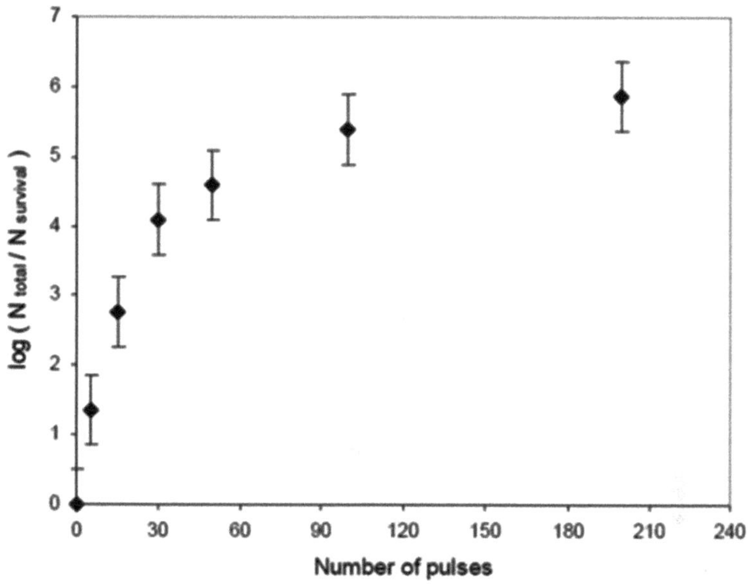

Fig. 6.1 Inactivation rate of *P. putida* after PEF treatment with rectangular pulses of 100 kV·cm^{-1} electric field strength and pulse duration of 200 ns. The conductivity of the medium was 2.0 mS·cm^{-1}, at 21 °C. The temperature increase due to electrical energy dissipation was below 7 K. With increasing number of pulses/treatment energy, the bacterial inactivation approaches a climax (6 log), which is below complete inactivation (8 log; correspond to 108 colony forming units (CFU) ·ml^{-1} initial cell density) [1]

pointed out that in this case the temperature rise was below 7 K in particular due to the fact that the heat could dissipate across the aluminum electrodes into the cooling medium surrounding the treatment chamber. Therefore, the observed bacterial inactivation reflects the impact due to the electric field and is not caused by electrically induced thermal heat. For orientation, the PEF treatment with a specific electrical energy input of 40 J·ml^{-1} causes heating with a temperature increase of 10 K, if adiabatic conditions are considered. For scalable systems, the temperature inside the treatment chamber will equilibrate according to treatment volume, pulse frequency, ambient temperature, and electrical energy.

In general, Gram-negative bacteria (such *as Escherichia coli* or *P. aeruginosa*) are much more sensitive to electric field treatment than Gram-positive bacteria (such as *Staphylococcus aureus* or *Enterococcus faecium*) [17, 39, 40]. This was demonstrated also for spiked hospital wastewater, which was treated by PEF at specific energies ranging from 80 to 190 J·ml^{-1} [2]. The PEF treatment was performed with a PEF system consisting of a 65-kV, 1-kJ, 10-Hz, two-stage pulse-forming generator. The pulse-forming network of the generator delivers a 1.2-μs-long rectangular output pulse. The field strength amplitude in the treatment chamber was 80 kV·cm^{-1}. A reduction efficiency of up to 5.5 decimal orders of magnitude was reached for the Gram-negative reference strain, *P. aeruginosa*, when the treatment energies exceeded 162 J·ml^{-1} (Table 6.2). In contrast to

Table 6.2 Reduction efficiency of the PEF treatment of hospital wastewater spiked with *Pseudomonas aeruginosa* depending on specific treatment energy [2]

	P. aeruginosa (strain 1095)		*P. aeruginosa* (strain 1071)	
Energy value ($J{\cdot}ml^{-1}$)	CFU ml^{-1}	Decimal order of magnitude reduction	CFU ml^{-1}	Decimal order of magnitude reduction
0	3 × 105		1 × 106	
84	2 × 103	2.2	1 × 103	3.0
117	4 × 102	2.9	1 × 102	3.0
137	1 × 103	2.5	2 × 103	2.7
146	1 × 102	3.5	2 × 102	3.7
162	0	5.5	20	4.7
190	0	5.5	20	4.7

Table 6.3 Reduction efficiency of the PEF treatment of hospital wastewater spiked with different *E. faecium* strains. Strains labeled with 2, 1265 and B7641 were isolated from hospital patients, whereas DSM strains originated from German Collection of Microorganisms

Bacterial strain	Energy value ($J{\cdot}ml^{-1}$)	Decimal order of magnitude reduction
E. faecium strain 2	84	1.3–2.1
Vancomycin resistant	162	1.7–3.0
E. faecium strain DSM 20478	124	1.0
E. faecium strain DSM 20477	121	0.3
E. faecium strain 1265	116	0.8
	126	0.7
E. faecium strain B7641	125	0.6
	5214	2.6

DSM Deutsche Sammlung von Mikroorganismen

P. aeruginosa, the reduction efficiency of the Gram-positive vancomycin-resistant *E. faecium* (strain 2) was lower (Table 6.3). Moreover, inactivation experiments on different *E. faecium* strains showed strongly varying inactivation efficiencies. In different independent experiments, different reduction patterns were observed. These differences are attributed, on the one hand, to a varying robustness of different species against PEF treatment and, on the other hand, to the time independent experiments and the resulting different bacterial populations in the wastewater matrices. Inactivation experiments on wastewater containing mixed bacterial populations reveal that the most robust bacteria strain dominates the resulting inactivation rate of the wastewater sample. At standard treatment conditions, $W = 120\ J{\cdot}ml^{-1}$, the inactivation rate varied between 1.5 and 3.4 log.

6.2.4 Sustainability of PEF Disinfection Method

In order to establish the PEF processing for disinfection of hospital wastewater, the hazards of this method with respect to environment and humans have to be analyzed. Consequently, two important issues have to be investigated: (i) the adaptation of bacteria to electric field treatment (*electro-tolerance*) and (ii) the mutagenicity of PEF-treated clinical wastewater.

In the first case, an adaptation of bacteria to PEF treatment would decrease the efficiencies of this disinfection method, with possibly hazardous consequences to environment and humans. Induced *electro-tolerance* in reference bacteria can be proved by consecutive treatment of already treated bacteria suspensions over several cycles. A selective enrichment of *electro-tolerant* bacteria would finally result in increased survival rates. For these investigations, surviving bacteria from the previous treatment were cultivated in growth medium for 3 days and treated again by PEFs. In case of *P. putida*, no significant deviation of the survival rate from the average could be observed over 30 treatment cycles (Fig. 6.2), indicating that bacteria do not develop an adaptation against PEF treatment [1].

Furthermore, the comparison of the band patterns of the variable intergenic spacer region (ISR) of ribosomal operon over 30 cycles of treatments reveals the same pattern for all samples. This result underlines that no alterations become visible within the ISR, which means that no phenotypic changes of the bacterial population were induced by repetitive PEF treatments. Thus, an enrichment of bacterial population with *electro-tolerant* bacteria after PEF treatment as well as the transmission of *electro-tolerance* from bacteria to bacterial descendants can be excluded.

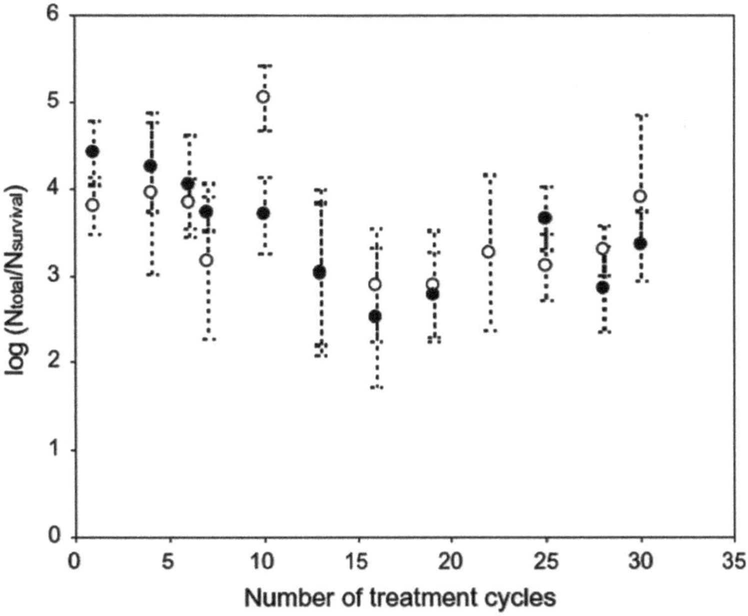

Fig. 6.2 Inactivation rate of samples of *P. putida* after consecutive PEF treatment. For each treatment cycle, surviving bacteria from the previous treatment were cultivated in growth medium for 3 days and treated again by PEF. An average inactivation of 3.6 log for a specific treatment energy of 120 J·ml^{-1} was obtained [1]

Another impediment of this method could be the generation of genotoxic by-products due to the electrolytic reaction of the wastewater compounds. The formation of such genotoxic compounds during PEF treatment might provoke mutations in the bacterial and human genomes, having unforeseeable consequences for aquatic environment and human population health. Usually, to assess the genotoxicity in water, the *umu*-test is employed as a standard method. The test is based on the ability of genetically modified bacteria strain (*Salmonella typhimurium*) to induce SOS response and to express the *umu*C gene induced by RNA lesions. Studies performed with SOS chromotest and *Salmonella* fluctuation test show that the different wastewater samples exhibit a genotoxic risk, which is not unusual for hospital wastewater [41]. This was also the case for wastewater from a sewer of the university hospital of the city of Mainz; all samples were positive when *umu*-test was performed [42]. However, PEF treatments with specific treatment energies of more than 250 $J{\cdot}ml^{-1}$ did not change the hospital wastewater genotoxicity. Moreover, in domestic water and phosphate buffer (PBS), no genotoxicity could be found before and after PEF treatment. Even PEF treatments with more than 480 $J{\cdot}ml^{-1}$ generated no genotoxic by-products in PBS.

6.2.5 Economic Feasibility of Combining PEF with Thermal Treatment

This section is dedicated to the economic feasibility of the PEF technology for wastewater disinfection. In order to achieve satisfactory bacterial inactivation (>4 log), an electric treatment energy between 120 and 240 $J{\cdot}ml^{-1}$ is necessary. This energy is enough to increase the fluid temperature by about 30–60 K, if adiabatic conditions are applied. Therefore, also the thermal treatment (*pasteurization*) might be considered as an alternative treatment.

Contrary to thermal disinfection or sterilization, PEF treatment reveals the important feature that the activity of nucleases is not affected by pulsed electric fields. Nucleases naturally digest free bacterial RNA fragments in water systems, thus preventing a spread of possibly antibiotic-resistant plasmids and transposons by uptake into other bacteria. It could be demonstrated that nuclease activity is not affected by PEF treatment [2, 42], whereas a thermal treatment to 72 °C already reduced DNA degradation by nucleases significantly. However, the degradation temperature of plasmids and DNA fragments is considerably higher ($T > 100$ °C).

It was shown that a combined treatment with PEF and heating increases bacterial inactivation of hospital wastewater significantly [42]. For comparison, the inactivation rate of *E. faecium* by PEF treatment with a specific energy input of 120 $J{\cdot}ml^{-1}$ at 25 °C is 2 log, whereas at a treatment temperature of 60 °C, a complete inactivation of the bacteria (>8 log) could be achieved at the same electrical energy expenditure. At treatment temperatures below 55 °C, the combined treatment does not improve the inactivation rate.

The effectiveness of the combined thermal and electric wastewater treatment was demonstrated on a prototype plant. Using heat exchangers for preheating and heat recovery, a disinfection of 4–5 log could be achieved with PEF-treatment energy of 40 J·ml^{-1}. Under this premise, the PEF processing costs of \$0.10 l^{-1} can be drastically reduced, resulting in a competitive price compared to conventional disinfection methods.

Considering all these results, it can be concluded that PEF treatment has no adverse effect on environment and humans, underlined by the facts that PEF processing does not increase the genotoxicity of hospital wastewater and also doesn't cause phenotypic changes, in terms of *electro-tolerance* induced by repetitive treatments. Further, PEF treatment became very effective in bacteria inactivation in combination with thermal treatment. Thus, PEF treatment is a sustainable and suitable disinfection method, which effectively reduces bacteria from hospital wastewater effluents without any negative impact on aquatic environment.

6.3 Algae (*Cyanobacteria*) Treatment in Dams and Lakes

Takashi Sakugawa

6.3.1 Introduction

Blue-green algal bloom is a rapid increase in the population of cyanobacterial cells in aquatic systems. In recent years, frequent occurrences of large-scale water blooms due to eutrophication of water threaten human health all over the world. Environmental problems caused by water blooms include clogging intake pipes and filter lines of water treatment facilities, loss of landscape beauty, and smell of drinking water. Moreover, when a bloom dies in a pond or shallow lake, severe oxygen depletion can produce objectionable odors and damage of fisheries.

The most serious problem is that some bloom-forming cyanobacterial genera produce toxic substances, which can cause a range of human health effects such as skin irritation, liver damage, stomach and intestinal illness, and so on [34]. Field investigations have reported that microcystins in drinking water are one major factor that resulted in locally high incidence of liver cancer. The current methods of water blooms treatment such as chemical compounds, ultrasonic, and microwave have been used to eliminate water blooms or to slow down cyanobacterial cell proliferation [46]. However, chemical compounds and ultrasonic can cause the cell membrane to be destroyed and the contents of the cells released into ambient water. And energy consumptions of ultrasonic treatment system and microwave options are too great to be used in large-scale treatments. Therefore, the development of an efficient method for large-scale water-bloom treatment becomes an urgent task of environmental protection.

It is well known that pulsed streamer discharges generate several physical phenomena and chemical phenomena and chemical reactions simultaneously. The use of pulsed streamer discharges in the gas to realize bacterial decontamination has been a firmly established method [47]. Accompanied with the development of applications of this kind of discharge in liquid, it has been used to realize water sterilization. A new method using underwater pulsed streamer-like discharges to prevent water blooms from large area generation is developed.

6.3.2 *Effects of Underwater Streamer-Like Discharge*

In recent years, pulsed power modulator for underwater streamer discharge has been developed. Figure 6.3 shows typical circuit of this modulator. A type of this modulator is magnetic pulse compressor (MPC). First-stage semiconductor switching devices are used such as thyristor and insulated-gate bipolar transistor (IGBT). Usually, the ferromagnetic core of the toroidal shape is used for the magnetic switch SI_1, SI_2, and SI_3. Input voltage and energy from the charger to C_0 are 1.2 kV and 5 (J/p). A maximum output voltage of the MPC is 30 kV with pulse width 4 μs. Maximum pulse repetition rate is 180 pulses per second (pps) [48]. Figure 6.4 shows typical waveforms of the discharge voltage and the current. The voltage risetime is about 800 ns. After 800 ns, the voltage is decreased and increasing again. In this time, the current flows. After 6 μs, it can be recognized that reverse phases of the voltage and the current are almost the same from Fig. 6.4 [48].

Photograph of underwater streamer-like discharge and effects of streamer discharge are shown in Fig. 6.5 [48]. It can be confirmed that streamer-like discharge branches have progressed radially from tip of the electrode. Distance from tip to tip of the streamer-like discharge branches is about 30 mm. Underwater streamer-like discharge is effective in the generation of radicals, high-electric fields, ultraviolet rays, microbubbles, and shock waves.

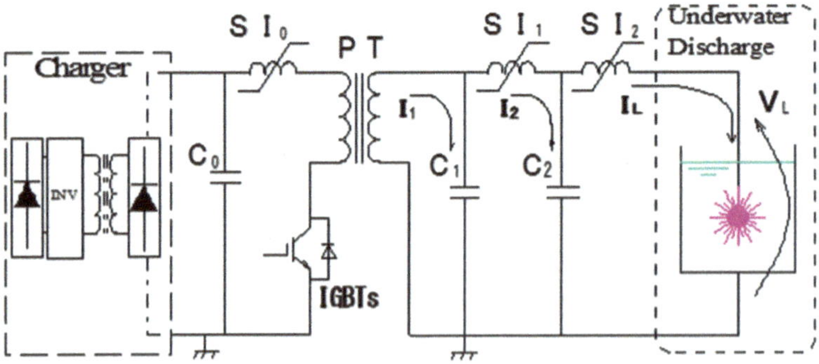

Fig. 6.3 Schematic diagram of pulsed power modulator for underwater streamer-like discharge

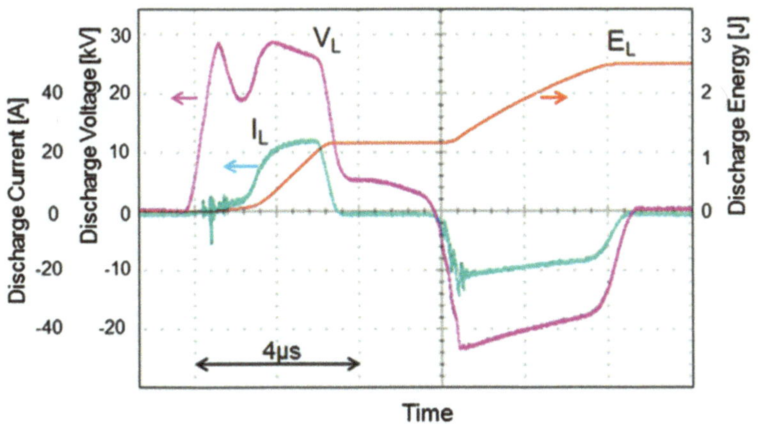

Fig. 6.4 Voltage and current waveforms of the underwater streamer-like discharge

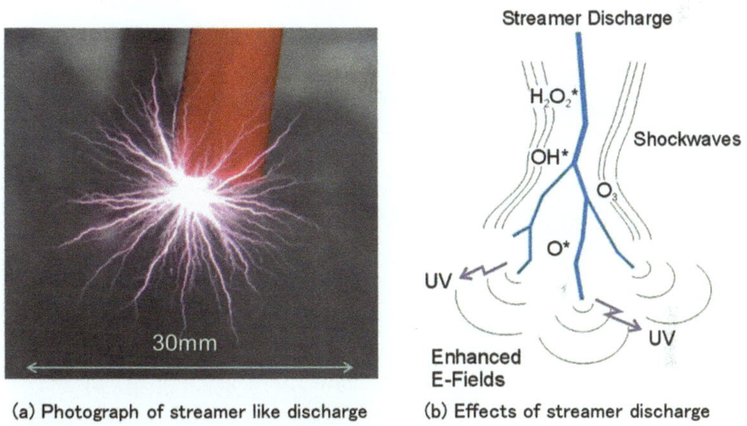

Fig. 6.5 Underwater streamer-like discharge

6.3.3 Method of Algae Treatment

Cyanobacteria, also called "blue-green algae," are relatively simple, primitive life forms closely related to bacteria. A diameter of the cyanobacterial cell is several micrometers. There are some honeycomb shape and hollow cylindrical structures inside the cytoplasm of cyanobacterial cells named as gas vesicles. Usually, cyanobacterial cells live in the surface of the water body. We have investigated that streamer-like discharge affects the cyanobacterial cells [48]. Figure 6.6 shows the effects of the streamer-like discharges for cyanobacterial cells. We have results of three kinds of experiments. Control is only stirred for 1 min. The second treatment is applied with 60 pulses and stirred for 1 min. The third treatment is applied with 600 pulses and stirred for 1 min. Volume of each water vessel is 1 l.

Fig. 6.6 Effects of the streamer-like discharges for cyanobacterial cells

A half day after treatment, the cells of the control rise to the surface of the water vessel. Some of the cells sink to the bottom of the water vessel by the second treatment. And almost all the cells sink to the bottom of the water vessel by the third treatment. We consider that the effect on gas vesicles of cell is assumed to be fast rising shock wave. We had results of laboratory experiments that UV and enhanced electric field do not affect the sedimentation of cyanobacterial cells [49]. Figure 6.7 shows TEM images of cyanobacterial cells. In Fig. 6.7b, gas vesicles are collapsed by applying streamer-like discharge [48].

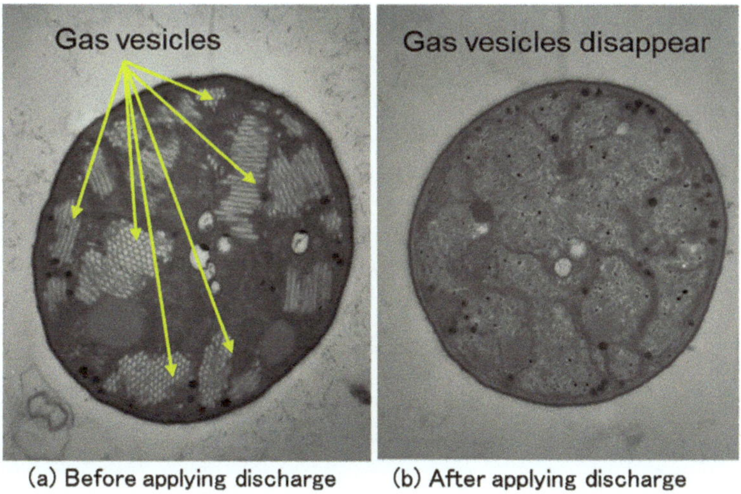

Fig. 6.7 TEM images of cyanobacterial cells

6.3.4 Algae Treatment Apparatus

Figure 6.8 shows a large-sized algae treatment apparatus constructed in 2006. This system becomes a product. The MPC is used as a pulsed power modulator [50–52]. Usually, pulsed power modulator for treatment of blue-green algae was using a voltage higher than 100 kV in order to increase volume of the streamer [51, 53, 54].

On the other hand, in recent study, we have developed a water intake mechanism that can treat the blue-green algae by the relatively small streamer-like discharge [48]. We were able to develop new practical blue-green algae treatment apparatus (named MIZUMORI-A2) using a low-voltage pulsed power modulator. The block diagram of the system is shown in Fig. 6.9. The MIZUMORI-A2 has photovoltaic cells as an independent power source. Maximum power generated by the photovoltaic cells is 518 W. Pulsed power system consists of the charger, IGBT switches, and MPC circuit. Pulse repetition rate of the modulator is 10 pps. Figure 6.10 shows the specifications and appearance of the MIZUMORI-A2. Photovoltaic cells are attached to the roof portion of the MIZUMORI-A2. There is a box of electrical components, including the MPC under the photovoltaic cells. In the center, streamer-like discharge generated place and algae intake of the MIZUMORI-A2 [48]. An anode of discharge electrode uses a tungsten diameter of 1 mm. It's covered with an insulating material with a diameter of 10 mm around the anode. A cathode of discharge electrode is made of stainless steel. In cylindrical shape, inner diameter is 104 mm. As the inner diameter of section of streamer-like discharge generation is 104 mm, water flow velocity of algae intake is 86 cm/s. Treatment flow rate is 26 m^3/h.

Fig. 6.8 A large-sized algae treatment system in 2006

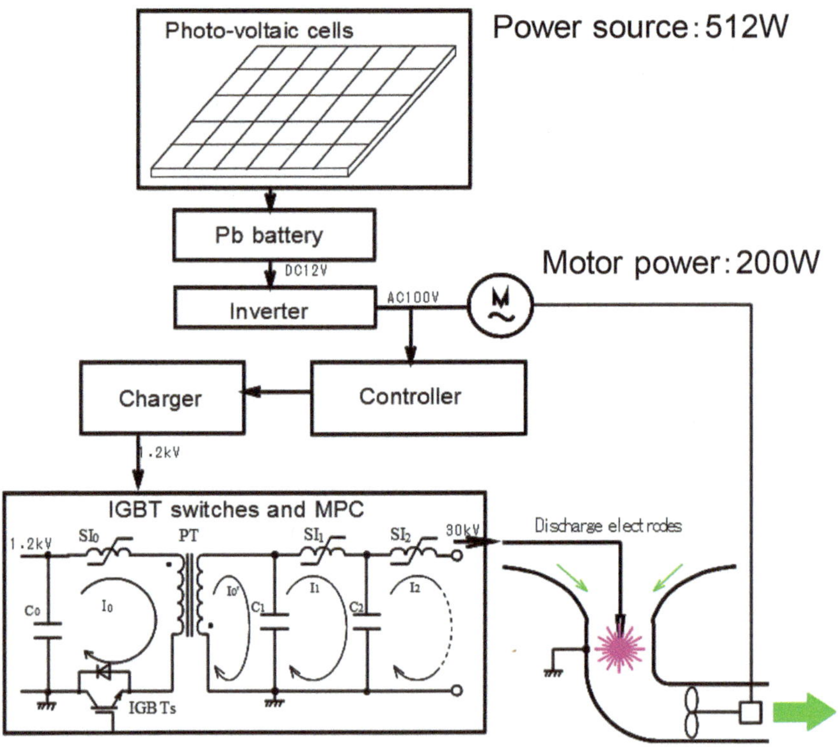

Fig. 6.9 Block diagram of algae treatment apparatus in 2012

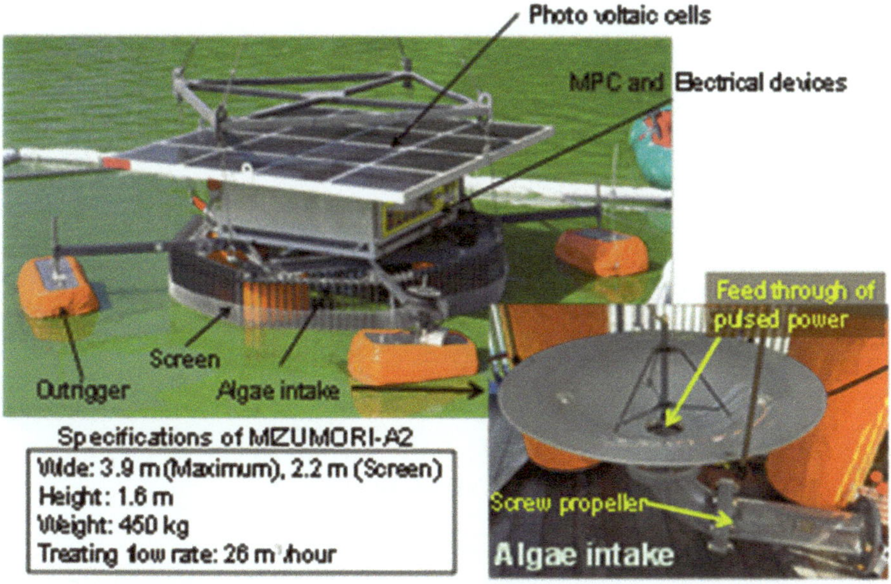

Fig. 6.10 Photograph of algae treatment apparatus in 2012

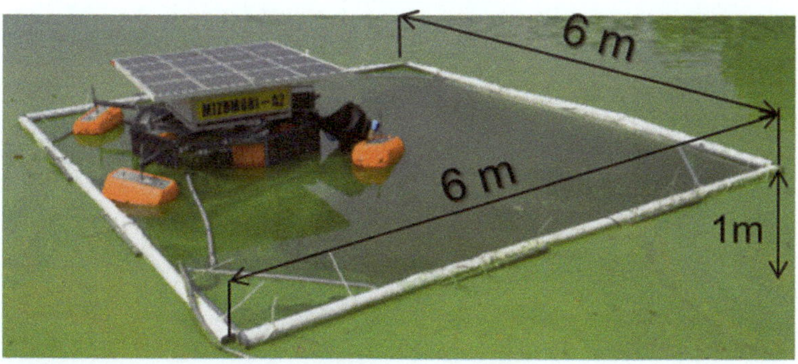

Fig. 6.11 Cyanobacterial bloom treatment at dam

We have demonstrated blue-green algae treatment by the MIZUMORI-A2 at the dam in Kyushu district, Japan [48]. Figure 6.11 shows the treating area by the MIZUMORI-A2. The treating area has a square of each side of 6 m. The area has been separated from the outside by a curtain fence of 1 m in depth. The MIZUMORI-A2 was carried out treating blue-green algae during 1 h at the treatment area. Pulse repetition rate of the modulator used in this experiment is 10 pps. We can recognize that blue-green alga has been lost from the surface of the water inside of the fence. Figure 6.12 shows experimental results of the treatment. In Fig. 6.12a, we found that chlorophyll concentration and turbidity decrease after the start of treatment by the MIZUMORI-A2. These results mean that *Cyanobacteria*

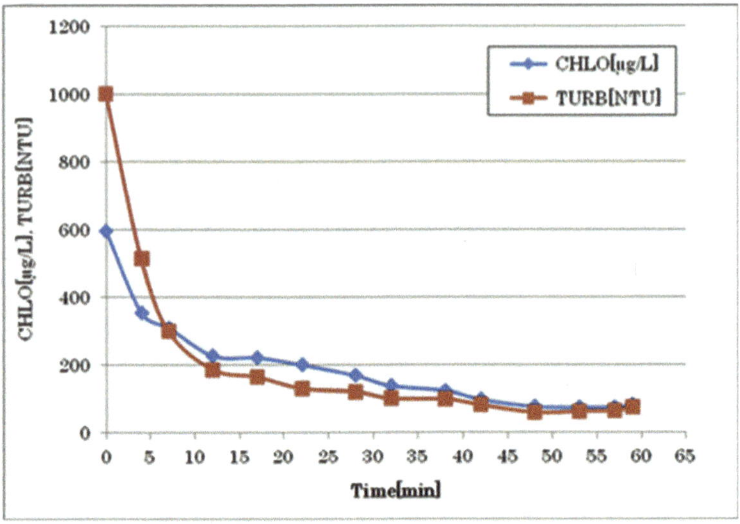

(a) Change of CHLO and TURB

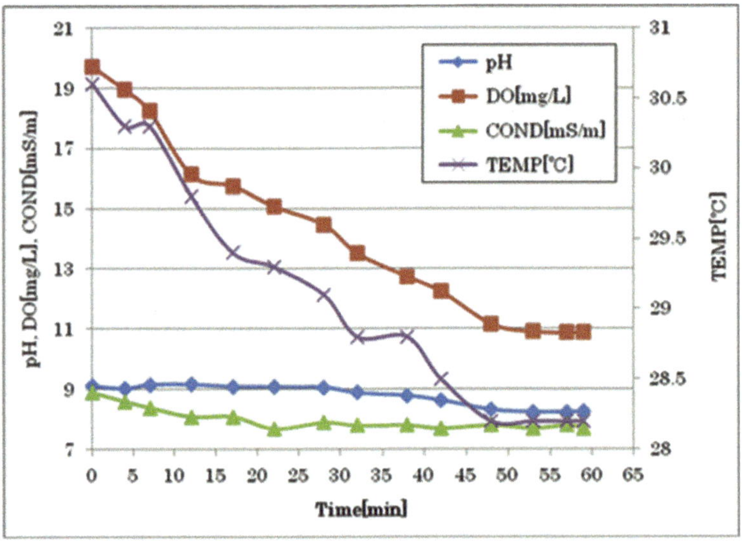

(b) Change of pH, DO, COND and TEMP

Fig. 6.12 The quality change of water by the cyanobacterial treatment apparatus (MIZUMORI-A2)

are precipitated under the influence of underwater streamer-like discharge. On the other hand, Fig. 6.12b shows potential of hydrogen (pH), dissolved oxygen (DO), conductivity (COND), and water temperature (TEMP). DO and COND are decreased gradually. We consider that reduction of DO indicates inactivated cyanobacterial cells or blue-green alga disappears from surface of water, because DO is increased by photosynthesis of the plant cell. Efficacy of *Cyanobacteria* treatment was confirmed by demonstrated experiments of the MIZUMORI-A2 at the dam.

6.4 Large-Scale PEF Processing of Sugar Beets and Grape Mash

Martin Sack, Juergen Sigler, and Georg Mueller

6.4.1 PEF Treatment of Sugar Beets

Pulsed electric field-assisted extraction of valuable substances from plant cells enables an energy-efficient processing compared to a conventional thermal process, fast processing, and a more complete extraction of ingredients.

For the production of sugar from sugar beets, saving energy is the main goal for the use of electroporation-assisted extraction. Other than for the production of sugar from sugar cane, the sugar production from sugar beets requires a considerable amount of energy covered by fossil fuels coal, gas, and oil. In the sugar beets, the sugar is stored inside vacuoles in the cells of the parenchyma tissue. The vacuole fills nearly the whole space of the cell, which has in average a diameter of 50 μm. Conventionally, the cells of the sugar beet tissue are opened thermally at a temperature of approximately 70–78 °C in order to prepare the tissue for the following extraction process. The thermal denaturation can be replaced by PEF treatment either of whole sugar beets or of cossettes [55–57]. For the extraction process, the cell membranes and the membranes of the vacuoles need to be opened. Figure 6.13 shows raw and permeabilized tissue in comparison. After PEF treatment, a part of the juice drains out of the cells due to the pressure inside the cells. The treatment of whole sugar beets has the advantage that the energy consumption for the subsequent slicing process and the wear of the blades inside the slicing machines are reduced. According to texture tests, cutting PEF-treated beets requires with 8 N approximately half of the mechanical force required for cutting fresh beets. A package of PEF-treated cossettes exhibits better perfusion properties. As single PEF-treated cossettes are more flexible than thermally treated cossettes, there is less fragmentation of the cossettes during transport and processing. The diffusion of sucrose molecules in an aqueous suspension is increased slightly. Moreover, the extracted juice exhibits a higher purity requiring less lime milk for purging. Hence, less lime stone and coke for the production of lime milk is required [57].

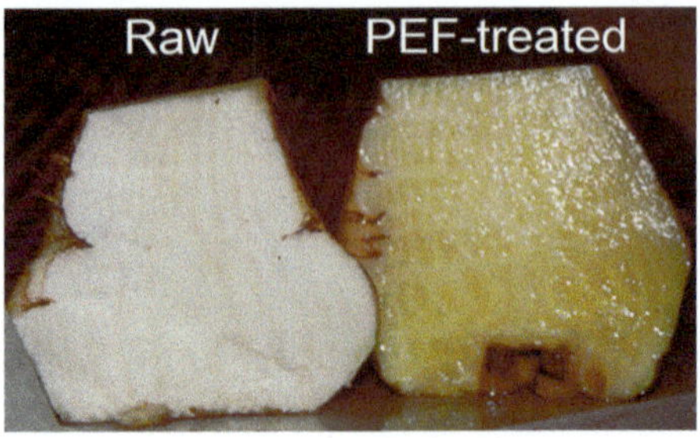

Fig. 6.13 Raw (*left*) and PEF-treated half of a sugar beet [346]

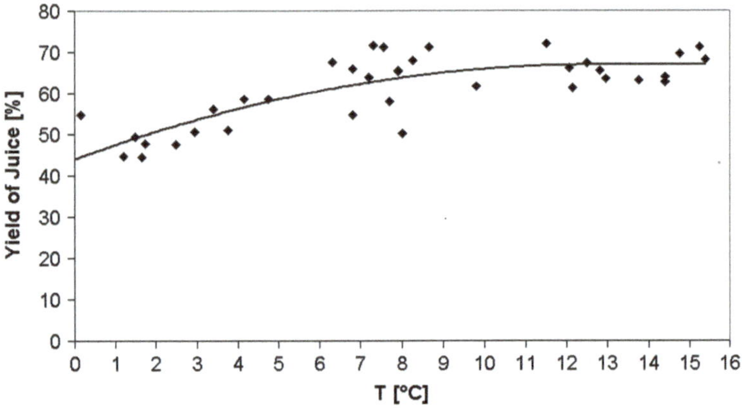

Fig. 6.14 Temperature dependency of the yield of juice after PEF treatment and subsequent pressing of slices of sugar beets ($\hat{E}=3$ kV/cm, 30 pulses) [77]

The combination of PEF treatment with an alkaline extraction enables more efficient dewatering. For alkaline extraction, lime milk is added to the PEF-treated cossettes before the extraction process. It causes strengthening of the cell walls and a chemical modification of the pectin of the beet tissue. Thus, the juice is able to drain better out of the cells. A dry matter content of approximately 40 % after pressing the cossettes has been achieved [57]. A typical value for thermally treated cossettes without alkaline extraction is 35 %. The increased dry matter content enables a reduction of energy required for a subsequent drying process of the cossettes.

However, the energy required for complete cell opening by pulsed electric fields depends on the temperature. Figure 6.14 shows the temperature dependence of the yield of juice after pressing for a PEF treatment with 30 pulses at $\hat{E}=3$ kV/cm. Experiments showed a significant decrease of the electroporation efficiency for a treatment temperature below 7 °C at an electric field strength of $\hat{E}=6$ kV/cm. In a sugar factory, thermal energy for warming up the sugar beets is available as waste heat. However, the surface-to-volume ratio of whole sugar beets is too small for a sufficiently fast heat transfer in the course of the production process. Hence, the PEF treatment of cossettes instead of whole sugar beets is preferred, as they can be warmed up in time. A processing temperature during PEF treatment of between approximately 10 °C and 30 °C for the cossettes is preferred, because with rising temperature also the conductivity of the suspension is increased causing a higher electric current during pulse application.

The temperature of the thin juice after extraction needs to be kept at approximately 60 °C in order to prevent the growth of mesophilic bacteria. In a conventional extraction process, the extraction starts at a temperature level of approximately 70 °C due to the requirements of the preceding thermal denaturation stage. The temperature is reduced during extraction due to thermal losses and increased again by adding hot water required for the extraction. For a combined process based on PEF treatment and alkaline extraction, an inverse temperature profile might be employed during the extraction process [58]. Starting with a temperature of approximately 45 °C, the cossettes are warmed up to approximately 60 °C during a countercurrent extraction process by the hot water required for the extraction. The alkaline extraction of PEF-treated cossettes requires less water. As a consequence, the draft, which is the ratio of juice to cossettes, can be reduced from approximately 105 % for the conventional process to less than 100 %. In experiments a draft of 90 % has been achieved [55, 57] Thus, less evaporation energy is required for the subsequent concentration process of the sugar solution. When using a combined process of PEF treatment and alkaline extraction, in total energy savings of more than 20 % of the primary energy consumption of a sugar factory can be expected. Moreover, the alkaline milieu due to the alkaline extraction prevents corrosion, which is an issue in the acid environment of pure sugar juice.

Apart from a conventional extraction process, a discontinuous process based on pressing of cossettes while a pulsed electric field is applied has been described [59]. An increased purity of the juice after PEF treatment has been observed.

Another approach for the sugar extraction is based on ultrafiltration [60]. After PEF treatment and pressing of the cossettes, the extracted juice is filtered by means of ultrafiltration. This process results in an undiluted juice without additional water from an extraction process. Moreover, there is no need for lime milk as for a conventional thermal process with countercurrent extraction and purging by lime milk.

Pulse shape, electric field strength, pulse length, and repetition rate of the pulse application vary depending on the employed technology. For the application of rectangular pulses with an electric field strength of 0.6 kV/cm and a pulse length between 7 and 10 ms at a temperature of 20 °C, a specific energy of 7.2–10.8 kJ/kg required for cell opening has been reported [79]. When applying an aperiodically damped pulse shape with an electric field strength of 5 kV/cm and a pulse length of approximately 2 μs at a temperature of 10–15 °C, a specific energy in the order of 4–5 kJ/kg is required for cell opening [57]. However, in both cases the cell membranes need to be charged to a value required for cell opening. Pulse length and a polarity reversal of the pulse have an influence on the electrochemical wear of the electroporation electrodes [61].

6.4.2 PEF Treatment of Grapes

For the production of wine from red and white grape varieties, valuable substances are extracted especially from the cells of the peel tissue. The pigments of most red grape varieties can be found in the grape skins only. They are stored in the cells inside vacuoles. Conventional methods for cell opening for red grape varieties comprise the fermentation on skins and thermovinification.

The fermentation on skins is a fermentation process, which takes at least several days up to 1, 2, or 3 weeks. The cells are opened by maceration, e.g., by enzymes. In the course of the fermentation, alcohol is produced, which supports the extraction of valuable substances to the must. In contrast to such an alcoholic extraction after thermovinification, an aqueous extraction takes place. For thermal denaturation, the destemmed and crushed grapes are heated up to a temperature of approximately 80 °C and kept at this temperature level for about 2 min and are subsequently cooled back to 45 °C or lower. The following extraction process is enhanced by the increased temperature. However, specific taste reminding of marmalade may occur due to the heating process. For white wines, often a light character is preferred. Therefore, gentle pressing and further processing are applied. However, added enzymes cannot always be removed completely in the course of the following production steps and can be found in the wine. The treatment of white grapes by pulsed electric fields enables fast and gentle processing at low temperature without additives.

Electroporation can be applied as self-contained method for cell permeabilization or as a supplementary processing step, especially in combination with fermentation on skins [62–65]. A comparative study of red wines produced from PEF treatment and thermovinification confirmed that equal contents of tannins and pigments for both disintegration methods could be achieved. PEF treatment of white grape varieties leads to an increased content of yeast available nitrogen, which helps in preventing the atypical aging defect in the course of the fermentation. Moreover, due to a better extraction of flavoring substances, the must has a higher content also of tannins, but less acidity compared with white wine produced by gently pressing the whole berries directly. The reason for less acidity is improved chemical buffering caused by a total increase of extracted substances [66].

A combination of PEF treatment of red grapes and subsequent fermentation on skins enables a more intense color of the must than it could be achieved by each method separately [65]. For such a combined process, the time for the fermentation on skins can be reduced to approximately 3 days without a substantial effect on the extraction result. Hence, a comparably fast and complete extraction is possible.

For an extraction of pigments and tannins from the grape skins, the cell membranes and vacuoles both need to be opened. Figure 6.15 shows a microscopic view of cells from the peel tissue of Lemberger grapes before and after cell permeabilization by pulsed electric fields. The extraction starts directly after the PEF treatment, so the must after PEF treatment is becoming red. The vacuoles of

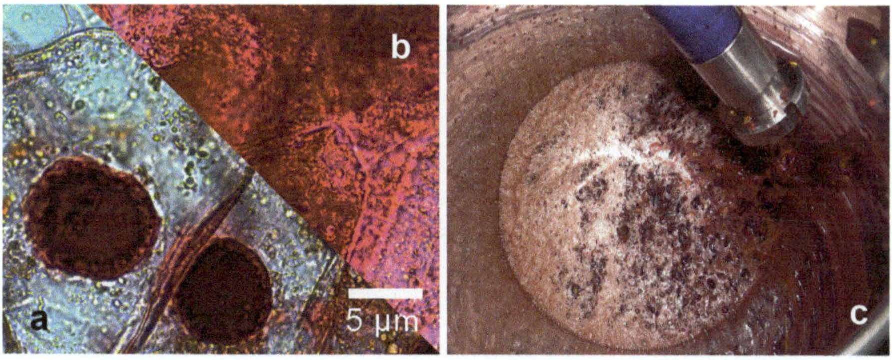

Fig. 6.15 Microscopic views of cells from the peel tissue of Lemberger grapes: (**a**) before PEF treatment; (**b**) after PEF treatment; (**c**) extraction of pigments to the must directly after PEF treatment [65, 347]

this grape variety have a diameter of approximately 5 μm. Whether a complete cell opening by PEF has been achieved can be evaluated by a comparison of the contents of pigments and tannins in the must of a PEF-treated sample with a sample denatured by thermovinification both with a subsequent extraction. The measurement of the disintegration index according to Angersbach [80] based on impedance measurements may not always correlate with the extraction result, because it is more sensitive to an opening of the cell membrane, while the extraction of pigments and tannins depends also on the degree of opening of the vacuoles [67, 68]. In a continuous process, cell opening is also caused to some extent by mechanical stress due to pumping. In order to distinguish this type of disintegration from the cell opening by pulsed electric fields, a sham treatment by pumping the crushed grapes through the treatment device with the pulse voltage being switched off should be included into the experiments [69]. Figure 6.16 shows extraction curves of must and the color of the wine for raw, untreated crushed grapes, grapes that have been pumped through the device only, and PEF-treated grapes. Processing of raw, untreated grapes results in a white wine made from a red grape variety (so-called blanc de noirs). For Pinot noir grapes, a complete cell opening has been achieved when applying an aperiodically damped pulse shape with an electric field strength of 20 kV/cm and a pulse length of approximately 1.5 μs at a temperature of 10–15 °C. It requires a specific energy in the order of 40 kJ/kg. For grape varieties in Spain, it has been reported that cell opening could be achieved with rectangular pulses at an electric field strength of $\hat{E} = 5$ kV/cm and a pulse length of 3 μs at an applied specific energy of 3.67 kJ/kg [348]. For grapes grown in Portugal, a treatment with rectangular pulses at an electric field strength of approximately $\hat{E} = 0.5$ kV/cm at pulse lengths between 610 μs and 1.53 ms at a flow rate between 11 and 28 t/h has been reported [70]. A specific energy of 0.4 kJ/kg has been applied to the grapes.

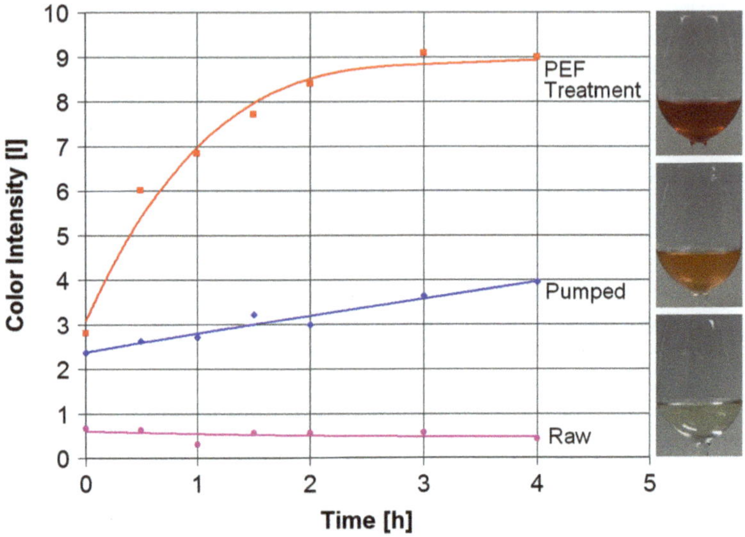

Fig. 6.16 Extraction curves for crushed Trollinger grapes and color of the wine: change of color intensity during extraction after PEF treatment, after pumping through the treatment device only, and untreated

6.4.3 Transport of Material

The transport of the material in large scale is performed as a package. The means for transport differ with the requirements of the material [71]. Whole sugar beets can be transported by means of a wheel equipped with rods made out of insulation material. Figure 6.17 shows such a wheel during assembly of the electroporation device. The distance of the rods is adapted to the size of the sugar beets in such a way that the package is moved as a whole rather than a few single beets only. A conveyor belt transports the beets to the top of the wheel. In order to establish a conductive connection to the electrodes for pulse application, the sugar beets are immersed into water, which fills the space around the lower half of the wheel. The wheel pushes the beets below the water surface and prevents them from floating.

Raw cossettes need to be transported gently in order to prevent them from breaking. Even in a vertical chute, the cossettes may block due to the formation of bridges, if too much pressure is applied. Therefore, the application of external pressure for the transport needs to be omitted.

Grapes also need to be transported gently through the PEF treatment device in order to keep mechanical damage as low as possible. Therefore, progressive cavity pumps or peristaltic pumps are used. Figure 6.18 shows an example for a flow scheme [72]. A minimum diameter of 40 mm is required for the grape transport. As progressive cavity pumps and peristaltic pumps do not convey the material continuously, an air chamber after the pump equalizes pressure variations and, hence, the transportation velocity. If during PEF treatment a high-electric field needs to be

Fig. 6.17 Wheel equipped with rods for the transport of whole sugar beets

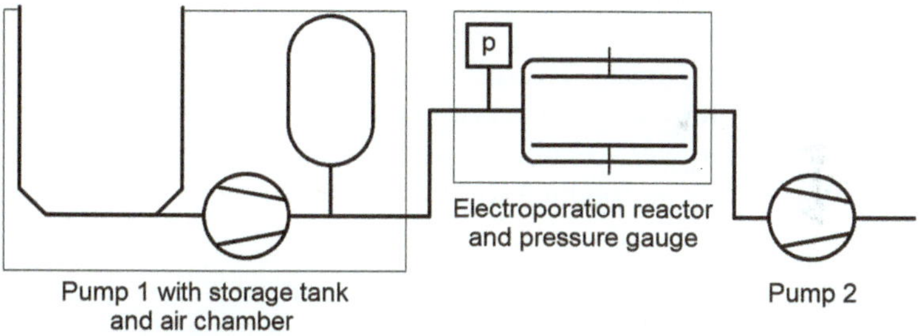

Fig. 6.18 Flow scheme of a PEF treatment device for crushed grapes [349]

applied, a flashover between the electrodes inside the treatment chamber has to be prevented. Air bubbles in the crushed grapes may initiate a flashover due to their lower breakdown strength. They might be removed partly by means of a degassing valve before the grapes enter the treatment chamber. Additionally, an increase of the pressure causes the bubbles to shrink and increases their breakdown strength. For pressure regulation, a second pump after the electroporation reactor might be used. A feedback control loop for the speeds of both pumps based on a pressure gauge keeps the pressure constant.

6.4.4 Electrode System for Pulsed Electric Field Application

The treatment chamber comprises an electrode system for applying the electric field to the material. For an energy-efficient operation to each volume element of material, approximately the same energy needs to be applied. If the material is transported through the electrode system as a package with equal velocity across the cross section, a substantially homogeneous electric field results in an equal energy distribution over the treatment volume. However, due to inhomogeneities at the borders of the electrodes near inlet and outlet of the treatment area, regions with low electric field outside the treatment area may occur, which cause additional losses [73]. A substantially homogeneous field might be established either by a plate electrode system according to Fig. 6.19a or by a collinear electrode arrangement as shown in Fig. 6.19b [74–76]. During the design, the size of the electrode system needs to be adapted, on one hand, to the requirements for the material flow and, on the other hand, to the other circuit elements of the pulse circuit to adjust the circuit to the required pulse shape. In the collinear electrode arrangement, the direction of the electric field is oriented in the direction of the material flow. So the electric resistance of the electrode system rises with the length of the PEF treatment area. For a plate electrode system, the electric resistance decreases inversely proportional to the length. When scaling the electrode system in both cases, the resistance varies inversely proportional with the scaling factor.

Due to the material supply at ground potential, the pulse circuit needs to be grounded at the treatment chamber in order to omit an electric current flow out of the treatment area. Additionally, for safety reasons, protective electrodes might be installed before and after the electroporation area, which are safely tied to ground potential [74, 75]. For a plate electrode system, a leakage current to ground can be prevented, if the voltage is supplied to both electrodes symmetrically to ground

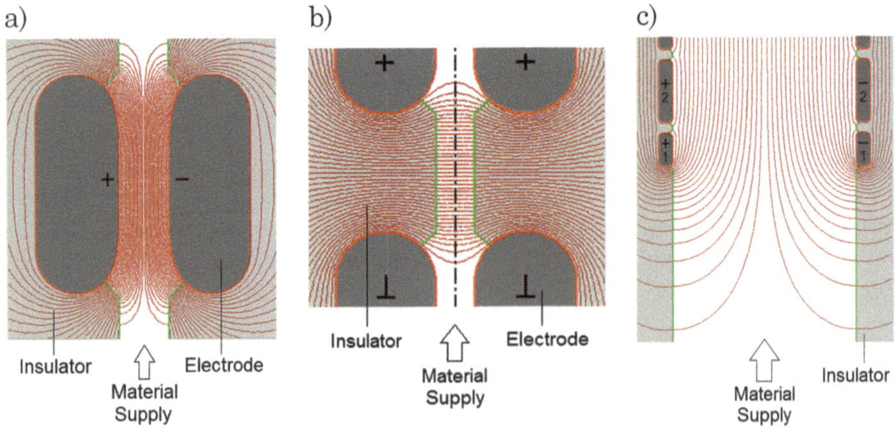

Fig. 6.19 (**a**) Plate electrode system, (**b**) collinear electrode system, (**c**) plate electrode system with separate pairs of electrodes (1, 2) each connected to one pulse generator

potential. So in the case of a homogeneous material distribution, the center of the treatment area will be at ground potential due to a resistive divider formed by both halves of the electrode system. Only in the case of an inhomogeneity in the material, there will be a considerable current flow out of the treatment zone to the protective electrodes. In order to achieve material supply at ground potential for a collinear electrode system, two electrode systems are put one after the other such that they share one common high-voltage electrode, while the other two electrodes are at ground potential. From an electric point of view, both electrode systems are switched in parallel. The voltage supply is done unsymmetrically to ground. So the insulation to ground needs to be designed for the total applied voltage. In the case of a supply symmetrically to ground, a more compact setup is possible, as the insulation to ground has to be designed for half of the total applied voltage only.

Especially for the application of higher electric field strength, the electrodes can be shaped such that the electric field at the edge is reduced. Usually, a radius or a combination of different radii is applied. The electrodes might be designed such that they provide a proper electric field distribution also inside the dielectric material surrounding the channel for material transport, where a capacitive electric field control is applied.

For large PEF treatment devices, the inductance of the pulse circuit might limit the current, which can be provided to the electrode system at a predefined pulse shape without considerable oscillations. In order to reduce the inductance pulse, generators might be connected in parallel configuration to the electrode system. Thus, the total inductance of the circuit is reduced according to the number of parallel current paths. However, if the switching time of at least one pulse generator exhibits a considerable jitter, energy oscillations between the generators may occur. In order to dampen such oscillations, the electrode system can be set up as separate pairs of electrodes with each pair connected to one pulse generator as shown in Fig. 6.19c. The pairs of electrodes are placed next to each other such that the electric field is still nearly homogeneous. The distance of two adjacent electrodes influences the electric resistance between both electrodes, which acts as damping element for the energy oscillations between different pulse generators. Each pair of electrodes has been designed for equal resistance. However, the pairs of electrodes at the inlet and at the outlet are much shorter, as they have to supply additionally the neighbored low-field regions.

Semiconductor-based pulse generators exhibit a negligibly small jitter of the switching moment. However, currently semiconductor switches for high switching power are still much more expensive than spark gap switches. Therefore, spark gap switches are for large-scale electroporation devices still a cost-effective alternative. If several Marx generators with spark gap switches are connected to a common PEF treatment chamber, they need to be synchronized to each other by triggering [73]. Marx generators equipped with spark gap switches might be advantageously triggered by means of overvoltage triggering of the first stage's spark gap. For overvoltage triggering, one or both charging coils between the first and second stage are replaced by pulse transformers, which are connected to semiconductor-based trigger pulse generators. For triggering by this arrangement, an additional pulse voltage is superimposed to the charging voltage of the spark gap. Subsequently, the remaining spark gaps ignite in the usual manner. For a low jitter, the first stage's spark gap is equipped with a corona wire. The high and inhomogeneous electric field in the vicinity of the wire causes the formation of a corona discharge, which emits ultraviolet light for seed electron generation in the electrode system. For an overvoltage-triggered seven-stage Marx generator, a jitter of the total switching time of less than 60 ns has been achieved [77].

6.5 Recovery of Valuable Components from Plants and Microorganisms

Eugene Vorobiev and Nikolai Lebovka

6.5.1 Basics of PEF-Induced Electroporation in Plants and Microorganisms

Application of pulsed electric field treatment (PEF) for recovery of valuable components from plants and microorganisms becomes more and more popular [81, 82]. The short-duration pulses (from several nanoseconds to several milliseconds) with amplitude from 100–300 V/cm to 10–50 kV/cm can affect the permeability and barrier properties of cell membranes. Usually, the effect of ohmic heating during PEF treatment is unessential, and there exists possibility of processing food materials at low temperatures. PEF treatment allows avoidance of undesirable changes in biological material, which is typical for other techniques, such as thermal, chemical, and enzymatic ones [83]. The supplementary advantage is possibility of microbial inactivation [84]. Modern PEF applications include intensification of separation, extraction, pressing, diffusion, and drying processes.

Impact of PEF on cells reflects losing of barrier functions by biomembranes. Traditionally, this phenomenon is called "electroporation," "electropermeabilization," or "electroplasmolysis" (see reference [85] and references cited therein). Efficient electroporation requires some threshold value of transmembrane potential, u_m, typically 0.5–1.5 V. Depending on conditions of treatment, the value of u_m, and PEF exposure time t_{PEF}, a temporary (reversible) or irreversible loss of barrier function can occur. The electroporation efficiency depends upon details of pulse protocol [81, 82]. The most important parameters are the electric field strength E and the total time of treatment t_{PEF}. For plant tissues, an important damage of rather large food tissue cells ($R \approx 30$–50 μm) can be observed at $E = 200$–1000 V/cm and treatment time within 10^{-4}–10^{-1} s [85], and for small microbial cells ($R \approx 1$–10 μm), larger field ($E = 20$–50 kV/cm) and smaller treatment time (10^{-5}–10^{-4} s) are required [84]. The experimental data evidence also that longer pulses are more effective and bipolar pulses are more advantageous [86]. Moreover, a complex protocol with adjustable long pause between the trains allows fine regulation of the disintegration of tissue without noticeable temperature elevation during the PEF treatment. The correlations between electroporation efficiency and electrical conductivity contrast $k = \sigma_d/\sigma_i$ (here, σ_d and σ_i are electrical conductivities of completely damaged and intact tissues, respectively) were also experimentally observed and reasonably explained on the base of electroporation theory [87].

In general, the electroporation effect does not require high-power consumption, and it stipulates the industrial attractiveness of PEF treatment. Experiments and theory evidence that minimal power consumption Q is observed at certain optimum electric field strength E_{op}. For vegetable and fruit tissues (apple, potato, cucumber, aubergine, pear, banana, and carrot), the typical values of E_{op} were within 200–700 V/cm and PEF treatment times t_{PEF}, required for effective damage, were within 1000 μs–0.1 s [82, 86]. The estimated power consumptions Q for PEF-treated tissues were found to be rather low and typically lying within 1–15 kJ/kg. It is rather low as compared to other methods of treatment like mechanical (20–40 kJ/kg), enzymatic (60–100 kJ/kg), and heating or freezing/thawing (>100 kJ/kg).

6.5.2 *Plants*

Among very promising potential applications of PEF are technologies of "cold" extraction and recovery of valuable components from fresh food plants and microorganisms, e.g., sucrose extraction from sugar beetroot, betanin extraction from red beet, inulin extraction from chicory, beta-carotene extraction from carrot, phenolics extraction from grapes, etc.

6.5.2.1 Potato and Apple

Potatoes and apples were frequently used as model systems for testing of different PEF-induced effects on plant tissues. For example, the effects of reversible electroporation, transient viscoelastic behavior, stress-induced effects, metabolic responses, and electrostimulated effects were intensively studied for these plants [87–91].

Potato was used in studies of textural and compressive properties of PEF-treated tissue [90]. PEF treatment allowed elimination of the textural strength to some extent, and the effects were more pronounced after mild thermal pretreatment at 45–55 °C. Experimental data for PEF-treated apple tissue were used for testing of the theoretical models of PEF-induced effects [89]. Potato and apple were widely used also in investigations of temperature and PEF protocol effects on characteristic damage time, dehydration, freezing, and drying [87–89, 91]. The effect of the synergy of PEF and thermal treatments on the textural properties of apple tissue and apple juice expression was demonstrated [91]. It was shown that mild thermal treatment allows increase of the damage efficiency of PEF treatment, and apple tissue preheated at 50 °C and treated by PEF at $E \approx 500$ V/cm exhibits a noticeable enhancement of juice extraction by pressing.

The effects of apple tissue anisotropy and orientation with respect to the applied electric field on electropermeabilization were reported [92]. It was shown that elongated cells (taken from the inner region of the apple parenchyma) responded to the electric field in a different manner than round cells, while no field orientation dependence was observed for round cells (taken from the outer region of parenchyma). The textural relaxation data support higher apple damage efficiency at longer pulse duration.

The attempts of practical application of PEF-assisted recovery of valuable components from potato and apple were also done. PEF treatment was used for facilitation of starch extraction from potato, and enhancement of the extractability of an anthocyanin-rich pigment was reported [93]. PEF treatment is also useful for production of the apple juice with increased pressing yield without using of enzymes [94]. PEF application improved noticeably the juice yield and soluble matter content in the juice. It was demonstrated that juice characteristics and yield are directly related to the size of slices [91]. Juice yield Y increased significantly after PEF treatment of large apple mash ($Y = 71.4$ %) at $E = 450$ V/cm as compared with the check sample with small apple mash (45.6 %) [94]. The PEF pretreatment was accompanied by a noticeable improvement of the apple juice clarity, an increase of the total soluble matter and the content of polyphenols, and intensification of the antioxidant capacities of juice.

6.5.2.2 Roots

6.5.2.2.1 Sugar Beet

Traditional recovery of sugar from sugar beet utilizes a power-consuming hot water diffusion of sugar from sliced cossettes at 70–75 °C and a very complex multi-staged juice purification using big quantity of lime [95]. PEF treatment has large potential for replacement or modification of the conventional thermal technology for sugar extraction and purification. Application of PEF treatment allowed recovery of sucrose even at ambient temperature, "cold" pressing of the PEF-treated sugar beet cossettes, preservation of thermal degradation of cell walls, and binding of pectins in the cellular matrix [96–109]. It was shown that up to 82 % of the overall yield could be achieved by two-stage pressing with an intermediate PEF application. The "cold" juices, expressed from the sugar beet slices after the intermediate PEF treatment, had higher purity (95–98 %) as compared to those before PEF application (90–93 %). Additionally, the quantity of pectin was noticeably lower, and the color of juice was systematically three to four times less intensive than the color of factory juices [21]. PEF-assisted "cold" extraction resulted in lower concentration of colloidal impurities (especially pectins), lower coloration, and better filterability of juice [109–111]. It is expected that PEF treatment will simplify (or even eliminate) the very complicated and polluting carbonic purification process existing today in sugar production [109–112]. The scale-up study of PEF-assisted aqueous recovery of sugar using a pilot countercurrent extractor with 14 extraction sections was reported [109]. The estimated energy gain for cold extraction with temperature reduction from 70 to 30 °C (i.e., by $\Delta T = 40$ °C) was ≈46.7 kW.h/t, and it was noticeably higher than the power consumption, required for PEF treatment, ≈5.4 kWh/t.

6.5.2.2.2 Red Beet

PEF treatment allowed achievement of the good recovery of betalains (≈90 %) from red beet (*Beta vulgaris* L.) [113, 114]. The combination of PEF (7 kV/cm) and pressing permitted 18-fold shortening of extraction time and ≈4-fold increase of betanin yield [113]. Effects of PEF (400–600 V/cm) and thermal (30–80 °C) treatments on degradation of colorants and recovery kinetics were also discussed [114, 115]. The positive effect of PEF treatment in improvement of extraction and decrease of degradation was observed. PEF-assisted "cold" recovery at 30 °C allowed reaching of the high yield of colorants (≈95 %) at lower level of colorant destruction (≈10 %).

6.5.2.2.3 Chicory

Traditional industrial production of inulin from chicory roots (*Cichorium intybus*) needs high-temperature diffusion. The effects of PEF treatment (100–600 V/cm) on efficiency of recovery of soluble matters (inulin, sucrose, proteins) from chicory were studied [116]. The benefits of the PEF application for "cold" recovery soluble matter extraction from chicory were demonstrated. It was shown that diffusion activation energy of the usual thermal damage was rather high (more than 200 kJ/mol). However, it could be noticeably reduced to about 20 kJ/mol by PEF treatment.

6.5.2.2.4 Rhizome

PEF treatment was applied to *Podophyllum peltatum* in order to enhance recovery of podophyllotoxin, which is valuable for the treatment of cancer and venereal warts [117]. Conventional extraction of this chemical is inefficient and involves mechanical fragmentation of dehydrated rhizomes followed by solvent extraction at warm temperatures. The dried rhizomes of *P. peltatum* were soaked in deionized water and then were PEF treated (17.7, 19.4 kV/cm). The color of the sample changed after PEF treatment from sand yellow to deep red. The data demonstrated significant increase (up to 47 %) of podophyllotoxin concentration as compared with control samples.

6.5.2.2.5 Ginseng

The effects of PEF on the drying of ginseng (at 55 °C) and hot water (95 °C) extraction of dried ginseng were studied [118]. PEF was applied to the fresh ginseng using 1- and 2-kV/cm electric field strengths, 30-μs pulse duration, 25- and 200-Hz frequencies, and a pulse number of 175. Such PEF treatment resulted in reduction (≈38 %) of the drying time. It also increased the soluble solid content and significantly reduced the sugar content of the extract as compared to samples that were not treated by PEF.

6.5.2.3 Grape

PEF treatment may be rather promising for different applications in winemaking industry. PEF-assisted pressing of wine grapes improved must expression [91] and enhanced compression kinetics [90]. For PEF-treated ($E = 750$ V/cm) white grapes, the final juice yield increased up to 73–78 % as compared to 49–54 % for the untreated grapes (Muscadelle, Sauvignon and Semillon) [91]. Moreover, the juices extracted from PEF-treated grapes were less turbid (clearer) and didn't require filtration through polluting filter aids (such as diatomite).

In the case of red grapes, application of PEF treatment is important for enhancement of the recovery of colorants, aromatic compounds, and phenolic bays [119, 120]. Phenolic compounds play an important role in enology owing to their contribution to the sensory properties of wine. The effects of different pretreatments (moderate thermal, ultrasound, and PEF) on the phenolics extraction from Cabernet Franc grapes were compared [121]. The results show that all pretreatments improve phenolic extraction (content of anthocyanins and tannins), color intensity, and scavenging activity of the samples during red fermentation. However, the PEF pretreatment was the most effective. For example, pretreatment at 0.8 kV/cm and 5 kV/cm increased the yield of phenolic extraction by 51 % and 62 %, respectively. The effects of pulsed ohmic heating (POH, 100–800 V/cm) on extraction of polyphenols from red grape pomace were studied [119]. The highest yields were obtained after POH pretreatment at 400 V/cm followed by 60 min diffusion at 50 °C and using solvent composed of 30 % of ethanol in water. PEF treatment of Merlot grapes (500–700 V/cm) has demonstrated positive impact on the evolution of color intensity and content of anthocyanins and phenolic during the alcoholic fermentation [122]. Sensory analysis indicates that PEF treatment contributes to the enhancement of the sensory attributes of wine.

The important potential application of PEF is also related with extraction of polyphenols from grape by-products (skins, seeds) [123]. The effects of PEF treatment on Cabernet Sauvignon grape skin histocytological structures and on the organization of skin cell wall polysaccharides and tannins were studied using two different protocols: PEF1 (4 kV/cm and 1 ms) and PEF2 (0.7 kV/cm and 200 ms). It was shown that PEF1 protocol had little effect on the polyphenol structure and pectic fraction, and PEF2 protocol profoundly modified the organization of skin cell walls [124, 125]. Application of PEF1 protocol resulted in alteration of the visual appearance of phenolic compounds in the skins and led to increased extraction of the anthocyanins (19 %). From the other hand, application of PEF2 protocol resulted in changes in the structure of grape skins, and produced wine was richer in tannins (34 %). It was demonstrated that changes in the operating parameters of the PEF treatment did not affect the alcohol content, total acidity, or volatile acidity in the finished wines compared to the control wine.

The polyphenol extractions from grape seeds after three different pretreatments by pulsed electric field (PEF) (8–20 kV/cm, 0–20 ms), high-voltage electrical discharges (HVED) (10 kA/40 kV, 1 ms), and grinding (180 W, 40 s) were compared [126]. The PEF efficiency was higher when the treatment was performed at 50 °C in the presence of ethanol. The subsequent solid–liquid separation was faster after PEF treatment as compared to ground and HVED treatments.

PEF treatment application is promising for reduction of the maceration time during vinification and production of wines with better characteristics [120]. The effect of PEF treatment on the cold maceration (6 days at 6 °C) of Cabernet franc and Cabernet Sauvignon grapes was investigated [127]. The wines obtained from PEF-treated musts had higher phenolic content and color intensity during the alcoholic fermentation period than wines obtained from the untreated musts. Recently, the potential of PEF treatment during winemaking was tested on industrial scale [128]. PEF treatment (4.3 kV/cm, 60 μs) was done in continuous mode using collinear treatment chamber (1900 kg/h). It was demonstrated that after 7 days of maceration, the color intensity, anthocyanin content, and polyphenol index in the tank, containing grapes treated by PEF, were higher by 12.5 %, 25 %, and 23.5 %, respectively, than in the tank containing untreated grapes. Finally, the capability of PEF to inactivate the wine spoilage microorganisms may be also important [129]. It allows the fine control of fermentation and enhancement of the quality of wines.

6.5.2.4 Mushroom

Application of PEF-assisted pressure and solvent extractions for recovery of total polyphenols, polysaccharides, and proteins from the mushrooms (*Agaricus bisporus*) was studied [130]. The traditional hot water or ethanol extractions resulted in cloudy extracts with low colloid stability. It was demonstrated that extracts produced by PEF-assisted pressure extraction (PE+PEF) were clear, and their colloid stability was high. In general, PE + PEF allowed production of mushroom extracts with high contents of fresh-like proteins and polysaccharides. PEF-assisted extraction technique was used to optimize conditions of extraction of exopolysaccharides (EPS) from Tibetan spiritual mushroom broth [131]. The results show that the optimal conditions of such extraction technique are 40 kV/cm electric field intensity, the number of pulses 8, and pH 7. The effect of different factors on EPS extraction increases in the following order: electric field intensity > pH > number of pulses. The optimal conditions increased the EPS extraction by 84.3 % compared to that of the control group.

6.5.2.5 Tea Leaves and Wine Shoots

Thin slices of the fresh tea leaves were subjected to PEF treatment (400–1100 V/cm) in order to recover polyphenols [132]. PEF treatment accelerated kinetics of extraction, and the maximum recovery yield (≈27 %) was reached at 900 V/cm. PEF-assisted extraction (tea leaves/water ratio of 1:16, 20 kV/cm, and pulse frequency of 125 Hz) was applied for production of high-aroma instant tea powder [132]. This method allowed avoiding of the tea aroma losses during thermal processing of tea beverage production. The product obtained using the proposed method had an excellent quality and good aromatic characteristics. The application of combined PEF and freeze concentration technology to tea soup was studied [133]. The optimal conditions of PEF extraction were obtained using electric field strength of 37 kV/cm and solid–liquid ratio of 1:30.

The effects of different physical treatments (PEF, high-voltage electrical discharges, and ultrasound US) on intensification of polyphenol and protein recovery from the vine shoots were studied [134]. It was demonstrated that HVED had the highest polyphenol and protein recovery yields with the lowest energetic prerequisite.

6.5.2.6 Herbal and Flowering Plants

The PEF-assisted extraction (20 kV/cm) by 90 % ethanol–water solution was applied for enhancement of recovery of an alkaloid (Guanfu base A, GFA) from Chinese medicinal herb *Aconitum coreanum* [135, 136]. It demonstrated the highest yield of GFA (3.94 mg g^{-1}) with the shortest extraction time (0.5–1 min) and the lowest energy costs as compared to other extraction methods (cold maceration extraction, percolation extraction, heat reflux extraction (HRE), and ultrasonic-assisted extraction (UE)). PEF treatment (5 kV/cm) resulted in enhancement of recovery of the major components (crocin, color; safranal, flavor; and picrocrocin, taste) from stigma and pomace of saffron (*Crocus sativus*) [137]. PEF treatment (30 kV/cm) was applied for reaching the optimized extraction of polysaccharides from the corn silk [138]. Corn silk is a traditional Chinese herbal medicine, which is rich in antioxidants, polyphenols, vitamins (vitamin K, C), and minerals. The yield of polysaccharides under the optimal extraction conditions was ≈7 %. The microwave and PEF-assisted extraction of polyphenols from defatted hemp seed cake by the mixed methanol, acetone, and water solvent (MAW, 7:7:6 v/v/v) was compared. It was suggested that microwave and PEF treatment can be integrated to enhance polyphenol extraction and maximize the yield [139]. A critical review on application of different pretreatments for the efficient extraction of bioactive compounds from herbal plants was recently presented [140].

6.5.2.7 Other Plants

PEF treatment was also applied for recovery of valuable components from other plants, e.g., carrot, red bell pepper and paprika, fennel, alfalfa, and red cabbage. A possibility of PEF-assisted selective recovery of water-soluble components (soluble sugars) and production of a "sugar-free" concentrate, rich in vitamins and carotenoids, was demonstrated for carrot [90]. The positive effect of PEF treatment (0.25–1.0 kV/cm) of the carrot purees on extraction of polyacetylene and sugars, color changes, and total carotenoid content was also shown [141]. The effect of PEF treatment (0.1–1 kV/cm and frequency of 5–75 Hz) on extraction of carotenoids from the carrot pomace was examined using different vegetable oils (sunflower, soya bean, and peanut) [142]. It was demonstrated PEF effects on the extractability of carotenoids from the carrot were dependent on the electric field strength, frequency, and type of vegetable oil. The stable microemulsions were used to extract β-carotene from PEF-treated carrot pomace [143]. The content of β-carotene, extracted from PEF-treated carrot pomace using microemulsions, was higher than when it was extracted from untreated pomace. The high recovery yield (up to 96.7 %) of lycopene from tomato residual was achieved using PEF treatment with electric field strength 30 kV/cm, liquid–stuff ratio 9 ml/g, temperature 30 °C, and ethyl acetate as an extraction solvent [144]. The proposed method was proved to be a fast way for lycopene extraction from the tomato residual. PEF treatment (2.5 kV/cm) applied to mashed red cabbage in a batch treatment chamber allowed enhancement of the total anthocyanin recovery by 2.15 times [145].

6.5.3 Microorganisms

6.5.3.1 Yeast Cells

The yeast cells were used in many works as a model microorganism for testing the effects of electrical stimulation, permeabilization of cell membranes, and cell lysis. Depending on the protocol of PEF treatment, inactivation [146] or partial electropermeabilization of yeast cells [147, 148] is possible. In addition, PEF treatment can be useful for recovery of the high-quality intracellular components (ions, saccharides, enzymes, proteins, and nucleic acids) from the yeast cells [149–151]. The most effective method for recovery of valuable substances is the intensive high-pressure homogenization (HPH); however, it produces not pure extracts. It was shown that PEF treatment combined with mild HPH can be used for more efficient extraction of proteins [151]. For example, PEF treatment (10 kV/cm) of the aqueous suspension of wine yeast allowed high extraction of ionic components and low extraction of high-molecular-weight components [150]. PEF treatment (40 kV/cm) of the aqueous suspension of the same wine yeast allowed extraction of the 70 % of ionic substances, 1 % of proteins, and 16 % of nucleic acids [151]. It was demonstrated that PEF and high-voltage electrical discharge (HVED) treatments always resulted in incomplete recovery from yeast cells, though efficiency of HVED was higher than that of PEF [151]. For example, treatment at $E = 40$ kV/cm allowed extraction of $\approx$80 % and $\approx$70 % of ionic substances, $\approx$4 % and $\approx$1 % of proteins, and $\approx$30 % and $\approx$16 % of nucleic acids in cases of HVED and PEF treatments, respectively. Recently, PEF treatment was applied for assistance of recovery of valuable components from waste brewing yeasts [152, 153]. For PEF treatment at $\approx$20 kV/cm, the rate of trehalose recovery was up to 2.635 % and was 15.96 times higher than that of extraction microwave and 34.08 times higher than for ultrasound technique. For PEF treatment at 50 kV/cm, recovery of RNA in ethanol extraction was 1.69 times higher than that of water extraction.

6.5.3.2 Microalgae

Microalgae have high content of lipids, proteins, polyunsaturated fatty acids, carotenoids, valuable pigments, and vitamins and can be used in the food, feed, cosmetics, pharmaceutical, and biofuel industries. Traditional methods for extraction of these components use environmentally toxic solvents. Recently, several groups have developed PEF-assisted techniques for extraction of valuable components from microalgae. PEF treatment (23–43 kV/cm) in a flow cell was applied to microalgae *Auxenochlorella protothecoides* suspension [154]. It was demonstrated that PEF-assisted extraction was highly selective and allowed release of the soluble intracellular matter, while extraction of lipids required application of solvents. The opportunity of using PEF treatment at the first step of extraction and solvents at the second step of extraction was demonstrated. Continuous PEF treatment was studied, and a flow technology was proposed for extraction of the total of cytoplasmic proteins from microalgae (*Nannochloropsis salina* and *Chlorella vulgaris*) [155]. The technology was developed on the preindustrial pilot scale that allows easy treatment of large volumes. Effective extraction was observed in the case when PEF was followed by a 24-h incubation period in a salty buffer. PEF treatment was applied to enhance lipid extraction from *Ankistrodesmus falcatus* wet biomass using the green solvent ethyl acetate. Application of PEF allowed significant enhancement of the rate of lipid recovery. It was noted that the increase in lipid recovery was due to the electroporation and not due to temperature effects [156]. Experimental data evidence that electrically based PEF (20 kV/cm)- and HVED (40 kV/cm)-assisted techniques allow selective recovery of water-soluble ionic components and microelements, small-molecular-weight compounds, and water-soluble proteins from microalgae *Nannochloropsis* sp. [157]. The increase in lipid recovery for PEF-treated microalgae combined with ethyl acetate was explained by electroporation [156]. PEF treatment of microalgae also resulted in significant increase of the yield of fatty acids [158]. PEF-assisted technology was used as an effective tool for extraction of proteins from microalgae [155]. The possibility of enhancement of the extraction yield of pigments (chlorophyll and carotenoid) from microalgae by PEF treatment was also shown [159]. PEF treatment was also proposed as an effective tool for control of predators in industrial scale microalgae cultures [160]. It was shown PEF can be used for selective elimination of such contaminant as protozoa that can highly jeopardize the productivity of the culture.

6.6 Bacterial Inactivation for Food Preservation

Ignacio Álvarez and Javier Raso

Pulsed electric field technology (PEF) is viewed as one of the most promising nonthermal methods for inactivating microorganisms in liquid foods. The local defects or pores created by the application of an external electric field led to the loss of the microbial membrane's integrity abolishing its capacity to maintain the microbial homeostasis. The treatment causes the inactivation of pathogenic and spoilage microorganisms, but it also results in the retention of flavor, nutrients, and the color of the food compared to thermal processing. As bacterial spores are resistant to PEF treatments, the main applications of this technology for food preservation must be focused on pasteurization. Commercialization of PEF technology as a pasteurization process requires the estimation of its efficacy against pathogenic and spoilage foodborne microorganisms. This chapter reviews the current state of the art in microbial inactivation by PEF. Particular attention is devoted to the microbial inactivation mechanisms and the different factors influencing the microbial inactivation by PEF.

6.6.1 Introduction

Most food products constitute a rich nutrient source for microbial development. Microbial growth of microorganisms causes food spoilage as a consequence of metabolic activities leading to the production of molecules that alter the sensorial attributes of foods. The microbial deterioration of food is evidenced by slime formation and changes in the appearance, texture, color, odor, and flavor [161]. More important, foods may be sources of foodborne pathogenic microorganisms. The consumption of foods contaminated with these poisoning microorganisms and without generally presenting any sign of food spoilage may cause serious illness outbreaks. While microbial food spoilage is a huge economic problem (it is estimated that about 25 % of the world's food supply is lost from microbial spoilage), foodborne illnesses are an enormous public health concern worldwide with severe direct and indirect economic consequences.

Food preservation technologies used in the food industry aim at combating the deleterious effects of microorganisms in foods and avoiding foodborne illnesses. These technologies act by preventing microbial growth or by microbial killing. Metabolic activity of the microorganisms may be inhibited or slowed through those factors such as temperature, water activity, preservatives, pH, or atmosphere that most effectively influence the growth and survival of microorganisms in foods. Chilling, freezing, drying, modified atmosphere packing, acidifying, and adding preservatives are examples of preservation techniques based on preventing or slowing microbial growth. Among the many techniques used to preserve foods, a much smaller number rely on killing microorganisms. These techniques that inactivate microorganisms are more effective to ensure that potentially hazardous levels of microorganisms are not present in foods at the time of consumption. Pasteurization and sterilization by heat are the methods of microbial destruction traditionally used in the food industry. These techniques are very effective for microbial inactivation. However excessive heat treatment may cause undesirable effects on foods such as protein denaturation, nonenzymatic browning, and loss of vitamins and volatile flavor compounds [162]. In order to reduce the negative effects of the heat treatments in foods, alternative technologies capable of inactivating microorganisms at temperatures below those used during thermal processing are being demanded by the food industry. These technologies, called nonthermal technologies, not only inactivate pathogenic and spoilage microorganisms but also result in the retention of flavor, nutrients, and color of foods compared to thermal processing [45].

Pulsed electric field technology (PEF) is viewed as one of the most promising nonthermal methods for inactivating microorganisms in liquid foods. Electroporation of microbial membranes has the ability to effectively inactivate vegetative forms of microorganisms, thus extending foods' shelf life and enhancing microbial food safety without compromising the nutritional and sensory characteristics of the foods.

6.6.2 Mechanism of Microbial Inactivation by PEF

Although the mechanism underlying microbial inactivation by PEF has not been fully elucidated, it is believed that the formation of local defects or pores that leads to an increment of the cell membrane permeability to ions and macromolecules is the main cause of microbial killing.

The basic function of the cell membrane is to protect the cell from its surroundings. The cell membrane separates the microbial cytoplasm from the outside environment, acting as a selectively permeable barrier to ions and organic molecules and controlling the movement of substances in and out of the cells. The maintenance of the microbial homeostasis requires that the cytoplasmic membrane acts as an intact semipermeable barrier under fluctuating external conditions. The local defects or pores created by the application of an external electric field lead to the loss of the membrane integrity, and uncontrolled molecular transport across the membrane may occur. These events may abolish the homeostatic capacity of the cells and will eventually lead to microbial death. Membrane damage presumably becomes lethal if the damage is of a nature that precludes resealing or if irreversible effects occur as a secondary consequence of the loss of permeability control, energy conservation, or other membrane functions [163].

However, in microorganisms including bacteria and yeast, the cytoplasmic membrane is not the only barrier that separates the cytoplasm from the environment. The cytoplasmic membrane of Gram-positive bacteria is surrounded by a thick cell wall made of peptidoglycans and tectonic acids. On the other hand, the cell wall in Gram-negative bacteria is thinner, but it is surrounded by an outer membrane that differs from typical biological membrane because the main molecular constituents of the external lipid bilayer are lipopolysaccharides. This outer membrane prevents the entrance of some molecules such as antibiotics, lytic enzymes, or bacteriocins but allows low-molecular-weight nutrients to diffuse into the periplasmic space. On the other hand, yeast cells are surrounded by a cell wall similar to Gram-positive bacteria. Important differences in the effect caused by PEF in Gram-positive and Gram-negative bacteria have been described. However, how the envelopes surrounding the cytoplasmic membrane influence electroporation is an aspect that requires further research.

Different techniques such as detection of leakage of intracellular material, measurement of osmotic response, or fluorescent dye exclusion assays have been used to evidence the electroporation of microbial cytoplasmic membrane caused by PEF.

The presence in the medium surrounding the microorganisms of ultraviolet-absorbing material such as nucleic acid, proteins, and adenosine triphosphate (ATP) is one of the most commonly used indicators of the leakage of intracellular material. Leakage of ultraviolet-absorbing material from different microorganisms treated by nonlethal PEF treatments has been observed, indicating that the temporary loss of permeability control is not necessarily lethal [164]. It was observed that increasing the severity of the treatment resulted in a greater leakage of ATP, nucleic acids, and proteins [40, 164, 165].

Bacterial plasmolytic response to osmotic stress is a physical indicator of membrane integrity. The ability of microbial cells to undergo plasmolysis indicates that a semipermeable membrane is present and functioning to maintain protoplast integrity. When intact microbial cells are suspended in a hypertonic medium, water diffuses from the cell, thus causing a strong condensation of the cytoplasmic content that can be determined by the increments in the optical density of the cell suspensions. Using this procedure, early studies on microbial electroporation demonstrated that cells of *Escherichia coli* treated by PEF lost their ability to plasmolyze in a hypertonic medium, which supports that the treatment affected the integrity of cytoplasmic membrane [17].

One of the most often used techniques to investigate the electroporation of the microbial membranes caused by PEF is the exclusion of dyes. The hydrophilic fluorescent molecule propidium iodide (PI) with a molecular weight of 660 Daltons has been the most commonly used probe for this purpose. PI that is only able to enter permeabilized microbial cells has been used to analyze the electroporation of individual cells with epifluorescent microscopy and flow cytometry, or of the whole population, by using spectrofluorometer procedures. Using flow cytometry, it has been observed that when the PI is added after the PEF treatment, there is a linear correlation between the number of permeabilized and inactivated cells for *E. coli*, *Listeria innocua*, and *Lactobacillus plantarum* [40, 164]. However, in the case of the yeast *Saccharomyces cerevisiae*, it was observed that some treatments that increased the permeability to PI did not necessarily cause a loss of viability. The reversible and irreversible electroporation can be detected by comparing the fluorescent intensity of a microbial suspension treated by PEF when the PI was added before or after the PEF treatment [166]. Reversible electroporation was detected when the fluorescent intensity of the microbial suspension that was in contact with PI during the PEF treatment was higher than the fluorescent intensity of the microbial suspension that was put in contact with PI after the PEF treatment. Results obtained using this technique supported that reversible and irreversible electroporation was involved in the microbial inactivation by PEF, depending on the treatment medium pH and the characteristics of the microbial envelopes surrounding the cytoplasmic membrane. At pH 7 and pH 4, the loss of viability for two Gram-positive bacteria (*Listeria monocytogenes* and *L. plantarum*) was correlated with an irreversible loss of membrane integrity. However, for the two Gram-negative bacteria (*E. coli* and *Salmonella senftenberg*), inactivation was correlated with the proportion of reversible and irreversible electroporated cells. Therefore, these results indicate that the reversible electroporation of the cytoplasmic membrane of Gram-negative bacteria may cause also microbial death. For Gram-negative bacteria treated at pH 4, no correlation was observed between the loss of viability and membrane permeabilization being the proportion of permeabilized cells lower than the inactivated ones. This behavior could be a consequence of that the size of the pores caused by PEF treatments in cells suspended in a medium of pH 4 was smaller than those required for the PI uptake but big enough to abolish the homeostatic capacity of the cells.

The ability of the microorganism to recover from the damage caused by PEF in the cytoplasmic membrane has been correlated with the occurrence of sublethal injury after the PEF treatment. The sublethally injured population fails to survive and multiply in harsh environments tolerated by the untreated cells. Comparison of cell counts of PEF-treated samples on selective (harsh environment) and nonselective media is the most conventional technique for detecting the occurrence of sublethal injury. Early studies on sublethal injury caused by PEF concluded that microbial inactivation by PEF was an all-or-nothing effect because after the treatment, alive or dead cells were detected but not sublethally injured ones

[165, 167]. However, at present it is well established that PEF causes sublethal injury depending on the microorganisms and pH of the treatment medium [168, 169]. Generally, a greater number of sublethally injured cells were detected in a population of Gram-negative bacteria when treated by PEF at pH 4 than at pH 7. In Gram-positive bacteria, the occurrence of sublethal injury was greater at pH 7 than at pH 4. The fact that the presence of NaCl in the recovery medium prevented the growth of sublethal injured cells of *E. coli* after PEF treatments and the demonstration that these damaged cells required energy and the synthesis of lipids for injury repair supports the involvement of the cytoplasmic membrane on the microbial inactivation by PEF [170].

Recovery media with bile salts added are generally used to evaluate the permeabilization of the outer membrane of Gram-negative microorganisms. Using this technique, several authors have reported that PEF did not affect the permeability barrier of the outer membrane of the bacteria surviving the treatments [17]. However, it was observed in cells of *Enterobacter sakazakii* treated by PEF in media of low pH the occurrence of sublethal injury in both cytoplasmic membrane and outer membrane [171]. This damage in the outer membrane facilitated the antimicrobial activity of citral in cells of this microorganism previously treated by PEF.

It can therefore be concluded that PEF may cause reversible or irreversible electroporation of the cytoplasmic membrane of microorganisms depending on the intensity of the applied treatment but also on the type of microorganism and pH of the treatment medium. Irreversible electroporation leads to microbial inactivation. However, reversible microbial electroporation may result in cells that are able to return to their original state by membrane resealing, in sublethally injured cells or dead cells. The presence of sublethally injured cells permits combining PEF with additional hurdles to improve the preservation effect of PEF. Such is the case of acid medium that inhibits the recovery of damaged cells [172] or the addition of antimicrobial substances that otherwise would not be effective in undamaged cells [173, 198, 199].

6.6.3 Factors Affecting Microbial Inactivation by Pulsed Electric Fields

The microbial inactivation by PEF has been found to depend on many factors [173, 174]. In order to define the processing conditions required to inactivate spoiling and pathogenic microorganisms, the influence of these factors must be understood. Critical factors affecting microbial inactivation can essentially be classified into three groups: processing parameters, microbial characteristics, and treatment medium characteristics (Table 6.4).

Table 6.4 Factors affecting microbial resistance by PEF. The main factors are indicated in bold letters

Process parameters	Microbial characteristics	Product parameters
Electric field strength	**Strain**	**Composition**
Treatment time	**Specie**	**pH**
Specific energy	Growth conditions:	Electrical conductivity
Temperature	Growth temperature	a_w
Pulse width	Growth phase	
Pulse shape	Recovery conditions:	
Frequency	**Medium composition**	
	Temperature	
	Recovery time	
	Oxygen concentration	

6.6.3.1 Processing Parameters

The most typical processing parameters that characterize PEF technology are electric field strength, pulse shape, pulse width, number of pulses, pulse-specific energy, frequency, and temperature. From among them, electric field strength, treatment time, specific energy, and temperature are the most critical for the effectiveness of microbial inactivation by PEF.

The distance between the electrodes of the treatment chamber and the voltage delivered defines the electric field strength that is generally reported as kV/cm. In general, microbial inactivation increases by augmenting the strength of the electric field over a threshold field strength called *critical electric field strength* (E_c). This E_c is the field strength required to exceed the transmembrane potential that varies from 0.5 to 1.0 V and is responsible for the membrane electroporation [175, 176]. The higher the field strength, the larger the electroporation phenomenon. Due to the size of microbial cells (1–10 μm), which is smaller than that of the eukaryote plant cells (40–200 μm), the critical electric field intensity to induce electroporation of microbial cells is much higher than for induce electroporation in eukaryote cells of plant or animal tissues (>5 kV/cm). This fact indicates that microbial inactivation requires more powerful equipment and higher energetic costs than the electroporation of eukaryote cells. Generally, studies on microbial inactivation by PEF have been conducted in the range of 10–30 kV/cm because the application of higher electric field strengths has technical limitations, especially at an industrial scale, and may cause the dielectric breakdown of the food material.

Treatment time is defined as a function of the duration of pulse width and the number of pulses applied. It is generally reported in μs. In square waveform pulses, pulse width corresponds to the duration of the pulse, but in exponential decay pulses, the time required for the input voltage to decay to 37 % of its maximum value has been adopted as the effective pulse width. The survival curves (Log_{10} of survivors along the time) at constant electric field strength are characterized by a fast inactivation in the first moments of the treatment, and then the number of

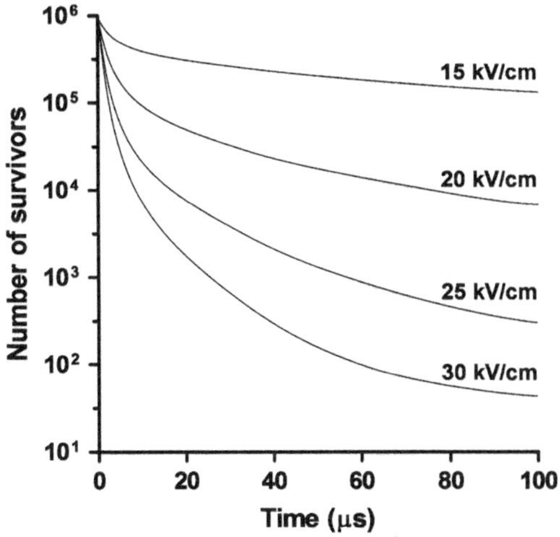

Fig. 6.20 Theoretical survival curves corresponding to microbial inactivation by PEF treatments at different electric field strengths

survivors slowly decreases as the number of pulses applied becomes longer (Fig. 6.20). Due to this kinetics of inactivation, many investigations have been carried out in order to define the most adequate mathematical model to describe it and to develop equations to predict the PEF microbial inactivation [177].

Specific energy of the treatment (energy applied per mass unit) depends on the applied voltage, pulse width, number of pulses, and resistance of the treatment chamber that varies according to its geometry and conductivity of the treated material. Usually it is reported in kJ/kg. This parameter permits the evaluation of the energy costs of the PEF process and, consequently, the comparison of the energy efficiency of PEF with other inactivation technologies. The specific energy has been proposed as a control parameter of the PEF process together with the electric field strength mainly when exponential decay pulses are used due to the lack of precision in the measurement of the pulse width [178]. Microbial inactivation by PEF increases with specific energy, but when different treatments of the same specific energy are compared in terms of microbial inactivation, those applied at higher electric fields are more effective [179, 180].

Microbial inactivation by PEF is usually enhanced when the temperature of the treatment medium is increased, even in ranges of temperatures that are not lethal for microorganisms (Fig. 6.21) [181–183]. The higher microbial PEF sensitivity when applying PEF at moderate temperatures is the basis of current PEF pasteurization treatments of fruit juices and smoothies [173]. The sensitizing effect of temperature has been attributed to changes in the phospholipid bilayer structure of the cell membranes, from a gel-like consistency to a liquid crystalline state that is caused by the temperature increase. The improved membrane fluidity reduces its stability and facilitates the PEF electroporation. However, further studies are required to demonstrate this fact.

It has to be pointed out that the combination of PEF with lethal temperatures has been recently investigated as a treatment to inactivate bacterial spores of *Bacillus*

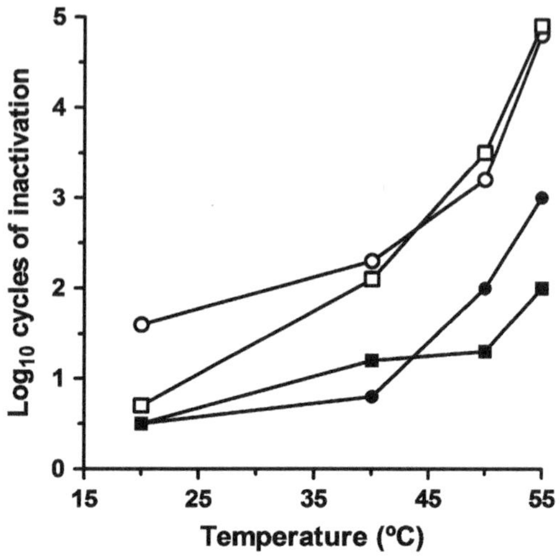

Fig. 6.21 Influence of the temperature on the inactivation of *Escherichia coli* O157:H7 (●); *Salmonella typhimurium* 878 (○); *Staphylococcus aureus* 4459 (■); and *Listeria monocytogenes* 5672 (□) by PEF treatments (30 kV/cm; 0.5 Hz; 1 pulse of 3 μs) in McIlvaine buffer of pH 3.5 (■)

subtilis [184]. In this case, a new processing concept has been proposed based on previous research of Heinz et al. [185] for the pasteurization of apple juice or milk by Guerrero-Beltrán et al. [186] on using the heating due to the Joule effect because of the electrical energy dissipation that occurs during the PEF processing. In this approach, treatment conditions were selected (9 kV/cm, 146–178 kJ/kg) to apply treatment temperatures over 80 °C but with extremely short residence times reducing up to 3 Log$_{10}$ cycles of the spore population. More research is necessary in order to identify the advantages of this process against traditional thermal processing in terms of energy requirements, impact on nutritional compounds, and sensorial properties.

Finally, there is some controversy concerning the influence of the pulse shape, width, and frequency on PEF microbial inactivation. It is generally accepted that square wave pulses are more efficient than exponential decay ones because the characteristic slow-decaying rate causes a long tail section that is ineffective in killing the microorganisms in the food material. Some authors have reported that when treatments of the same duration are applied with pulses of different width or at different frequencies, longer pulses and higher frequencies are more effective [187, 188]. However, these two parameters apparently do not exert an influence on microbial inactivation when the temperature rise of the medium caused by the application of longer pulses of higher frequencies is avoided [189].

6.6.3.2 Microbial Characteristics

Microbial inactivation by PEF depends on microbial properties such as the type of microorganism, characteristics of the cell envelopes (Gram-positive or Gram-negative), cell size and shape, growth conditions (growth temperature and phase), and recovery conditions (medium composition, temperature, recovery time, oxygen concentration, etc.). Generally, it has been reported that bacteria are more PEF resistant than yeast; Gram-negative microorganisms are more sensitive than Gram-positive microorganisms; and cocci are more resistant than rods. However, it seems that the intrinsic microbial resistance is more important than the effect of the microbial characteristics in determining the microbial sensitivity to PEF. When the PEF resistance of different microorganisms is compared under the same experimental conditions, it is observed that some yeast cells are more PEF resistant than some bacteria, some Gram-positive microorganisms are more sensitive than Gram-negative microorganisms, and some yeast species and some rod bacteria are more resistant than some coccus bacteria [177, 183].

The PEF resistance of different strains of bacterial species may vary greatly. It has been observed that depending of the strain and pH of the treatment medium, the inactivation of different strains of the same microorganism may range from hardly any inactivation to more than 4.0 Log_{10} CFU/ml [190]. As the PEF resistance of the different strains depended on the pH of the treatment medium, the target microorganisms to define treatment conditions for PEF pasteurization could be expected to be different for foods, depending on their pH (Fig. 6.22). This can be a limitation of the technology for industrial application.

Culture conditions of microorganisms also influence microbial inactivation by PEF. Generally, it has been reported that microorganisms at the exponential phase of growth are more PEF sensitive than those at the stationary phase [191, 192]. The higher size of the cells in the exponential phase could explain this difference in resistance. On the other hand, microbial resistance to PEF depends on the cultivation temperature. It has been reported that cells grown at temperatures lower than the optimal one are more PEF sensitive than those grown at optimal temperature. Variations in the lipid composition of cultures grown at different temperatures could be the reason of this behavior.

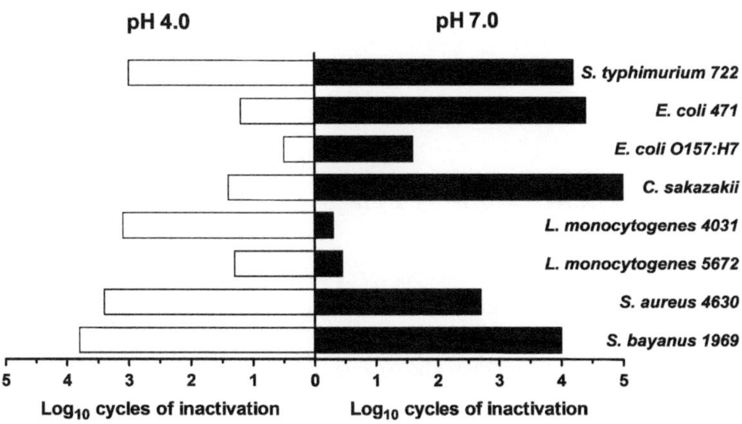

Fig. 6.22 Influence of the pH of the treatment medium on the resistance of different microbial species and strains to a PEF treatment (30 kV/cm; 50 pulses of 3 μs) (Adapted from [45, 205, 206])

6.6.3.3 Treatment Medium Characteristics

Generally, studies on microbial inactivation by PEF have been conducted with microorganisms suspended in liquid media, receiving less attention the microbial inactivation in solid media. The influence of the pH and the electrical conductivity of the substrate on microbial inactivation have been the main investigated treatment medium parameters. It has been observed that microbial PEF resistance varies considerably, depending of the pH of the treatment medium. An increment, decrease, or no effect on the PEF microbial inactivation has been reported when varying the pH of the treatment medium when applying the same PEF treatment [177]. Generally, Gram-positive microorganisms are more PEF resistant in media of neutral pH than in acidic conditions, and Gram-negative ones are more resistant in media of acidic pH than in neutral conditions [193] (Fig. 6.22). This effect of the pH on microbial resistance has been confirmed in both buffers and liquid foods. The mechanism that explains these differences seems to be related to the occurrence of sublethal membrane damage by PEF as it has been previously described [193]: when Gram-positive bacteria are treated in neutral media, their ability for repairing sublethal injury caused by PEF is higher than when treated in low pH media. On the contrary, the higher PEF ability for repairing sublethal injury in Gram-negative bacteria occurs when they are treated in acidic pH media.

Several studies have reported that the electrical conductivity of the treatment medium affects microbial inactivation. However, it is unclear if the conductivity influences electroporation or if the observed effect is a consequence of the influence of conductivity on the intensity of the PEF treatment applied. A change in conductivity modifies the resistance of the treatment chamber, and as a consequence, it may cause changes in the electric field strength and the pulse width and total specific energy of the pulses. Studying the lethality in a range from 0.05 S/m up to 0.4 S/m, which corresponds to the conductivity of most of liquid foods, it has been observed that the conductivity did not affect microbial inactivation when the input voltage and input pulse width were modified in order to obtain the same treatment (electric field strength and treatment time) in media of different conductivities [179, 180].

6.6.4 Food Pasteurization by PEF

Inactivation of vegetative cells of bacteria and yeast by PEF has widely been demonstrated. However, the few studies conducted on the inactivation of bacterial spores by PEF at moderate temperatures describe these structures as resistant to PEF treatments [194, 195]. Therefore, currently practical applications of PEF processing aim at replacing thermal pasteurization as a means of killing vegetative microorganisms rather than sterilization. Pasteurization refers to a treatment used for food preservation that aims to inactivate pathogenic forms of vegetative microorganisms. Although the main objective of PEF pasteurization is to guarantee food safety, a large proportion of the population of vegetative spoilage microorganisms is also inactivated by the treatment. Therefore, PEF treatments may extend the shelf life of foods [196, 197]. However, the treatment is not capable of achieving commercial sterility because spores or other nonpublic health significant microorganisms can survive the treatment; thus other preservation techniques, such as refrigeration, atmosphere modification, the addition of preservatives, or a combination of these techniques, will be required to preserve the quality and stability of the food during its distribution and storage [173, 198, 199].

PEF is gaining interest as a gentle method of food preservation of heat-sensitive foods such as fruit juices. Demands by consumers for fresh-like and natural taste foods have promoted the introduction in the market of non-pasteurized fruit juices produced from fresh fruit and distributed under refrigeration. As psychotropic microorganisms are able to grow at refrigeration temperatures, the shelf life of these fruit juices is very short (around 7 days). On the other hand, several outbreaks associated with the consumption of unpasteurized juices have demonstrated that these products can be a vehicle of foodborne illnesses caused by pathogens such as *Salmonella* spp. or *E. coli* O157:H7 [200].

Commercial exploitation of PEF for food pasteurization requires proof that PEF promotes a level of microbial safety that is equal to that made possible via traditional processing. A 5-Log_{10} reduction of the most resistant microorganism of public health significance has been established by the US Food and Drug Administration (FDA) concerning fruit juice pasteurization [201]. Studies to evaluate the application of PEF for microbial decontamination at room temperature have shown that to obtain these levels of reduction, it is necessary to apply long treatments (i.e., >100 μs) at high-electric field strengths (i.e., ≥ 30 kV/cm) [173, 177, 179, 180, 202]. At a commercial scale, technical and economical limitations exist in applying these intense treatments in a continuous flow. However, several studies confirmed that the application of PEF at moderate temperatures provides the possibility of obtaining substantial microbial inactivation of pathogenic microorganisms that are particularly PEF resistant with a short residence time (less than 1 s) at moderate electric field strengths (≤ 25 kV/cm). Under these treatment conditions, the shelf life of fruit juices and smoothies is extended up to 21 days, while maintaining a fresh-like taste and product quality. Currently such products are commercialized in the Netherlands, Germany, and the UK, using PEF equipment with a capacity in the range of 1500–8000 l/h. The total processing costs, including investment and operation, could vary in a range from 0.01 to 0.03 euros/l [203, 204].

Although the first commercial applications of PEF are available in the market, more multidisciplinary research efforts are required to identify the most PEF-resistant pathogens of concern for each specific food, to define process criteria for PEF pasteurization, to get a better mechanistic understanding of the critical parameters affecting microbial inactivation, and to combine all this knowledge with engineering aspects involving the distribution of the electric field strength in continuous flow treatment chambers or to develop suitable sensors to assess the

PEF process. A deeper knowledge of these aspects is needed to satisfy regulatory agencies and to enhance the safety and stability of minimal processed foods of the future.

6.7 Microbial Decontamination Using Atmospheric-Pressure Plasma

Loree C. Heller and Muhammad A. Malik

6.7.1 Introduction

Adding energy to a gas breaks bonds between atoms and ionizes the atoms creating positively charged particles (ions) and negatively charged particles (electrons). When a significant number of atoms in the gas ionize, the resulting overall electrically neutral medium of free positive and negative particles can be described as plasma. When the temperature of neutral particles (gas temperature, T_g) is the same as electron temperature (T_e) and ion temperature (T_i), it is called equilibrium plasma. In nonequilibrium plasma, electron temperature is much higher than gas temperature, i.e., $T_e >> T_g \approx T_i$. The plasma can be formed at reduced pressure or at atmospheric pressure. Atmospheric-pressure nonthermal plasma relevant to microbial decontamination is described in the following chapter.

Plasma is usually generated by high-voltage electrical discharges between electrodes with the interelectrode gap filled with gas. The gas discharges are usually classified on the basis of electrode configuration and power source/excitation frequencies. The main groups of discharges relevant of microbial decontamination are corona discharges powered by direct current (dc) or pulsed dc, dielectric barrier discharges (DBD) powered by alternating voltages of low frequency to several megahertz, and atmospheric-pressure plasma jets (APPJs) powered by dc to some gigahertz [207–209].

Corona discharge is formed in a diverging electrode geometry, such as needle to plate or wire to cylinder [210]. An intense electric field is formed at the needle tip or thin wire upon application of high voltage. Electrical breakdown in the gas occurs in the form of initiation of several thin plasma channels distributed along the electrode that propagate toward the counter electrode. Arcing is avoided by either limiting the current in the case of dc power or by applying pulsed dc to cut off the electric field before arcing. Inserting a dielectric layer in the discharge gap between electrodes changes the corona discharge into sliding discharge at the solid–gas interface [214, 215]. Covering at least one of the electrodes also avoids arcing in the

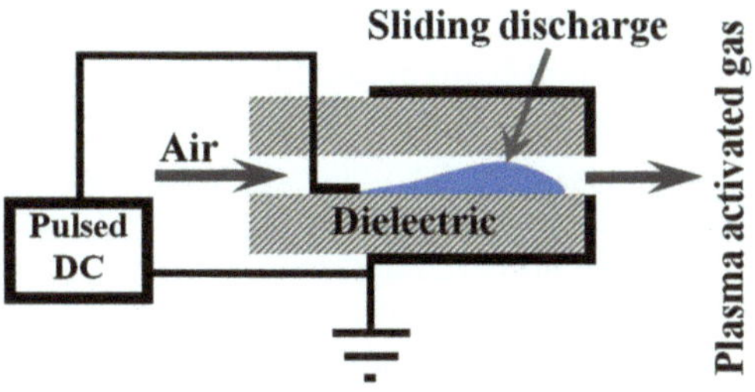

Fig. 6.23 Cross-section view of a sliding discharge-based large area plasma device

case of DBD [213]. The electrodes in DBD can be large area plate-to-plate configuration where a large number of micro-plasma channels are formed, distributed in time and space along the electrode. Accumulation of charges at the dielectric reduces local electric field and extinguishes the discharge in a few nanoseconds. New discharges initiate and continue the process as long as the alternating voltage continues. Removing the air gap between the electrodes, i.e., placing a wire or strip electrode on the dielectric and a plate-type electrode covering the opposite side of the dielectric, transforms the volume DBD into surface DBD [214, 215].

Several versions of atmospheric-pressure plasma jets have been developed which generate plasma in open space [207, 208]. The plasma in these cases is formed in a dielectric tube with attached electrode(s) and is expelled in the open space by fast flowing gas and/or by redistribution of electric fields driving the plasma into the open space. The target to be treated with plasma may be one of the electrodes responsible for the redistribution of the electric fields. The APPJs have been extended into a brush-shaped plasma, e.g., by using an array of the jets [215] or an edge electrode to generate large area plasmas [216]. A cross-section view of a large area plasma device used in bacterial decontamination on surfaces [212] is illustrated in Fig. 6.23. It forms sliding discharges in the chamber, activating air. The activated air is expelled and targeted on the surface to be decontaminated. It can be enlarged further by stacking and operating multiple discharge chambers in parallel [211].

6.7.2 *What Makes Plasma Reactive?*

Nonthermal plasmas are usually formed by high-voltage electrical discharges. Any free electrons in the interelectrode gap are accelerated under the influence of strong electric fields. Inelastic collisions between high-energy electrons and ambient gas molecules causes ionization, dissociation, or excitation of ambient gas molecules producing ions, more free electrons, reactive free radicals, and excited states like O, N, H, OH, O_2*, and N_2*. A major fraction of the input electrical energy ultimately ends up as heat. Some of the excited state species emit photons that may fall in UV or VUV spectrum range. The primary reactive species ultimately produce secondary reactive species, like O_3, H_2O_2, and nitrogen oxides (NOx). All of them affect the cell being treated by the plasma to a varying extent.

What reactive component exists in the plasma and which one plays a dominant role in microbial decontamination depends on the feed gas, experimental conditions, and type of cell being treated. For example, noble gas plasma is easier to form and sustain and efficiently produces and transmits VUV and UV radiations, but has a low concentration of reactive oxygen species (ROS) and reactive nitrogen species (RNS) [217]. Addition of a reactive gas such as oxygen in the noble gas can generate ROS-like O, metastable oxygen, and O_3. Addition of water vapor results in production of H, OH, and H_2O_2. Therefore, air, particularly humid air, is an excellent option because it is easily available and produces a wide range of ROS and RNS. Particularly the strongest oxidizing agents, i.e., OH and $ONOO^-$, are produced in humid air and are considered to play a major role in microbial decontamination [218]. UV radiations are found to play in synergy with free radicals in the case of reduced pressure plasmas [209]. In atmospheric-pressure plasmas, free radicals and reactive neutrals play a dominant role, while most of the UV radiations are absorbed in air [219].

Direct exposure to plasma is generally more effective than remote exposure. In the case of direct exposure of the target to the plasma, both short-lived species like O and N and long-lived species like OH, O_3, H_2O_2, and NOx affect the cells being treated. Heat [220, 221] and electric fields [222, 223] that are often associated with the direct plasma treatment are also known to play a significant role. In the case of remote plasma exposure, only long-lived ROS and RNS, like O_3, OH, H_2O_2, and $ONOO^-$, are primarily involved in the microbial decontamination.

6.7.3 The Effects of Plasma Reactive Components on Microbes

The observation that atmospheric-pressure plasma reduced vegetative bacterial numbers [224] led to a large number of studies confirming inactivation of a variety of organisms, including vegetative bacteria, bacterial spores, bacteriophages, and viruses, fungi, and fungal spores with a variety of direct and indirect plasma types in proof-of-principle or translational studies. Several excellent review articles are available for more detail [209, 225–228]; this chapter presents an overview of mechanisms and recent translational studies.

Some species of bacteria and fungi are protected by a mucoid polysaccharide capsule, which may shield the exterior of these organisms from plasma components. All bacterial and fungal cells possess a cell wall composed primarily of peptidoglycan or chitin respectively that protects the cell from mechanical damage and osmotic rupture. Microscopically observable cell surface lesions [229] commonly described as "etching" are observed microscopically after plasma application.

Although bacteria can generally be categorized as Gram-negative or Gram-positive based on the specific cell wall composition, this category does not necessarily predict a cell's plasma sensitivity [212, 221, 230–237]. Although the cell wall may protect the cell from lysis, damage to lipids in the plasma membrane can produce loss of membrane potential [238] and leakage of cellular contents [229, 231, 239]. Plasma exposure acidifies lipid films [240] and membrane lipids [238], although this is not a primary antibacterial component [241]. Viral genomes are encased in a protein capsid that may or may not be enclosed in a protective membrane, so the sensitivity of various free viral species to plasma application should vary. Interestingly, while several groups have demonstrated free viral inactivation, intracellular virus is plasma resistant [236], an indication of the limitations of plasma inactivation.

Proteins may be imbedded in membranes or free in the cytoplasm. ROS may generate many temporary and permanent amino acid modifications, the peptide backbone can be cleaved, and the protein can aggregately react with the oxidation products fatty acids and carbohydrates. Several groups have demonstrated protein degradation [242] or a loss in enzymatic activity after plasma application [242–244]. Proteins within cells may be more sensitive to plasma application than on surfaces [245].

The interaction between superoxide and nitric oxides produces a wide array of reactive molecular species that can deaminate DNA, cause abasic sites, strand breaks, and cross-link and induce reactions with proteins and with metal cofactors [246]. Direct exposure to plasma degrades DNA [242, 245]. In addition, oxidative DNA damage markers significantly increase with plasma exposure [238].

All cells possess protective enzymes such as catalases, peroxidases, superoxide dismutase, scavengers including glutathione, and stress response genes to combat reactive species. A number of pathways are devoted to DNA repair by reversal, removal, or repair of damaged bases. Genetic studies in *Escherichia coli* [247], the yeast *Saccharomyces cerevisiae* [248], and *Bacillus subtilis* [245] elucidated the important antimicrobial constituents. Oxidative stress appears to be the major antimicrobial component, but osmotic stress and heat also contribute to inactivation. Damage markers for nearly all cell components, the membrane, cell wall, DNA, and protein are upregulated. Clearly, multiple plasma components participate in cell inactivation.

Potential applications of the antimicrobial properties of direct or indirect plasma delivery include surface, air, and liquid decontamination and medical and dental antimicrobial treatments. The next section highlights a small number of recent microbial decontamination studies from a considerable body of literature.

6.7.4 Microbial Decontamination on Surfaces

One primary application of surface inactivation is environmental decontamination. In healthcare environments, an important reservoir of nosocomial pathogens related to transmission is inanimate surfaces, including hospital room surfaces and medical plastics. Many planktonic clinically relevant bacteria, yeasts, and one viral model, a bacteriophage, have been inactivated by plasma application on plastics, glass, laminate, linoleum, cloth, and steel [212, 237, 249, 250]. Bacterial cells in biofilms are protected by a bacterially produced polymer matrix that makes them more environmentally resistant than planktonic cells. In spite of this, these cells are significantly inactivated by plasma application on plastics, glass, and stainless steel [234, 251–257]. Spores are particularly difficult to inactivate, producing varying results [250, 258]. The mechanisms of plasma microbial inactivation appear to be independent of bacterial antibiotic resistance status [229, 237, 255, 256, 258, 259].

Biological residues on medical instruments and consumables may also pose an environmental hazard. An interesting potential use of plasma is the decontamination of biological residues on medical instruments and consumables. Prions, or infectious misfolded proteins [260], normal protein [261], and amyloid [262] are reduced substantially by plasma application.

Other applications of plasma decontamination are reviewed elsewhere in this book. A burgeoning area of research for microbial inactivation by plasmas is in food preservation, particularly in fresh produce and meats [228, 261, 263, 264]. Since plasma application to food can cause discoloration and other undesirable effect, one group has explored "gentle" plasma application parameters [265].

Bacteria, yeasts, and molds on biological surfaces such as corneas, dentin, and skin are inactivated by various plasmas [219, 221, 233, 236, 259, 266–269]. Dental and medical applications are discussed elsewhere in this book including wound decontamination, which has reached clinical trials [270]. Decontamination of biological surfaces may be more challenging than inanimate surfaces since plasma modification of the surface itself may be more detrimental.

6.7.5 Microbial Decontamination in Air

Experimental design to confirm air decontamination is a daunting task. Aerosolized bacteria, fungal spores, bacteriophages, or viruses can be inactivated after capture on a HEPA filter [271], contained within a chamber [272] or in closed circulation systems [273–275]. In a fascinating study, nebulized airborne human pathogenic respiratory viruses were inactivated to a higher level by a combination of gas plasma and UV than by either component alone [276].

6.7.6 Microbial Decontamination in Liquids

Vegetative bacteria [277–280], bacterial spores [281], and bacteriophages [275, 282] have been significantly inactivated in medium, broth, or water. Enteropathogenic *E. coli* O157:H7 was in activated in apple juice [283] and an *E. coli* model in milk [284]. One concern of plasma inactivation in liquids is potential for a significant pH decrease with exposure, which would potentially affect beverage flavor or modify a liquid's composition.

6.7.7 *Conclusions*

Multiple studies confirm that plasma has many biological and biomedical applications, including microbial inactivation on surfaces, in food and beverages, in air, and in medical and dental practices. Differences in antimicrobial sensitivity may not only be due to differences between organisms but also be due to differences in the employed plasmas and in the medium or surface used as a support for the inactivation. Inactivation on both biological and nonbiological surfaces can be less effective than on smooth agar plates; an obvious hypothesis for this lower inactivation is that a surface quality such as topography protects the organisms. For each plasma device, it is important to characterize the components that play a role in decontamination. A major future development area for plasma translation is the application-specific device design and scaling of plasma delivery devices.

6.8 Downstream Processing of Microalgae for Energetic Use

Wolfgang Frey and Christian Eing

6.8.1 *Introduction*

The depletion of petroleum-based fossil resources, the urgent need to reduce greenhouse gas emissions, and the increasing demand for fuels for the transportation and energy sector call for alternatives based on renewable sources. During the past decades, research and development has been intensified to replace fossil fuels like diesel and gasoline by biomass-derived products.

The production of *first-generation biofuels*, i.e., biodiesel and bioethanol, from oil- and carbohydrate-rich, agriculturally grown biomass turned out to evoke a strong competition with food industry for arable land resources. Moreover, the current amount of worldwide produced vegetable oil and fat might only cover about 10 % of the global diesel demand [285]. *Second-generation biofuels* are based on the conversion of energy crops, agriculture and forestry waste streams, and food industry residuals to fuels. This approach can indeed alleviate competition problematics, but also not satisfy the future demand for transportation fuels to any remarkable extent [286].

During the last decade, research and development efforts have been intensified on microalgae biomass, referred to as *third-generation biomass*. Although not being demonstrated on industrial scale for fuel production yet, it exhibits advantageous properties over conventional biomass that justifies its consideration as a sustainable feedstock for future biofuel production.

6.8.2 Microalgae: Third-Generation Biomass

Microalgae exhibit a high biomass yield per footprint which is two to five times higher compared to agriculturally grown biomass. Typical yields of efficient energy crops, e.g., miscanthus, are 20 $t{\cdot}ha^{-1}{\cdot}year^{-1}$ [287], whereas microalgae cultivation in closed systems, i.e., photobioreactors [288], can yield up to 70–100 $t{\cdot}ha^{-1}{\cdot}year^{-1}$ [289]. Cultivation in photobioreactors saves resources, such as water and nutrients, and enables biomass production on arid or barren lands, thus avoiding competition with food production and preventing degradation of soil as evident for intensive energy crop agriculture.

Microalgae can accumulate high amounts of lipids. The lipid content of some species, e.g., *Botryococcus braunii*, can exceed 70 % of weight [290] when cultivated under nitrogen depletion. Unfortunately, for these cultivation conditions, biomass production rates are poor. The lipid content of microalgae still exhibiting satisfying biomass growth is on the order of 20 % of weight. For comparison, the oil content of rape is only 5 % in maximum related to the entire biomass [291]. Thus, conservatively estimated, the possible oil yield per footprint from microalgal biomass 0.2 70 = 14 $t{\cdot}ha^{-1}{\cdot}year^{-1}$ is more than ten times higher compared to rapeseed, which allows lipid yields of 1.2 $t{\cdot}ha^{-1}{\cdot}year^{-1}$. Optimistic scenarios even estimate oil yields of more than 50 $t{\cdot}ha^{-1}{\cdot}year^{-1}$ [292].

Besides lipids, microalgae produce other valuable components as proteins, carbohydrates, antioxidants, and vitamins [293, 294]. Altogether the content of value-added products (including lipids) of microalgae biomass exceeds 80 %, which is a value much higher than for conventional biomass. Nevertheless, multiple component recovery from microalgae is still a biotechnological challenge, since all these components are stored intracellularly and they are protected by a rigid cell wall.

6.8.3 Some Energetic and Economic Numbers for Microalgae Production and Processing

Compared to agriculturally grown biomass, a photobioreactor-based production of microalgae biomass is more expensive. For northern irradiation conditions and based on production systems already operating at commercial intermediate scale, the costs for 1 kg_{dw} of dry (dw, dry weight) microalgae biomass range between 4 and 5 € [295]. Important cost factors [296, 297] are light transfer, mixing of the microalgae suspension in the photobioreactor to prevent sedimentation, $C0_2$ supply, nutrients, and dewatering. Investment costs for photobioreactors and labor costs for maintenance and operation are noticeable and have to be kept on a minimum extent [316]. On industrial scale, for a 100 ha production site, these costs are forecasted to be reduced to 0.68 €/kg_{dw}, which appears to be reasonable for an economic microalgae-based biofuel production [295]. This scenario already involves the use of flue gas as $C0_2$ source, wastewater for nutrient supply, and irradiation conditions being available in southern European countries.

From the energetic point of view, the production of 1 kg_{dw} of microalgal biomass with flat panel photobioreactors consumes lowest 10 MJ of energy [298]. Depending on the lipid content, the energy (enthalpy of combustion) stored in microalgae ranges between 20.6 and 26.7 MJ/kg_{dw} [299]. The lower value was obtained from microalgae exhibiting a lipid content of about 20 %. In consequence, almost half of stored energy of microalgae biomass is consumed by cultivation.

As a first step for efficient downstream processing, microalgae have to be concentrated from a biomass density out of the photobioreactor, i.e., 2–5 g_{dw}/l, to a density of 150–200 g_{dw}/l. Dewatering of cell suspensions is commonly done by batch or continuous centrifugation methods on bench scale. Some energy values for centrifugation of microalgae biomass given in literature are 0.95 kWh/kg_{dw} [www.evodos.eu] and 8 kWh per m^3 of centrifuged microalgae suspension [300], which corresponds to 6 MJ/kg_{dw} (1 kWh = 3.6 MJ) for a cell density of 5 g_{dw}/l and can reach up to 10 MJ/kg_{dw} [289]. They additionally depend on strain, processing scale, targeted final density of the biomass wet paste, and, last but not least, the tolerable content of small debris in the supernatant. This makes clear that the number of biomass washing steps and concentration steps has to be restricted to a minimum extent along the complete microalgae processing chain. Furthermore, for large-scale energetic use of microalgae biomass, centrifugation is agreed to be too energy intensive [301]. Biofuel production scenarios pursue flocculation combined with pressing, optimistically requiring less than 0.5 MJ/kg_{dw} on a 100 ha production scale [302].

After harvesting and dewatering of microalgae to a microalgae paste, which then exhibits a dry biomass content of about 200 g_{dw}/l, subsequent processing alternatives split up into dry-route and wet-route processing. In the first case, the energy for evaporation of water, i.e., 7 MJ/kg_{dw}, has to be expended for biomass drying. Already at this point, it immediately can be realized that wet-route processing of microalgae paste has to be the most appropriate choice for a targeted energetic use.

For efficient recovery of intracellularly stored components, a cell disintegration step is needed. Among microwave or enzymatic treatment, ultrasound sonication, and ball milling, high-pressure homogenization and ball milling were demonstrated to be efficient in conventional microalgae processing. High-pressure homogenization has developed to a standard cell disruption method for ranking the efficiency of alternative methods. Recently bead milling was reported to require less than 2.5 kWh/kg_{dw} [303]. Energy consumption values from literature for high-pressure homogenization vary over several orders of magnitude, starting from some kWh/kg_{dw} and ending up at values of more than 100 kWh/kg_{dw} [304]. It has to be emphasized that the cell disruption efficiency of high-pressure homogenization and bead milling highly depends on biomass density in the suspension to be treated. For constant treatment conditions, the fraction of disrupted cells decreases at higher cell densities.

After cell disruption, microalgae-based fuel production requires a solvent extraction step for lipid recovery followed by a refinement step and subsequent transesterification to biofuel [305]. Processing data on larger scale for lipid conversion into biofuels are rare in literature. Nevertheless, fatty acid profile determination revealed that common species are suitable for refining to biodiesel allowing the use of currently available refining technologies [292]. Nowadays, the direct transesterification of microalgae biomass is discussed to be a promising conversion route and has been demonstrated on pilot scale [306]. Under realistic conditions on pilot scale, the yield on fatty acid methyl esters (FAME, biodiesel) was determined to be 15 % related to initial microalgae biomass exhibiting a lipid content of 17 %wt [307]. On pilot scale, realistic overall production costs for biodiesel by direct transesterification of microalgae biomass at current state of technology are close to 30 €/kg [300]. Projections into near future allow expecting a cost reduction by a factor of 10 for microalgae-based fuel [308, 317]. Without doubt, microalgal-based biofuels will be a sustainable future source for transportation fuels, but nowadays can only become feasible if additional products and components except lipids can be valorized [309]. Therefore and as agreed in the community, energy-efficient and fractionating downstream processing methods are mandatory.

6.8.4 PEF Processing of Microalgae Biomass

Pulsed electric field treatment, involving plasma membrane permeabilization as basic biophysical process, is an efficient wet-route processing technique exhibiting fractionating properties. When treating microalgae suspensions of *A. protothecoides*, pre-concentrated from cultivation density of 5 to 100 g_{dw}/l, with rectangular 1 μs pulses of an electric field strength of 34 kV/cm, an increase of suspension conductivity due to ion release by a factor of 1.6–2 can be observed immediately at the outlet of the flow treatment chamber, indicating efficient membrane permeabilization. In this case, 15 % of the total biomass could be released into the extracellular medium right after PEF treatment [310]. This water-soluble fraction contains salts, sugars, amino acids, and water-soluble proteins. Without washing step between microalgae harvesting and PEF treatment, the conductivity of the pre-concentrated microalgae suspension was 1 mS/cm. Under these conditions, an increase in component release could already be obtained for specific treatment energy values of 50 kJ per kg of treated suspension (kJ/kg_{sus}). A saturation of the yield of water-soluble components was observed for energies higher than 150 kJ/kg_{sus}.

An additional waiting time of 2 h after PEF treatment further increased the yield of water-soluble components up to 20 % [313]. After that time, no further increase could be obtained for *A. protothecoides*.

Intracellularly stored oil consists in bodies with an average diameter of ~1 μm [311]. Mainly due to their size and their hydrophobic character, oil bodies cannot pass cell wall and permeabilized membrane and remain intracellular. After separation of the water-soluble fraction, lipids were extracted with ethanol from the residual, lipid-rich biomass fraction. At a specific treatment energy of 150 kJ/kg_{sus}, the lipid yield from the PEF-treated residual fraction was three to four times higher, compared to the untreated sample, recovering more than 80 % of the stored lipids on average [312]. This indicates that PEF treatment not only improves component transport to the extracellular medium but also solvent access into the cell by a single PEF-treatment step.

Contrary to the curve progression of the yield of water-soluble products over treatment energy, which shows a steep initial increase at low energies and saturation behavior at values above 150 kJ/kg_{sus} [310], the energy dependence of the lipid yield is different. Up to values of 75 kJ/kg_{sus}, the yield increase is small and can be neglected. A steep increase in lipid yield can be observed at 100 kJ/kg_{sus} and higher [312]. These results indicate that an improvement of intracellular solvent access requires higher treatment energies, whereas water-soluble components already can be released under milder treatment conditions.

Measurements on the influence of pulse protocol on ethanolic extraction yield of intracellular components, i.e., carotenoids and chlorophylls, from *Chlorella vulgaris*, revealed that pulse shape and duration can impact extraction efficiency [313]. Comparable extraction yields could be achieved with monopolar 3 μs pulses at a specific treatment energy of 17 kJ/kg_{sus}, whereas for 1 ms bipolar pulses, 150 kJ/kg_{sus} were required. This compares well with our findings on *C. vulgaris*. Compared to *A. protothecoides*, the required energy for increased lipid recovery from *C. vulgaris* is lower and amounts to 25 kJ/kg_{sus}. In general, the required specific energy for effective PEF treatment depends on the algae strain under consideration and has to be determined individually. For instance, PEF treatment of *Neochloris oleoabundans* within an energy range of 25 kJ/kg_{sus} and 150 kJ/kg_{sus} did not influence component yield at all, neither in lipid yield nor for water-soluble component release.

Treatment energies for microalgae processing are related to one kilogram of dry biomass. In the case that the required treatment energy is 150 kJ per kilogram of treated suspension (kJ/kg_{sus}) and the biomass density of the pre-concentrated suspension amounts to 100 g_{dw}/l, the specific energy value related to dry biomass is 1.5 MJ/kg_{dw}. Since treatment energy is coupled to the suspension conductivity, the specific energy consumption can be reduced by increasing the cell density in the suspension to be treated, provided that the extraction efficiency is not affected. Experiments on the influence of biomass density in the suspension to be treated on the component yield have shown that the extraction efficiency did not decrease at higher biomass densities up to 160 g_{dw}/l [310]. Consequently, microalgae suspensions of 200 g_{dw}/l, which are still well pumpable, require a specific treatment energy of 0.75 MJ/kg_{dw}, which is considerably lower compared to conventional processing methods. Moreover, PEF treatment does not produce cell debris, which facilitates subsequently required separation processes.

These merits of PEF processing can satisfy the demand for a low-energy-consuming technology for cascade valorization of microalgae biomass for energetic use. PEF-assisted fractionating component recovery allows for partial compensation of the comparatively high-energy demand for cultivation by simultaneous valorization of both higher-value water-soluble products and lipid-rich residual biomass. Furthermore, the fractionating character of PEF processing opens several energetic processing alternatives for microalgae [314] to be investigated in the future.

As a first valorization step, water-soluble proteins can be separated as already demonstrated for *Nannochloropsis* [155], *C. vulgaris*, and *H. pluvialis* [315]. In subsequent steps, either lipids can be extracted from the residual lipid-rich biomass or the lipid-rich biomass fraction may be fed into other energetic valorization pathways, i.e., thermochemical conversion or biogas production. The challenge for future R&D work will be to identify the appropriate pathway exhibiting the highest economic and energetic gain. Since energy consumption of PEF treatment is very low, it undoubtedly will advance to an important tool promoting the energetic valorization of microalgae.

6.9 Protein Electroextraction from Microorganisms (Bacteria, Yeasts, Microalgae)

Justin Teissie

It will describe that long (ms) pulses affect the cell membrane and cell wall in such a way that a slow release of proteins is induced. A flow technology is available and is described. The cell remains macroscopically intact allowing an easy separation without the formation of debris.

6.9.1 Introduction

Walled microorganisms are cell factories. Among the many systems available for heterologous protein production, the Gram-negative bacterium *Escherichia coli* remains one of the most attractive ones because of its ability to grow rapidly and, at high density on inexpensive substrates, its well-characterized genetics and the availability of a large number of cloning vectors and mutant host strains [318, 319].

Yeasts are widely used for industrial production of homologous proteins, most of them with intracellular location. Nowadays, *Saccharomyces cerevisiae* as well as yeasts from genera *Kluyveromyces*, *Pichia*, and *Hansenula* have become a suitable host for industrial production of recombinant proteins with high biotechnological and pharmacological values [320, 321]. Their secretion from the cell is the optimal way for isolation, but often this is impossible or of low efficiency. Thus, the newly synthesized heterologous proteins remain accumulated in the cell cytoplasm.

The simple growth requirements of microalgae made these microorganisms attractive bioreactor systems for the production of high-value heterologous proteins. The cultivation could be achieved in photobioreactors (PBRs) which provided safe control of culture environment. It was possible to produce recombinant proteins either in cytosol or chloroplast of microalgae [322]. A promising development was given by the production of unique immunotoxin cancer therapeutics in algal chloroplasts [323]. Extraction of proteins was hindered by the cell wall barrier. Only a slow release was observed on the microalgae having a rigid cell wall, and therefore, disrupting the rigid cell wall of C. vulgaris was required to obtain a complete protein release after extraction [324].

The main methods utilized for liberation of intracellular enzymes are mechanical disintegration and chemical extraction [325, 326]. Although applicable on a large scale, these are relatively drastic procedures, which affect the stability of the proteins or introduce additional impurities to be removed in the associated downstream processes. The running costs are high. Formation of debris is a problem for the downstream purification. Thus alternative methods for enzyme extraction are investigated to ensure a higher selectivity of release and mild experimental condition preserving in maximal level the enzyme activity.

Pulsed electric fields are described as one of the most promising approaches. Long pulses (ms long) appear to permeabilize the plasma membrane and to induce a structural change in the wall. As a final consequence, a slow release of soluble cytoplasmic proteins is obtained. Field conditions can be adjusted in such a way to leave the vacuole intact, to prevent the release of proteolytic enzymes. The proof of concept of the flow process protocol to treat industrially significant volumes was previously validated [149] (Fig. 6.24).

Flow process electroextraction was indeed patented to the CNRS (FR # 0013415; Euro/PCT # 1982525.6). A preindustrial pilot was developed in our group during the FP7 "Electroextraction" project ([FP7-SME-2007-1], Grant agreement n°222220) and is now commercially available.

6.9.2 *Electroextraction by a Flow Process*

6.9.2.1 Description

The basic concept was to apply calibrated electric pulses with a delivery delay which was linked to the flow rate (Figs. 6.24 and 6.25). The desired number of

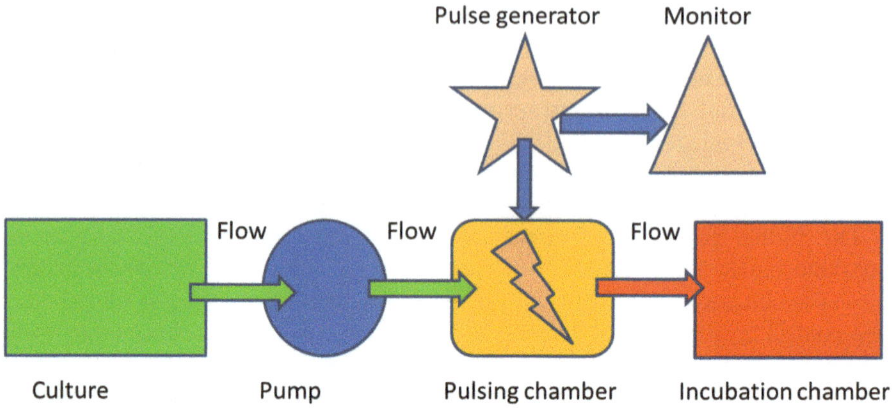

Fig. 6.24 Flow process for electroextraction. The microorganisms are washed and pumped across the pulsing chamber. A train of pulses are delivered by the pulse generator, and their voltage and current are monitored on line. The pulsed flow is collected and incubated in the appropriate incubation buffer. The extracted proteins are treated downstream from the supernatant

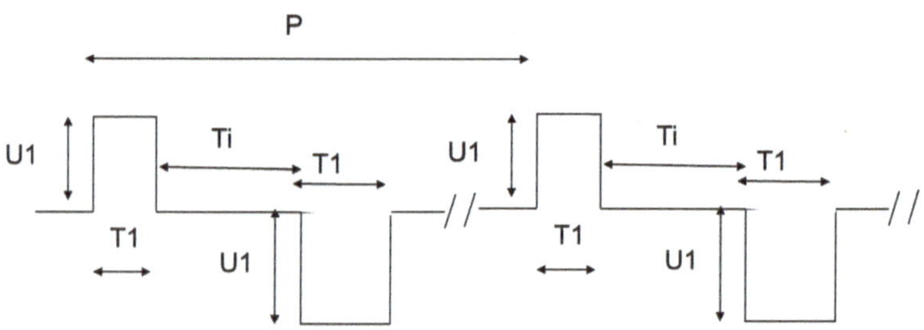

Fig. 6.25 Train of pulses. A positive voltage pulse (intensity U1, duration T1) is followed after an interval Ti by a negative voltage pulse (intensity U2, duration T2). This is repeated at a period P to limit electrochemical reactions, U1T1 = U2T2. In practice U1 = U2 and T1 = T2

pulses was actually delivered on each cell during its residency in the pulsing chamber. The geometry of the chamber (flat parallel electrodes) was chosen to give a homogeneous field distribution. Therefore, the residency time Tres of a given cell in the chamber was

$$\mathrm{Tres} = \mathrm{Vol}/\mathrm{Q} \tag{6.1}$$

where Vol was the volume of the pulsing flow chamber and Q, the flow rate. The number of pulses delivered per cell was

$$\mathrm{N} = \mathrm{Vol}/\mathrm{QP} \tag{6.2}$$

P being the period of the pulses. The field strength was

$$E = U/d \tag{6.3}$$

d being the width between the two electrodes and U the voltage.This average power associated to the train of pulses was

$$\langle Pow \rangle = UIT/P \tag{6.4}$$

T being the single-pulse duration and I the current.As the chamber resistance, when filled by the sample, could be approximated by

$$R = d/(\Lambda S) \tag{6.5}$$

where Λ is the conductance of the sample and S the section of the electrodes, then

$$\langle Pow \rangle = TE^2 \Lambda Vol/P \tag{6.6}$$

From Eq. 6.2, an increase in the flow rate Q while keeping the number of applied pulses N constant needs to increase Vol/P, i.e., either increase in Vol or decrease in P (or both). From Eq. 6.6, more power is needed.

One other physical problem is that Λ increases with the temperature, meaning that due to the Joule heating, the current increases during the pulse and from Eq. 6.6 more power than predicted is needed.

6.9.2.2 Specifications of the Pulse Generator

Long square wave pulses, not an accumulation of short pulses, are needed to obtain cytoplasmic soluble protein extraction [157, 327]. The voltage to be delivered must be adjusted to the microorganism and to the pulsing chamber to obtain a field strength critical to trigger the plasma membrane permeabilization. To keep the required voltage at a high current during the pulse duration (Eq. 6.4), a significant electrical charge must be stored in the capacitors to be partly delivered in the biological sample. To obtain a high frequency in the delivery of pulses in the train needed with the flow process, the power supply that provides the charge to the capacitor needs to be designed large enough.

The use of long electric pulses is associated to a technical drawback: electrochemical reactions are occurring at the surface of the electrode. Formation of bubbles of H_2 and Cl_2 are observed. Electro-erosion of the electrode surface is present (details are given in Chap. 2). This can be prevented to delivering trains of pulses with alternating polarities with a short (about 10 ms or less) delay between each.

The DeexBio pilot was therefore designed to provide trains of bipolar pulses (Betatech, France). The electric system was made of two S20u generators each able to deliver up to 2 kV under 10 A (maximum for safety reason), with adjustable pulse duration of a few milliseconds. Delays between pulses were down to 30 ms.

An analog switch working at 15 ms was used to connect the two generators with the pulsing chamber. All electrical settings (T1, Ti, U1, and P) were selected from the touch screen. Current and voltage were monitored online.

6.9.2.3 Pulsing Chambers (Applicators)

The pulsing chamber was built of several parts, the most important being the discharge chamber. A parallel plate configuration was chosen to obtain a homogeneous field. Two geometries were designed to obtain fields up to 6 kV/cm.

For microalgae or yeast, the distance between electrodes was 6 mm, the height was 6 mm, the length was longer than 1 cm, and the volume was up to 1.08 ml. The other applicator, used for bacteria or smaller microalgae, had a distance between electrodes reduced to 3 mm, the width was 5 mm, and the volume was 150 μl.

A laminar flow was present where the velocity of the microorganisms is not uniform. To obtain a more homogeneous treatment, several chambers are connected in series (see Fig. 6.26) with a remixing of the microorganism suspension. This feature limits the Joule heating [328].

Fig. 6.26 The Deex Bio. The two pulse generators (*yellow*) deliver voltage with alternating polarities on the pulsing chambers (*gray*, two in series). The flow of cells is pumped at a controlled flow rate (pump on the right). The pulsed cells are collected in the reservoir on the left. Temperatures are controlled at different levels to avoid overheating. The pulses are monitored online on the PC (in the back) (Picture by courtesy of Beta tech)

6.9.3 Electroextraction Protocols: A General Approach

To limit the current delivered along the pulse application, there is a need to work with a suspension in a low conductivity buffer. Cells are therefore washed in distilled (or at least detergent-free) water to a final conductance of about 0.2 mS/cm.

The cell concentration is limited by the electropermeabilization-induced release of the cytoplasmic ions that induces a fast increase in the suspension conductivity. This increase is a linear function of the cell concentration.

After the pulse train delivery, the cell suspension is diluted in a salty solution containing DTT (PB/DTT) to weaken the S-S bonds and the electrostatic interactions present in the wall.

The suspension is incubated at room temperature during several hours. Under the microscope, it is observed that the morphology of the pulsed cells is not affected when compared to controls. Separation of the supernatant where the released proteins are from the pulsed cells is easy as no debris are present. This can be obtained by a mild centrifugation and does not affect the downstream separation by chromatography [329].

Permeabilization was assayed by the propidium uptake and from the increase of the conductivity of the pulsed solution.

Protein extraction was assayed by the Coomassie blue assay. Due to the excretion process, protein release was observed in controls when the postpulse incubation was long (overnight).

SDS-PAGE of protein samples was performed on 12 % acrylamide slab gel as described by Laemmli [330]. The silver staining of gels as described by Nesterenko et al. [331] was made with a protein molecular weight marker (such as Page-RulerTM Prestained Protein Ladder, Thermo Scientific). Most of the bands appeared between 35 and 170 kDa. The same bands appeared both on negative controls (due to excretion) and samples, with a higher intensity for the samples incubated in water or phosphate buffer (PB/DTT)(PB 105 mM, 0.3 M Glycerol, 1 mM DTT, pH = 7). Several new proteins appeared after electroextraction in comparison to the negative controls.

Protein activity was assayed by routine-specific protocols. 3-phosphoglycerate kinase (PGK)(45 kDa) activity was determined according to Kulbe and Bojanovski [332], glyceraldehyde-3-phosphate dehydrogenase (GAPDH) (37 kDa) according to Kirshner and Voigt [333], hexokinase (a dimer of 100 kD) according to McDonald [334], and β-D-galactosidase (a 464-kDa homotetramer) according to Mbuyi-Kalala et al. [335]. Protease activity was determined according to Meussdoerffer et al. [336].

6.9.4 Electroextraction Protocols for Bacteria

Electroextraction has been proposed as a method to recover cytoplasmic content from bacteria [337–339]. The following protocols are given for *E. coli*.

6.9.4.1 Microorganisms and Cultivation Conditions

Escherichia coli BL21 and DH5α strains were used.

Cells were grown in LB, 0.5 % (w/v) glucose. After reaching appropriate growth phase, the cells were collected by centrifugation at 4000 × g for 10 min, washed once with deionized water, and resuspended to a final concentration corresponding to 5 % wet weight.

Viability after the different PEF treatments was obtained by counting colony-forming units (CFU). After discharge delivery, the suspensions were removed and incubated for 1 h at 37 °C. The bacterial suspensions were diluted, and 100 μl samples from selected dilutions were spread onto Petri dishes for incubation overnight at 37 °C.

6.9.4.2 Results

Extraction of GAPDH from cells of strain BL21 in exponential growth phase was obtained when cells were treated with 15 pulses, 4 Hz, 0.5, and 1 ms duration, followed by a postpulse incubation at 30 °C. Maximal release was obtained for GAPDH at field intensity of 7 kV/cm and pulse duration of 0.5 ms. In control cells, the incubation in buffer provoked the liberation of only 25 % of GAPDH (half less than in pulsed cells).

The upper limits in the field strength were detected by formation of precipitates and a decrease of GAPDH activity tested. This may be due to an increased Joule heating during electric treatment. Similar precipitation was observed with the application of pulses with 1 ms duration at field intensities over 5.75 kV/cm or of pulses of 0.5 ms duration but over 7.25 kV/cm.

Extraction was strain dependent. The optimal conditions for electroextraction from DH5α were the same as those for BL21. But cells of this strain were less sensitive to electric fields, in comparison with BL21. A lower extraction (GAPDH and PGK) was obtained.

6.9.4.2.1 Influence of Incubation Buffer Content

The buffer content for postpulse incubation was found to be crucial for protein recovery. The common buffer for cell lysis and protein extraction with *E. coli* is Tris buffer pH = 8–8.5. Very often additional components such as EDTA and DTT are present.

6.9.4.2.2 Influence of the Growth Phase

The cell wall porosity strongly depends on growth phase. We checked the influence of growth phase on the release of intracellular enzymes.

The release of PGK in the middle and late exponential phase cells was considerably higher than for GAPDH. This might be due to its lower molecular weight, thus demonstrating some sieving selectivity of protein electroextraction. The electro-induced release of enzymes from cells in late exponential growth phase was only 50 % of what was obtained in the middle exponential phase. When the cells were in stationary growth phase, the electric field was completely inefficient.

6.9.4.2.3 Increase in Sensitivity to Lytic Enzymes

Electric field pulse in bacteria not only permeabilized plasma membrane but also affected other cell envelop components. Pulsed and control cells were incubated for 30 min in medium containing lysozyme, in concentrations considerably lower than that used for cell lysis . As a result of this mild lytic treatment, the release of GAPDH and PGK of pulsed cells was up to 85–86 % of total activity for both enzymes. The effect on control, unpulsed cells was present but distinctly lower—only 28 % of total activity.

6.9.5 Electroextraction Protocols for Yeasts

Cytoplasmic proteins can be extracted from yeasts by electropulsation (pulsed electric field technology, PEF) [149, 327, 340–344].

Protein release from the cells is a slow process occurring along several hours during the incubation in the specific buffer.

Flow process treatment of yeasts (*Saccharomyces cerevisiae*), with high-intensity electric field pulses, allows the release of the intracellular protein content on large culture volumes [149, 343].

The extraction yield is dependent on the field strength (to obtain plasma membrane permeabilization), pulse duration (several ms is more effective than subms), and the number of successive pulses. Optimization is cell strain dependent. DTT brings a significant increase in extraction due to the effect on the wall [345].

Extraction from *Saccharomyces cerevisiae* was obtained with 15 pulses of 2 ms at 6 Hz. Maximal yield for glyceraldehyde-3-phosphate dehydrogenase (GAPDH, 145 kDa) (85 %) was obtained at 3.2 kV/cm where all cells were permeabilized. About 58 % from cell activity was liberated within 1 h after pulsation. The maximal release of two other cytoplasmic enzymes 3-phosphoglycerate kinase (PGK, 45 kDa) and hexokinase (HK, 100 kDa) was obtained during the same time period.

6.9.5.1 Growth Phase

Cells in stationary phase are frequently used as source for production of enzymes. Previous data with a batch process showed that the growth phase determined the effect of electric pulsation. Electroextraction of protein using cells in stationary phase (5×10^8cells/ml) was obtained. When applying 15 pulses (2 ms duration, 6 Hz), maximal yield of GAPDH 4 h after pulsation (85 %) was obtained at 3.5 kV/cm. While at 3.2 kV/cm, which was optimal for cells in exponentially phase, the activity of GAPDH in the supernatant was only 65 % from the total.

6.9.5.2 Protein Stability

The extractable total protein for different pulse length/intensity/frequency combinations remained limited to about 37–45 % from the total content. But the specific activity of three electroextracted enzymes (GAPDH, PGK and HK) was about two times higher than that was obtained in cell extracts, from either after enzymatic lysis or mechanical grinding.

An increase of the intensity above the optimal value (established for a given cell concentration) provoked a decrease of enzyme activity—probably a thermal denaturation. This is a critical event as an increase of intensity from 2.7 to 3 kV/cm on a 20 % wet weight suspension resulted in about 90 % decrease of extracted GAPDH activity.

The proteins from crude extracts and supernatants of electrically treated cells were analyzed by SDS–polyacrylamide gel electrophoresis under reducing conditions. Most of the bands above 29 kDa present in the lysate, even those with highest molecular weights (proteins over 97 kDa), were quite similar in both electropulsated and mechanically broken cells.

6.9.5.3 Vacuole Preservation

In yeast the main protease activity is concentrated in the vacuole, which is affected by classical extraction methods. But when using electrical conditions giving 100 % permeabilization, and isotonic postpulse medium, the liberated protease activity at 4 and 6 h after pulsation was only about 15 % and 20 % from what was present in the extract after classical lysis. The preservation of vacuole integrity was checked by a classical Lucifer yellow (LY) staining of pulsed cells. Electropermeabilized cells incubated in isotonic medium showed fluorescence only in the cytoplasm, the vacuole remaining unstained—showing that the vacuolar membrane was intact.

6.9.5.4 Cell Concentration

The possibility for treatment of more concentrated suspensions—up to 20 % wet weight (2.2×10^9 cells/ml)—was checked by assaying GAPDH extraction. Results were that when using shorter pulses, i.e., 1 ms, one could obtain 85 % extraction at lower field intensity—2.7 kV/cm. At this intensity (15 pulses, 1 ms duration, 6 Hz), only about 40 % IP could be achieved when pulsing the more diluted 4.5 % wet weight (5×10^8 cells/ml) suspension. At the high concentration, the optimal intensity was 4.3 kV/cm, i.e., the five time increase of cell concentration is associated to an about 1.6 reduction of optimal field intensity.

These effects could be attributed to the increased conductivities of the suspensions during the pulse application, as a result of higher inorganic ion outflow. The conductivity of 4.5 % and 20 % wet weight suspensions before pulsation was about 40 and 100 μS/cm, respectively. Immediately after pulsation (30s), these values were ten times greater. Thus, the concentrated yeast suspensions were treated with higher currents. The technical limit is that the current that was delivered is higher (reaching the limits of the specifications of the pulse generator) and the Joule effect is larger (risks in thermal denaturation).

6.9.5.5 Postpulse Conditions

It was observed that the incubation of pulsed cells at room temperature did not influence the efficiency of extraction (only about 3 % decrease in GAPDH and PGK leakage as compared to 30 °C). This is a positive advantage in running costs.

The presence of glycerol and DTT in the postpulse incubation medium contributed to higher GAPDH (about 15 %) and PGK (about 20 %) activities in the supernatant of electrically treated cells but did not influence hexokinase activity, 80 % of HK activity being recovered within 4 h (data not shown).

6.9.6 Electroextraction Protocols for Microalgae

6.9.6.1 Panel of Strains

C. vulgaris and *H. pluvialis* are growing in freshwater, while *Dunaliella salina* and *N. salina* are marine species. All have a rigid cell wall.

Culture conditions were in photoreactors. Growth was checked by measuring the density of algae under an inverted microscope with a Malassez slide.

6.9.6.2 Results

It should be kept in mind that microalgae are indeed photosynthetic yeasts. Electroextraction protocols were rather similar. Protein release was a slow process (h) following the very fast pulse delivery (<s) [317].

6.9.6.3 Influence of the Electric Field on Protein Electroextraction

The extraction increased with the increase in the field strength (always with a single-pulse duration of 2 ms). Extraction efficiency was dependent on the field strength in a specie-dependent manner. Fields of 3 kV/cm were efficient on *C. vulgaris* and *H. pluvialis*, while 6 kV/cm was needed for *N. salina* due to its smaller size.

6.9.6.3.1 Influence of the Number of Pulses and of the Delay Between the Train of Pulses

A systematic evaluation of the effect of accumulating pulses showed that the amount of released proteins was increased with the number of successive pulses, whatever the field strength applied by the applicator.

A highly efficient protein release was obtained with one cycle of 15 bipolar pulses. There was no improvement in extraction when cycling again the microalgae (*C. vulgaris*) suspension.

6.9.6.3.2 Pulse Duration

Pulse duration is clearly a leading factor to obtain protein extraction from walled species. Microsecond long pulses were not inducing the protein release even if the cumulated application time lasts several ms. A slight modification of the single-pulse duration, while keeping the cumulated pulse duration constant, did not affect the extraction efficiency. Thirty pulses of 1 ms gave the same amount of released proteins from *Chlorella* as 15 pulses of 2 ms. The 30 pulses were obtained by doing two passages, while such a protocol did not improve the extraction efficiency for pulses of 2 ms. The temperature increase linked to the Joule heating was reduced.

6.9.6.3.3 Post-Zap Dilution

In our previous results obtained on yeasts, there was a need for postpulse incubation in a salty buffer containing DTT. The pulsed sample (in a low conductivity buffer) was routinely diluted in 4 volumes of an incubation buffer (dilution factor of 5). But further industrial development, a one to one dilution (with a 1.7× concentrated extraction buffer), was observed to be as efficient with *C. vulgaris* and *N. salina*, after treating under their respective optimized conditions.

6.9.6.3.4 Kinetics of Protein Leakage After PEF

Protein leakage from microalgae happened in a time-dependent manner. A massive leakage was obtained during the first 30 min following the postpulse dilution, while no release of protein was observed in the control samples during that time. To obtain a more complete extraction, an overnight postpulse incubation was required. Due to endogenous excretion processes, a substantial amount of proteins was released during that long incubation period that was run at room temperature.

6.9.6.3.5 Postpulse Temperature

No obvious difference was observed between incubation at 4 °C and at 20 °C, whatever the duration of incubation.

6.9.6.3.6 Influence of the Cell Concentration

C. vulgaris at 105 cells/ml and 106 cells/ml were exposed to nine pulses of 2 ms at 3 kV/cm. After 24 h incubation at 20 °C, results showed that the increase of cell concentration increased linearly the extracted protein concentration. Pulsing more concentrated suspension brought the problem of the increase in conductivity due to the release of cytoplasmic ions.

6.9.7 Conclusions and Tips

The effect of electric pulsation is size dependent, and stronger electric treatment should be applied to obtain the same effect on smaller cells.

Pulse duration should be long (ms). Pulses should be delivered in a low conductivity buffer. The release of ions during the pulse delivery is a limit in the cell concentration to be treated. Joule heating (Temperature increase) can be reduced by using a series of pulsing chambers and coolers, using a larger number of shorter pulses with more successive chambers.

Postpulse incubation should be operated in an ionic solution containing DTT. The release of proteins during the incubation is a slow process (h.) but occurs at room temperature. Separation of the supernatant where the released proteins are from the pulsed cells is easy as no debris are present. The downstream separation by chromatography is not subject to clogging. No ultrastructural damages are observed.

Acknowledgments The authors appreciate the support from the COST Action TD1104 (EP4Bio2Med - European network for development of electroporation-based technologies and treatments).

References

1. Gusbeth, C., Frey, W., Volkmann, H., Schwartz, T., Bluhm, H.: Pulsed electric field treatment for bacteria reduction and its impact on hospital wastewater. Chemosphere **75**, 228–233 (2009)
2. Rieder, A., Schwartz, T., Schön-Hölz, K., Marten, S.M., Süss, J., Gusbeth, C., Kohnen, W., Swoboda, W.: Molecular monitoring of inactivation efficiencies of bacteria during pulsed electric field treatment of clinical wastewater. J. Appl. Microbiol. **105**, 2035–2045 (2008)

3. Schwartz, T., Kohnen, W., Jansen, B., Obst, U.: Detection of antibiotic resistant bacteria and their resistance genes in wastewater, surface water, and drinking water biofilms. Microbiologica **43**, 325–335 (2003)
4. Schwartz, T., Volkmann, H., Kirchen, S., Kohnen, W., Schön-Hölz, K., Jansen, B., Obst, U.: Real-time PCR detection of Pseudomonas aeruginosa in clinical and municipal wastewater and genotyping of the ciprofloxacin-resistant isolates. FEMS Microbiol. Ecol. **57**, 158–167 (2006)
5. Volkmann, H., Schwartz, T., Bischoff, P., Kirchen, S., Obst, U.: Detection of clinically relevant antibiotic-resistance genes in municipal wastewater using RealTime PCR (TaqMan). J. Microbiol. Methods **56**, 277–286 (2004)
6. Davies, J.: Inactivation of antibiotics and the dissemination of resistance genes. Science **264**, 375–382 (1994)
7. Davies, J.: Origins, Acquisition, and Dissemination of Antibiotic Resistance Determinants. Wiley, Chichester (1997)
8. Environmental Protection Agency.: Design Manual: Municipal Wastewater Disinfection. EPA/625/1-86/021. Cincinnati (1986)
9. Hülsheger, H., Pottel, J., Niemann, E.G.: Killing of bacteria with electric pulses of high field strength. Radiat. Environ. Biophys. **20**, 53–65 (1981)
10. Neumann, E., Rosenheck, K.: Permeability changes induced by electric impulses in vesicular membranes. J Membr. Biol **14**, 194–196 (1972)
11. Min, S., Evrendilek, G.A., Zhang, H.Q.: Pulsed electric fields: processing system, microbial and enzyme inhibition, and shelf life extension of foods. IEEE Trans. Plasma Sci. **35**, 59–73 (2007)
12. Schoenbach, K.H., Joshi, R.P., Stark, R.H.: Bacterial decontamination of liquids with pulsed electric fields. IEEE Trans. Dielectr Electr Insul. **7**, 637–645 (2000)
13. Castro, A.J., Barbosa-Canovas, G.V., Swanson, B.G.: Microbial inactivation of foods by pulsed electric fields. J. Food Process Preserv. **17**, 47–73 (1993)
14. Guderjan, M., Elez-Martínez, P., Knorr, D.: Application of pulsed electric fields at oil yield and content of functional food ingredients at the production of rapeseed oil. Innov. Food Sci. Emerg. **8**, 55–62 (2007)
15. Heinz, V., Alvarez, L., Angersbach, A., Knorr, D.: Preservation of liquid foods by high intensity pulsed electric fields-basic concepts for process design. Trends Food Sci. Technol. **12**, 103–111 (2002)
16. Rivas, A., Rodrigo, D., Martínez, A., Barbosa-Cánovas, G.V., Rodrigo, M.: Effect of PEF and heat pasteurization on the physical–chemical characteristics of blended orange and carrot juice. LWT Food Sci. Technol. **39**, 1163–1170 (2006)
17. Hamilton, W.A., Sale, A.J.H.: Effects of high electric fields on microorganisms: II. Killing of bacteria and yeast. Biochim. Biophys. Acta **148**, 789–800 (1967)
18. Environmental Protection Agency: Wastewater Technology Fact Sheet: Chlorine Disinfection. EPA 832-F-99-062. Office of Water, Washington, DC (1999)
19. Dietrich, J., Loge, F., Ginn, T., Basagaoglu, H.: Inactivation of particle associated microorganisms in wastewater disinfection: modelling of ozone and chlorine reactive diffusive transport in polydispersed suspensions. Water Res. **41**, 2189–2201 (2007)
20. Emmanuel, E., Keck, G., Blanchard, J.M., Vermande, P., Perrodin, Y.: Toxicological effects of disinfections using sodium hypochlorite on aquatic organisms and its contribution to AOX formation in hospital wastewater. Bull. Environ. Contam. Toxicol. **30**, 891–900 (2004)
21. Environmental Protection Agency: Wastewater Technology Fact Sheet: Ozone Disinfection. EPA 832-F-99-063. Office of Water, Washington, DC (1999)
22. Sohn, J., Amy, G., Cho, J., Lee, Y., Yoon, Y.: Disinfectant decay and disinfectant by-products formation model development: chlorination and ozonation by-products. Water Res. **38**, 2461–2478 (2004)
23. Gordon, G., Grunwell, J.: Comparison of analytical methods for residual ozone. In: Municipal Wastewater Disinfection – Proceedings of Second National Symposium. U. S. Environmental Protection Agency. EPA-600/9-83-009: pp. 26–245 (1983)

24. Wojtenko, I., Stinson, M.K., Field, R.: Performance of ozone as a disinfectant for combined sewer overflow. Crit. Rev. Environ. Sci. Technol. **31**(4), 295–309 (2001)
25. Field, R.: Combined sewer overflows: control and treatment. In: Control and Treatment of Combined Sewer Overflows. Van Nostrand Reinhold, New York (1990)
26. Jacangelo, J.G., Patania, N.L., Reagan, K.M., Aieta, E.M., Krasner, S.W., McGuire, M.J.: Ozonation: assessing its role in the formation and control of disinfection by-products. Res. Technol. J. Am. Water Works Assoc. **81**, 74–84 (1989)
27. White, C.: Handbook of Chlorination and Alternative Disinfectants, 4th edn. Wiley, New York (1999)
28. Martin, S., Gagnon, T., Gagnon, G.A.: Impact of ozonation on water quality in marine recirculation systems. Aquac. Eng. **29**, 125–137 (2003)
29. Venosa, A.D., Meckes, C.M.: Control of ozone disinfection by exhaust gas monitoring. In: Municipal Wastewater Disinfection – Proceedings of Second National Symposium. U. S. Environmental Protection Agency. EPA-600/9-83-009: pp. 246–254 (1983)
30. Environmental Protection Agency: Wastewater Technology Fact Sheet: Ultraviolet Disinfection. EPA 832-F-99-064. Office of Water, Washington, DC (1999)
31. Jungfer, C., Schwartz, T., Obst, U.: UV-induced dark repair mechanisms in bacteria associated with drinking water. Water Res. **41**, 188–196 (2007)
32. Environmental Protection Agency.: Wastewater Treatment by Pulsed Electric Field Processing, Washington, DC (2003)
33. Tokuşoğlu, Ö., Swanson, B.G.: Improving food quality with novel food processing technologies. CRC Press, ISBN 9781466507241 (2014)
34. Codd, G.A., Bell, S.G., Kay, K., Ward, C.J., Metcalef, J.S.: Cyanobacterial toxins, exposure routes and human health. Eur. J. Phycol. **34**(04), 405–415 (1999)
35. Gaudreau, M., Hawkey, T., Kempkes, M., Petry, J.: Scale up of PEF systems for food and waste streams. Proceedings of the 3rd Innovative Food Centre Conference, Bedford (2006)
36. Kotnik, T., Mir, L.M., Flisar, K., Puc, M., Miklavčič, D.: Cell membrane electropermeabilization by symmetrical bipolar rectangular pulses: part I. Increased efficiency of permeabilization. Bioelectrochemistry **54**, 83–90 (2001)
37. Kotnik, T., Miklavčič, D., Mir, L.M.: Cell membrane electropermeabilization by symmetrical bipolar rectangular pulses: part II. Reduced electrolytic contamination. Bioelectrochemistry **54**, 91–95 (2001)
38. Frey, W., Gusbeth, C., Schwartz, T.: Inactivation of Pseudomonas putida by pulsed electric field treatment: a study on the correlation of treatment parameters and inactivation efficiency in the short-pulse range. J. Membr. Biol. **246**(10), 769–781 (2013)
39. Wouters, P.C., Alvarez, I., Raso, J.: Critical factors determining inactivation kinetics by pulsed electric field food processing. Trends Food Sci. Technol. **12**, 112–121 (2001)
40. Wouters, P.C., Bos, A.P., Ueckert, J.: Membrane permeabilization in relation to inactivation kinetics of Lactobacillus species due to pulsed electric fields. Appl. Environ. Microbiol. **67**, 3092–3101 (2001)
41. Jolibois, B., Guerbet, M.: Evaluation of industrial, hospital and domestic wastewater genotoxicity with the Salmonella fluctuation test and the SOS chromotest. Mutat. Res. **565**, 151–162 (2005)
42. Gusbeth, C., Frey, W., Schwartz, T., Rieder, A.: Critical comparison between the pulsed electric field and thermal decontamination methods of hospital wastewater. Acta Phys. Pol. A **115**, 1092–1094 (2009)
43. Davison, J.: Genetic exchange between bacteria in environment. Plasmid **42**, 73–91 (1999)
44. Volkmann, H., Schwartz, T., Kirchen, S., Stofer, C., Obst, U.: Evaluation of inhibition and cross-reaction effects on real-time PCR applied to the total DNA of wastewater samples for the quantification of bacterial antibiotic resistance genes and taxon-specific targets. Mol. Cell. Probes **21**, 125–133 (2007)
45. Zhang, H.Q., Barbosa-Cánovas, G.V., Balasubramaniam, V.M., Dunne, C.P., Farkas, D.F., Yuan, J.T.C.: Pulsed Electric Field Processing Nonthermal Processing Technologies for Food. Blackwell Publishing Ltd. ISBN: 978-0-813-81668-5 (2011)

46. Zhang, G., Zhang, P., Liu, H., Wang, B.: Ultrasonic damages on cyanobacterial photosynthesis. Ultrason. Sonochem. **13**(6), 501–505 (2006)
47. Akiyama, H., Sakugawa, T., Namihira, T., Takaki, K., Minamitani, Y., Shimomura, N.: Industrial applications of pulsed power technology. IEEE Trans. Dielectr. Electr. Insul. **14** (5), 1051–1064 (2007)
48. Sakugawa, T., Aoki, N., Akiyama, H., Ishibashi, K., Watanabe, M., Kouda, A., Suematsu, K.: A method of cyanobacteria treatment using underwater streamer-like discharge. IEEE Trans. Plasma Sci. **42**(3), 794–798 (2014)
49. Gnapowski, S., Akiyama, H., Sakugawa, T., Akiyama, M.: Effect of pulse power discharges in water on algae treatment. IEEJ Trans. Fundam. Mater **133**(4), 198–204 (2013)
50. Choi, J., Yamaguchi, T., Yamamoto, K., Namihira, Sakugawa, T., Katsuki, S., Akiyama, H.: Feasibility studies of EMTP simulation for the design of the pulsed-power generator using MPC and BPFN for water treatments. IEEE Trans. Plasma Sci. **34**(5), 1744–1750 (2006)
51. Li, Z., Ohno, T., Sato, H., Sakugawa, T., Akiyama, H., Kunitomo, S., Sasaki, K., Ayukawa, M., Fujiwara, H.: A method of water-bloom prevention using underwater pulsed streamer discharge. J. Environ. Sci. Health Part A **43**, 529–551 (2008)
52. Akiyama, H., Sakai, S., Sakugawa, T., Namihira, T.: Environmental applications of repetitive pulsed power. IEEE Trans. Dielectr. Electr. Insul. **14**(4), 825–833 (2007)
53. Sakugawa, T., Yamaguchi, T., Yamamoto, K., Kiyan, T., Namihira, Katsuki, S., Akiyama, H.: All solid state pulsed power system for water discharge. Proceedings of IEEE International Pulsed Power Conference 2005, pp. 1057–1060 (2006)
54. Akiyama, H., Li, Z., Ohno, T., Lin, X. F., Sato, H., Namihira, Sakugawa, T., Katsuki, S., Ayukawa, M., Fujiwara, H., Kunitomo, S., Sasaki, K.: Water-bloom treatment by underwater pulsed streamer-like discharges. Proceedings of IEEE International Pulsed Power Conference 2007, pp. 324–327 (2007)
55. Eshtiaghi, M.N., Knorr, D.: High electric field pulse pretreatment: potential for sugar beet processing. J. Food Eng. **52**, 265–272 (2002)
56. Schultheiss, C., Bluhm, H., Mayer, H.G.: Processing of sugar beets with pulsed electric fields. 2nd Symp. Nonthermal/Medical/Biological Treatments using Electromagnetic Fields and Ionized Gases, Portsmouth, Va., May 21–23, (2001)
57. Frenzel, S., Michelberger Th., Arnold J., Bluhm H., Sack M., Kern M.: Entwicklung und Bau einer Elektroimpuls-Pilotanlge zum nicht-thermischen Aufschluss pflanzlicher Zellen zur Schonung von Energieressourcen am Beispiel der Zuckerindustrie, Abschlussbericht, Fkz. 0330434, TIB-Hannover, (2005)
58. Arnold, J., Frenzel, S., Michelberger, Th., Scheuer, T.: Extraktion von Inhaltsstoffen aus Rübenschnitzeln, Patentschrift DE 10 2004 028 782 B4, Anmeldetag 16.06.2004.
59. Jemai, A.B., Vorobiev, E.: Pulsed electric field assisted pressing of sugar beet slices: towards a novel process of cold juice extraction. Biosyst. Eng. **93**(1), 57–68 (2006). doi:10.1016/j.biosystemseng.2005.09.008
60. Mhemdi, H., Bals, O., Grimi, N., Vorobiev, E.: Alternative pressing/ultrafiltration process for sugar beet valorization: impact of pulsed electric field and cossettes preheating on the qualitative characteristics of juices. Food Bioprocess. Technol. **7**, 795–805 (2014). doi:10.1007/s11947-013-1103-y
61. Morren, J., Roodenburg, B., De Haan, S.W.H.: Electrochemical reactions and electrode corrosion in pulsed electric field (PEF) treatment chambers. Innov. Food Sci. Emerg. Technol. **4**(2003), 285–295 (2003)
62. Donsi, F., Ferrari, G., Fruilo, M., Pataro, G.: Pulsed electric field-assisted vinification of aglianico and piedirosso grapes. J. Agric. Food Chem. **58**, 11606–11615 (2010). doi:10.1021/jf102065v
63. Grimi, N., Lebovka, N.I., Vorobiev, E., Vaxelaire, J.: Effect of a pulsed electric field treatment on expression behavior and juice quality of chardonnay grape. Food Biophys. **4**, 191–198 (2009). doi:10.1007/s11483-009-9117-8

64. Sigler, J., Schultheiss, C., Kern, M.: Maischeporation – ein neuer Weg der Weinbereitung. Schweiz. Z. Obstet. Weinbau **16**, 14–16 (2005)
65. Sigler, J., Sack, M., Eing, C., Müller, G., Waidelich, G.: Elektroporation zum verbesserten Aufschluss von Trauben und Maische. Intervitis Interfructa : Internat. Technologiemesse für Wein, Obst, Fruchtsaft und Spirituosen, Stuttgart, 24.-28.März 2010
66. Sigler, J., Schultheiss, C., Mayer, H.G., Kern, M.: Zellporation in der Weinbereitung. INTERVINIS – INTERFRUCTA 2004 : Internat. Technologiemesse für Wein, Obst und Fruchtsaft, Stuttgart, 11.-15.Mai 2004
67. Sack, M., Sigler, J., Stukenbrock, L., Eing, C., Mueller, G.: Lab-scale experiments on PEF-treatment of grape mash at different pulse parameters. 4th Euro-Asian Pulsed Power Conf. (EAPPC 2012), 19th Internat.Conf.on High-Power Particle Beams (BEAMS 2012), Karlsruhe, September 30–October 4, 2012
68. Sack, M., Sigler, J., Stukenbrock, L., Eing, C., Mueller, G.: On the variation of pulse parameters for PEF treatment of grape mash. Proc.of the Internat.Conf.on Bio and Food Electrotechnologies (BFE 2012), Fisciano, I, September 26-28, 2012, Salerno : ProdAl S.c.a. r.l, 2012 S.223–227
69. Schmidt, O.., Schick, A., Sack, M., Sigler, J.: Sesam öffne Dich – Elektroporation von Trauben, das deutsche weinmagazin • 6/24. März 2012
70. Redondo, L.M., Andrade, J., Santos, J.O., Barros, F., Pereira, M.T.: Industrial processing of red and white grapes assisted by Pulsed Electric Fields, Proc. EAPPC 2012, Karlsruhe, 30.9.-04.10.2012
71. Müller, G., Frey, W., Sack, M., Schultheiss, C., Mayer, H.-G., Sigler, J., Kern, M., Günther, U.: Karlsruher Elektroporationsanlage KEA – Die Erfolgsgeschichte eines Technologietransfers in die Industrie, NACHRICHTEN – Forschungszentrum Karlsruhe Jahrg. 39 3/2007 S. 153–158 (2007)
72. Sack, M., Müller, G.: Optimisation of an electroporation device for mash. Cernat, M. [Hrsg.] Proc.of the 11th Internat.Conf.on Optimization of Electrical and Electronic Equipment (OPTIM '08), Brasov, R, May 22–24, (2008)
73. Sack, M., Schultheiss, C., Bluhm, H.: Triggered Marx generators for the industrial scale electroporation of sugar beets. IEEE Trans. Ind. Appl. **41**, S.707–S.714 (2005). doi:10.1109/TIA.2005.847307
74. Sack, M., Attmann, F., Müller, G.: EMV-Aspekte beim Entwurf einer Elektroporationsanlage. Internat.Fachmesse und Kongress für elektromagnetische Verträglichkeit (EMV), Düsseldorf, 9.-11.März 2010
75. Sack, M., Müller, G.: Electrical design of electroporation reactors
76. Yongguang Yin, Quinghua Howard Zhang, Sudhir Kartikeya Sastry: High-voltage pulsed electric field treatment chamber for the preservation of liquid food, Patent US 5 690 978, 30. Sept. 1996
77. Sack, M., Bluhm, H.: New measurement methods for an industrial-scale electroporation facility for sugar beets. IEEE Trans. Ind. Appl. **44**, 1074–83 (2008). doi:10.1109/TIA.2008.926222
78. Loginova, K.V., Vorobiev, E., Bals, O., Lebovka, N.I.: Pilot study of countercurrent cold and mild heat extraction of sugar from sugar beets, assisted by pulsed electric fields. J. Article J. Food Eng. **102**(4), 340–347 (2011)
79. Kseniia Loginova, Eugène Vorobiev, Nikolai Lebovka: Qualitative characteristics of sugar beet juices obtained in pilot extractor with pulsed electric field (PEF) pre-treatment. 11th International Congress on Engineering and Food (ICEF), May 22–26, 2011 Athens Greece
80. Angersbach, A., Heinz, V., Knorr, D.: Electrophysical model of interact and processed plant tissues: Cell disintegration criteria. Biotechnol. Prog. **15**(1999), 753–762 (1999)
81. Raso, J., Heinz, V. (eds.): Pulsed Electric Field Technology for the Food Industry. Fundamentals and Applications. Springer, New York (2006)
82. Vorobiev, E.I., Lebovka, N.I. (eds.): Electrotechnologies for Extraction from Food Plants and Biomaterials. Springer, New York (2008)

83. Knorr, D., Engel, K.-H., Vogel, R., Kochte-Clemens, B., Eisenbrand, G.: Mol. Nutr. Food Res. **52**, 1539 (2008)
84. Barbosa-Canovas, G.V., Gongora-Nieto, M.M., Pothakamury, U.R., Swanson, B.G.: Preservation of Foods with Pulsed Electric Fields. Academic, London (1998)
85. Lebovka, N., Vorobiev, E., In: Pakhomov, A.G., Miklavcic, D., Markov, M.S. (eds.) CRC Press, pp. 463–490 (2010)
86. Vorobiev, E.I., Lebovka, N.I. In: Lebovka, N., Vorobiev, E., Chemat, F. (eds.) CRC Press, Taylor & Francis LLC,, pp. 25–83 (2011)
87. Ben Ammar, J.: Etude De L'effet Des Champs Electriques Pulses Sur La Congelation Des Produits Vegetaux. Universite de Technologie de Compiegne, France, Compiegne (2011)
88. Shynkaryk, M.: Influence De La Permeabilisation Membranaire Par Champ Electrique Sur La Performance De Sechage Des Vegetaux. Universite de Technologie de Compiegne, France, Compiegne (2006)
89. Bazhal, M.: Etude Du Mécanisme D'électropermeabilisation Des Tissus Végétaux. Application à L'extraction Du Jus Des Pommes. Universite de Technologie de Compiegne, France, Compiegne (2001)
90. Grimi, N.: Vers L'intensification Du Pressage Industriel Des Agroressources Par Champs Electriques Pulses: Etude Multi-echelles. Universite de Technologie de Compiegne, France, Compiegne (2009)
91. Praporscic, I.: Influence Du Traitement Combine Par Champ Electrique Pulse Et Chauffage Modere Sur Les Proprietes Physiques Et Sur Le Comportement Au Pressage De Produits Vegetaux. Universite de Technologie de Compiegne, France, Compiegne (2005)
92. Chalermchat, Y., Malangone, L., Dejmek, P.: Biosyst. Eng. **105**(3), 357 (2010)
93. Toepfl, S.: Pulsed Electric Fields (PEF) for Permeabilization of Cell Membranes in Food- and Bioprocessing – Applications, Process and Equipment Design and Cost Analysis, Institut für Lebensmitteltechnologie und Lebensmittelchemie, (2006)
94. Turk, M.: Vers Une Amelioration Du Procede Industriel D'extraction Des Fractions Solubles De Pommes à L'aide De Technologies Electriques. Universite de Technologie de Compiegne, France, Compiegne (2010)
95. der Poel, P.W., Schiweck, H., Schwartz, T.: Sugar Technology: Beet and Cane Sugar Manufacture, p. 1118. Verlag Dr Albert Bartens KG, Berlin (1998)
96. Bouzrara, H., Vorobiev, E.I.: Int. Sugar J. **102**(1216), 194 (2000)
97. Bouzrara, H., Vorobiev, E.I.: Zucker **126**, 463 (2001)
98. Bouzrara, H., Vorobiev, E.I.: Chem. Eng. Process: Process Intensification **42**, 249 (2003)
99. Eshtiaghi, M. N., Knorr, D. (1999)
100. Jemai, A.B., Vorobiev, E.: J. Food Eng. **59**, 405 (2003)
101. Jemai, A.B., Vorobiev, E.: Biosyst. Eng. **93**, 57 (2006)
102. El-Belghiti, K., Vorobiev, E.I.: Food Bioprod. Process. **82**, 226 (2004)
103. El-Belghiti, K., Rabhi, Z., Vorobiev, E.: J. Sci. Food Agric. **85**, 213 (2005)
104. El-Belghiti, K., Rabhi, Z., Vorobiev, E.: J. Food Process Eng. **28**, 346 (2005)
105. El-Belghiti, K.: Effets D'un Champ Électrique Pulsé Sur Le Transfert De Matière Et Sur Les Caractéris-tiques Végétales. Universite de Technologie de Compiegne, France, Compiegne (2005)
106. Lopez, N., Puertolas, E., Condon, S., Raso, J., Alvarez, I.: LWT Food Sci. Technol. **42**, 1674 (2009)
107. Vorobiev, E.I., Jemai, A.B., Bouzrara, H., Lebovka, N.I., Bazhal, M.I.: In: Barbosa-Canovas, G., Tapia, M.S., Cano, M.P. (eds.) Novel Food Processing Technologies, pp. 105–130. CRC Press, New York (2005)
108. Lebovka, N.I., Shynkaryk, M.V., El-Belghiti, K., Benjelloun, H., Vorobiev, E.: J. Food Eng. **80**, 639 (2007)
109. Loginova, K., Loginov, M., Vorobiev, E., Lebovka, N.I.: J. Food Eng. **106**, 144 (2011)
110. Loginov, M., Loginova, K., Lebovka, N., Vorobiev, E.: J. Membr. Sci. **377**, 273 (2011)

111. Loginova, K., Loginov, M., Vorobiev, E., Lebovka, N.I.: LWT Food Sci. Technol. **46**, 371 (2012)
112. K. Loginova (Sereda), Mise En Oeuvre De Champs Electriques Pulses Pour La Conception D'un Procede De Diffusion a Froid a Partir De Betteraves a Sucre Et D'autres Tubercules Alimentaires (etude Multi-echelle), Compiegne: Universite de Technologie de Compiegne, France, (2011)
113. Lopez, N., Puertolas, E., Condon, S., Alvarez, I.: J. Food Eng. **90**, 60 (2009)
114. Loginova, K.V., Lebovka, N.I., Vorobiev, E.: J. Food Eng. **106**, 127 (2011)
115. Shynkaryk, M.V., Lebovka, N.I., Vorobiev, E.: Drying Technol. **26**, 695 (2008)
116. Loginova, K.V., Shynkaryk, M.V., Lebovka, N.I., Vorobiev, E.: J. Food Eng. **96**, 374 (2010)
117. Abdullah, S.H., Zhao, S., Mittal, G.S., Baik, O.-D.: Sep. Purif. Technol. **93**, 92 (2012)
118. Lim, J.H., Shim, J.M., Lee, D.U., Kim, Y.H., Park, K.-J.: Korean J. Food Sci. Technol. **44**, 704 (2012)
119. El Darra, N., Grimi, N., Vorobiev, E., Louka, N., Maroun, R.: Food Bioprocess Technol. **6**, 1281 (2013)
120. Puertolas, E., Lopez, N., Condon, S., Alvarez, I., Raso, J.: Trends Food Sci. Technol. **21**(5), 247 (2010)
121. El Darra, N., Grimi, N., Maroun, R.G., Louka, N., Vorobiev, E.: Eur. Food Res. Technol. **236**, 47 (2013)
122. Delsart, C., Ghidossi, R., Poupot, C., Cholet, C., Grimi, N., Vorobiev, E., Milisic, V., Peuchot, M.M.: Am. J. Enol. Vitic. **63**, 205 (2012)
123. Nakagawa, A., Hatayama, H., Takaki, K., Koide, S., Kawamura, Y.: IEEJ Trans. Fundam. Mater. **133**, 32 (2013)
124. Cholet, C., Delsart, C., Petrel, M., Gontier, E., Grimi, N., L'Hyvernay, A., Ghidossi, R., Vorobiev, E., Mietton-Peuchot, M., Gény, L.: J. Agric. Food Chem. **62**, 2925 (2014)
125. Delsart, C., Cholet, C., Ghidossi, R., Grimi, N., Gontier, E., Gény, L., Vorobiev, E., Mietton-Peuchot, M.: Food Bioprocess Technol. **7**(2), 424 (2014)
126. Boussetta, N., Vorobiev, E., Le, L.H., Cordin-Falcimaigne, A., Lanoisellé, J.-L.: LWT Food Sci. Technol. **46**, 127 (2012)
127. El Darra, N., Grimi, N., Vorobiev, E., Maroun, R.G., Louka, N.: Am. J. Enol. Vitic. **64**, 476 (2013)
128. Luengo, E., Franco, E., Ballesteros, F., Álvarez, I., Raso, J.: Food Bioprocess Technol. **7**, 1457 (2014)
129. Luengo, E., Puértolas, E., López, N., Álvarez, I., Raso, J.: Stewart Postharvest Rev. **8**(2), 1 (2012)
130. Parniakov, O., Lebovka, N.I., Van Hecke, E., Vorobiev, E.: Food Bioprocess Technol. **xx**, 1 (2013)
131. Zhang, T.-H., Wang, S.-J., Liu, D.-R., Yuan, Y., Yu, Y.-L., Yin, Y.-G.: Jilin Daxue Xuebao (Gongxueban) J. Jilin Univ. Eng. Technol. Ed. **41**, 882 (2011)
132. Ye, D., Zhang, L., Sun, S., Chen, J., Fang, T.: J. Food Process Eng. **37**, 273 (2014)
133. Chen, J., Li, Y., Sun, S., Fang, T.: Nongye Gongcheng Xuebao Trans. Chin. Soc. Agric. Eng. **30**, 260 (2014)
134. Rajha, H.N., Boussetta, N., Louka, N., Maroun, R.G., Vorobiev, E.: Food Res. Int. **xx**, xx (2014)
135. Bai, Y., Li, C., Zhao, J., Zheng, P., Li, Y., Pan, Y., Wang, Y.: Chromatographia **76**, 635 (2013)
136. Bai, Y.-Z., Li, C.-Y., Zhao, J.-H., Wang, Y.-P.: Chin. Tradit. Herb. Drugs **44**, 1267 (2013)
137. Pourzaki, A., Mirzaee, H., Hemmati Kakhki, A.: J. Food Process. Preserv. **37**, 1008 (2013)
138. Zhao, W., Yu, Z., Liu, J., Yu, Y., Yin, Y., Lin, S., Chen, F.: J. Sci. Food Agric. **91**, 2201 (2011)
139. Teh, S.-S., Niven, B.E., Bekhit, A.E.-D.A., Carne, A., Birch, E.J.: Food. Bioprocess Technol. **7**(11), 3064 (2014)
140. Zhao, S., Baik, O.-D., Choi, Y.J., Kim, S.-M.: Crit. Rev. Food Sci. Nutr. **54**, 1283 (2014)

141. Aguiló-Aguayo, I., Hossain, M.B., Brunton, N., Lyng, J., Valverde, J., Rai, D.K.: Innov. Food Sci. Emerg. Technol. **23**, 79 (2014)
142. Roohinejad, S., Everett, D.W., Oey, I.: Int. J. Food Sci. Technol. **49**, 2120 (2014)
143. Roohinejad, S., Oey, I., Everett, D.W., Niven, B.E.: Food Bioprocess Technol. **7**(11), 3336 (2014)
144. Jin, S., Yin, Y.: Nongye Gongcheng Xuebao Trans. Chin. Soc. Agric. Eng. **26**, 368 (2010)
145. Gachovska, T., Cassada, D., Subbiah, J., Hanna, M., Thippareddi, H., Snow, D.: J. Food Sci. **75**, E323 (2010)
146. Schrive, L., Grasmick, A., Moussiere, S., Sarrade, S.: Biochem. Eng. J. **27**, 212 (2006)
147. El Zakhem, H., Lanoisellé, J.L., Lebovka, N.I., Nonus, M., Vorobiev, E.: J. Colloid Interface Sci. **300**, 553 (2006)
148. El Zakhem, H., Lanoiselle, J.-L., Lebovka, N.I., Nonus, M., Vorobiev, E.: Colloids Surf. B: Biointerfaces **47**, 189 (2006)
149. Ganeva, V., Galutzov, B., Teissié, J.: High yield electroextraction of proteins from yeast by a flow process. Anal. Biochem. **315**, 77–84 (2003)
150. Shynkaryk, M.V., Lebovka, N.I., Lanoisellé, J.L., Nonus, M., Bedel-Clotour, C., Vorobiev, E.: J. Food Eng. **92**, 189 (2009)
151. Liu, D., Lebovka, N.I., Vorobiev, E.: Food Bioprocess Technol. **6**, 576 (2013)
152. Jin, Y., Wang, M., Lin, S., Guo, Y., Liu, J., Yin, Y.: Afr. J. Biotechnol. **10**, 19144 (2011)
153. Liu, J.-B., Yu, Y.-D., Wang, M., Lin, S.-Y., Wang, Q., Gao, L.-X., Sun, P.: Jilin Daxue Xuebao (Gongxueban) J. Jilin Univ. Eng. Technol. Ed. **40**, 1171 (2010)
154. Goettel, M., Eing, C., Gusbeth, C., Straessner, R., Frey, W.: Algal Res. **2**, 401 (2013)
155. Coustets, M., Al-Karablieh, N., Thomsen, C., Teissié, J.: Flow pProcess for eElectroextraction of tTotal pProteins from mMicroalgae. J. Membr. Biol. **246**, 751–60 (2013)
156. Zbinden, M.D., Sturm, B.S., Nord, R.D., Carey, W.J., Moore, D., Shinogle, H., Stagg-Williams, S.M.: Biotechnol. Bioeng. **110**(6), 1605 (2013)
157. Grimi, N., Dubois, A., Marchal, L., Jubeau, S., Lebovka, N.I., Vorobiev, E.: Selective extraction from microalgae Nannochloropsis sp. using different methods of cell disruption. Bioresour. Technol. **153**, 254 (2014)
158. Sheng, J., Vannela, R., Rittmann, B.E.: Environ. Sci. Tech. **45**, 3795 (2011)
159. Luengo, E., Condon-Abanto, S., Alvarez, I., Raso, J.: J. Membr. Biol. **xx**, 1 (2014)
160. Rego, D., Costa, L., Navalho, J., Paramo, J., Geraldes, V., Redondo, L.M., Pereira, M.T. In: Plasma Science (ICOPS), 2013 Abstracts IEEE International Conference On, p. 1 (2013)
161. Remenant, B., Jaffrès, E., Dousset, X., Pilet, M.F., Zagorec, M.: Bacterial spoilers of food: behavior, fitness and functional properties. Food Microbiol. **45**(A), 45–53 (2015)
162. Arnoldi, A.: Thermal processing and nutritional quality. In: Henry, C.J.K., Chapman, C. (eds.) The Nutrition Handbook for Food Processing, pp. 265–286. Woodhead, Cambridge (2002)
163. Mackey, B.M.: Injured bacteria. In: Lund, B.M., Baird-Parker, T.C., Gould, G.W. (eds.) The Microbiological Safety and Quality of Food, pp. 1–16. Aspen Publishers, Gaithersburg (2000)
164. Aronsson, K., Rönner, U., Borch, E.: Inactivation of *Escherichia coli*, *Listeria innocua* and *Saccharomyces cerevisiae* in relation to membrane permeabilization and subsequent leakage of intracellular compounds due to pulsed electric field processing. Int. J. Food Microbiol. **99**, 19–32 (2005)
165. Simpson, R.K., Whittington, R., Earnshaw, R.G., Russell, N.J.: Pulsed high electric field causes all or nothing membrane damage in *Listeria monocytogenes* and *Salmonella typhimurium*, but membrane H^+-ATPase is not a primary target. Int. J. Food Microbiol. **48**, 1–10 (1999)
166. García, D., Gómez, N., Mañas, P., Condón, S., Raso, J., Pagán, R.: Pulsed electric fields cause bacterial envelopes permeabilization depending on the treatment intensity, the treatment medium pH and the microorganisms investigated. Int. J. Food Microbiol. **113**, 219–217 (2006)

167. Russell, N.J., Colley, M., Simpson, R.K., Trivett, A.J., Evans, R.I.: Mechanism of action of pulsed high electric field (PHEF) on the membranes of food-poisoning bacteria is an "all-or-nothing" effect. Int. J. Food Microbiol. **55**, 133–136 (2000)
168. García, D., Gómez, N., Mañas, P., Condón, S., Raso, J., Pagá, R.: Pulsed electric fields cause sublethal injury in *Escherichia coli*. Lett. Appl. Microbiol. **36**, 140–144 (2003)
169. García, D., Gómez, N., Mañas, P., Condón, S., Raso, J., Pagán, R.: Occurrence of sublethal injury after pulsed electric fields depending on the microorganism, the treatment medium pH and the intensity of the treatment investigated. J. Appl. Microbiol. **99**, 94–104 (2005)
170. García, D., Gómez, N., Mañas, P., Condón, S., Raso, J., Pagán, R.: Biosynthetic requirements for the repair of sublethal membrane damage in *Escherichia coli* cells after pulsed electric fields. J. Appl. Microbiol. **100**, 428–435 (2006)
171. Arroyo, C., Somolinos, M., Cebrián, G., Condón, S., Pagán, R.: Pulsed electric fields cause sublethal injuries in the outer membrane of *Enterobacter sakazakii* facilitating the antimicrobial activity of citral. Lett. Appl. Microbiol. **51**, 525–531 (2010)
172. García, D., Hassani, M., Mañas, P., Condón, S., Pagán, R.: Inactivation of *Escherichia coli* O157:H7 during the storage under refrigeration of apple juice treated by pulsed electric fields. J. Food Saf. **25**, 30–42 (2005)
173. Saldaña, G., Álvarez, I., Condón, S., Raso, J.: Microbiological aspects related to the feasibility of PEF technology for food pasteurization. Crit. Rev. Food Sci. Nutr. **54**(11), 1415–1426 (2014)
174. Jin, T.Z., Guo, M., Zhang, H.Q.: Upscaling from benchtop processing to industrial scale production: more factors to be considered for pulsed electric field food processing. J. Food Eng. **146**, 72–80 (2015)
175. Ho, S.Y., Mittal, G.S.: Electroporation of cell membranes: a review. Crit. Rev. Biotechnol. **16**, 349–362 (1996)
176. Weaver, J.C., Chizmadzhev, Y.A.: Theory of electroporation: a review. Bioelectrochem. Bioenerg. **41**, 135–160 (1996)
177. Álvarez, I., Condón, S., Raso, J.: Microbial inactivation by pulsed electric fields. In: Raso, J., Heinz, V. (eds.) Pulsed Electric Fields Technology for the Food Industry: Fundamentals and Applications, pp. 97–129. Springer, New York (2006)
178. Heinz, V., Álvarez, I., Angersbach, A., Knorr, D.: Preservation of liquid foods by high intensity pulsed electric fields – basic concepts for process design. Trends Food Sci. Technol. **12**, 103–111 (2001)
179. Álvarez, I., Pagán, R., Condón, S., Raso, J.: The influence of process parameters for the inactivation of *Listeria monocytogenes* by pulsed electric fields. Int. J. Food Microbiol. **87**, 87–95 (2003)
180. Álvarez, I., Raso, J., Sala, F.J., Condón, S.: Inactivation of *Yersinia enterocolitica* by pulsed electric fields. Food Microbiol. **20**, 691–700 (2003)
181. Saldaña, G., Puértolas, E., Álvarez, I., Meneses, N., Knorr, D., Raso, J.: Evaluation of a static treatment chamber to investigate kinetics of microbial inactivation by pulsed electric fields at different temperatures at quasi-isothermal conditions. J. Food Eng. **100**, 349–356 (2010)
182. Sharma, P., Bremer, P., Oey, I., Everetta, D.W.: Bacterial inactivation in whole milk using pulsed electric field processing. Int. Dairy J. **35**(1), 49–56 (2014)
183. Timmermans, R.A.H., Groot, M.N., Nederhoff, A.L., van Boekelb, M.A.J.S., Matsera, A.M., Mastwijka, H.C.: Pulsed electric field processing of different fruit juices: impact of pH and temperature on inactivation of spoilage and pathogenic micro-organisms. Int. J. Food Microbiol. **173**, 105–111 (2014)
184. Siemer, C., Toepfl, S., Heinz, V.: Inactivation of *Bacillus subtilis* spores by pulsed electric fields (PEF) in combination with thermal energy – I. Influence of process- and product parameters. Food Control **39**, 163–171 (2014)
185. Heinz, V., Toepfl, S., Knorr, D.: Impact of temperature on lethality and energy efficiency of apple juice pasteurization by pulsed electric fields treatment. Innov. Food Sci. Emerg. Technol. **4**(2), 167–175 (2003)

186. Guerrero-Beltrán, J.A., Sepulveda, D.R., Góngora-Nieto, M.M., Swanson, B., Barbosa-Cánovas, G.V.: Milk thermization by pulsed electric fields (PEF) and electrically induced heat. J. Food Eng. **100**(1), 6–60 (2010)
187. Wouters, P.C., Dutreux, N., Smelt, J.P.P., Lelieveld, H.L.M.: Effects of pulsed electric fields on inactivation kinetics of *Listeria innocua*. Appl. Environ. Microbiol. **65**, 5354–5371 (1999)
188. Aronsson, K., Ronner, U.: Influence of pH, water activity and temperature on the inactivation of *Escherichia coli* and *Saccharomyces cerevisiae* by pulsed electric fields. Innov. Food Sci. Emerg. Technol. **2**, 105–112 (2001)
189. Raso, J., Álvarez, I., Condón, S., Sala, F.J.: Predicting inactivation of *Salmonella senftenberg* by pulsed electric fields. Innov. Food Sci. Emerg. Technol. **1**, 21–30 (2000)
190. Saldaña, G., Puértolas, E., López, N., García, D., Álvarez, I., Raso, J.: Comparing the PEF resistance and occurrence of sublethal injury on different strains of *Escherichia coli*, *Salmonella* Typhimurium, *Listeria monocytogenes* and *Staphylococcus aureus* in media of pH 4 and 7. Innov. Food Sci. Emerg. Technol. **10**, 160–165 (2009)
191. Álvarez, I., Pagán, R., Raso, J., Condón, S.: Environmental factors influencing the inactivation of *Listeria monocytogenes* by pulsed electric fields. Lett. Appl. Microbiol. **35**, 489–493 (2002)
192. Rodrigo, D., Ruíz, P., Barbosa-Cánovas, G.V., Martínez, A., Rodrigo, M.: Kinetic model for the inactivation of *Lactobacillus plantarum* by pulsed electric fields. Int. J. Food Microbiol. **81**, 223–229 (2003)
193. García, D., Gómez, N., Raso, J., Pagán, R.: Bacterial resistance after pulsed electric fields depending on the treatment medium pH. Innov. Food Sci. Emerg. Technol. **6**, 388–395 (2005)
194. Pagán, R., Esplugas, S., Góngora-Nieto, M.M., Barbosa-Cánovas, G.V., Swanson, B.G.: Inactivation of *Bacillus subtilis* spores using high intensity pulsed electric fields in combination with other food conservation technologies. Food Sci. Technol. Int. **4**, 33–44 (1998)
195. Pol, I.E., Mastwijk, H.C., Slump, R.A., Popa, M.E., Smid, E.J.: Influence of food matrix on inactivation of *Bacillus cereus* by combinations of nisin, pulsed electric field treatment and carvacrol. J. Food Prot. **64**, 1012–1018 (2001)
196. El-Hag, A.H., Jayaram, S.H., Griffiths, M.W.: Inactivation of naturally grown microorganisms in orange juice using pulsed electric fields. IEEE Trans. Plasma Sci. **34**(4 II), 1412–1415 (2006)
197. Min, S., Jin, Z.T., Min, S.K., Yeom, H., Zhang, Q.H.: Commercial-scale pulsed electric field processing of orange juice. Food Chem. Toxicol. **64**, 1265–1271 (2003)
198. Raso, J., Barbosa-Cánovas, G.V.: Nonthermal preservation of foods using combined processing techniques. Crit. Rev. Food Sci. **43**, 265–285 (2003)
199. Monfort, S., Saldaña, G., Condón, S., Raso, J., Álvarez, I.: Inactivation of *Salmonella* spp. in liquid whole egg using pulsed electric fields, heat, and additives. Food Microbiol. **30**(2), 393–399 (2012)
200. Bull, M.K., Szabo, E.A., Cole, M.B., Stewart, C.M.: Toward validation of process criteria for high-pressure processing of orange juice with predictive models. J. Food Prot. **68**, 949–954 (2005)
201. Food and Drug Administration (FDA).: Hazard analysis and critical control point (HACCP) procedures for the safe and sanitary processing and importing of juice; final rule. Federal Register 66 FR 6137–6202 (2001)
202. Sampedro, F., Rivas, A., Rodrigo, D., Martínez, A., Rodrigo, M.: Effect of temperature and substrate on PEF inactivation of Lactobacillus plantarum in an orange juice-milk beverage. Eur. Food Res. Technol. **223**, 30–34 (2006)
203. Toepfl, S.: Pulsed electric field food processing – industrial equipment design and commercial applications. Stewart Posthar Rev. **2**, 1–7 (2012)
204. Sampedro, F., McAloon, A., Yee, W., Fan, X., Geveke, D.J.: Cost analysis and environmental impact of pulsed electric fields and high pressure processing in comparison with thermal pasteurization. Food Bioprocess Technol. **7**(7), 1928–1937 (2014)

205. Puértolas, E., López, N., Condón, S., Raso, J., Álvarez, I.: Pulsed electric fields inactivation of wine spoilage yeast and bacteria. Int. J. Food Microbiol. **130**(1), 49–55 (2009)
206. Arroyo, C., Cebrián, G., Pagán, R., Condón, S.: Resistance of *Enterobacter sakazakii* to pulsed electric fields. Innov. Food Sci. Emerg. Technol. **11**(2), 314–321 (2010)
207. Lu, X., Laroussi, M., Puech, V.: On atmospheric-pressure non-equilibrium plasma jets and plasma bullets. Plasma Sources Sci. Technol. **21**(3), 034005 (2012). doi:10.1088/0963-0252/21/3/034005
208. Ehlbeck, J., Schnabel, U., Polak, M., Winter, J., von Woedtke, T., Brandenburg, R., von dem Hagen, T., Weltmann, K.D.: Low temperature atmospheric pressure plasma sources for microbial decontamination. J. Phys. D-Appl. Phys. **44**(1), 013002 (2011). doi:10.1088/0022-3727/44/1/013002
209. De Geyter, N., Morent, R.: Nonthermal plasma sterilization of living and nonliving surfaces. Annu. Rev. Biomed. Eng. **14**(1), 255–274 (2012). doi:10.1146/annurev-bioeng-071811-150110
210. Chang, J.S., Lawless, P.A., Yamamoto, T.: Corona discharge processes. IEEE Trans. Plasma Sci. **19**(6), 1152–1166 (1991). doi:10.1109/27.125038
211. Malik, M.A., Xiao, S., Schoenbach, K.H.: Scaling of surface-plasma reactors with a significantly increased energy density for NO conversion. J. Hazard. Mater. **209**, 293–298 (2012). doi:10.1016/j.jhazmat.2012.01.024
212. Edelblute, C.M., Malik, M.A., Heller, L.C.: Surface-dependent inactivation of model microorganisms with shielded sliding plasma discharges and applied air flow. Bioelectrochemistry **103**, 22–27 (2015). doi:10.1016/j.bioelechem.2014.08.013
213. Kogelschatz, U.: Dielectric-barrier discharges: their history, discharge physics, and industrial applications. Plasma Chem. Plasma Process. **23**(1), 1–46 (2003). doi:10.1023/A:1022470901385
214. Morfill, G.E., Shimizu, T., Steffes, B., Schmidt, H.U.: Nosocomial infections-a new approach towards preventive medicine using plasmas. New J. Phys. **11**, 115019 (2009). doi:10.1088/1367-2630/11/11/115019
215. Cao, Z., Walsh, J.L., Kong, M.G.: Atmospheric plasma jet array in parallel electric and gas flow fields for three-dimensional surface treatment. Appl. Phys. Lett. **94**(2), 021501 (2009). doi:10.1063/1.3069276
216. Lu, X., Wu, S., Chu, P.K., Liu, D., Pan, Y.: An atmospheric-pressure plasma brush driven by sub-microsecond voltage pulses. Plasma Sources Sci. Technol. **20**(6), 065009 (2011). doi:10.1088/0963-0252/20/6/065009
217. Brandenburg, R., Lange, H., von Woedtke, T., Stieber, M., Kindel, E., Ehlbeck, J., Weltmann, K.D.: Antimicrobial effects of UV and VUV radiation of nonthermal plasma jets. IEEE Trans. Plasma Sci. **37**(6), 877–883 (2009). doi:10.1109/Tps.2009.2019657
218. Brisset, J.L., Hnatiuc, E.: Peroxynitrite: a re-examination of the chemical properties of non-thermal discharges burning in air over aqueous solutions. Plasma Chem. Plasma Process. **32**(4), 655–674 (2012). doi:10.1007/s11090-012-9384-x
219. Kolb, J.F., Mattson, A.M., Edelblute, C.M., Hao, X.L., Malik, M.A., Heller, L.C.: Cold DC-operated air plasma jet for the inactivation of infectious microorganisms. IEEE Trans. Plasma Sci. **40**(11), 3007–3026 (2012). doi:10.1109/Tps.2012.2216292
220. Camp, J.T., Jing, Y., Zhuang, J., Kolb, J.F., Beebe, S.J., Song, J.H., Joshi, R.P., Xiao, S., Schoenbach, K.H.: Cell death induced by subnanosecond pulsed electric fields at elevated temperatures. IEEE Trans. Plasma Sci. **40**(10), 2334–2347 (2012). doi:10.1109/Tps.2012.2208202
221. Heller, L.C., Edelblute, C.M., Mattson, A.M., Hao, X., Kolb, J.F.: Inactivation of bacterial opportunistic skin pathogens by nonthermal DC-operated afterglow atmospheric plasma. Lett. Appl. Microbiol. **54**(2), 126–132 (2012). doi:10.1111/j.1472-765X.2011.03186.x
222. Schoenbach, K.H., Xiao, S., Joshi, R.P., Camp, J.T., Heeren, T., Kolb, J.F., Beebe, S.J.: The effect of intense subnanosecond electrical pulses on biological cells. IEEE Trans. Plasma Sci. **36**(2), 414–422 (2008). doi:10.1109/Tps.2008.918786

223. Sysolyatina, E., Mukhachev, A., Yurova, M., Grushin, M., Karalnik, V., Petryakov, A., Trushkin, N., Ermolaeva, S., Akishev, Y.: Role of the charged particles in bacteria inactivation by plasma of a positive and negative corona in ambient air. Plasma Processes Polym. **11** (4), 315–334 (2014). doi:10.1002/ppap.201300041
224. Laroussi, M.: Sterilization of contaminated matter with an atmospheric pressure plasma. IEEE Trans. Plasma Sci. **24**(3), 1188–1191 (1996)
225. Morfill, G.E., Kong, M.G., Zimmermann, J.L.: Focus on plasma medicine. New J. Phys. **11**, 115011 (2009). doi:10.1088/1367-2630/11/11/115011
226. Kong, M.G., Kroesen, G., Morfill, G., Nosenko, T., Shimizu, T., van Dijk, J., Zimmermann, J. L.: Plasma medicine: an introductory review. New J. Phys. **11**, 115012 (2009). doi:10.1088/1367-2630/11/11/115012
227. Morent, R., Geyter, N.D.: Inactivation of bacteria by non-thermal plasmas. Biomed. Eng. Front. Chall. InTech. (2011). doi:10.5772/18610
228. Plasma for Bio-Decontamination, Medicine and Food Security. In: Machala, Z.H. K., Akishev, Y. (eds.) (2012)
229. Kvam, E., Davis, B., Mondello, F., Garner, A.L.: Nonthermal atmospheric plasma rapidly disinfects multidrug-resistant microbes by inducing cell surface damage. Antimicrob. Agents Chemother. **56**(4), 2028–2036 (2012). doi:10.1128/AAC.05642-11
230. Marsili, L., Espie, S., Anderson, J.G., MacGregor, S.J.: Plasma inactivation of food-related microorganisms in liquids. Radiat. Phys. Chem. **65**(4–5), 507–513 (2002)
231. Digel, I., Temiz Artmann, A., Nishikawa, K., Cook, M., Kurulgan, E., Artmann, G.M.: Bactericidal effects of plasma-generated cluster ions. Med. Biol. Eng. Comput. **43**(6), 800–807 (2005)
232. Scholtz, V., Julak, J., Kriha, V., Mosinger, J., Kopecka, S.: Decontamination effects of low-temperature plasma generated by corona discharge. Part II: new insights. Prague Med. Rep. **108**(2), 128–146 (2007)
233. Ermolaeva, S.A., Varfolomeev, A.F., Chernukha, M.Y., Yurov, D.S., Vasiliev, M.M., Kaminskaya, A.A., Moisenovich, M.M., Romanova, J.M., Murashev, A.N., Selezneva, I.I., Shimizu, T., Sysolyatina, E.V., Shaginyan, I.A., Petrov, O.F., Mayevsky, E.I., Fortov, V.E., Morfill, G.E., Naroditsky, B.S., Gintsburg, A.L.: Bactericidal effects of non-thermal argon plasma in vitro, in biofilms and in the animal model of infected wounds. J. Med. Microbiol. **60**(Pt 1), 75–83 (2011). doi:10.1099/jmm.0.020263-0
234. Alkawareek, M.Y., Algwari, Q.T., Gorman, S.P., Graham, W.G., O'Connell, D., Gilmore, B. F.: Application of atmospheric pressure nonthermal plasma for the in vitro eradication of bacterial biofilms. FEMS Immunol. Med. Microbiol. **65**(2), 381–384 (2012). doi:10.1111/j.1574-695X.2012.00942.x
235. Klampfl, T.G., Isbary, G., Shimizu, T., Li, Y.F., Zimmermann, J.L., Stolz, W., Schlegel, J., Morfill, G.E., Schmidt, H.U.: Cold atmospheric air plasma sterilization against spores and other microorganisms of clinical interest. Appl. Environ. Microbiol. **78**(15), 5077–5082 (2012). doi:10.1128/AEM.00583-12
236. Brun, P., Brun, P., Vono, M., Venier, P., Tarricone, E., Deligianni, V., Martines, E., Zuin, M., Spagnolo, S., Cavazzana, R., Cardin, R., Castagliuolo, I., Valerio, A.L., Leonardi, A.: Disinfection of ocular cells and tissues by atmospheric-pressure cold plasma. PLoS One **7** (3), e33245 (2012). doi:10.1371/journal.pone.0033245
237. Cahill, O.J., Claro, T., O'Connor, N., Cafolla, A.A., Stevens, N.T., Daniels, S., Humphreys, H.: Cold air plasma to decontaminate inanimate surfaces of the hospital environment. Appl. Environ. Microbiol. **80**(6), 2004–2010 (2014). doi:10.1128/AEM.03480-13
238. Joshi, S.G., Cooper, M., Yost, A., Paff, M., Ercan, U.K., Fridman, G., Friedman, G., Fridman, A., Brooks, A.D.: Nonthermal dielectric-barrier discharge plasma-induced inactivation involves oxidative DNA damage and membrane lipid peroxidation in Escherichia coli. Antimicrob. Agents Chemother. **55**(3), 1053–1062 (2011)
239. Lu, H., Patil, S., Keener, K.M., Cullen, P.J., Bourke, P.: Bacterial inactivation by high-voltage atmospheric cold plasma: influence of process parameters and effects on cell leakage and DNA. J. Appl. Microbiol. **116**(4), 784–794 (2014). doi:10.1111/jam.12426

240. Helmke, A., Hoffmeister, D., Mertens, N., Emmert, S., Schuette, J., Vioel, W.: The acidification of lipid film surfaces by non-thermal DBD at atmospheric pressure in air. New J. Phys. **11**, 115025 (2009). doi:10.1088/1367-2630/11/11/115025
241. Korachi, M., Gurol, C., Aslan, N.: Atmospheric plasma discharge sterilization effects on whole cell fatty acid profiles of Escherichia coli and Staphylococcus aureus. J. Electrostat. **68** (6), 508–512 (2010). doi:10.1016/j.elstat.2010.06.014
242. Kim, S.M., Kim, J.I.: Decomposition of biological macromolecules by plasma generated with helium and oxygen. J. Microbiol. **44**(4), 466–471 (2006)
243. Pankaj, S.K., Misra, N.N., Cullen, P.J.: Kinetics of tomato peroxidase inactivation by atmospheric pressure cold plasma based on dielectric barrier discharge. Innov. Food Sci. Emerg. **19**, 153–157 (2013). doi:10.1016/j.ifset.2013.03.001
244. Surowsky, B., Fischer, A., Schlueter, O., Knorr, D.: Cold plasma effects on enzyme activity in a model food system. Innov. Food Sci. Emerg. **19**, 146–152 (2013). doi:10.1016/j.ifset.2013.04.002
245. Lackmann, J.W., Schneider, S., Edengeiser, E., Jarzina, F., Brinckmann, S., Steinborn, E., Havenith, M., Benedikt, J., Bandow, J.E.: Photons and particles emitted from cold atmospheric-pressure plasma inactivate bacteria and biomolecules independently and synergistically. J. R. Soc. Interface Roy. Soc. **10**(89), 20130591 (2013). doi:10.1098/rsif.2013.0591
246. Jena, N.R.: DNA damage by reactive species: mechanisms, mutation and repair. J. Biosci. **37** (3), 503–517 (2012)
247. Perni, S.: Probing bactericidal mechanisms induced by cold atmospheric plasmas with Escherichia coli mutants. Appl. Phys. Lett. **90**(7), 073902 (2007)
248. Feng, H., Wang, R., Sun, P., Wu, H., Liu, Q., Fang, J., Zhu, W., Li, F., Zhang, J.: A study of eukaryotic response mechanisms to atmospheric pressure cold plasma by using Saccharomyces cerevisiae single gene mutants. Appl. Phys. Lett. **97**(13). DOI:DOI 10.1063/1.3491180 (2010)
249. Montie, T.C., Kelly-Wintenberg, K., Roth, J.R.: An overview of research using the one atmosphere uniform glow discharge plasma (OAUGDP) for sterilization of surfaces and materials. IEEE Trans. Plasma Sci. **28**(1), 41–50 (2000)
250. Klämpfl, T.G., Shimizu, T., Koch, S., Balden, M., Gemein, S., Li, Y.-F., Mitra, A., Zimmermann, J.L., Gebel, J., Morfill, G.E., Schmidt, H.-U.: Decontamination of nosocomial bacteria including clostridium difficile spores on dry inanimate surface by cold atmospheric plasma. Plasma Processes Polym. **11**(10), 974–984 (2014). doi:10.1002/ppap.201400080
251. Vleugels, M., Shama, G., Deng, X.T., Greenacre, E., Brocklehurst, T., Kong, M.G.: Atmospheric plasma inactivation of biofilm-forming bacteria for food safety control. IEEE Trans. Plasma Sci. **33**(2), 824–828 (2005). doi:10.1109/Tps.2005.844524
252. Kamgang, J.O., Briandet, R., Herry, J.M., Brisset, J.L., Naitali, M.: Destruction of planktonic, adherent and biofilm cells of Staphylococcus epidermidis using a gliding discharge in humid air. J. Appl. Microbiol. **103**(3), 621–628 (2007). doi:10.1111/j.1365-2672.2007.03286.x
253. Abramzon, N., Joaquin, J.C., Bray, J., Brelles-Marino, G.: Biofilm destruction by RF high-pressure cold plasma jet. IEEE Trans. Plasma Sci. **34**(4), 1304–1309 (2009)
254. Zelaya, A.J., Stough, G., Rad, N., Vandervoort, K., Brelles-Marino, G.: Pseudomonas aeruginosa biofilm inactivation: decreased cell culturability, adhesiveness to surfaces, and biofilm thickness upon high-pressure nonthermal plasma treatment. IEEE Trans. Plasma Sci. **38**(12), 3398–3403 (2010). doi:10.1109/TPS.2010.2082570
255. Joshi, S.G., Paff, M., Friedman, G., Fridman, G., Fridman, A., Brooks, A.D.: Control of methicillin-resistant Staphylococcus aureus in planktonic form and biofilms: a biocidal efficacy study of nonthermal dielectric-barrier discharge plasma. Am. J. Infect. Control **38** (4), 293–301 (2010). doi:10.1016/j.ajic.2009.11.002
256. Cotter, J.J., Maguire, P., Soberon, F., Daniels, S., O'Gara, J.P., Casey, E.: Disinfection of meticillin-resistant Staphylococcus aureus and Staphylococcus epidermidis biofilms using a

remote non-thermal gas plasma. J. Hosp. Infect. **78**(3), 204–207 (2011). doi:10.1016/j.jhin.2011.03.019

257. Pei, X., Lu, X., Liu, J., Liu, D., Yang, Y., Ostrikov, K., Chu, P.K., Pan, Y.: Inactivation of a 25.5 mu m Enterococcus faecalis biofilm by a room-temperature, battery-operated, handheld air plasma jet. J. Phys. D-Appl. Phys. **45**(16), 165205 (2012). doi:10.1088/0022-3727/45/16/165205
258. Galvin, S., Cahill, O., O'Connor, N., Cafolla, A.A., Daniels, S., Humphreys, H.: The antimicrobial effects of helium and helium–air plasma on Staphylococcus aureus and Clostridium difficile. Lett. Appl. Microbiol. **57**(2), 83–90 (2013). doi:10.1111/lam.12091
259. Maisch, T., Shimizu, T., Li, Y.F., Heinlin, J., Karrer, S., Morfill, G., Zimmermann, J.L.: Decolonisation of MRSA, S. aureus and E. coli by cold-atmospheric plasma using a porcine skin model in vitro. PLoS One **7**(4), e34610 (2012). doi:10.1371/journal.pone.0034610
260. Baxter, H.C., Campbell, G.A., Whittaker, A.G., Jones, A.C., Aitken, A., Simpson, A.H., Casey, M., Bountiff, L., Gibbard, L., Baxter, R.L.: Elimination of transmissible spongiform encephalopathy infectivity and decontamination of surgical instruments by using radio-frequency gas-plasma treatment. J. Gen. Virol. **86**(Pt 8), 2393–2399 (2005). doi:10.1099/vir.0.81016-0
261. Deng, X.T., Shi, J.J., Chen, H.L., Kong, M.G.: Protein destruction by atmospheric pressure glow discharges. Appl. Phys. Lett. **90**(1), 013903 (2007). doi:10.1063/1.2410219
262. Bayliss, D.L., Walsh, J.L., Shama, G., Iza, F., Kong, M.G.: Reduction and degradation of amyloid aggregates by a pulsed radio-frequency cold atmospheric plasma jet. New J. Phys. **11**, 115024 (2009). doi:10.1088/1367-2630/11/11/115024
263. Niemira, B.A.: Cold plasma decontamination of foods. Ann. Rev. Food Sci. Technol. **3**(1), 125–142 (2012). doi:10.1146/annurev-food-022811-101132
264. Misra, N.N., Tiwari, B.K., Raghavarao, K.S.M.S., Cullen, P.J.: Nonthermal plasma inactivation of food-borne pathogens. Food Eng. Rev. **3**(3-4), 159–170 (2011). doi:10.1007/s12393-011-9041-9
265. Baier, M., Görgen, M., Ehlbeck, J., Knorr, D., Herppich, W.B., Schlüter, O.: Non-thermal atmospheric pressure plasma: screening for gentle process conditions and antibacterial efficiency on perishable fresh produce. Innov. Food Sci. Emerg. **22**, 147–157 (2014). doi:10.1016/j.ifset.2014.01.011
266. Jiang, C.Q., Chen, M.T., Schaudinn, C., Gorur, A., Vernier, P.T., Costerton, J.W., Jaramillo, D.E., Sedghizadeh, P.P., Gundersen, M.A.: Pulsed atmospheric-pressure cold plasma for endodontic disinfection. IEEE Trans. Plasma Sci. **37**(7), 1190–1195 (2009). doi:10.1109/Tps.2009.2014870
267. Hammann, A., Huebner, N.O., Bender, C., Ekkernkamp, A., Hartmann, B., Hinz, P., Kindel, E., Koban, I., Koch, S., Kohlmann, T., Lademann, J., Matthes, R., Muller, G., Titze, R., Weltmann, K.D., Kramer, A.: Antiseptic efficacy and tolerance of tissue-tolerable plasma compared with two wound antiseptics on artificially bacterially contaminated eyes from commercially slaughtered pigs. Skin Pharmacol. Physiol. **23**(6), 328–332 (2010)
268. Rupf, S., Lehmann, A., Hannig, M., Schafer, B., Schubert, A., Feldmann, U., Schindler, A.: Killing of adherent oral microbes by a non-thermal atmospheric plasma jet. J. Med. Microbiol. **59**(Pt 2), 206–212 (2010)
269. Jiang, C., Schaudinn, C., Jaramillo, D.E., Webster, P., Costerton, J.W.: In vitro antimicrobial effect of a cold plasma jet against Enterococcus faecalis biofilms. ISRN Dent. **2012**, 295736 (2012). doi:10.5402/2012/295736
270. Isbary, G., Morfill, G., Schmidt, H.U., Georgi, M., Ramrath, K., Heinlin, J., Karrer, S., Landthaler, M., Shimizu, T., Steffes, B., Bunk, W., Monetti, R., Zimmermann, J.L., Pompl, R., Stolz, W.: A first prospective randomized controlled trial to decrease bacterial load using cold atmospheric argon plasma on chronic wounds in patients. Br. J. Dermatol. **163**(1), 78–82 (2010)
271. Kelly-Wintenberg, K., Sherman, D.M., Tsai, P.P.Y., Ben Gadri, R., Karakaya, F., Chen, Z.Y., Roth, J.R., Montie, T.C.: Air filter sterilization using a one atmosphere uniform glow discharge plasma (the Volfilter). IEEE Trans. Plasma Sci. **28**(1), 64–71 (2000)

272. Nojima, H., Park, R.-E., Kwon, J.-H., Suh, I., Jeon, J., Ha, E., On, H.-K., Kim, H.-R., Choi, K., Lee, K.-H., Baik-Lin, S., Jung, H., Kang, S.J., Namba, S., Takiyama, K.: Novel atmospheric pressure plasma device releasing atomic hydrogen: reduction of microbial-contaminants and OH radicals in the air. J. Phys. D Appl. Phys. **40**(2), 501 (2007)
273. Vaze, N.D., Gallagher, M.J., Park, S., Fridman, G., Vasilets, V.N., Gutsol, A.F., Anandan, S., Friedman, G., Fridman, A.A.: Inactivation of bacteria in flight by direct exposure to nonthermal plasma. IEEE Trans. Plasma Sci. **38**(11), 3234–3240 (2010). doi:10.1109/Tps.2010.2072788
274. Ye, S.Y., Song, X.L., Liang, J.L., Zheng, S.H., Lin, Y.: Disinfection of airborne spores of Penicillium expansum in cold storage using continuous direct current corona discharge. Biosyst. Eng. **113**(2), 112–119 (2012). doi:10.1016/j.biosystemseng.2012.06.013
275. Wu, Y., Liang, Y., Wei, K., Li, W., Yao, M., Zhang, J., Grinshpun, S.A.: MS2 virus inactivation by atmospheric pressure cold plasma using different gas carriers and power levels. Appl. Environ. Microbiol. (2014). doi:10.1128/AEM.03322-14. in press
276. Terrier, O., Essere, B., Yver, M., Barthelemy, M., Bouscambert-Duchamp, M., Kurtz, P., VanMechelen, D., Morfin, F., Billaud, G., Ferraris, O., Lina, B., Rosa-Calatrava, M., Moules, V.: Cold oxygen plasma technology efficiency against different airborne respiratory viruses. J. Clin. Virol.: Off. Publ. Pan Am. Soc. Clin. Virol. **45**(2), 119–124 (2009). doi:10.1016/j.jcv.2009.03.017
277. Abou-Ghazala, A., Katsuki, S., Schoenbach, K.H., Dobbs, F.C., Moreira, K.R.: Bacterial decontamination of water by means of pulsed-corona discharges. IEEE Trans. Plasma Sci. **30**(4), 1449–1453 (2002). doi:10.1109/Tps.2002.804193
278. Martines, E., Zuin, M., Cavazzana, R., Gazza, E., Serianni, G., Spagnolo, S., Spolaore, M., Leonardi, A., Deligianni, V., Brun, P., Aragona, M., Castagliuolo, I., Brun, P.: A novel plasma source for sterilization of living tissues. New J. Phys. **11**(11), 115014 (2009)
279. Sohbatzadeh, F., Hosseinzadeh, C.A., Mirzanejhad, S., Mahmodi, S.: E. coli, P. aeruginosa, and B. cereus bacteria sterilization using afterglow of non-thermal plasma at atmospheric pressure. Appl. Biochem. Biotechnol. **160**(7), 1978–1984 (2010)
280. Ziuzina, D., Patil, S., Cullen, P.J., Keener, K.M., Bourke, P.: Atmospheric cold plasma inactivation of Escherichia coli in liquid media inside a sealed package. J. Appl. Microbiol. **114**(3), 778–787 (2013). doi:10.1111/jam.12087
281. Kuo, S.P., Tarasenko, O., Nourkbash, S., Bakhtina, A., Levon, K.: Plasma effects on bacterial spores in a wet environment. New J. Phys. **8**, 41 (2006). doi:10.1088/1367-2630/8/3/041
282. Alshraiedeh, N.H., Alkawareek, M.Y., Gorman, S.P., Graham, W.G., Gilmore, B.F.: Atmospheric pressure, nonthermal plasma inactivation of MS2 bacteriophage: effect of oxygen concentration on virucidal activity. J. Appl. Microbiol. **115**(6), 1420–1426 (2013). doi:10.1111/jam.12331
283. Montenegro, J., Ruan, R., Ma, H., Chen, P.: Inactivation of E. coli O157: H7 using a pulsed nonthermal plasma system. J. Food Sci. **67**(2), 646–648 (2002)
284. Gurol, C., Ekinci, F.Y., Aslan, N., Korachi, M.: Low temperature plasma for decontamination of E. coli in milk. Int. J. Food Microbiol. **157**(1), 1–5 (2012). doi:10.1016/j.ijfoodmicro.2012.02.016
285. de Boer, K., Moheimani, N.R., et al.: Extraction and conversion pathways for microalgae to biodiesel: a review focused on energy consumption. J. Appl. Phycol. **24**(6), 1681–1698 (2012)
286. Eisentraut, A.: Sustainable production of second-generation biofuels. OECD/IEA 2010: International Energy Agency. http://www.iea.org/topics/biofuels/publications/: 1–221 (2010)
287. Lewandowski, I., Heinz, A.: Delayed harvest of miscanthus – influences on biomass quantity and quality and environmental impacts of energy production. Eur. J. Agron. **19**(1), 45–63 (2003)
288. Dillschneider, R., Posten, C.: Closed bioreactors as tools for microalgae production. In: Lee, J.W. (ed.) Advanced Biofuels and Bioproducts, pp. 629–649. Springer, New York (2013)

289. Schenk, P.M., Thomas-Hall, S.R., et al.: Second generation biofuels: high-efficiency microalgae for biodiesel production. Bioenergy Res. **1**(1), 20–43 (2008)
290. Chisti, Y.: Biodiesel from microalgae. Biotechnol. Adv. **25**(3), 294–306 (2007)
291. Amaro, H.M., Guedes, A.C., et al.: Advances and perspectives in using microalgae to produce biodiesel. Appl. Energy **88**(10), 3402–3410 (2011)
292. Gouveia, L., Oliveira, A.: Microalgae as a raw material for biofuels production. J. Ind. Microbiol. Biotechnol. **36**(2), 269–274 (2009)
293. Pulz, O., Gross, W.: Valuable products from biotechnology of microalgae. Appl. Microbiol. Biotechnol. **65**(6), 635–648 (2004)
294. Rosello-Sastre, R.: Products from microalgae: an overview. Microalgal biotechnology: integration and economy, pp. 13–50. C. Posten and C. Walter, de Gruyter, Berlin/Boston (2012)
295. Norsker, N.-H., Barbosa, M.J., et al.: Microalgal production – a close look at the economics. Biotechnol. Adv. **29**(1), 24–27 (2011)
296. Jacobi, A., Posten, C.: Energy considerations of photobioreactors. In: Borowitzka, M.A., Moheimani, N.R. (eds.) Algae for Biofuels and Energy, vol. 5, pp. 223–232. Springer, Dordrecht/Heidelberg/New York/London (2013)
297. Morweiser, M., Kruse, O., et al.: Developments and perspectives of photobioreactors for biofuel production. Appl. Microbiol. Biotechnol. **87**(4), 1291–1301 (2010)
298. Sierra, E., Acien, F.G., et al.: Characterization of a flat plate photobioreactor for the production of microalgae. Chem. Eng. J. **138**(1-3), 136–147 (2008)
299. Dillschneider, R., Steinweg, C., et al.: Biofuels from microalgae: photoconversion efficiency during lipid accumulation. Bioresour. Technol. **142**, 647–654 (2013)
300. Molina, E., Acien, G., et al.: Downstream processing of cell mass and products. In: Richmond, A., Hu, Q. (eds.) Handbook of Microalgal Culture, 2nd edn, pp. 267–309, Wiley Blackwell, Chicester (2013)
301. Lardon, L., Helias, A., et al.: Life-cycle assessment of biodiesel production from microalgae. Environ. Sci. Technol. **43**(17), 6475–6481 (2009)
302. Xu, L.X., Brilman, D.W.F., et al.: Assessment of a dry and a wet route for the production of biofuels from microalgae: energy balance analysis. Bioresour. Technol. **102**(8), 5113–5122 (2011)
303. Postma, P.R., Miron, T.L., et al.: Mild disintegration of the green microalgae Chlorella vulgaris using bead milling. Bioresour. Technol. **184**, 297–304 (2015)
304. Günerken, E., D'Hondt, E., et al.: Cell disruption for microalgae biorefineries. Biotechnol. Adv. **33**(2), 243–260 (2015)
305. Halim, R., Danquah, M.K., et al.: Extraction of oil from microalgae for biodiesel production: a review. Biotechnol. Adv. **30**(3), 709–732 (2012)
306. Macías-Sánchez, M.D., Robles-Medina, A., et al.: Biodiesel production from wet microalgal biomass by direct transesterification. Fuel **150**, 14–20 (2015)
307. Molina, E., González, M., et al.: Solvent extraction for microalgae lipids. In: Borowitzka, M. A., Moheimani, N.R. (eds.) Algae for Biofuels and Energy, vol. 5, pp. 187–205. Springer, Netherlands (2013)
308. Slade, R., Bauen, A.: Micro-algae cultivation for biofuels: cost, energy balance, environmental impacts and future prospects. Biomass Bioenergy **53**, 29–38 (2013)
309. Wijffels, R.H., Barbosa, M.J.: An outlook on microalgal biofuels. Science **329**(5993), 796–799 (2010)
310. Goettel, M., Eing, C., et al.: Pulsed electric field assisted extraction of intracellular valuables from microalgae. Algal Res. Biomass Biofuels Bioprod. **2**(4), 401–408 (2013)
311. Gusbeth, C., Eing, C., et al.: Fluorescence diagnostics for lipid status monitoring of microalgae during cultivation. Int. J. Renew. Energy Biofuels: **2016**, 12, Article ID 899698 (2016). IBIMA Publishing. http://www.ibimapublishing.com/journals/IJREB/ijreb.html. doi:10.5171/2016.899698

312. Eing, C., Goettel, M., et al.: Pulsed electric field treatment of microalgae-benefits for microalgae biomass processing. IEEE Trans. Plasma Sci. **41**(10), 2901–2907 (2013)
313. Luengo, E., Martínez, J., et al.: A comparative study on the effects of millisecond- and microsecond-pulsed electric field treatments on the permeabilization and extraction of pigments from chlorella vulgaris. J. Membr. Biol. **248**(5), 1–9 (2015)
314. Kotnik, T., Frey, W., et al.: Electroporation-based applications in biotechnology. Trends Biotechnol. **33**(8), 480–488 (2015)
315. Coustets, M., Joubert-Durigneux, V., et al.: Optimization of protein electroextraction from microalgae by a flow process. Bioelectrochemistry **103**, 74–81 (2015)
316. Acién, F.G., Fernández, J.M., et al.: Production cost of a real microalgae production plant and strategies to reduce it. Biotechnol. Adv. **30**(6), 1344–1353 (2012)
317. Brownbridge, G., Azadi, P., et al.: The future viability of algae-derived biodiesel under economic and technical uncertainties. Bioresour. Technol. **151**, 166–173 (2014)
318. Chen, R.: Bacterial expression systems for recombinant protein production: E. coli and beyond. Biotechnol Adv **30**(5), 1102–1107 (2012)
319. Huang, C.J., Lin, H., Yang, X.: Industrial production of recombinant therapeutics in Escherichia coli and its recent advancements. J. Ind. Microbiol. Biotechnol. **39**(3), 383–399 (2012)
320. Gellissen, G., Hollenberg, C.P.: Application of yeasts in gene expression studies: a comparison of Saccharomyces cerevisiae, Hansenula polymorpha, and Kluyveromyces lactis – a review. Gene **190**, 87–97 (1997)
321. Cereghino, J., Cregg, J.M.: Heterologous protein expression in the methylotrophic yeast Pichia pastoris. FEMS Microbiol. Rev. **24**, 45–66 (2000)
322. Walker, T.L., Purton, S., Becker, D.K., et al.: Microalgae as bioreactors. Plant Cell Rep. **24**, 629–641 (2005)
323. Tran, M., Van, C., Barrera, D.J., Pettersson, P.L., Peinado, C.D., Bui, J., Mayfield, S.P.: Production of unique immunotoxin cancer therapeutics in algal chloroplasts. Proc. Natl. Acad. Sci. U. S. A. **110**(1), E15–E22 (2013)
324. Safi, C., Charton, M., Pignolet, O., et al.:Influence of microalgae cell wall characteristics on protein extractability and determination of nitrogen-to-protein conversion factors, J. Appl. Phycol. **25**(2), 1–7 (2012)
325. Schuttle, H., Kula, M.R.: Pilot and process-scale techniques for cell disruption. Biotechnol. Appl. Biochem. **12**, 559–620 (1990)
326. Naglak, T.J., Hettwer, D.J., Wang, H.J.: Chemical permeabilization of cells for intracellular product release. Bioprocess Technol. **9**, 177–205 (1990)
327. Ganeva, V., Galutzov, B.: Electropulsation as an alternative method for protein extraction from yeast. FEMS Microbiol. Lett. **174**(2), 279–284 (1999)
328. Barbosa-Canovas, G., Gongora-Nieto, M., Pothakamury, U., Swanson, B.: Preservation of Foods with Pulsed Electric Fields. Academic, San Diego (1999)
329. Simental-Martınez, J., Reddy Vennapusa, R., Benavides, J., et al.: A novel process for the recovery of superoxide dismutase from yeast exploiting electroextraction coupled to direct sorption. J. Chem. Technol. Biotechnol. **88**, 1498–1505 (2013)
330. Laemmli, U.K.: Cleavage of structural proteins during the assembly of the head of bacteriophage T4. Nature **227**, 680–685 (1970)
331. Nesterenko, M.V., Tilley, M., Upton, S.J.: A simple modification of Blum's silver stain method allows for 30 minute detection of proteins in polyacrylamide gels. J. Biochem. Biophys. Methods **28**, 239–242 (1994)
332. Kulbe, K., Bojanovski, M.: 3-Phosphoglycerate kinase from bovine liver and yeast. Methods Enzymol. **90**, 115–120 (1982)
333. Kirschner, V.B.: Purity criteria for the crystallizable isoenzyme of D-glyceraldehyde-3-phosphate dehydrogenase from baker's yeast, Hoppe Seylers Z. Physiol. Chem. **349**, 632–644 (1968)
334. McDonald, M.: Yeast hexokinase. Methods Enzymol. **1**, 269–272 (1955)

335. Mbuyi-Kalala, A., Schnek, A.G., Leonis, J.: Separation and characterization of four enzyme forms of b-galactosidase from Saccharomyces lactis. Eur. J. Biochem. **178**, 437–443 (1988)
336. Meussdoerffer, F., Tortora, P., Holzer, H.: Purification and properties of proteinase A from yeast. J. Biol. Chem. **25**, 12087–12093 (1980)
337. Ohshima, T., Hama, Y., Sato, M.: Releasing profiles of gene products from recombinant Escherichia coli in a high-voltage pulsed electric field. Biochem. Eng. J. **5**, 149–155 (2004)
338. Shiina, S., Ohshima, T., Sato, M.: Extracellular release of recombinant alpha-amylase from Escherichia coli using pulsed electric field. Biotechnol. Prog. **20**, 1528–1533 (2004)
339. Coustets, M., Ganeva, V., Galutzov, B., et al.: Millisecond duration pulses for flow-through electro-induced protein extraction from E. coli and associated eradication Bioelectrochemistry in press (2014)
340. Suga, M., Goto, A., Hatakeyama, T.J.: Electrically induced protein release from Schizosaccharomyces pombe cells in a hyperosmotic condition during and following a high electropulsation. Biosci. Bioeng. **103**(4), 298–302 (2007)
341. Suga, M., Hatakeyama, T.: Gene transfer and protein release of fission yeast by application of a high voltage electric pulse. Anal. Bioanal. Chem. **394**(1), 13–16 (2009)
342. Ganeva, V., Galutzov, B., Eynard, N., et al.: Electroinduced extraction of beta-galactosidase from Kluyveromyces lactis. Appl. Microbiol. Biotechnol. **56**(3-4), 411–413 (2001)
343. Ganeva, V., Galutzov, B., Teissié, J.: Flow process for electroextraction of intracellular enzymes from the fission yeast, Schizosaccharomyces pombe. Biotechnol. Lett. **26**, 933–937 (2004)
344. Zakhartsev, M., Momeu, C., Ganeva, V.: High-throughput liberation of water-soluble yeast content by irreversible electropermeation (HT-irEP). J. Biomol. Screen. **12**(2), 267–275 (2007)
345. Zlotnik, H., Fernandes, M.P., Blowers, B., et al.: Increased cell wall porosity in Saccharomyces cerevisiae after treatment with dithiothreitol or EDTA. J. Gen. Microbiol. **135**, 2077–2084 (1984)
346. Schultheiss, C., Bluhm, H., Mayer, H.G., Kern, M., Michelberger, T., Witte, G.: Processing of sugar beets with pulsed-electric fields. IEEE Trans. Plasma Sci. 30(4), August 2002, 1547–1551 (2002)
347. Sack, M., Sigler, J., Frenzel, S., Eing, C., Arnold, J., Michelberger, T., Frey, W., Attmann, F., Stukenbrock, L., Mueller, G.: Research on industrial-scale electroporation devices fostering the extraction of substances from biological tissue. Food Eng. Rev. **2**,147–156 (2010). doi:10.1007/s12393-010-9017-1
348. Puértolas, E., Hernández-Orte, P.E., Saldaña, G., Álvarez, I., Raso, J.: Improvement of winemaking process using pulsed electric fields at pilot-plant scale. Evolution of chromatic parameters and phenolic content of Cabernet Sauvignon red wines. Food Res. Int. **43**(2010), 761–766 (2010)
349. Sack, M., Sigler, J., Eing, C., Stukenbrock, L., Stängle, R., Wolf, A., Müller, G.: Operation of an electroporation device for mash. Proc. IEEE PPC. (2009) June 28 2009-July 2 2009, ISBN 978-1-4244-4064-1, doi:10.1109/PPC.2009.5386165

Index

A
Accidental cell death (ACD), 238
AC-driven plasma jets, 114
Acidified nitrites, 125
Actin, 213–214
Actin Arrays, 198–199
Aggregation of platelets, 27
Air bubbles, 415
Algae treatment system, 406
Algal bloom, 401
Alkaline extraction, 410
Animal responses, 359–360
Antennas, pulse delivery systems, 31
Anthocyanins, 422
Antibiotic-resistant bacteria, 391
Anticancer agents, 290
Apoptosis, 25, 241
Apple, 420
Apple juice, 420

B
B16 melanoma, 330
B16f10 tumors, 321
Bead milling, 445
Biocidal agents, 115–116
Bioelectrics, 320
Biomass yield, 444
Bipolar pulses, 25
Bleomycin, 279, 291
Blood coagulation, 339–340
Blood vessels, 300
Blumlein line, 63
Bone, 303
Brain, 303–304
Brain tumors, 297
Breast cancer, 292
Brown antenna, 98

C
Ca^{2+}, 325
  entry, 195
  uptake, 191
Calcium mobilization, 26–27
Cancer, 280, 317–318, 320
Cancer therapy, 313
Capacitive dividers, 78
Capacitive energy storage, 48–50
Cardiovascular, 280
Carotenoids, 425, 427
Carrot, 425
Caspase, 240–245
Cats, 354–360
Cell alterations, 185–186
Cell disintegration, 445
Cell membrane, 290
Cell models, 170–171
Cell stimulation, 340–342
Cell volume, 192–193
Cell wall cytoskeletal continuum, 197, 200
Chemical thermodynamic concept, 2
Chemotherapeutic agents, 290
Chemotherapeutic drug, 279
Chemotherapy, 347–348
Chicory, 421
Chlorophyll, 427
CLC energy transfer circuit, 50–51
Clinical data, 302
Clinical studies, 333

H. Akiyama, R. Heller (eds.), *Bioelectrics*, DOI 10.1007/978-4-431-56095-1

Clinical studies of GET, 316–319
Clinical trials, 296–297
Coccus bacteria, 435
Collinear electrode arrangement, 416
Common mode, 88
Computational models, 166
Conductivity, wastewater, 396
Contact wire electrodes, 350
Continuum modeling, 166–176
Corona discharges, 111, 438
Cortical microtubules, 197, 201, 203
CT treatment, 349–352
Cyanobacteria, 403
Cytoplasmic membrane, 429
Cytoskeleton, 196–197

D
DBD plasma, 342
DC-driven plasma jets, 114
D-dot probe, 82–84
D-dot sensors, 78
Deeper chronic wounds, 324–327
Delivery of nucleic acids, 20
Delivery to muscle, 313
Delivery to skin, 314
Delivery to tumors, 314
Dental treatment, 342–343
Dielectric barrier discharges (DBD), 116, 438
Differentiating-integrating (D–I) circuits, 78
Direct transesterification, 446
Discone antenna, 98
Disintegration index, 413
DLCO, 296
DNA vaccines, 280
Dogs, 354–360
Downstream processing, 445
Drug delivery, 283–285
Ductal systems, 301

E
Ebola, 318
eIF2α, 226
Electric field distribution, 417
Electric field effects on proteins, 31
Electric field parameters, 182–183
Electric fields, 127
Electrical energy storage, 45
Electrochemotherapy (ECT), 20, 279, 290, 292, 346–362
Electrode array, 311
Electrode selection, 311–312
Electrodes, 350
Electro-erosion, 451
Electroextraction, 21, 449
Electrofusion, 18
Electromagnetic interference (EMI), 333, 335
Electromagnetic radiation, 92
Electropermeabilization, 180, 280
Electropermeabilization measurement, 184
Electroporation-based therapies (EBTs), 228
Electro-tolerance, 399
Ellipsoidal reflector, 98
EMC, 88
EMI. *See* Electromagnetic interference (EMI)
Endocytosis-like processes, 186
Endocytotic uptake, 289
Endodontic treatment, 344–345
Endoscopic electrodes, 297
Endothelial cells, 341
Energy oscillations, 417
Entracellular trafficking, 213–219
Enzyme, 291
Experimental measurement, 163–164
Experimental protocol, 181–182
Extracorporeal shock wave lithotripsy (ESWL), 118

F
Far-infrared (FIR) gas lasers, 129
Fermentation, skins, 412
Fluorescent dye uptake, 189–191
Food preservation, 436
Fractionating component recovery, 448
Free electron-based sources, 130–135
Free radicals, 439
Frequency, 359–360

G
Gas vesicles, 404
Gene delivery, 286–288
Gene electrotransfer (GET), 279, 307–329
Gene therapy, 292
Genotoxicity, 400
Glyceraldehyde-3-phosphate dehydrogenase (GAPDH), 455
Gram-negative bacteria, 397, 429
Gram-positive bacteria, 397, 429
Grape transport, 414
Ground potential, 416
Guanfu base A, 424

H
Heart failure, 320
Heat, 116
Hepatitis B, 318

Heterogeneous conductivity, 230
High-power ultrawideband electromagnetic pulses, 127
High-pressure homogenization, 445
HIV vaccines, 318
Horse sarcoids, 360–362
Human brain, 101
Hydrogen peroxide, 119
Hydrophilic drugs, 291
Hydrophilic molecules, 290
Hydrophilic pore, 290

I
Immune response, 305
Immunotherapy, 298
Impulse generators, 51–52
In vivo delivery
  of genes, 18
  of plasmid DNA, 18
Inactivation of bacteria, 118
Induced membrane field, 11
Induced transmembrane voltage, 156–158
Inductive energy storage, 46, 67
Infectious disease, 318–319
Influence of the temperature, 434
Influenza, 318
Intracellular calcium release, 26–27
Intracellular signaling, 25
IRE therapy, 228–229, 305
Irreversible electroporation, 298–307, 430
Irreversible electroporation (IRE), 228, 304
Irreversible permeabilization, 17
Ischemia, 280

J
Joule heating, 452

K
Kidney, 302
Klystrons, 128

L
Laser-driven THz emitters, 129
LC generator, 53–54
Lemberger grapes, 412
Lipid electropores, 173–179
Lipid membrane, 290
Lipid yield, 447
Lipids, 444
Lipophilic drugs, 290
Liver, 303
Low jitter, 418

M
Macropinocytotic pathways, 289
Magnetic pulse compression, 63–66
Magnetic pulse compressor (MPC), 402
Magnetic switches, 70–72
Malignant tumors, 347–348
Mammalian cells, 290
MAPK pathways, 221–222
Marx generator, 52–53
Matching, 58
Membrane sensor, 201–205
Metabolic disorders, 280
Metal–oxide–semiconductor field-effect transistor (MOSFET), 73, 74
Microalgae, 426, 449
Micro-hollow cathode discharges (MHCDs), 112
Microsecond pulse electric fields (μsPEF), 276–278
Microtubular arrays, 197–198
Microtubule, 215–216
Microvessel density (MVD), 321
Microwave discharges, 113
Microwave-driven plasma jets, 115
Minor trauma, 296
Mitochondria membrane potential, 29
Mitomycin C, 291
Mixed bacterial populations, 398
Molecular dynamic models, 175–176
Molecular dynamic simulations, 177–179
Molecular mechanical (MM), 175
MOSFET. *See* Metal–oxide–semiconductor field-effect transistor (MOSFET)
Muscle cells, 313
Myocardial ischemia, 327–329

N
Nanoelectroablation, 332–333
Nanoelectropores, 187–189
Nanoporation, membranes, 24–25
Nanopores, 188
Nanosecond pulse application, 320–337
Nanosecond pulse electric fields (nsPEF), 276, 277, 330
Nanosecond pulses, 329–330
Necroptosis, 245–246
Needle electrodes, 352

Nerves, 300
Nonequilibrium transport, 167
Noninvasive electrode, 312
Nonpenetrating electrodes, 313
Nonthermal atmospheric-pressure plasmas (NTAPs), 278, 338
Nonthermal plasmas, 111–126, 438
nsPEF. *See* Nanosecond pulse electric fields (nsPEF)
Nuclear entry of plasmids, 217–218
Nuclease activity, 400
Nucleic acid delivery, 279–280
Numerical computation, 162–163

O
Opening switch, 67
Operating modality, 358
Overvoltage triggering, 417
Oxidative stress, 441
Ozone, 119

P
Pain medication, 296
Pancreas, 301–302
Pancreatic cancer, 297
Parthanatos, 246–247
Patch clamp, 193–194
pDNA. *See* Plasmid DNA (pDNA)
PEF resistance, 435
Penetrating electrodes, 314
Periodontal treatment, 343–344
Peroxynitrites, 119
pH, treatment medium, 431
Phosphoinositide signaling pathway, 224–225
Photobioreactor, 445
Photolysis, 126
Physiological effects, 194
Pigments, 412
Pinot noir grapes, 413
Plant actin, 207–211
Plant cells, 196–197
Plant cytoskeleton, 198–199, 205–207
Plasma, 111
Plasma jets, 113
Plasma medicine, 337–346
Plasma sensitivity, 441
Plasma-activated water (PAW), 124
Plasmid, 213–219
  design, 312–313
  movement, 218–219
Plasmid DNA (pDNA), 288, 313
Plasmolysis, 429
Plate electrodes, 350
Platelet, 320, 324–327
Platelet-rich plasma (PRP), 325
Podophyllotoxin, 421
Polyphenols, 423, 424
Pore behavior, 169
Pore density, 16
Pore density fraction, 12
Pore energy landscape, 169
Pore formation, 4, 7
Pore longevity, 6
Pore models, 167–168
Pore polarization volume, 8
Pore resealing, 7
Post-discharge reactions, 123
Post-field relaxation, 4
Potato, 419
Power junction field-effect transistors, 76
Power semiconductor device, 67–69
Preclinical data, 301
Preclinical studies, 313–316, 330–332
Production, wine, 412
Programmed cell death (PCD), 27
Prostate, 304–305
Prostate-specific antigen (PSA), 318
Protein, 448
  activity, 453
  extraction, 453
  phosphorylation, 219
Protein kinase, 219
Protein–DNA complexes, 216–217
Protocol development, 308–313
Pulsed current measurements, 85–88
Pulsed electromagnetic fields, 110
Pulsed electromagnetic radiation, 110
Pulsed power, 110
Pulsed power-driven plasma jets, 114
Pulsed streamer discharges, 402
Pulse-forming line, 61
Pulse-forming networks, 59–61
Pulsing procedure, 359–360

Q
Quantum mechanical (QM), 175

R
Reactive chemical species, 119–120
Reactive nitrogen species (RNS), 116
Reactive oxygen species (ROS), 119, 327, 338, 342, 441

Reactive species, 439
Red beet, 421
Regulated cell death (RCD), 238, 240–245
Reperfusion injury, 327–329
Resealing, 185
Resistive dividers, 78
Resting potential, 157
Resting voltage, 157
Retreatment, 295
Reverse drift, 25
Reversible electroporation, 430
Reversible permeabilization, 17
RF-driven plasma jets, 115
Rogowski coil, 86
ROS. *See* Reactive oxygen species (ROS)

S
Saturable inductors, 65
Scaling of nanosecond pulse effects, 25
Secondary biological effects, 123
Semiconductor opening switches (SOS), 72–73
Shock waves, 118–119
Side effects, 296
Signal transduction, 219–220
Simple pulsed power generator, 61
Simplest cell model, 167
siRNA, 213–219
Skin cancer, 321–322
Skin tumors, 295
Sleeve antenna, 98
Solid tumors, 346–362
Solute transport, 171
Solvent extraction, 446
Specific energy, 433
Spheroids, 186–187
Squamous cell carcinoma, 279
Standard operating procedures, 293
Static induction transistors (SIT), 76
Streamer-like discharge, 402
Stress response, 225–227
Subcellular membrane effects, 22–24
Sugar beets, 409–411, 420
Superficial skin, 324–327
Surface inactivation, 441
Susceptibility to IRE, 235–237

T
Tandem shock waves, 119
Tannins, 412, 422
Thermal effects, 229
Thermally assisted electroeffects, 32
Thermally based therapies, 299
Thermodynamic concept, 3
Thermovinification, 412
3D multicellular spheroids, 186
Thyristors, 75–76
Tl^+ ions, 191
Transdermal delivery, 18
Translational suppression, 225
Transmembrane potentials, 180–181, 290
Transmission line, 55–59
Treatment chamber, 416
Treatment energy, 447
Treatment medium, 436
Treatment time, 432
Tumor therapy, 329–337
TUNNEL, 321
Turn-off thyristor, 75–76

U
Ultrasound transducer, 335
Ultraviolet radiation, 116

V
Vacuole integrity, 456
Veterinary oncology, 346–362
VUV photolysis, 117

W
Wine grapes, 422
Wound care, 338–339
Wound disinfection, 339
Wound healing, 280, 320

Y
Yeast, 425, 449, 455
YO-PRO-1, 189

FSC
www.fsc.org
MIX
Papier | Fördert
gute Waldnutzung
FSC® C083411

Kolibri 360 GmbH
Ferdinand-Jühlke-Straße 7
99095 Erfurt, Deutschland
produktsicherheit@kolibri360.de